D1164099

THE ONTOGENY OF VERTEBRATE BEHAVIOR

CONTRIBUTORS

P. P. G. Bateson

Douglas K. Candland

Joseph Church

Maurice Hershenson

D. S. Lehrman

Peter Marler

Barton Meyers

Howard Moltz

Paul Mundinger

J. S. Rosenblatt

Leonard A. Rosenblum

Richard E. Whalen

W. I. Welker

THE ONTOGENY OF VERTEBRATE BEHAVIOR

Edited by HOWARD MOLTZ
Department of Psychology
The University of Chicago
Chicago, Illinois

ACADEMIC PRESS New York and London 1971

ACADEMIC PRESS, INC.
111 Fifth Avenue, New York, New York 10003

United Kingdom Edition published by
ACADEMIC PRESS, INC. (LONDON) LTD.
24/28 Oval Road, London NW1 7DD

LIBRARY OF CONGRESS CATALOG CARD NUMBER: 79-159616

PRINTED IN THE UNITED STATES OF AMERICA

CONTENTS

LIST OF CONTRIBUTORS

Numbers in parentheses indicate the pages on which the authors' contributions begin.

P. P. G. BATESON, Sub-Department of Animal Behaviour, University of Cambridge, Cambridge, England (369)

DOUGLAS K. CANDLAND, Department of Psychology, Bucknell University, Lewisburg, Pennsylvania (95)

JOSEPH CHURCH, Department of Psychology, Brooklyn College and the University Graduate Center of the City University of New York, Brooklyn, New York (451)

MAURICE HERSHENSON, Department of Psychology, Brandeis University, Waltham, Massachusetts (29)

D. S. LEHRMAN, Institute of Animal Behavior, Rutgers University, Newark, New Jersey (1)

PETER MARLER, The Rockefeller University, New York, New York (389)

BARTON MEYERS, Department of Psychology, Brooklyn College of the City University of New York, Brooklyn, New York (57)

HOWARD MOLTZ, Department of Psychology, The University of Chicago, Chicago, Illinois (263)

PAUL MUNDINGER, The Rockefeller University, New York, New York (389)

J. S. ROSENBLATT, Institute of Animal Behavior, Newark, New Jersey (1)

LEONARD A. ROSENBLUM, Primate Behavior Laboratory, State University of New York, Downstate Medical Center, Brooklyn, New York (315)

RICHARD E. WHALEN, Department of Psychobiology, University of California, Irvine, California (229)

W. I. WELKER, Laboratory of Neurophysiology, The University of Wisconsin Medical School, Madison, Wisconsin (171)

PREFACE

This volume is more than a collection of chapters devoted to current research in the area of comparative psychology (or biopsychology, to use the more fashionable term). It is a volume focused on development, specifically the development of vertebrate behavior. Ontogenetic determinants are emphasized, and the questions treated are those related to the differentiation of selected response systems. The systems themselves range widely, with Chapter 6, for example, treating sexual behavior in the rat and Chapter 11 language acquisition in the child. The vertebrates included are birds, rodents, lagamorphs, monkeys, and men.

The audience for which this work is intended, of course, will have an interest in behavior development, but will come from a variety of disciplines: some from psychology, others from ethology, endocrinology, and behavioral biology.

The problems relating to ontogeny are many, and the solutions require a spectrum of different technical and conceptual skills. It is to the total effort of understanding the development of vertebrate behavior that this present volume is offered.

HOWARD MOLTZ

CHAPTER 1

THE STUDY OF BEHAVIORAL DEVELOPMENT*

D. S. Lehrman and J. S. Rosenblatt

In the study of behavioral development, as in the study of other aspects of behavioral biology, it is neither possible nor necessary to agree about a single formulation of *the* major problems, for the purpose of defining and delimiting the paths to be followed by scientific investigation. The diversity of conceptual and methodological approaches (and of investigative techniques) is limited only by the ability of investigators to perceive new relationships and ask new questions about them.

It is possible, however, in surveying the actual activities of existing groups of scientists, to distinguish a limited variety of major approaches to the groups of problems making up the study of behavioral development. What we propose to do in this chapter is to briefly characterize some of the main groups of problems and thus define different ways of approaching the study of the development of behavior. We do not intend to present either an exhaustive catalog of problems or a detailed analysis of any one problem. Rather, it is our intention to convey an impression of the variety of problems formulated by scientists with very different theoretical and technical inter-

*Contribution #114. The preparation of this paper was supported by a research career award (DSL) and by research grants (DSL and JSR) from the National Institute of Mental Health, which are gratefully acknowledged.

ests, who work with methods which (though very dissimilar) complement each other to build up a growing understanding of the phenomena of behavioral development.

I. Aspects of Individual Development

Many aspects of animal function, properly regarded as basic equipment for the development of social behavior, have developmental histories which must be understood before one can fully appreciate the ways in which social behavior develops. A few selected examples will be given.

The Role of Experience in the Development of Basic Functions

1. Sensorimotor Functions

The cerebral cortex of the cat contains cells which "fire" in response to various kinds of light stimulation of the animal's eye. Each of these cells is connected to a group of light sensitive retinal cells in a specific area of the retina. This area, connected to a single cortical neuron in the visual area of the cat's brain, is called a "receptive field." (Adjacent receptive fields may overlap to some extent.) Some of these neurons do not respond merely to the appearance and disappearance of light, but respond rather specifically to the movements of contours (boundaries between light areas and dark areas) across the receptive field (Hubel & Wiesel, 1959). For example, some cortical cells may respond most intensely to a vertical line moving across the receptive field, while other neurons may respond most intensely to a horizontal line moving up and down in the visual field. Some of these cortical neurons are binocularly represented—that is, they are fired equally well (or almost equally well) upon appropriate stimulation of corresponding receptive fields in each eye separately (Hubel & Wiesel, 1962).

The binocular character of the response of these cortical neurons, representing the coordination of the activities of both eyes in interpreting the visual field, is already present to some degree in kittens tested shortly after their eyes have opened (i.e., before they have had any visual experience), although it is not as well developed in newborn kittens as it will be in adulthood. That is, in the kitten the cortical cells respond less symmetrically to stimulation of corresponding receptive fields in both eyes than they do in the adult (Hubel & Wiesel, 1963).

If the kitten is reared with 1 eye deprived of pattern vision until it is 2 or 3 months old, then the binocular character of the response of the cortical visual neurons is lost, and they respond only to stimulation of the eye that has not been deprived (Wiesel & Hubel, 1963a, 1963b). This means that a sensory organization characteristic of the adult eye-brain relationship exists already at birth in a less developed form, and is maintained and further developed through the influence of visual experience made possible by the degree of organization present at birth. Similar conclusions about the relationship between visual experience and the development of basic visual capacities have been reached through behavioral studies of visual discrimination in cats reared in the dark (or with one or both eyes covered) during the first few months of life (Riesen, 1960).

For investigators concerned with the question of the role played by experience in the development of behavioral capacities, data like these could be described in two ways, which can be made to seem contradictory to each other. One can say that the organization of the cat's visual system develops independently of experience, but degenerates in the absence of practice. By contrast, one could say that the cat's visual experience since birth is in part responsible for the organization observed in the visual system of the adult cat. These seem to us to be two different, equally accurate, ways of describing the same process with different emphases. Our own feeling is that dichotomous formulations of experience *against* structure, or nature *against* nurture, or maturation *against* learning are rather artificial, and generally tend to deflect attention from the fact that the adult organization is the end-product of a complex developmental process in which the processes of growth and of experience are intricately interrelated in ways elucidated by experiments of the sort just described.

A similar pattern emerges when we turn from electrophysiological studies of the visual system itself to behavioral studies of vision-dependent responses to the complex environment. If a cat learns to discriminate between 2 patterns using only 1 eye, it can immediately make the same discrimination using only the other eye. This indicates that visual experience has some sort of common representation in the central nervous system regardless of which eye receives the visual impressions which are the basis of the experience. Experiments involving the rearing of kittens in the dark, or with 1 or both eyes covered, show that the ability of the cat to transfer experience readily from 1 eye to the other depends in part upon the fact that both eyes have, during early development, participated simultaneously in identical visual experience (Riesen, Kurke, & Mellinger, 1953).

The act of reaching for a visually perceived object represents a complex coordination of visual and motor functions. Many years ago, Riesen (1947) showed that a young chimpanzee reared in the dark failed to reach for an approaching nursing bottle until the nipple actually touched his lips. When he was kept in the light after 18 months or so of dark rearing, visually guided reaching developed only very slowly and inadequately.

This demonstration that the corrdination of visually guided reaching depended to some degree upon visual experience has been greatly extended and confirmed by the work of Held and Hein (1963) on the development of visuomotor coordination in kittens. Their work has shown that the development of visually coordinated reaching behavior depends upon experience of the visual consequences of motor acts. That is, in order to be able to reach for something under visual guidance, the cat must have had the opportunity to learn how movements of parts of its body look, or how the visual world is changed by locomotor movements of its body through space.

In a most striking demonstration, Held and Hein (1963) reared 2 kittens together in a device in which they were mechanically linked to each other, 1 of the kittens ("active") having its feet in contact with the floor, while the other ("passive") was suspended slightly above the floor. Every time the active kitten walked, both the active and the passive kittens moved through visual space in the same way. Both kittens could freely make "locomotor" movements; both kittens had simultaneous and equivalent visual experience. However, only in the case of the active kitten did the animal experience the visual consequences of its *own* locomotor movements. And the outcome of the experiment was that the active kitten developed active visual reaching ability, while the passive kitten did not!

Similar results can be found in other types of visually guided behavior. For example, the placing response of the cat is a response to a visually approaching surface, such as the top of the table toward which the cat is being lowered by an experimenter holding it under its forelegs, or a narrow ledge towards which the cat leaps from a tree above. The response consists of extending the limbs so as to meet the visually oncoming surface, and it includes both a component, or aspect, of extension of the limbs from the body, and of orientation in the plane perpendicular to the body; the cat, in making the placing response, can guide its limbs appropriately with respect to irregularities or gaps in the surface to which it is responding. Hein and Held (1967) have shown that if a kitten is reared in the dark and then tested for the placing response, it will extend the limbs appropriately, but will not be able to guide them to the details of the oncoming surface as does a normally reared animal. Here, again, experimentally definable features of the animal's visual capacities are seen to depend in experimentally definable ways upon particular aspects of its experience, interacting with growth processes in the central nervous system and in the animal as a whole.

The utterance of a song, characteristic of the species, is a characteristic motor ability of birds (Thorpe, 1958a, 1958b). The fact that all members of a given species sing songs that closely resemble each other and that different species, even closely related ones living in the same neighborhood, sing songs that are distinctly and recognizably different from each other would suggest that the characteristic song of a bird might be rigidly determined by the structure of the bird's nervous system and motor apparatus with experience playing little or no necessary role in its development. In some species of birds, this indeed seems to be the case for at least some major parts of the song repertoire (Blase, 1960; Mulligan, 1966; Nottebohm, 1971). There are a number of species, however, in which ingenious experiments have revealed very substantial and unexpected influences of individual experience upon the development of species-characteristic patterns.

As one example among several which have been studied, the White-Crowned Sparrow (*Zonotrichia leucophrys*) of the Pacific coast of North America has local "song dialects"; that is, birds of this species coming from different local populations sing recognizably different song patterns. If birds of this species are reared in isolation from the first few days of life, they develop a song which is quite different in many respects from that of any normal White-Crowned Sparrow. Further, although there are striking individual differences among the songs of such isolation-reared birds, they cannot be related to the group differences in the populations from which the different individuals came. However, if the birds are allowed to remain in their normal environment until after they leave the nest and are then placed in isolation only after they have acquired their feathers and begun to be independent of the parents, they then, during isolation, develop not only normal White-Crowned Sparrow songs, but the particular dialect of the population from which they came (Marler & Tamura, 1962, 1964)!

These experiments indicate rather clearly that the normal, universal, species-specific song of the White-Crowned Sparrow, by which any ornithologist (and presumably any White-Crowned Sparrow) can recognize the bird as a member of the species, develops in part through the auditory experience of the individual bird during a rather restricted period of its early life. It is possible to analyze the nature of this experience further by allowing birds isolated from an early (nestling) age to hear prerecorded songs of different kinds during various periods of the birds' development, and to then analyze the kind of song that the experimental bird eventually develops (Thorpe, 1958b). By this technique, a rather startling feature of the "learning" of song in the White-Crowned Sparrow can be demonstrated: although the birds can quite precisely imitate White-Crowned Sparrow songs (with respect to which there is a great deal of individual variability), their song development is not at all affected by hearing the songs of other closely related species of sparrows! When they hear the songs of several species, they imitate only the White-Crowned Sparrow song, copying the individual details of the specific example used as a prerecorded "tutor," while ignoring the songs of the other species. If they hear *only* the songs of other species, their song development is no different from that of White-Crowned Sparrows which have heard no bird songs during their development (Marler, 1967).

The role of auditory experience can be further elucidated by deafening the birds during different stages of the development of their song and by limiting their access to auditory experience. If a White-Crowned Sparrow is deafened after it has had the auditory experience on the basis of which it would be expected to develop a normal song but before the song has actually developed, it does not develop normal song; the songs which such birds eventually produce are like those of birds which have not had the necessary auditory experience (Konishi, 1965). Furthermore, when a Song Sparrow (*Melospiza melodia*), which does *not* need to hear Song Sparrow song in order to develop the normal song, is deafened before its song has developed, it likewise fails to develop a normal song (Mulligan, 1966). This means that the role of hearing in the development of song is not simply a matter of imitating a song heard in the outer environment. In developing the motor patterns of its own song, the bird must be matching the variable and plastic songs which it sings while it is developing the definitive species song to an "internalized" auditory pattern. This appears to be true both in species such as the White-Crowned Sparrow, in which the internal auditory pattern to which the bird's own song must be matched has developed through hearing an appropriate song from another member of the species, and in species such as the Song Sparrow, in which the internal auditory pattern ("template") must develop in the course of the growth of the bird's central nervous system, without reference to specifically relevant auditory experience.

Birds of this kind do not sing the full species song until the spring when they are almost 1 year old, when they are becoming sexually mature, and when they are brought into condition to sing by male hormones secreted by their maturing testes. They thus use auditory experience acquired at an age when they are physiologically incapable of "practice" as a contribution to the development of their behavior at a much later stage of development.*

2. *Emotion and Motivation*

All normally reared dogs will vigorously try to escape from a painful stimulus, and will learn to avoid the sources of such painful stimulation after only very few experiences.

When dealing with such a basic and universal form of response and of learning capacity which all dogs have in common and which they share with many other kinds of animals, one is tempted to think that the behavior must be impervious to environmental influence, and to regard the qualification "normally reared" as a mere *pro forma* concession to the remote possibility that the behavior might be indirectly affected by some sort of extreme environmental manipulation. It is quite surprising to find that this behavior is strikingly affected by a realm of individual experience which is very easy to control. Melzack and Scott (1957) reared Scottish terriers from weaning (4 weeks of age) until they were mature (8 months of age) in box cages constructed so that the dogs could be kept in isolation, could not see anything in the environment around the box (although light was permitted to get into the box), and had no opportunity to see or to interact with the people who cleaned their cages. These dogs, and a group of normal dogs reared as pets, were tested for their response to electric shock, to a heated match held near the nose, and

*More detailed discussion of the development of bird song may be found in Thorpe (1958) and in the book edited by Hinde (1969), particularly the chapter by Konishi and Nottebohm (1969) from which we have borrowed heavily.

to a pin prick. The normal dogs reacted with obvious pain to the stimuli and very quickly learned to avoid them. By contrast, the dogs reared in the restricted environment showed a reduced capacity to respond to the "painful" stimulation, some of them not withdrawing from the stimuli at all; in addition, they required many more experiences before they had learned to avoid the painful objects, some of them not learning to avoid at all during the entire experiment. As Melzack and Scott pointed out, it seems that the capacity to respond appropriately to painful stimuli, in spite of its universality, cannot be explained in terms of imperative reflexes alone without regard to the earlier experience of the animals.

Observations like this cannot be explained merely as conventional learning by saying that the animals did not have previous opportunity to learn about sources of painful stimulation. They responded very poorly to all sorts of stimulation other than pain (Melzack & Thompson, 1956). In addition, the restricted-environment animals showed frequent, intense excitement in all of their behavior (Melzack, 1954) and seemed unable to respond in the coordinated, well-organized way in which normally reared dogs respond both to familiar and novel objects. Melzack (1968) offers the plausible explanation that the lack of early perceptual experience led to an inability to distinguish among different objects and to respond differently to novel objects and familiar objects, or to acquire familiarity with the objects and the immediate environment. Thus everything that stimulates them stimulates them too much, arousing intense general activity, without appropriate orientation toward or reaction to the specific characteristics of the stimulus. They even lack (or have lost) the ability to react differently to strong and weak stimuli.

This work of Melzack demonstrates that emotional characteristics, and the relationship of emotional behavior to perception of the emotion-arousing events in the environment, although they seem so universal as strongly to suggest that they might be "built-in" to the animal's structure, nevertheless require early experience, even though the early experience may be simply experience of perceiving and reacting to a constantly changing environment. This sort of experience is, of course, something which all dogs have in common.

This demonstration of the importance of early experience which all normally reared animals of a species universally pass through is a salutary warning against regarding the universality, or species-typical character, of any kind of behavior as evidence of the fact that experience did not play a role in its development.

Another striking example of the role of early experience in the development of species-specific, universally observed characteristics of behavior may be found in the social and sexual behavior of the Rhesus monkey. Monkeys of this species, reared in social isolation from other monkeys during their first year, are subsequently unable to develop any of the normal social relationships that characterize the interactions of normally reared Rhesus monkeys with other members of their species. They cannot play as monkeys "spontaneously" and "naturally" do, they are drastically unable to engage in normal sexual behavior, etc. Even partial isolation, in which the monkey is able to see and hear other monkeys but without being able to make any physical contact with them, is almost as destructive to the development of normal behavior patterns as is complete isolation. Although interaction with the mother is important to the young monkey, it appears that lack of opportunity to play with other monkeys of his own age is an even more significant aspect of the isolation experiments (Harlow, 1966; Harlow, Dodsworth, & Harlow, 1965).

Even in animals not having such a long socialization period, and in which there is not such an elaboration of complex social interactions within a group as normally occurs in monkeys, early experience often plays a significant role in the development of basic aspects of the behavior of the species. For example, the level of sexual "motivation" in guinea pigs is substantially affected by whether the animals have been able to associate with other guinea pigs just after they are weaned from their parents (Lehrman, 1962; Young, 1961).

Nonsocial motivated behavior (such as feeding behavior) also undergoes developmental change not only with respect to the actual movements but with respect to the relationship between various elements of the behavior pattern and the motivational causes. Sucking, which slightly later in the life of an animal appears clearly related to the need for food, appears at first to be a response which is relatively independent of nutritional status or of food intake. Puppies appear to have a "need" for a certain amount of sucking, even if an experimenter arranges the situation so that they get adequate food intake with much less sucking (James, 1957; Levy, 1934; Ross, 1950). If some kittens are allowed to suck at a female who gives milk while others suck from an anesthetised female which is not producing milk, both groups of kittens do the same amount of sucking (adequate food was provided directly into the stomach by the experimenter for the first 3 weeks) (Koepke & Pribram, 1971). This means that, early in life, sucking is not stimulated by hunger, nor is it terminated by the filling of the stomach (James & Rollins, 1965), although preloading the stomach with large amounts of milk does reduce the amount of sucking in puppies (Satinoff & Stanley, 1963).

After kittens are a few weeks old, it *is* important whether sucking does or does not lead to the getting of milk into the stomach. At this age, kittens whose sucking does not secure milk gradually suck less and less, while those whose sucking succeeds in obtaining milk continue to suck vigorously (Koepke & Pribram, 1971). If the kitten is not allowed to suck at all during these first 3 weeks, it will then be too late for it to develop the ability to suck from the mother (Kovach & Kling, 1967). In a normal mother-kitten relationship, the mother takes the initiative to approach the kitten and initiate the feeding episodes during the early days of the kitten's life, but after about 3 weeks the kitten takes the initiative, seeking the mother out when it is hungry (Rosenblatt, Turkewitz, & Schneirla, 1961).

All these facts taken together suggest that, although we see the same relationships between sucking and food intake during the whole period through weaning (i.e., the infant sucks the mother's nipple at regular intervals, the intervals being appropriate for the provision of an adequate supply of food for the infant), the internal relationships between the animal's behavior and the needs which it serves are constantly changing through some interaction between the growth of the infant and the experience it has in the interaction with its mother made possible by that growth. At first the sucking behavior is reflexly initiated by stimuli offered by the initiative of the mother, or simply by the proximity of the mother to the infant. The growth of the infant and its experience in the situation described by this interaction lead to the emergence of an internal connection between (a) the sucking behavior by which the infant gets food, (b) the felt need for food, and (c) the external source of food, so that the infant becomes able to perceive the source of food as such, to orient to it at its own initiative, and to learn to distinguish many subtle details of the food source. It is important to recognize that complex and dramatic changes may be taking place in the organization of the young animal's behavior during a period when, to simple observation, the behavior does not seem to be changing very drastically.

II. Aspects of Social Development

The materials which we have just summarized, and many other observations, indicate that many aspects of the young animal's psychological organization depend upon the experience during early life. This is by no means the same as saying that they are arbitrarily learned in the sense that a dog could just as well learn to be a cat, or that a kitten could learn to get food perfectly well without going through the stage of sucking not related to hunger. The animal has predispositions, which are not necessarily based upon specific previous experience, to take advantage of particular kinds of experience at particular stages of its development.

The behavior of most higher animals develops in a social setting in which it is constantly interacting with other members of its species, particularly its parents. The development of behavior in such social animals cannot be fully understood unless the social setting, and particularly the interaction with the mother, is taken into account.

A. Development in a Social Setting

1. Species Recognition

The development of appropriate social behavior implies the development not only of the basic sensorimotor and emotional processes that we have discussed above but also of the animal's ability to direct its social behavior selectively toward, and to respond selectively to stimuli from, members of its own species. A number of observations suggest that some sorts of experiential interaction between the developing young animal and other members of its species play a role in the development of the behavioral tendencies that we take as evidence that the animal "knows" to what species it belongs.

Domestic sheep move in tightly knit flocks in which the animals remain physically close to one another, move more or less in unison across their fields, approach each other closely when alarmed, and in general demonstrate continuous awareness of each other and continuous adjustment to each other's movements. Scott (1945) isolated at birth a lamb born in such a flock, and maintained it by hand until it was 10 days old, after which it was returned to its mother. This animal never achieved anything like the normal integration into the flock. Even as an adult, it was usually comparatively physically distant from the other animals, did not move about in coordination with them, and generally appeared to the observer like an outsider in the group.

Schneirla and Rosenblatt (1961) raised domestic kittens from birth in an artificial brooder in which the kitten was totally isolated from other animals, receiving milk from an artificial nipple, and contact comfort and warmth from felt-covered solid forms. When such an animal was returned to its litter as early as 7 days or as late as 45 days, it, like Scott's lamb, failed to act like a normal nursing infant.

At 7 days or 25 days it did not orient appropriately to the mother's body in such a way as to find the nipple (Rosenblatt *et al.*, 1961) and at later ages (e.g., 45 days) it did not approach the mother to feed and it did not engage in the playful interactions of its siblings.

The young of many species of birds are able to walk, and to follow their parents about, almost immediately after hatching from the egg. Lorenz (1935) showed that, in some such birds (e.g. ducks and geese), the recently hatched young will at first follow any of a very wide variety of stimuli, but that the following response quickly becomes restricted to those stimuli that the bird has already followed. Lorenz coined the now widely used term "imprinting" (*Prägung*) for this phenomenon. By the term, he meant a special form of learning characterized by the (assumed) facts that it (a) required only 1 or a very few trials, (b) occurred only during a sharply defined and time-limited "critical period" during early life, and (c) was irreversible. Lorenz believed that these characteristics defined a phenomenon that was not identical, or even continuous, with any other form of learning.

Recent work suggests that these characteristics are not so sharply defined as was once thought. The critical period is not necessarily inherently determined or sharply defined, but may be altered by altered conditions of rearing (Moltz & Stettner, 1961). Chicks learn discriminations better when the patterns to be discriminated in a test situation include some which have become familiar in the home cage (Bateson, 1964a, 1964b), and they avoid strange patterns more than they do patterns with which they have become familiar by apparently passive cohabitation. This suggests that familiarity with the imprinted object, and the consequent development of avoidance of other, now "strange" objects, may account for the end of the imprinting period (Sluckin, 1962). If so, imprinting could be regarded as the outcome of well-known learning processes in a special developmental situation, rather than as a qualitatively unique or special form of learning itself (Bateson, 1966; Hinde, 1962; Moltz, 1963, 1968).

The assertion that "imprinting" may be continuous with other forms of learning should not be allowed to obscure the very interesting fact that Lorenz was pointing out in his discussion: in many species of birds, social reactions become directed to members of the bird's own species as a result of the developing individual's experience in a normal social setting during a specific stage in its development.

In introducing this discussion, we used the notion that an animal may learn through experience to "know" to what species it belongs. When we examine the problems of development analytically, however, it soon becomes apparent that this is a misleading formulation. It assumes in advance that the animal has a global, integrated cognitive appreciation of the fellow member of the species (conspecific) as an entity, and perceives it more or less as we do. It assumes that the process of learning to recognize and react to conspecifics is a unitary one, and that "recognition" is a single item in the organization of the animal's behavior. None of these assumptions is necessarily true.

In many cases, different social reactions of an animal to the same conspecific may be elicited by different features of the stimulating animals. For example, Immelmann (1959) found that male zebra finches would perform various social responses to rather crude artificial "zebra finch" models. A very simple, unbirdlike model would induce courtship behavior, provided only that it had a red bill. Aggressive responses, however, required the presence of malelike markings on the head and body of the model. Similarly, ter Pelkwijk and Tinbergen (1937) showed that the aggressive responses of the male three-spined stickleback are elicited primarily by the red belly characteristic of a male, irrespective of shape, while the courtship responses depend on the characteristic shape of the belly of the gravid female.

These observations do not, by themselves, elucidate problems of behavioral development, since they may be made with animals which have never seen a conspecific (stickleback:Cullen, 1960).* Their significance for our discussion lies in the fact that the animal's social behavior *as a whole* cannot be said to be oriented to a conspecific "recognized" as such in a unitary or global way. Careful analysis is required to determine which aspects of the conspecific are rele-

*Caution is required in interpreting the overall developmental significance of such observations of complex behavioral responses occurring in the absence of relevant experience. The hostile reaction of the male stickleback to the vertical posture of another male (ter Pelkwijk & Tinbergen, 1937; Tinbergen, 1942) undoubtedly occurs without previous experience of conspecifics. This reaction of a territory-holding male can be demonstrated by confining a second stickleback in a test tube inside the tank; when the stimulus fish is forced into a vertical position by the experimenter's manipulation of the test tube, it elicits an attack. After a series of such tests, the *empty* test tube alone will now elicit the attack, but only if it is held in a vertical position (Verplanck, 1955). This means that even though an inexperienced animal will respond to this posture the identical reaction of an *experienced* fish to the same posture cannot be said to be free of any influence of the previous experience.

vant to which aspects of the behavior of the acting animal, and the developmental processes must be separately considered for each aspect of the behavior.

These considerations bear upon the use by Scott (1962) of the concept of "critical period" or "sensitive period" to mean a specific time in the life of the animal when all its social responses are especially labile and predeterminedly subject to the effects of experience. He uses the term "socialization" to refer to the development, at a particular period, of the dog's major orientation to other animals (e.g. to human handlers or to other dogs). Schneirla and Rosenblatt (1963) point out that a kitten kept in isolation from mother and siblings for short periods will, when returned to the litter situation, behave differently than the littermates, and that this type of effect is not limited to a particular stage of development, although the specific effects will be different at different ages. This is quite different from the notion of a "sensitive period" at which the major socialization of the animal occurs, preceded and followed by periods in which the animal is resistent to the effects of social experience. Similarly, Klinghammer (1967; Klinghammer & Hess, 1964a) showed that the age at which doves require experience of human beings in order to become "tame" is not identical with the age at which they require experience of doves in order to mate with doves, and that they retain for a very long time (perhaps throughout life) the ability to take advantage of experience with conspecifics, by increasing their readiness to mate with them.

The processes by which animals become able to react appropriately to members of their own species cannot be easily generalized across species, since the ecological circumstances, and the overall behavioral organization, of the particular species may impose constraints upon the possible roles of experience. For example, doves of many species, reared by foster parents of other species of dove may, when adult, readily mate with members of the same species as the foster parents, rather than of their own species (Whitman, 1919). It would be premature, however, to conclude from this that all birds are dependent on early experience with parents to establish mating preferences. The American cowbird is a brood parasite, which lays its eggs (1 to a nest) in the nests of hosts of other species (Friedmann, 1929). Birds of this species are thus *always* reared by foster parents of other species. Yet they always flock together in one-species flocks, and mate only with cowbirds. Obviously, natural selection is exerting strong pressure against any characteristics that would permit these birds to base their mating preferences on early experience with the adults that reared them.

The differences between animals that are, and those that are not, influenced by experience in their species identification is not an all-or-none one. Presumably the features of the cowbird that make it possible for it to react selectively to other cowbirds without specific experience may be represented to some degree in other birds, as well. Doves that have been reared by human hands, and that court humans, will gradually come to court doves if later allowed to associate with them, and will prefer doves if hand reared in pairs; doves reared by doves will not come to prefer humans if later allowed to associate with them (Klinghammer, 1967). Ducklings can be imprinted on a wide variety of objects (Hess, 1959). However, wood ducks, which nest in holes in trees, from which the young are eventually called by their mother, can be imprinted to specific sounds, while mallard ducks, which hatch in ground nests from which they follow their mothers, can be imprinted to sounds only when the sounds are associated with an object that the ducklings can follow visually (Gottlieb, 1963, 1965).

The development of species identification and of species recognition must be seen as occurring in a social setting defined by the overall pattern of the animal's ecology, social organization, and behavior, and the analysis of this development must take this specific setting into consideration before a generalized account can be evolved.

2. *Individual Recognition*

In many—perhaps most—higher animals many elements of social behavior are selectively oriented toward particular individuals, and consequently the recognition of individuals plays an important role in social organization. The distinction between species recognition and individual recognition is, of course, only relative. One cannot say that there is a sharp separation, either in time or in content, between the processes of development of what *we* distinguish as these two kinds of recognition. It is, nevertheless, helpful to make this relative distinction.

Any observer of social groups of adult birds or mammals quickly becomes convinced that they recognize each other individually (Altmann, 1958; Lorenz, 1931, 1935; Tinbergen, 1953; van Lawick-Goodall, 1969; and many others). The stability of dominance hierarchies in small flocks of domestic fowl, which survives the transfer of the flock to new quarters, undoubtedly depends upon individual recognition (Collias, 1944). Domestic fowl apparently recognize each other in part by characteristics of the shape and color of the head (Collias, 1944).

Birds also recognize each other by voice. Ovenbirds (Weeden & Falls, 1959) and White-Throated Sparrows (Falls, 1969) sing in response to hearing the songs of other birds of their own species. They will also respond to recorded songs played back through loudspeakers placed near their territories. When tested with various recorded songs, they respond best to the songs of males from territories distant from their own (hence unfamiliar), and least to songs of neighboring (hence familiar) males. A particularly convincing bit of evidence for individual recognition of the song is that the White-Throated Sparrow responds characteristically to the song of a familiar male even if it is played from an unfamiliar direction!

In mammals, individual recognition undoubtedly often depends partly on olfaction (Beach and Jaynes, 1956a, 1956b; LeMagnen, 1953), but visual and auditory cues are also important, no doubt to different degrees in different species. Sexual preferences in monkeys depend on both olfactory and nonolfactory stimuli (Michael & Saayman, 1967; Michael, Saayman, & Zumpe, 1967), and monkeys in social groups respond discriminatively to the voices of other individuals (E. H. Hansen, personal communication).

Studies of the development of individual recognition in young animals are few, but illuminating. Using playbacks of recorded adult vocalizations, Beer (1969, 1970) has shown that Laughing Gull chicks, removed from the nest and immediately tested in the laboratory, can distinguish the voices of their parents from those of neighboring adults at the age of 6 days. The chicks consistently approach the loudspeaker from which the parent's voice is heard, while avoiding or ignoring the calls of other adults.

Tschanz (1968), in similar experiments, showed that a young guillemot (a bird related to the gulls) would actually touch and peck at a loudspeaker playing the voice of its parent, while avoiding it when the voice was that of another adult. Tschanz found that the learning of the parents' calls occurs exceedingly early in the life of the chick. Incubator-hatched chicks responded preferentially to recorded calls which the experimenter had played to them while they were *in the egg* before hatching!

Experiments of comparable elegance are not yet available for mammals, but any one who has watched domestic cats or horses is aware that the young animal very quickly learns selectively to follow its mother and that this specific individual association, physically very close, lasts for a very long time in horses and other herd-living animals (Altmann, 1958).

Kittens learn during the first few days of life to orient to the mother, and to the area in which the mother spends her time, on the basis of their olfactory characteristics (Rosenblatt, Turkewitz, & Schneirla, 1969). It is not yet clear to what extent this involves the learning of the specific characteristics of the individual mother, but there is strong evidence that this is the case since at 6 days of age kittens, even though very hungry, will not readily suckle from a strange mother (Ewer, 1968; Rosenblatt, personal observation). Infant rhesus monkeys clearly recognize their mothers from an early age, staying close to them, and running to them when alarmed; the fact that they are several weeks old before they spend any time farther than arm's reach from the mother makes it difficult to be precise about when the individual recognition develops.

It is clear that the development of social behavior in the social setting involves not only the development of responsiveness to, and orientation toward, members of the animal's own species, but also the integration of recognition of, and relationships with, particular individuals, into the pattern of the animal's behavior.

3. *Mother-Young Synchrony*

The mother (and, in some cases, the father) is the central and crucial focus of the developing infant's environment. But she is not merely a stable, passive part of the environment. A major key for the understanding of the psychosocial development of young mammals (Schneirla, 1946, 1951; Schneirla & Rosenblatt, 1961) lies in the facts that the mother actively reacts to the infant, that her physiological and psychological state changes as a consequence of association with the infant, just as his does as a consequence of association with her, that her changing behavior toward the infant reflects her changing internal state as well as her reaction to his changing characteristics, and that, consequently, the social development of the infant is one expression of a reciprocal developmental interaction between infant and mother.

The changes in the mother that prepare her to participate in the postnatal development of the infant begin early in pregnancy. We use the terms "maternal" or "maternal condition" to refer to the condition of a rat that reacts to rat pups by retrieving them, licking them, building a nest for them, and assuming the nursing position over them. A virgin rat is ordinarily not maternal (Wiesner & Sheard, 1933), although she can be induced to become maternal by several days of enforced cohabitation with pups (Rosenblatt, 1967). A rat that has just given birth is, however, immediately maternal (Rosenblatt & Lehrman, 1963). This maternal condition is brought about by the termination of pregnancy, and the hormonal events of pregnancy bring about the development of the condition in which the rat can be made maternal by terminating the pregnancy. There are three lines of evidence supporting this conclusion: (a) Caesarian delivery of the fetuses facilitates the induction of the maternal condition, and it does so more effectively, the later in pregnancy it is carried out (Rosenblatt, 1969); (b) blood plasma from a parturient, but not a late pregnant, rat facilitates the induction of the maternal condition (Terkel and Rosenblatt, 1967); and (c) a course of hormone treatment approximately imitating the endocrine events of pregnancy and parturition induces maternal behavior in nonpregnant rats (Moltz, Lubin, Leon, & Numan, 1970).

Mammals that have just given birth act, differently in different species, in such a way as to facilitate access to the nipples by the young. The rat licks the helpless, nonmobile pups, builds up the nest under and around them, tucks them under her, and crouches over them in the nursing position (Wiesner & Sheard, 1933). The cat lies quietly on her side, occasionally licking the kittens, so that the kittens' activity stimulated by her licking, and the kittens' tendency to press against the furry surface of the mother's body, can lead them to find the nipples (Schneirla, Rosenblatt, & Tobach, 1963). Goats and sheep lick the infant until it stands up and continue licking it while it is standing just after birth. Gradually, in the course of licking and thus remaining in contact with the infant, the mother comes into an orientation to it which brings the infants head near the nipples (Blauvelt, 1955; Collias, 1956). The rhesus monkey pulls the infant against the ventral side of the mother's body where the infant grasps and makes a connection which may not be broken for many days (Hinde & Spencer-Booth, 1968; Tinklepaugh & Hartman, 1932).

In all these cases, it is clear that the initial reactions of the infant are adapted to making contact with the mother, but that specific behavior on the part of the mother is also necessary in order for functional contact to be made. Further, these actions are different in different species; in each case the characteristic postparturitive behavior of the mother and the characteristic initial behavior of the neonate are adapted to each other, and in each case the behavior of the mother includes active and positive elements of orientation and attention to the neonate.

We have already pointed out that these interactions between neonate and mother may be the occasion for important experiential effects on the infant's development. The mother may also learn in this situation. For example, if a neonatal goat or sheep is separated from its mother for even an hour or less immediately after parturition, the mother remains indifferent thereafter to the infant (Hersher, Moore, & Richmond, 1958; Moore, 1968). By the use of artificial odors, Klopfer and Gamble (1966) showed that in this situation the mother learns the individual smell of the infant, and that this provides the initial orientation of the mother to the neonate through which the infant is first brought into the normal social situation.

Individual recognition of the young is apparently not at all as significant in rats as it seems to be in the other species we have described. Rats readily retrieve, lick, cuddle, and nurse strange foster young. Indeed, the usual techniques for testing the maternal condition in rats depend upon this fact (Rosenblatt & Lehrman, 1963). Even in these animals, however, it is possible to demonstrate, by suitable tests, that mothers respond preferentially to their own young. Maternal rats tend to retrieve their own young before strange ones, and spend more time sniffing at the odor of their own young than of others (Beach & Jaynes, 1956a).

The connection thus established between mother and neonate creates the social situation in which the infant grows physically and develops its social integration with other members of its species. Both the physical growth and the behavioral development are, in one way or another, created by the interchange with the mother.

But the mother is also changing. If the neonate is removed from the mother at birth, the maternal condition and the lactational state, both of which spontaneously occur as a consequence of parturition, will quickly disappear (Bruce, 1961; Rosenblatt & Lehrman, 1963); if the infants remain with their mother, both lactation and the maternal condition persist. (The hormonal basis of the maternal condition, while it may not be identical with that underlying lactation, certainly has a great deal in common with it.) This means that the physiological, and consequently behavioral, condition of the mother is influenced by the presence of the young in a way that is responsible for the further maintenance and development of her parental behavior toward them, and consequently for the maintenance and development of the situation in which the normal development of the young is created.

As the young animals grow, the situation between mother and young changes. The young become more mobile, more interested in aspects of the environment other than the mother, and less attached to her. In both cat (Schneirla *et al.*, 1963) and rat (Rosenblatt & Lehrman, 1963) the mother initiates most nursing episodes during early postparturitive life, and there is a transition, sometime before weaning, to a period of nursings initiated by the young. Thus, as the young get older, and approach a stage at which they will be able to get food elsewhere than from the mother, the mother is also changing in such a way that she becomes less interested in nursing the young. This kind of change in the mother depends partly upon the changing character of the young. Mother mice or rats can be maintained in a maternal condition for periods of up to 3 or 4 times as long as the normal weaning period simply by repeatedly removing the young and replacing them with younger foster pups so that the mother continuously tends infants that do not become old enough to make her lose interest in maternal care (Bruce, 1961; Nicoll & Meites, 1959; Wiesner & Sheard, 1933). When her young reach weaning age and the mother and young naturally abandon each other, the mother is not simply ceasing to react to the young because their character as stimuli change; she loses the maternal condition, as measured by her responses to *young* pups.

What the pup experiences, then, is not a stable, unchanging mother who responds appropriately to the infant, and therefore changes her behavior as his stimulus characteristics change. He has, rather, a mother whose physiological and motivational condition is itself changing partially in response to changes in the young, just as the infant's physiological and behavioral state changes partly through experience and partly through the growth changes made possible by the nurturing behavior that he elicits from the mother. The changing social situation as the infant develops thus represents a synchronization between the physiological and behavioral states of mother and young.

Rhesus monkeys have also been shown to develop in a social situation in which there is an intricate interlacement of reciprocal influences between developing behavioral states in mother and young. Hinde and Spencer-Booth (1968) observed mother-infant interactions over many months, counting such items as the number of occasions on which the infant was off or was on the mother, the occasions when the infant moved away from the mother, and when the mother approached or left the infant. Using a number of ingenious measures based on these observations, they were able to demonstrate that in these monkeys, as in cats and rats, the immediate cause of the increasing independence of the growing infant lies in a decrease in the mother's initiatives toward the infant and in her attentiveness to him. As they point out, this increasing indifference of the mother may depend upon changes, based on growth and on experience, in the physical and behavioral characteristics of the young which elicit her attention.

A similar conception of physiological and behavioral synchrony can be usefully applied to the development of many birds. In the ring dove, the act of incubating the eggs appears to be the source of stimulation that induces the secretion of prolactin by the birds' pituitary glands (Friedman & Lehrman, 1968; Patel, 1936). This hormone induces growth of the crop, which secretes a milky substance that the parents regurgitate to the young when the eggs hatch (Riddle, 1963); it also induces the parents to feed the young (Lehrman, 1955). The crop is well developed by the time the eggs hatch. The pituitary gland continues to secrete prolactin and thus the crop is stimulated to produce crop milk for 12 or 15 days if the young remain with the parents; if the young are removed at hatching prolactin secretion declines and new eggs are produced after 8 or 9 days instead of after the normal period of 20 days or so (Kobayashi, 1953). The crop-milk secretion gradually declines starting a few days after hatching so that the crop contains, day by day, more grain (swallowed by the adults) and less crop milk (Beams & Meyer, 1931); thus by the time the crop is almost inactive after 15–20 days, the parents, when they regurgitate to the young, are regurgitating mostly grain. This change in crop content appears to parallel the developing ability of the young doves' digestive system to handle food other than crop milk (Klinghammer & Hess, 1964b).

In the ring dove, the situation is reminiscent of that in the mammals we have been discussing: the physiological condition of the parent changes after hatching in ways that are relevant to the changing behavior and changing requirements of the young. The parents take the initiative, when the young are about 15 days old, in refusing to regurgitate food when the young beg from them, and it is at about this time that the young make the transition from begging from the parent to pecking at grain independently. By testing 14- to 15-day-old young doves with strange adults whose own young had been hatched at various times before the test, Wortis (1969) demonstrated that the readiness of the parent doves to feed young depends partly upon the time since the parents' eggs have hatched, and that the pecking at food of the parent dove when it refused to feed a young by regurgitation contributes to the young bird's ability to feed independently.

The examples we have given from several widely different types of mammal and from a bird indicate, we think, that synchrony between the changing psychobiological states of mother and young, based in part on preparturitional (or prehatching) changes in the parent, partly on growth changes in the young, partly on cyclically endogenous postparturitional changes in the mother, and partly on mutual influences of the two on each other, is a centrally important conception for the analysis of the development of social behavior in higher animals.

III. Conclusion

We have attempted to survey a number of problems related to, and a number of conceptual approaches to the study of aspects of behavior relevant to social integration. It will be apparent to the reader that we do not present a coherent, formal, or specifically articulated "theory" of development. At the present stage of development of our field (or at least of our knowledge of it), we do not feel any impulse toward such theorizing. We hope we have expressed our view of the importance of some orienting attitudes: respect for diversity both in nature and in the activities of our fellow scientists; attention to the relationship between morphological and experiential influences in development; appreciation of continuity, as well as discontinuity, in development; consideration of the setting in which developmental processes take place; and the perception of interaction as the central source of developmental dynamics.

References

Altmann, M. Social integration of the moose calf. *Animal Behaviour*, 1958, **6**, 155–159.

Bateson, P. P. G. Effect of similarity between rearing and testing conditions on chick's following and avoidance response. *Journal of Comparative and Physiological Psychology*, 1964, **57**, 100–103. (a)

Bateson, P. P. G. Relation between conspicuousness of stimuli and their effectiveness in the imprinting situation. *Journal of Comparative and Physiological Psychology*, 1964, **58**, 407–411. (b)

Bateson, P. P. G. The characteristics and context of imprinting. *Biological Reviews of the Cambridge Philosophical Society*, 1966, **41**, 177–220.

Beach, F. A., & Jaynes, J. Studies of maternal retrieving in rats. I. Recognition of young. *Journal of Mammalogy*, 1956, **37**, 177–180. (a)

Beach, F. A., & Jaynes, J. Studies of maternal retrieving in rats. III. Sensory cues involved in the lactating female's response to her young. *Behaviour*, 1956, **10**, 104–125. (b)

Beams, H. W., & Meyer, R. K. The formation of pigeon "milk." *Physiological Zoölogy*, 1931, **4**, 486–500.

Beer, C. G. Laughing gull chicks: Recognition of their parents' voices. *Science*, 1969, **166**, 1030–1032.

Beer, C. G. Individual recognition of voice in the social behavior of birds. *Advances in the Study of Behavior*, 1970, **3**, 27–74.

Blase, B. Die Lautäusserungen des Neuntoters (*Lanius c. collurio* L.). Freilandbeobachtungen und Kasper-Häuser-Versuche. *Zeitschrift für Tierpsychologie*, 1960, **17**, 293-344.

Blauvelt, H. Dynamics of the mother-newborn relationship in goats. In B. Schaffner (Ed.), *Group processes. Transactions of the first conference.* New York: Josiah Macy, Jr. Found., 1955. Pp. 221-258.

Bruce, H. M. Observations on the suckling stimulus and lactation in the rat. *Journal of Reproduction and Fertility*, 1961, **2**, 17-34.

Collias, N. E. Aggressive behavior among vertebrate animals. *Physiological Zoölogy*, 1944, **17**, 83-123.

Collias, N. E. The analysis of socialization in sheep and goats. *Ecology*, 1956, **37**, 228-239.

Cullen, E. Experiment on the effect of social isolation on reproductive behaviour in the three-spined stickleback. *Animal Behaviour*, 1960, **8**, 235.

Ewer, R. F. *Ethology of mammals.* London: Logos Press, 1968.

Falls, J. B. Functions of territorial song in the white-throated sparrow. In R. A. Hinde (Ed.), *Bird vocalizations in relation to current problems in biology and psychology.* London and New York: Cambridge Univ. Press, 1969. Pp. 207-232.

Friedman, M., & Lehrman, D. S. Physiological conditions for the stimulation of prolactin secretion by external stimuli in the male ring dove. *Animal Behaviour*, 1968, **16**, 233-237.

Friedmann, H. *The cowbirds. A study in the biology of social parasitism.* Springfield, Ill.: Thomas, 1929.

Gottlieb, G. A naturalistic study of imprinting in wood ducklings (*Aix sponsa*). *Journal of Comparative and Physiological Psychology*, 1963, **56**, 86-91.

Gottlieb, G. Imprinting in relation to parental and species identification avian neonates. *Journal of Comparative and Physiological Psychology*, 1965, **59**, 343-356.

Harlow, H. F. The primate socialization motives. *Transactions & Studies of the College of Physicians of Philadelphia*, 4th ser., 1966, **33**, 224-237.

Harlow, H. F., Dodsworth, R. O., & Harlow, M. K. Total social isolation in monkeys. *Proceedings of the National Academy of Sciences of the United States*, 1965, **54**, 90-97.

Hein. A., & Held, R. Dissociation of the visual placing response into elicited and guided components. *Science*, 1967, **158**, 390-392.

Held, R., & Hein, A. Movement-produced stimulation in the development of visually guided behavior. *Journal of Comparative and Physiological Psychology*, 1963, **56**, 872-876.

Hersher, L., Moore, A. U., & Richmond, J. B. Effect of postpartum separation of mother and kid on maternal care in the domestic goat. *Science*, 1958, **128**, 1342-1343.

Hess, E. H. Imprinting. *Science*, 1959, **130**, 133-141.

Hinde, R. A. Some aspects of the imprinting problem. *Symposia of the Zoological Society of London*, 1962, **8**, 129-138.

Hinde, R. A. (Ed.) *Bird vocalizations in relation to current problems in biology and psychology.* London and New York: Cambridge Univ. Press, 1969.

Hinde, R. A., & Spencer-Booth, Y. The study of mother-infant interaction in captive group-living rhesus monkeys. *Proceedings of the Royal Society, Ser. B*, 1968, **169**, 177-201.

Hubel, D. H., & Wiesel, T. N. Receptive fields of single neurons in the cat's striate cortex. *Journal of Physiology*, (*London*), 1959, **148**, 574.

Hubel, D. H., & Wiesel, T. N. Receptive fields, binocular interaction and functional architecture in the cat's visual cortex. *Journal of Physiology (London)*, 1962, **160**, 106-154.

Hubel, D. H., & Wiesel, T. N. Receptive fields of cells in striate cortex of very young, visually inexperienced kittens. *Journal of Neurophysiology*, 1963, **26**, 994-1002.

Immelmann, K. Experimentelle Untersuchungen über die biologische Bedeutung artspezifischer Merkmale beim Zebrafinken (*Taeniopygiacastanotis* Gould). *Zoologische Jahrbücher*, 1959, **86**, 437-592.

James, W. T. The effect of satiation on the sucking response in puppies. *Journal of Comparative and Physiological Psychology*, 1957, **50**, 375-378.

James, W. T., & Rollins, J. Effect of various degrees of stomach loading on the sucking response in puppies. *Psychological Reports*, 1965, **17**, 844-846.

Klinghammer, E. Factors influencing choice of mate in altricial birds. In H. W. Stevenson (Ed.), *Early behavior: Comparative and developmental approaches*. New York: Wiley, 1967. Pp. 5-42.

Klinghammer, E., & Hess, E. H. Imprinting in an altricial bird: The Blond Ring Dove (*Streptopelia risoria*). *Science*, 1964, **146**, 265-266. (a)

Klinghammer, E., & Hess, E. H. Parental feeding in ring doves (*Streptopelia roseogrisea*): Innate or learned? *Zeitschrift für Tierpsychologie*, 1964, **21**, 338-347. (b)

Klopfer, P. H., & Gamble, J. Maternal "imprinting" in goats: The role of chemical senses. *Zeitschrift für Tierpsychologie*, 1966, **23**, 588-592.

Kobayashi, H. Studies on molting in the pigeon. IV. Molting in relation to reproductive activity. *Japanese Journal of Zoology*, 1953, **11**, 11-20.

Koepke, J., & Pribram, K. Effect of milk on the maintenance of sucking in kittens. *Journal of Comparative and Physiological Psychology*, 1971, in press.

Konishi, M. Effects of deafening on song development in American robins and Blackheaded Grosbeaks. *Zeitschrift für Tierpsychologie*, 1965, **22**, 584-599.

Konishi, M., & Nottebohm, F. Experimental studies in the ontogeny of avian vocalizations. In R. A. Hinde (Ed.), *Bird vocalizations in relation to current problems in biology and psychology*. London and New York: Cambridge Univ. Press, 1969. Pp. 29-48.

Kovach, J. K., & Kling, A. Mechanisms of neonate sucking behaviour in the kitten. *Animal Behaviour*, 1967, **15**, 91-101.

Lehrman, D. S. The physiological basis of parental feeding behaviour in the ring dove (*Streptopelia risoria*). *Behaviour*, 1955, **7**, 243-286.

Lehrman, D. S. Interaction of hormonal and experiential influences on development of behavior. IN E. L. Bliss (Ed.), *Roots of behavior*. New York: Harper (Hoeber), 1962. Pp. 142-156.

LeMagnen, J. L'olfaction: Le fonctionnement olfactif et son intervention dans les regulations psycho-physiologiques. *Journal de Physiologie (Paris)*, 1953, **45**, 285-326.

Levy, D. M. Experiments on the suckling reflex and social behavior of dogs. *American Journal of Orthopsychiatry*, 1934, **4**, 203-224.

Lorenz, K. Beiträge zur Ethologie sozialer Corviden. *Journal für Ornithologie*, 1931, **79**, 67-120.

Lorenz, K. Der Kumpan in der Umwelt des Vogels. *Journal für Ornithologie*, 1935, **83**, 137-213, 289-413.

Marler, P. Animal communication signals. *Science*, 1967, **157**, 769-774.

Marler, P., & Tamura, M. Song "dialects" in three populations of white-crowned sparrows. *Condor*, 1962, **64**, 368-377.

Marler, P., & Tamura, M. Culturally transmitted patterns of vocal behavior in sparrows. *Science*, 1964, **146**, 1483-1486.

Melzack, R. The genesis of emotional behavior: An experimental study of the dog. *Journal of Comparative and Physiological Psychology*, 1954, **47**, 166-168.

Melzack, R. Early experience: A neuropsychological approach to heredity-environment interactions. In G. Newton & S. Levine (Eds.), *Early experience and behavior*. Springfield, Ill.: Thomas, 1968. Pp. 65-82.

Melzack, R., & Scott, T. H. The effects of early experience on the response to pain. *Journal of Comparative and Physiological Psychology*, 1957, **50**, 155-161.

Melzack, R., & Thompson, W. R. Effects of early experience on social behaviour. *Canadian Journal of Psychology*, 1956, **10**, 82-90.

Michael, R. P., & Saayman, G. S. Individual differences in the sexual behaviour of male rhesus monkeys (*Macaca mulatta*) under laboratory conditions. *Animal Behaviour*, 1967, **15**, 460-466.

Michael, R. P., Saayman, G. S., & Zumpe, D. Sexual attractiveness and receptivity in rhesus monkeys. *Nature* (*London*), 1967, **215**, 554-556.

Moltz, H. Imprinting: An epigenetic approach. *Psychological Review*, 1963, **70**, 123-137.

Moltz, H. An epigenetic interpretation of the imprinting phenomenon. In G. Newton & S. Levine (Eds.), *Early experience and behavior*. Springfield, Ill.: Thomas, 1968. Pp. 3-41.

Moltz, H., Lubin, M., Leon, M., & Numan, M. Hormonal induction of maternal behavior in the ovariectomized nulliparous rat. *Physiology & Behavior*, 1970, **5**, 1373-1377.

Moltz, H., & Stettner, L. J. The influence of patterned-light deprivation on the critical period for imprinting. *Journal of Comparative and Physiological Psychology*, 1961, **54**, 279-283.

Moore, A. U. Effects of modified maternal care in the sheep. In G. Newton & S. Levine (Eds.), *Early experience and behavior*. Springfield, Ill.: Thomas, 1968. Pp. 481-529.

Mulligan, J. A. Singing behavior and its development in the song sparrow *Melospiza melodia*. *University of California, Berkeley, Publications in Zoology*, 1966, **81**, 1-76.

Nicoll, C. S., & Meites, J. Prolongation of lactation in the rat by litter replacement. *Proceedings of the Society for Experimental Biology and Medicine*, 1959, **101**, 81-82.

Nottebohm, F. Vocalizations and breeding behaviour of surgically deafened ring doves. *Animal Behaviour*, 1971, in press.

Patel, M. D. The physiology of the formation of the pigeon's milk. *Physiological Zoölogy*, 1936, **9**, 129-152.

Riddle, O. Prolactin or progesterone as key to parental behaviour. *Animal Behaviour*, 1963, **11**, 419-432.

Riesen, A. H. The development of visual perception in man and chimpanzee. *Science*, 1947, **106**, 107-108.

Riesen, A. H. Effects of stimulus deprivation on the development and atrophy of the visual sensory system. *American Journal of Orthopsychiatry*, 1960, **30**, 23-36.

Riesen, A. H., Kurke, M. I., & Mellinger, J. C. Interocular transfer of habits learned monocularly in visually naive and visually experienced cats. *Journal of Comparative and Physiological Psychology*, 1953, **46**, 166-172.

Rosenblatt, J. S. Nonhormonal basis of maternal behavior in the rat. *Science*, 1967, **156**, 1512-1514.

Rosenblatt, J. S. The development of maternal responsiveness in the rat. *American Journal of Orthopsychiatry*, 1969, **39**, 36-56.

Rosenblatt, J. S., & Lehrman, D. S. Maternal behavior of the laboratory rat. In Harriet Rheingold (Ed.), *Maternal behavior in mammals.* New York: Wiley, 1963. Pp. 8-57.

Rosenblatt, J. S., Turkewitz, G., & Schneirla, T. C. Early socialization in the domestic cat as based on feeding and other relationships between female and young. In B. M. Foss (Ed.), *Determinants of infant behaviour.* Methuen: London, 1961. Pp. 51-74.

Rosenblatt, J. S., Turkewitz, G., & Schneirla, T. C. Development of home orientation in newly born kittens. *Transactions of the New York Academy of Sciences,* 2nd ser., 1969, **31**, 231-250.

Ross, S. Sucking behavior in neonate dogs. *Journal of Abnormal and Social Psychology,* 1950, **46**, 142-149.

Satinoff, E., & Stanley, W. C. Effect of stomach loading on sucking behavior in neonatal puppies. *Journal of Comparative and Physiological Psychology,* 1963, **56**, 66-68.

Schneirla, T. C. Problems in the biopsychology of social organization. *Journal of Abnormal and Social Psychology,* 1946, **41**, 385-402.

Schneirla, T. C. A consideration of some problems in the ontogeny of family life and social adjustments in various infrahuman animals. IN M. J. E. Senn (Ed.), *Problems of infancy and childhood.* New York: Josiah Macy, Jr. Found., 1951. Pp. 81-124.

Schneirla, T. C., & Rosenblatt, J. S. Behavioral organization and genesis of the social bond in insects and mammals. *American Journal of Orthopsychiatry,* 1961, **31**, 223-253.

Schneirla, T. C., & Rosenblatt, J. S. "Critical periods" in the development of behavior. *Science,* 1963, **139**, 1110-1115.

Schneirla, T. C., Rosenblatt, J. S., & Tobach, E. Maternal behavior in the cat. In Harriet Rheingold (Ed.), *Maternal behavior in mammals.* New York: Wiley, 1963. Pp. 122-168.

Scott, J. P. Social behavior, organization and leadership in a small flock of domestic sheep. *Comparative Psychology Monographs,* 1945, **18**, 1-29.

Scott, J. P. Critical periods in behavioral development. *Science,* 1962, **138**, 949-958.

Sluckin, W. Perceptual and associative learning. *Symposia of the Zoological Society of London,* 1962, **8**, 193-198.

ter Pelkwijk, J. J., & Tinbergen, N. Eine reizbiologische Analyse einiger Verhaltensweisen von *Gasterosteus aculeatus* L. *Zeitschrift für Tierpsychologie,* 1937, **1**, 193-200.

Terkel, J. & Rosenblatt, J. S. Maternal behavior induced by maternal blood plasma injected into virgin rats. *Journal of Comparative and Physiological Psychology,* 1967, **65**, 479-482.

Thorpe, W. H. Further studies on the process of song learning in the Chaffinch (*Fringilla coelebs gengleri*). *Nature* (*London*), 1958, **182**, 554-557.(a)

Thorpe, W. H. The learning of song patterns by birds, with especial reference to the song of the Chaffinch *Fringilla coelebs. Ibis,* 1958, **100**, 535-570.(b)

Tinbergen, N. An objectivistic study of the innate behaviour of animals. *Bibliotheca Biotheoretica, Leiden,* 1942, **1**, 1-98.

Tinbergen, N. *The herring gull's world.* London: Collins, 1953.

Tinklepaugh, O. L., & Hartman, C. G. Behavior and maternal care of the newborn monkey (*Macaca mulatta* = "*M. rhesus*"). *Journal of Genetic Psychology,* 1932, 40, 257-286.

Tschanz, B. Trottellummen: Die Entstehung der personlichen Beziehungen zwischen Jungvögel und Eltern. *Zeitschrift für Tierpsychologie, Suppl.* 1968, **4**, 1-103.
van Lawick-Goodall, J. The behaviour of free-living chimpanzees in the Gombe Stream Reserve. *Animal Behaviour Monographs*, 1969, **1**, 161-311.
Verplanck, W. S. Since learned behavior is innate, and vice versa, what now? *Psychological Review*, 1955, **62**, 139-144.
Weeden, J. S., & Falls, J. B. Differential responses of male ovenbirds to recorded songs of neighboring and more distant individuals. *Auk*, 1959, **76**, 343-351.
Whitman, C. O. The behavior of pigeons. *Carnegie Institution of Washington Publication*, 1919, **257**, 1-161.
Wiesel, T. N., & Hubel, D. H. Effects of visual deprivation on morphology of cells in the cat's lateral geniculate body. *Journal of Neurophysiology*, 1963, **26**, 978-993. (a)
Wiesel, T. N., & Hubel, D. H. Single-cell responses in striate cortex of kittens deprived of vision in one eye. *Journal of Neurophysiology*, 1963, **26**, 1003-1017. (b)
Wiesner, B. P., & Sheard, N. M. *Maternal behaviour in the rat.* Edinburgh: Oliver & Boyd, 1933.
Wortis, R. P. The transition from dependent to independent feeding in the young ring dove. *Animal Behaviour Monographs*, 1969, **2**, 3-54.
Young, W. C. The hormones and mating behavior. In W. C. Young (Ed.), *Sex and internal secretions.* Vol. II. Baltimore, Md.: Williams & Wilkins, 1961. Pp. 1173-1239.

CHAPTER 2

THE DEVELOPMENT OF VISUAL PERCEPTUAL SYSTEMS*

Maurice Hershenson

The attempt to understand ontogenetic changes in perceptual abilities is becoming at once more simple and more complex than had been thought. This peculiar state of affairs results from the clarification of conceptual systems leading to more precise formulations of the possible mechanisms involved, while at the same time making it inevitable that more complex mechanisms are invoked for explanation. It is necessary, therefore, in attempting to understand the ontogenesis of perception, to recast the modes of thought of an earlier era into those which are compatible with modern methods and data. The

*Preparation of this chapter was supported, in part, by a grant from the United States Public Health Service, MH 18092.

first section of this chapter will be concerned with this task. It will then be possible to evaluate some of the recent research on very young organisms—particularly the human infant—with respect to the altered conceptions.

I. Setting the Problem

At least three different views of perceptual development have been suggested: (a) organization of the perceptual system manifest in an increasing ability to perceive; (b) elaboration of more distinctive percepts manifest in an increasing ability to differentiate aspects of stimulus information; and (c) integration of perceptual and motor systems, as in acquiring motor skills. The first view is concerned with the construction of a perceptual system, the other two require an intact and functional system from which the processes of development proceed. Thus the major conceptual difference is between acquiring a perceptual system and improving it or integrating it with another system. Since a functional perceptual system is a prerequisite for improvement or integration, the problem of development in the sense of construction may be said to approximate most closely the essence of the old nativism-empiricism debate. As Gibson (1966, p. 268) has noted, however, neither camp ever doubted the assumption that sensations were innate. The modern version of this debate is phrased in a new terminology, asking about the relative plasticity of the perceptual system—the degree to which it responds to extrinsic stimulation—and the relative organization of the nervous system—the degree to which the system contains analyzers which signal specific environmental invariants.

It should be apparent that these questions can be meaningfully answered only with respect to particular species and in relation to particular perceptual acts, systems, or abilities. The necessity for relating the discussion to species is a function of the comparative evolution of perceptual systems which have evolved in response to needs for different kinds of information from the environment (see Section III). Thus, comparison across species becomes meaningful only in the light of the differences among them. The particular system in question can also be understood as having evolved to deal with particular organism-environment problems, requiring detection of particular subsets of the large number of invariants available in stimulation. Thus ontogenetic development in one system in a particular species need not reflect on that system in another species or on another system in that same species.

The distinctions become clear when comparing two very different perceptual processes in the human being: the perception of space and the perception of meaningful pattern. Many environmental invariants underlying the perception of space have been identified by investigators with different orientations, but the specifications of the informational aspects of these cues and their relation to neural mechanisms have certainly not been simple and have not led to general agreement. This is especially true when asking about mechanisms in ontogenesis. In contrast, there is little doubt that the stimuli which comprise meaningful patterns, e.g., lines and angles, are analyzed or decoded by mechanisms whose organization is genetically determined. Recent physiological evidence demonstrates the existence of many different analyzers in many different species, and psychophysical evidence strongly suggests their existence in man (Weisstein, 1969). Nevertheless, the arbitrary relationship between the physical configuration and the meaning of the stimulus is indisputable.

In light of the great differences in the nature of the mechanisms involved and all of the possible differences across species, attempts to write a general theory of perceptual development make little sense. Thus, the approaches to be outlined below serve mainly to supply conceptual alternatives which may be applied within the more limited problem set by species and specific perceptual acts.

II. Major Conceptual Alternatives

The four major conceptual schemes that have been suggested as mechanisms of perceptual development are: (a) maturation, (b) construction, (c) learning and differentiation, and (d) adaptation.

A. Maturation

The classical nativism which viewed the genetic inheritance of a species as the sole determinant of the perceptual abilities of its adults has been replaced by the maturational view that seeks causal explanations in terms of genetic and environmental determinants acting in concert. This view emphasizes the genetic contribution to the changes observed over time. Thus, it sees the infant as possessing the necessary organizational blueprints for using environmental information in the development of a perceptual system—time and the appropriate environment, then, are all that is necessary for the predetermined organization to be manifest. It should be clear, moreover, that the "organ of perception" (the eye-brain-muscle combination) sets the limits on the kind of stimulus information that can be registered. This view also raises the possibility that adult abilities are based on the learned use of certain informational aspects of the environment, the learning being based on an innately organized and functional perceptual system.

B. Construction

In the constructionist view, the perceptual system is built up from basic sensory elements which are present at birth. The developmental consequence of the constructionist position is an increasing ability to perceive as the newborn grows into infancy and the infant grows into childhood and adulthood. While this view is an elementaristic one with respect to development, it is important to keep in mind the distinction between an elementarism in a developmental theory of perception and an elementarism in a theory of perception in general (such as the kind proposed by the structuralists). The modern version of a general theory of construction (Neisser, 1967) is microgenetic not ontogenetic—a particular percept is constructed from a particular input. In contrast, elementarism in a theory of perceptual development does not imply an elementaristic approach to perception in general. It implies nothing about the nature of the structures which are constructed and in that sense does not undermine the understanding of perception in terms of Gestalt psychology. Theories of perceptual development of this sort are attempts to explain the Gestalt, not invalidate it.

The constructionist view has become increasingly attractive to students of development because of the recent data which suggest that mechanisms which detect invariants of stimulation such as contours, angles, lines of particular orientation, etc., can be found in the visual system of species as different as the frog, the cat, and the monkey (Weisstein, 1969). The possibility that elements or stimulus analyzers of a similar sort exist in the human newborn has lent credence to the "building-block" approach. This view is also supported by reports of qualitative differences in the "looking" of younger and older human infants: the younger infant seems to be "captured by the stimuli" (Ames & Silfen, 1965), to show "obligatory attention" (Stechler & Latz, 1966) in a "vigilant-like state" (Stechler, Bradford, & Levy, 1966); the older infant appears to be "capturing stimuli with his visual behavior" (Ames & Silfen, 1965), to become "aroused" or "excited" by visual stimulation (Stechler & Latz, 1966). Such differences in looking could reflect qualitatively different experiential states mediated by different neuromechanisms and, indeed, theories encompassing this view have been suggested (e.g., Dodwell, 1964; Hebb, 1949; Sackett, 1963).

The conceptual difference between the elements and the larger structures they will be used to synthesize reflects their differential functions as components of the perceptual act—the difference between reception (invariance detection) and organization (synthesis, interpretation, judgment)—as well as their differential roles in development. Thus, the elements, sensory analyzers, are thought to be complex functional units "tuned" to specific invariants of stimulation such as contours, angles, and lines of particular orientation. Activation of a unit signals detection of a particular invariant in the stimulus array—the units are "invariance-bound." In general, they extract and process information from circumscribed regions of the visual field and do not alter their function over time as a result of experience. These elementary sensory mechanisms, then, differ from perceptual mechanisms in their specificity, rigidity, and restricted range of analysis (Hershenson, 1967).

In contrast, perceptual mechanisms reflect the unity and plasticity which are the hallmarks of perception. The perceptual system is plastic in the sense that it adapts readily to systematic variation of the input; it has unity in the sense that perceptual organization encompasses larger portions of the visual field. Thus, perceptual processes involve both spatial and temporal integration. One speaks, for example, of response to "pattern" as a perceptual ability, emphasizing both its global nature in the processing of information from different parts of the visual field, and its unity in integrating the information into a single percept. Moreover, the use of the term "pattern" implies that the information is represented by the relationships among the various parts of the stimulus—by comparison and integration of sensory information over the entire field. But this is simply to reiterate Gestalt psychology. It is these very properties—integrity, unity, and plasticity—which make the problem of perceptual development a meaningful one (Hershenson, 1967).

A constructionist view of perceptual development, then, would have simulus information conveyed through the newborn's visual system to some central locus where perceptual development could take place. The newborn's visual behavior would be determined primarily by the sensory input; the older infant would be able to integrate such input and therefore would enjoy a more flexible relationship with his environment. The change from "obligatory attention" to active searching would reflect the shift from control by sensory mechanisms to mediation by the newly organized perceptual system.

C. Learning and Differentiation

There is little doubt that the perceptual system is intimately involved in developmental changes which take place as a result of learning. The ability to read and to speak, the ability to understand and manipulate symbols, are prime examples of this. What does remain doubtful, however, is the nature and location of the effects. Does the learning take place within the perceptual system so that the actual percept is altered or is it acting within memory or in the response system entirely? Or perhaps the learning is related to the type of contact made between perception and memory. These questions are not easily answered and have been the central issue of much recent research (e.g., Kolers, 1969; Posner, 1969).

On the other hand, there is little doubt that perceptual learning does take place in the sense of a refined ability to make discriminations, i.e., to discover higher-order invariances of stimulation (Gibson, 1966; Gibson & Gibson, 1955). This developmental trend is manifest in all sense modalities: the wine taster's learned ability to differentiate among wines, the musician's ability to listen for subtle tonal quality, the adult's ability to read innuendo in facial expressions. These all represent a similar type of refinement of an ability to discriminate which results in an altered perception. But it is not the kind of learning that is usually meant by this term: it is not an accrual of associations, an attaching of responses, or an accumulation of memories. As a mechanism for perceptual development, this model would start with a system which is capable of a few global discriminations and would then be able progressively to differentiate finer and finer gradations of stimulus structure. In addition, the span of attention would increase with practice by being enlarged in scope and extended in time. Thus the "process is one of learning what to attend to, both overtly and covertly. For the perception of objects, it is the detection of distinctive features and the abstraction of general properties. This almost always involves the detection of invariants under changing stimulation" (Gibson, 1966, p. 270).

D. Adaptation

Since adaptation also implies change over time, it may easily be mistaken for profound developmental change unless the change is carefully followed. The conception of the perceptual system as an adaptive system means that, indeed, the perceptual system is responding to experience, even to past experience, but that there is no lasting effect. It has been viewed as a "self-tuning" system with a nervous system "resonating to the stimulus information" rather than storing images or connecting nerve cells (Gibson, 1966). In this view, the system acts like a monitor—taking note of present or immediately past stimulation and adjusting its baseline or coordinates accordingly. Thus, the major distinction which sets off an adaptive system from all of the others is the nature of the changes. A system which is "constructed" retains its basic structure for as long as the organism lives, as does a system which has been modified by learning. Therefore, theories of perception which relate to adaptation level (e.g., Helson, 1964) or to perception as a result of a weighted average of experience (Ittelson, 1960), are dealing with changes *from* baseline rather than changes *in* baseline. The vast literature on adaptation to distortion (with prisms) usually can be seen as a response to such change.

The distinction between the response to the environment of an adaptive system and one which produces lasting structural change is easier to understand if one thinks of the effects of experience as taking place within a limited time frame—say a week or a month. There are two major considerations for suggesting such a model for the perceptual system—or at least that part of it which deals with the perception of orientation in space, of up and down, of left and right, and of tilt. First, the perceptual response to distortion or displacement is adaptive both in the sense that there is rapid change and also that this change is not lasting. If the system were merely one which responds to the environment by incorporating new information into the already present structures which were built in response to past information, and if this new information were added in with equal weight, then the rapid adaptation which is found would not be expected (Kohler, 1964). Certainly even adaptation over a few months must be considered rapid since the past experience involved here is that which has been occurring over the entire life span of the organism. Thus, given equal weights for each moment of experience over the life span, a few months for adaptation to distorted input is indeed a short time.

Second, the very nature of the change itself suggests that the system is not simply altering its structure in the same way it had been formed. There is no theory of learning or of development in which a gradual buildup of structure over a lifetime is able to be altered in a few months. The conceptualization necessary to describe such a system is one which integrates its input over a limited time span, and is thus modifiable.

Thus, the usual characteristics of adaptive change are readily discernable and can be seen to be very different from those of a system whose basic structure has changed. Adaptation is rapid—it can usually be measured in hours, days, or weeks—and it usually involves changes in perception which could not occur if the system had been built up over many decades of learning or of construction. These will prove to be powerful arguments for the adaptation view. Further, the effects of past experience must be severely limited in this view. A monitoring system will take into account stimuli which occurred over some interval in the recent past, but since the changes are not lasting, they cannot be too remote. Such a system would act like a thermostat with a slight drag.

III. The Evolution of Perceptual Systems

Building upon Walls' *The Vertebrate Eye* (1942), Gibson has traced the evolution of the visual system in terms of the information it supplies to the perceiving organism about his environment. Gibson assumes that ambient light, not the retinal image, is the stimulus for the visual system, and that the information in the light falling upon the organism is the situation to which animals have adapted in the evolution of ocular systems. "The realization that eyes have evolved to permit perception . . . is the clue to a new understanding of human vision itself" (Gibson, 1966, p. 155).

Thus, the essential question of Gibson's analysis is "What is vision used for?" His answer is that vision supplies the organism with three main classes of information: (1) the layout of the surround, (2) the detection of change, and (3) the detection and control of locomotion. The differences picked up by the visual system within each of these three categories may vary from gross to subtle differences in stimulation.

Perhaps an example will clarify Gibson's approach. In analyzing the response to a simple invariant of nature—the fact that strong light comes from above and weak light comes from below—Gibson notes that an

> . . . animal with mere photoreceptive spots on its outer skin can orient to the sky-earth differential by keeping its upper eyespots more strongly stimulated than its lower ones . . . this is already a response to a difference of light in different directions, to an array, not simply a response to 'light.' It is detection of an invariant, not a response to energy; a *sky*tropism, not a *photo*tropism; and it is registered by the relation between stimulated eyespots, not by the stimulation of an eye spot (Gibson, 1966, p. 156).

Gibson sees the evolution of a visual system as a consequence of the unlimited specifying ability of the "superstimulus" provided by the structure of an optic array in transformation from one station point to another. The superstimulus is an

> inexhaustible reservoir of potential information about the world and about the individual's behavior in it . . . When *sequence* is combined with *scene,* vision makes possible the achieving of geographical orientation, the feats of navigation over time, and the cognitive mapping of the environment. It also makes possible the control of skilled movements and the coordination of hand with eye in primates. In man, the ability to control the movements of the hand by vision has led to picture-making and even to ideographic or phonetic writing, from which a new level of cognition emerges. There is visual feedback at all levels of activity: upright posture, locomotion, homing, and the control of vehicles; manipulation, tool-using, mechanical problem solving, and graphic representation. As the information feedback becomes differentiated the skill can become learned (1966, p. 163).

Gibson's detailed analysis of the evolution of visual systems provides a deeper understanding of comparative studies of perception as well as new and different insight into the human visual system. For example, the problem of postural stability of the eyes is understood only when the relation between their functional anatomy and the stimulus array is understood. Vertebrate eyes are set in sockets of the skull so that eye and head movements may be reciprocal. This relationship enables the eyes to stabilize independently of the head and body and to be closely anchored to the array of ambient light. Thus the eyes are now able to maintain their orientation to the surroundings even when the head and body move.

In a similar fashion, the analysis of the position of the eyes with respect to one another can be informative about their evolved function. Lateral eyes (e.g., in fish) afford a panoramic view of the visual field, whereas frontal eyes assure a double input of information about the field ahead. But a frontal position for the eyes does not necessarily mean that the eyes are fixated when turned forward. Fixation requires compulsory convergence (the coordinated pointing of both eyes) of the centers of both eyes, a characteristic of primates. This ability probably provides a different kind of information about the layout of things, namely, the disparity of the overlapping arrays. Thus, primates can rely on both the perspective of motion and the perspective of disparity (the difference between the information from the two converged eyes) for spatial information. Perhaps even more important, compulsory convergence allows primates to look at their hands and at objects in their hands.

With compulsory convergence, the eyes have lost their independence, but they gain the advantage of supplying double assurance of seeing with the central retina. Thus the development of foveas—small areas where receptors are densely concentrated—in the frontal eyes must be accompanied by the development of the ability to explore the optic array by scanning. An ability to look around must compensate for a restriction of panoramic vision. The development of an eye that could fixate carried with it a special advantage even though it had to jump around in order to explore—this eye could fixate a moving object. Further differentiation of functional anatomy within the retina, the duplex retina containing both rods and cones, could also improve the vision of primates by permitting greater differentiation of intensities in both the high and low ranges and by permitting color vision—the detection of the pigmentation of substances in the environment.

The single ability which remains a puzzle for Gibson is accommodation, the ability to focus a definite image on the retina. The functional significance of accommodation is unclear because Gibson does not view the retinal image as the stimulus for the ocular system. "The *structural* properties of the retinal image are what count, not its *visible* properties . . . *The structure of a retinal image is a sample of the structure of ambient light, obtained indirectly from an inverse ray-sheaf*" (1966, p. 172).

IV. Perceptual Abilities of the Human Infant

The resurgence of interest in the human infant as a subject for perceptual research takes its impetus from the pioneering work of Fantz (1958, 1963, 1965; Fantz, Ordy, & Udelf, 1962). The problems and methods have been expanded to such an extent over the past 10 years that it is now feasible to draw some general conclusions from evidence. For example, it seems reasonable to assume that soon after birth the infant is able to extract and utilize complex visual information from the environment. Certainly the components of space and form—edges and angles—are available to the infant visual system. This is not to say that perceptual abilities do not improve with experience or to deny that the developing organism may learn to use different kinds of information to supplement or replace its repertoire of innate mechanisms. Thus, the visual discriminations of the adult and child may be vastly different from those of the infant. In this sense, perhaps Fantz (1965) was correct in his belief that perception is "innate in the neonate and learned in the adult." The following sections will attempt to document these conclusions.

A. Analysis of Experiments on Infants

The empirical test of the various conceptions of perceptual development requires an analysis of the perceptual act of the newborn organism. This analysis must be so conceived that the function of "stimulus feature analyzers" could be differentiated from the activity of "hypercomplex" organizations underlying a global perceptual achievement (Weisstein, 1969). Unfortunately, the design of perceptual experiments has rarely been in response to developmental problems of this sort. For example, Garner, Hake, and Eriksen (1956), in demonstrating the need for differentiating perceptual from response processes, implicitly assumed that the perceptual system is organized and functional [see Fig. 1(a)]. Clearly this proposition cannot be assumed in the present context since it is the very problem under study. Alternative hypotheses which permit analysis of the input side of the perceptual act are needed to complement Garner, Hake, and Eriksen's analysis of the output side. Weisstein (1969) has recently

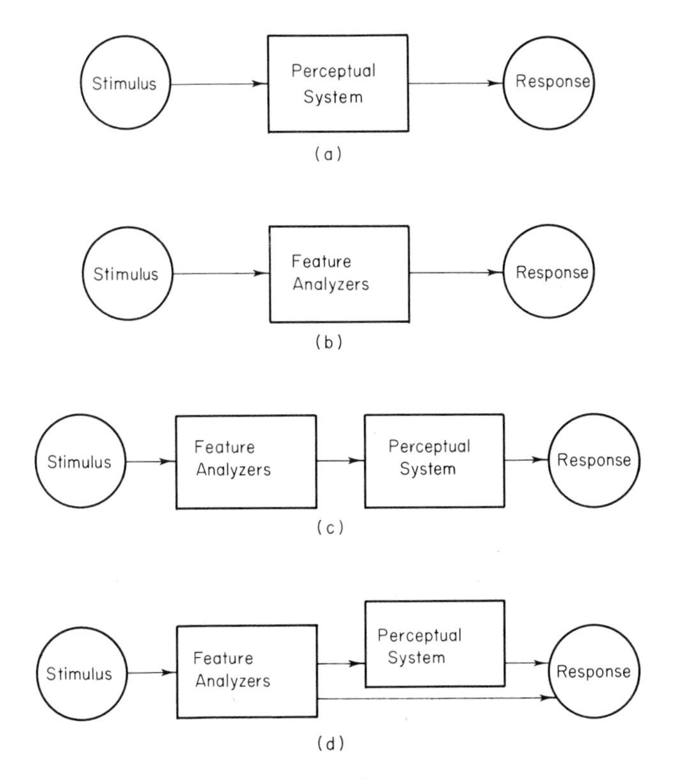

Fig. 1. Alternative schematizations of information flow from stimulation through "feature analyzers" or perceptual systems to response.

provided a methodological analysis of the input side but it is not clear that the psychophysical methods she proposed are applicable to developmental problems.

Specifically to meet this need, Hershenson (1967) provided an analysis of the possible empirical outcomes. Two simplifying assumptions were proposed to clarify the terminology: (a) that feature analyzers are functional, and (b) that the analyzers supply information to the perceptual system, i.e., the analyzers are always prior to the perceptual system in the flow diagram. The three contingencies of analyzer functioning with respect to the problem of development can be seen in Fig. 1. They show (a) analyzers operating in the absence of perceptual organization [Fig. 1(b)], (b) analyzers feeding information to an organizing (perceptual) system [Fig. 1(c)], and (c) an operating perceptual system which, for one reason or another, does not contribute to the observed response in the particular experimental situation. This contingency [Fig. 1(d)] incorporates both the series operation of contingency (b) and the alternative of contingency (c)—the "parallel" connection which enables the sensory analyzers to mediate responses without interference from an otherwise functional perceptual system. The determination that observed responses were contingent upon manipulation of a perceptual dimension suggests the series linkage of Fig. 1(c). It is the necessary and sufficient condition for describing the attributes of the perceptual system while simultaneously describing perceptual experience.

B. Prerequisites for Perception

A recent assessment of the functional integrity of the newborn's visual system (Hershenson, 1967) suggests that a high degree of organization exists at birth. For example, the data from studies of the electroretinogram of the newborn clearly suggest that both photopic and scotopic systems are functional at birth. Studies of the electroencephalogram and of the visual evoked response in the human newborn leave little doubt that the primary sensory pathways are intact and functional at birth.

Conjugate movement of the eyes has been reported for newborns who located and followed a moving black dot (Dayton, Jones, Steele, & Rose, 1964b). The infants' eyes moved simultaneously in the direction of the target over the same period of time, suggesting well-coordinated, purposeful ocular control. Convergence and conjugation in the presence of stationary targets are less clear. Hershenson (1964) photographed ocular orientation to stationary targets and found that both eyes were directed at the same target during fixations, and also that the eyes moved conjugately from target to target. Wickelgren (1967) reported that convergence varied greatly, depending upon the stimulus: the eyes converged 70% of the time for a striped-gray pair; 42% of the time for stimuli differing in color and brightness; and only 9% of the time for a blinking light. In reviewing this work, Hershenson (1967) suggested that the studies may not be in conflict: Wickelgren sampled ocular orientation at one frame per second over 4 minutes whenever the infant's eyes were open; Hershenson photographed at 24 frames per second for short bursts of about 2 seconds, when only one eye was oriented in the direction of a stimulus. Thus Hershenson probably measured the convergence and conjugation which Wickelgren found more or less frequent over longer periods of time, depending upon the nature of the stimulus.

The question remaining, then, is the relative frequency of convergence per opportunity to converge. It raises, first, the experimental problem of interpreting ocular orientation as an index of perception and, second, the problem of describing the processes of perceptual development. The experimental problem arises when it is assumed that convergence is present throughout an experimental session. If the frequency of convergence is not known, then orientation of one eye (e.g., Fantz, 1963) or of two eyes independently (e.g., Hershenson, 1964) is difficult to relate to the experience of the subject. The descriptive problem arises from comparison of newborn visual behavior with that of the normal adult. The experience of a single unified image in the mature perceptual system is thought to result from integration of correlated information from converged eyes. While it is possible, perhaps, for fusion to occur in the immature system with partially correlated input, it is also possible that stimulus information from one eye is suppressed or that information from the two eyes is processed alternately. The relaxation or disappearance of suppression or alteration in experience would then become important developmental problems.

Even with proper alignment, improper focal adjustment would result in blurred images on the retina. The impact of the resulting retinal blur on perception is not clearly understood, thereby complicating the study of accommodation in development. An inefficient accommodative system may, nonetheless, provide an inexperienced organism with sufficient sensory information for the development of perception. On the average, a newborn would require a far greater amount of accommodative effort than would an adult, merely to overcome his refractive error. The most recent and comprehensive surveys of refractive errors indicate hyperopia in about 75% of newborn eyes. The probability of ametropia is also far greater than in the adult eye. Thus, in a sense, the newborn starts with an accommodative handicap.

Using dynamic retinoscopy, Haynes, White, and Held (1965) measured the newborn's ability to accommodate actively to a patterned stimulus placed at the front of the retinoscope. Accommodative responses of infants under 1 month of age did not adjust to changes in target distance. The system appeared to be locked at one focal distance whose median value was 19 cm (about 8 inches). Radial tracking of a target moving in depth was not observed in infants under 1 month of age.

Nevertheless, Gorman, Cogan, and Gellis (1957), using the reflexive ocular-following response known as optokinetic nystagmus (OKN) as an index of visual acuity, found that 93 of 100 infants under 5 days of age followed stripes of 33.3 minutes of arc (Snellen equivalent = 20/670). Also using OKN, Dayton, Jones, Aiu, Rawson, Steele, and Rose (1964a) found that half of their subjects responded to targets with a Snellen equivalent of 20/150, 14 of 18 responded to a target equivalent of 20/190.

Riesen (1960) and Fantz *et al.* (1962) have argued, however, that this evidence, although convincing, is based upon a response mediated by neural mechanisms which may not be involved in pattern perception. To eliminate this possibility, Fantz *et al.* measured duration of fixation on stationary stimulus pairs of stripes and neutral grays. They assumed that neural pathways which mediate voluntary attention to pattern would then be directly involved. Infants under 1 month of age were able to resolve lines of ⅛ inch at 10 inches (20/200), a reasonable corroboration of the OKN data.

Thus, the evidence suggests the presence of the requisite muscular control and detector apparatus to bring the image of an object onto the fovea and to hold it there reasonably well, even if the object is moving. This can be done with both eyes in a coordinated fashion, at least for short durations. While some question remains as to the effect of the possibly limited accommodative capacity, the image is at least sharp enough for manifestation of a fair amount of resolving power for the newborn eye. Add to this the evidence for the apparent functional integrity of the neural pathways necessary for mediation of sensory information and a very good case could be made for the potential presence of perception. Clearly, there is no evidence that argues very strongly against it.

C. Response to Changes in Intensity

The development of brightness sensitivity has been measured by Doris (Doris, Casper, & Poresky, 1967; Doris & Cooper, 1966) in infants up to 113 days of age. Thresholds were measured by optokinetic nystagmus to a moving field of alternate light- and dark-gray. The contrast between the light and dark stripes was varied on successive trials by changing the brightness of the lighter stripes, and the disappearance of nystagmoid movements was taken as an index of threshold. The results indicated rapid development of brightness sensitivity in the first 2 months of life.

D. Response to Simple Stimulus Features

The importance of "feature analyzers" for building-block theories of perceptual development has already been discussed. This view has gained strength in recent years from experiments which have shown, by direct physiological measurement, the existence of many different kinds of feature analyzers. Single units have been found (a) in the retina of the frog responding only to moving, convex dark objects (Lettvin, Maturana, Pitts, & McCulloch, 1961); (b) in the ganglion cell layer of the retina of the rabbit responding primarily to movement in a particular direction at a particular velocity (Barlow & Hill, 1963; Barlow, Hill, & Levick, 1964; Barlow & Levick, 1965); (c) in the visual cortex of the cat responding to edges, bars, or slits of particular widths and lengths, in particular orientations, and in certain locations in the visual field (Hubel & Wiesel, 1962, 1965); (d) in the lateral geniculate of the rhesus monkey responding differentially to wavelength, to achromatic spatial arrangement, and to chromatic spatial arrangement (Wiesel & Hubel, 1966); and (e) in the visual cortex of the spider monkey responding to edges, slits, and bars in complex ways (Hubel & Wiesel, 1968). Thus, even in subhuman vertebrates, it is clear that the information reaching the higher integrative centers already contains a complex coded representation of many aspects of the stimulus.

The possible existence of similar organized feature analyzers in the human must be investigated by less direct methods (Weisstein, 1969); however, there too organization appears to be the rule. For example, Kessen, Salapatek, and Haith (1965) presented newborn humans with a single contour in either vertical or horizontal orientation, situated 3 inches from the center of the visual field. Fixations on the blank control field were distributed fairly normally around the center of the field. The distributions of eye fixations for horizontal contour were virtually identical to that for the blank screen. The vertical contours, on the other hand, were clearly attractive to the Ss. For the left-of-center contour, the distribution of eye fixations showed a sharp peak in the region of the contour. Fixation tended to remain within 1½ inches of the contour a good deal of the time. For the right-of-center contour, the distributions of fixations shifted to the right, but the peaking occurred to the right of the contour region. Thus, infants appear to be attracted to those contours which can be easily crossed. This implies that retinal stimulation occurs when the edge is crossed, that this stimulation is implicated in the mechanism which maintains fixation around the contour, and that it is easier for an infant to move his eyes horizontally across a vertical edge than vertically across a horizontal edge.

When figures are introduced into the visual field, a slightly different kind of visual behavior occurs. Salapatek and Kessen (1966) presented newborns with a large black triangle while photographing their eye fixations. Fixation points were plotted with respect to the real triangle presented to the experimental group or with respect to an imaginary triangle for the control group. Figure 2 shows responses to a blank homogeneous field. Each dark point represents a fixation and these are connected in sequence. The eye moved almost every second, and much more in the horizontal than in the vertical direction.

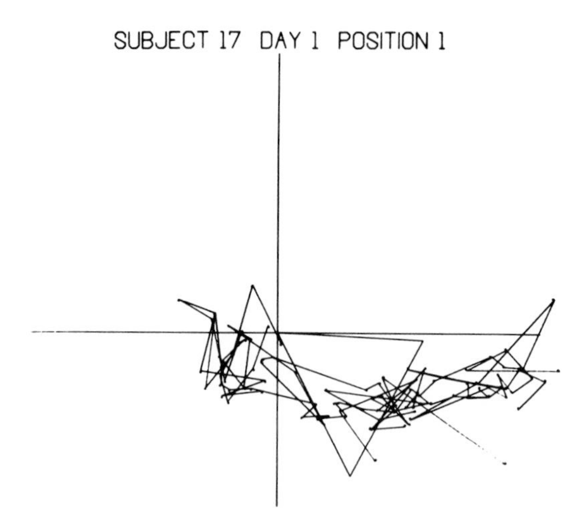

Fig. 2. Visual scanning pattern of human newborn of a homogeneous field. Each dot represents one time sample of eye orientation. (From Salapatek & Kessen, 1966.)

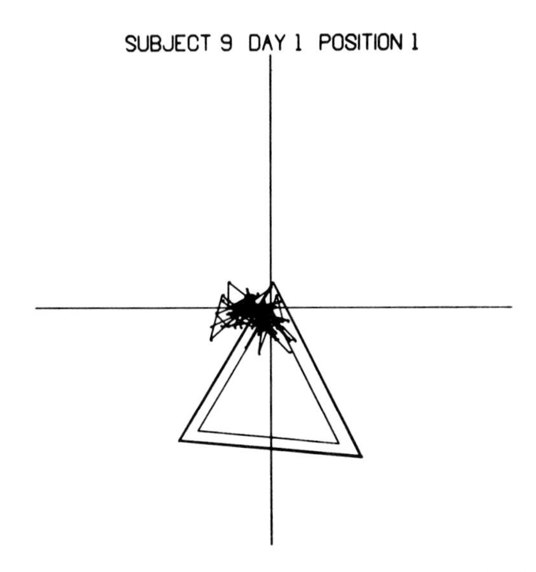

Fig. 3. Visual scanning pattern of human newborn of large black equilateral triangle. Each dot represents one time sample of eye orientation. (From Salapatek & Kessen, 1966.)

Figure 3 shows the response of a typical experimental subject. Eye movements were much shorter and the fixations tended to cluster around one of the three vertices. Thus, by fixating fairly closely to one of the corners, the infants appeared to be responding to "angle" rather than to total "form." Three alternative explanations for this pattern of fixations were offered by Salapatek and Kessen: that the infants were responding to (a) the angle itself—the "angle-detector" hypothesis; (b) a brightness transition, there being two at the vertex, or (c) an optimal level of brightness which might occur near the vertex. The angle-detector hypothesis was supported by a subsequent experiment in which three aspects of the triangular array were compared (Nelson & Kessen, 1969). Newborns were shown three stimuli: a complete outline triangle, only the sides of this triangle, and only the angles of this triangle. The infants typically looked toward a single angle component; the angular elements attracted the infants' gaze whether the sides of the triangle were present or not.

Salapatek (1968) varied size, angularity, figure-ground contrast, and figure type (solid or outline) in an attempt to separate the important aspects of stimulation which were attracting the newborns' gaze. The ocular response was subjected to detailed analysis by measuring (a) location of gaze, (b) dispersion of gaze, (c) number of shifts in gaze, (d) direction of shifts, (e) length of shifts, and (f) time spent looking at (1) total figure, (2) part of figure, (3) contour *vs.* center of circles, and (4) sides *vs.* angles of triangles.

The results were not unambiguous with respect to the specific stimuli used. They did not indicate that either specific figures as wholes, or specific parts of figures were more attractive than others. Nevertheless, the experiment did supply firm evidence in support of the earlier studies, leading Salapatek (1968, 1969) to characterize the visually naive newborn's response to patterned two-dimensional surfaces as follows: (a) Visual scanning is more dispersed in the horizontal than in the vertical dimension of the visual field under all stimulus conditions; (b) Introduction of a geometric figure into the newborn's visual field results in a decrease in the horizontal dispersion of visual scanning and a decreased difference between horizontal and vertical dispersion of scanning; (c) Newborns modify the pattern of visual scanning to fit the physical features of the stimulus. The broad automatic scan described by Salapatek (1968) occurred reliably in the vast majority of infants whenever they were placed before a homogeneous field with their eyes open. The scan occurred for extended periods of time and did not appear to require ambient illumination since a similar response has been reported to occur in darkness (Salapatek, 1969).

This general pattern of scansion is markedly altered when a figure or pattern is introduced into the visual field. Within a very short time, the majority of newborns "cease scanning broadly and fixate the figure's contour, typically some limited segment or feature of the figure's contour . . . Localization occurs for the majority of subjects within three or four seconds, and often involves an eye movement across some portion of the figure (e.g., linear contour) during the approach to the particular figure-segment, or feature, which is selected for prolonged fixation" (Salapatek, 1969). This pattern of behavior suggested to Salapatek that the newborn is equipped with "directionally-appropriate localizing tendencies towards peripheral pattern stimulation" and also that he is capable of "some pattern discrimination in the peripheral visual field, which determines the particular pattern feature that is approached." And, as Salapatek pointed out, angles appear to be particularly salient in this respect.

The mechanism for maintenance of feature localization is much more difficult to conceive and to understand. The majority of newborns maintained line of sight on a feature or segment for prolonged periods of time once it had been localized on the central portion of the retina. The scan appeared to be loosely centered over the feature rather than a steady sustained fixation or a compact series of fixations. Salapatek suggested and subsequently discounted a number of possible interpretations for this loose scan, including binocular rivalry and the correction of localizing errors. The most plausible explanation is suggested by the relative lack of development of the newborn macula. If, for example, the peripheral retina provides some reasonable degree of pattern acuity, and if there are relatively strong tendencies to fixate salient peripheral patterns which are less acute when fixated, then one might expect scanning patterns similar to those obtained.

In two recent experiments, Salapatek (1969) studied the development of visual scanning through the first 10 weeks of life. In the first experiment, two groups of infants were studied: one group ranging from 4 to 6 weeks of age and the other between 8 and 10 weeks. The stimuli used were a circle, a square, a triangle, a regular 12-turn shape, and an 11-turn random shape. Only in the presence of a very finely textured homogeneous field did all infants tend to fixate the extreme portions of their visual fields, become fussy, or fall asleep. When a geometric figure was introduced into the field, however, the infants rapidly fixated and scanned the figure for an extended period of time. In general, the scan was directed toward an angle, if one was present in the figure.

The differences between infants 5 weeks of age and those 9 weeks of age can be summarized as follows: (a) Young infants concentrated most of their visual scanning on a small portion of the stimulus to a greater degree than the older ones did. (b) Younger infants showed a more focalized scan; older infants directed their scan more loosely. (c) Although all infants fixated the contour rather than the center of the geometric patterns, the younger infants limited their gaze to a very small segment of the contour whereas the older infants showed a tendency to direct some fixations toward the centers of the figures.

In the second study, Salapatek presented 33 infants between 4 and 10 weeks of age with five different geometric patterns: (a) a ¾-inch outline square, (b) two ¾-inch outline squares, (c) a 6-inch outline square, (d) a 6-inch outline square in which a single ¾-inch outline square was positioned, and (e) a 6-inch outline square in which two ¾-inch outline squares were centrally positioned. Some of these stimuli have internal and external contours—they might even be thought to have internal features similar to those of a face. The data suggest a shift in attention toward internal as compared to external features of a geometric pattern during this stage of development.

Older infants fixated more on internal features than did younger ones. The infant between 8 and 10 weeks of age tended to select an internal feature of a geometric figure for prolonged visual regard more than the infant 4 to 6 weeks of age did.

E. Response to Complex Organization: Space

For the young infant, "perceiving space" means that a three-dimensional world of surfaces, edges, and solid objects is represented in his awareness. Presumably it is a world which contains objects devoid of meaning, whose apparent size and shape are determined by their intersections with the metric of the enclosed visual space. Experimental manipulation of spatial parameters with newborn infants has not been attacked systematically. Fantz (1963) has reported that infants look longer at three-dimensional reproductions of faces and objects than their two-dimensional representations; however, this evidence is merely suggestive.

The demonstration that the young infant can perceive objects in a meaningless Euclidean space, then, presupposes that a particular kind of information is being processed. For Gibson (1950, 1966) the perception of space depends upon the accurate reception of information about surfaces and edges. Gibson begins his analysis of space perception from the point of view of the receiving eye. An eye, being at a particular point in space at a particular time, receives a converging bundle of light rays—the optic array. Because adjacent elements in a particular surface absorb and reflect light differentially, the reflected rays of light have different intensities and the optic array possesses a texture—the optical texture. The optic array may be composed of combinations or of hierarchies of optical textures.

Each class of substances in the world reflects a unique pattern of light. Therefore the optical texture in this reflected light contains repetitive, congruent luminous patterns corresponding to the properties of the different substances. This makes it possible to assume that there can be, in the optic array, repetitive congruent luminous patterns, and allows Gibson to define an optical texture gradient as the rate of change of the density of elements. For Gibson, then, the optical texture gradients supply the stimuli for the detection and perception of surfaces at a slant, and for the specification of their intersection as edges. The young organism would need the mechanisms appropriate for the processing of information relative to optical texture gradients to have a volumetric space determined.

Before turning to the evidence on space perception in human infants, it might be interesting to speculate about the possibility of an innately organized hierarchical system for detecting surface and edge. The possibility is worth considering in light of the kinds of analyzers suggested above. If, for example, these analyzers are understood in relation to the kind of information necessary for the specification of space, then it should be clear that "line-at-a-slant" and "angle" information can be integrated at a higher level to yield surface at a slant. The detection of the frontoparallel plane projection of a surface is merely a simultaneous firing of lines at different, but particular, slants. Continued scanning of a plane could produce a higher order invariant of stimulation and of neural discharge to which the human could have adapted in evolution. Combinations of these invariances could form the basis of an innate volumetric visual space.

In a series of experiments, T. G. R. Bower has investigated the human infant's ability to respond to spatial invariants. He has found that infants are able to respond to objects which are changing their spatial positions in a manner which is independent of the spatial change. This ability has been called perceptual constancy and rests upon the specific spatial invariants discussed above.

In order to test the ability of premotor infants to discriminate depth, Bower (1964) trained subjects to turn their heads to the left to get a "peek-a-boo" reinforcement from the experimenter. A cube served as the conditioned stimulus. Four generalization tests were used which pitted projective size against object size. This is an important experiment since empiricist theory suggests that the discrimination of depth is dependent upon prior action in space which enables the organism to use the information to order objects in space. The results indicated that at least some of the information specifying distance and size-at-a-distance was received and used by the premotor infants. Thus infants who have had no opportunity for action in space discriminated depth on the basis of information other than the projective size of objects displaced in depth.

In a subsequent study, Bower (1965) showed that motion parallax was the necessary variable for perceptual discrimination in infants between 1 and 2 months of age. These infants were also trained to turn their heads to one side to see the experimenter saying "peek-a-boo." There were three groups: a monocular group, a binocular group, and a group which saw only a projection of the real scene. All groups were able to make a discrimination based on changes in the size of the retinal image. Only the first two groups, however, could discriminate among presentations where the retinal size was not changed. This response can be interpreted to mean that the infants could not only detect spatial position but could also detect size-at-a-distance, i.e., they manifested size constancy. Moreover, since the rich array of pictorial cues presented in the projected array was not sufficient for discrimination, this information seems not to have been involved. The ability to perceive size as invariant with distance transformed appears to have been based on information due to motion parallax. For both the binocular and monocular groups, head movements could produce relative displacements of objects and spaces in the visual field sufficient for the judgment.

In a similar experiment, Bower (1965) trained 2-month-old infants to turn their heads to the left (the conditioned response) whenever they were presented with a particular wooden block in a particular orientation (the conditioned stimulus) in order to be able to see the experimenter's head pop up from under the table and say "peek-a-boo" (the reinforcement). The infants were then tested for generalization with four stimuli: (a) the CS at its original 45° orientation, (b) the CS in the frontoparallel plane, (c) a trapezoid in the frontoparallel plane whose retinal projection was then equal to that of the CS at 45°, and (d) this trapezoid at an angle of 45°. In terms of difference from the CS, the four presentations may be classified as follows: (a) no change, (b) same objective shape, different projective shape, different orientation, (c) different objective shape, same projective shape, different orientation, (d) different objective shape, different projective shape, same orientation. The infants responded approximately twice as much to the original CS as they did to the stimuli which projected a retinal shape identical with it. They responded almost as much to the same objective shape in the different orientation as they did to the original CS. Thus Bower argued that the infants had not learned to respond to a projective or retinal shape but to an objective shape which could be recognized in a new orientation. To this extent, they showed shape constancy.

This experiment clearly eliminated the strong empiricist hypothesis; however, the possibility that constancy develops out of the correlation between two perceptually prior variables—projective shape and apparent slant—is still viable. This view requires that projective shape and perceived orientation precede shape constancy in perceptual development, and that projective shape, orientation, and real shape are separable attributes of an object in space. To test this hypothesis, Bower performed another experiment in which infants received similar training, but were given, in subsequent sessions, each of the following tasks: (a) the original block was exposed in four orientations: 5, 15, 30, and 45 degrees counterclockwise; (b) the projective equivalents of the positions in (a) shown in the frontoparallel plane; and (c) an opaque cardboard screen placed in front of the block. An aperture was cut in the screen so that the body of the block was visible but its edges were not. Thus, in this condition, only orientation was available to differentiate the presentations of the block in its four positions. These conditions presented the infant with: (a) a form in space having objective shape, projective shapes and orientations with only the first variant; (b) a set of projective shapes with orientation invariant; and (c) a set of orientations with projective shape invariant. The developmental hypothesis being tested in this experiment predicts that discrimination performance should be best under condition (a) where the infant had both projective shape and orientation. This prediction was not confirmed. Moreover, if the two variables were registering separately, then the fact that there were two variables in (a) should have made that discrimination easier to learn. That it did not suggests that the variables were not separately registered.

A third experiment performed to make certain that all training contingencies had been taken into account showed that variations in orientation of the same object, with projective shape and orientation variant and only real shape invariant, produced a higher degree of identity or "sameness" response than did variations in projective shape alone, with orientation invariant, or variations in orientation, with projective shape invariant.

Bower summed up this series of experiments by suggesting that the human infant possesses the capacity for shape constancy, i.e., he can detect the invariants of shape under transformational rotation in the third dimension. Moreover, the response to a shape invariant appears to be more primitive than response to simple variables such as orientation. Thus it would appear that the perceptual system of the infant responds to higher order invariants directly; that shape constancy is not learned or computed from projective shape and orientation.

One final experiment of Bower's (1967) is important because it studied the psychophysics of existence constancy—the fundamental mode of apprehending objects and events in the world reflecting a belief in the continued existence of objects which have disappeared, and in the preexistence of objects which have just appeared. In terms of its formal logic, this belief is similar to the constancy problem in perception. Using both operant conditioning techniques and the startle response in a series of experiments, Bower found that existence constancy as a perceptual phenomenon appears very early.

With one exception, the responses to different types of disappearance transformations did not differ between infants 5 and 59 weeks of age. Conceptual constancy, which concerns objects rather than events, developed later. Moreover, conceptual constancy did not seem to be an extension of perceptual structuring but appeared to develop in opposition to it.

The apparently surprising fact that human infants show perceptual constancy should have been anticipated. Size and shape constancy have been found in many species which are not capable of higher functions usually reserved for man. Size constancy has been reported in the monkey (Locke, 1937), cat (Freeman, 1968; Gunter, 1951), rat (Heller, 1968), and duck (Pastore, 1958). Thus, the constancy response may be mediated by simpler mechanisms, or better, by mechanisms which appear early in evolutionary history. This would suggest that there has been ample time for the evolutionary development of a complex constancy (volumetric space) mechanism in man.

F. Response to Complex Organization: Configurational Meaning

Configurational meaning can refer both to invariant and to arbitrary relations between a stimulus and its meaning. The arbitrary meanings of symbols come into play during the preschool years in humans. The understanding of this process involves an analysis beyond the scope of this discussion. However, there is evidence about the degree to which the meaning of a stimulus is related to its configurational characteristics and, moreover, the degree to which these invariant relations can be understood to contain an hereditary component. For example, in studies where patterned (or figural) stimuli are paired with plain stimuli, it seems clear that patterned stimuli are invariably given the greater number of fixations and the greater amount of visual attention by human infants (Fantz, 1963, 1966; Fantz & Nevis, 1967). Although differential responding was not obtained from human newborns in the first week of life when exposed to paired stimuli showing a human face and two variations of alteration of its elements (Hershenson, Kessen, & Munsinger, 1967), Watson (1966) has shown that infants 8 to 26 weeks of age can respond to the orientation of a face, and Lewis (1969) has shown that as early as 3 months of age, infants fixate facelike stimuli.

Sackett (1966) makes a stronger case for the implication of complex innate mechanisms in visually mediated behavior in monkeys. He reared monkeys in isolation from birth to 9 months of age. On the fourteenth day of their lives, the monkeys were allowed visual stimulation controlled either by the experimenter or by their own activity. The stimuli included pictures of other monkeys in various acts and expressing various "emotions": they included pictures of threatening, playing, fearful, withdrawing, exploring, and "sexing" monkeys, as well as pictures of infants, a mother and an infant together, and monkeys doing "nothing." Control pictures included a living room, a red sunset, an outdoor scene with trees, an adult female human, and various geometric patterns. Exploration, play, vocalization and disturbance occurred most frequently with pictures of monkeys threatening and pictures of infants. From 2½ to 4 months of age, threat pictures yielded a high frequency of disturbance level—monkeys initiated activity to expose the pictures at a very low rate during this period. Pictures of infants and of threat thus appear to have nonlearned, prepotent general activating properties for socially naive infant monkeys, while pictures of threat appear to "release" a maturationally determined, innate fear response.

The implications are clear: at least certain aspects of complex social communication in primates

> may lie in innate recognition mechanisms, rather than in acquisition through social learning processes during interactions with other animals. Although the maintenance of responses to socially communicated stimuli may well depend on learning and some type of reinforcement process, the initial evocation of such complex responses may have an inherited, species-specific structure. Thus . . . innate releasing mechanisms such as those identified by ethologists . . . for insects and avian species may also exist in some of the more complex behaviors . . . in . . . primates (Sackett, 1966, p. 1473).

V. Conclusion

There are four main points which should be stressed in conclusion. First, in general, psychologists are, once again, attempting to understand man within his biological heritage. Even his cognitive abilities are now seen to reflect his peculiar heredity. Second, perceptionists are not only taking this view of the human's general perceptual abilities, but the physiological mechanisms underlying first-stage information processing are becoming known and are supplying models for more complex processing. Third, the more detailed our understanding gets, the more the similarities and differences among species become apparent. This, along with the fourth point, the greater conceptual differentiation of perceptual tasks, allows the modern perceptionist much greater flexibility in his theorizing. Unburdened by the demand to produce a single grand theory, many different models emerge as viable proposals within more circumscribed domains.

References

Ames, E. W., & Silfen, C. K. Methodological issues in the study of age differences in infants' attention to stimuli varying in movement and complexity. Paper presented at the meeting of the Society for Research in Child Development, Minneapolis, March, 1965.

Barlow, H. B., & Hill, R. M. Evidence for a physiological explanation of the waterfall phenomenon and figural after-effects. *Nature* (*London*) 1963, **200**, 1345–1347.

Barlow, H. B., Hill, R. M., & Levick, W. R. Retinal ganglion cells responding selectively to direction and speed of image motion in the rabbit. *Journal of Physiology* (*London*), 1964, **173**, 377-407.

Barlow, H. B., & Levick, W. R. The mechanisms of directionally selective units in the rabbit's retina. *Journal of Physiology* (*London*), 1965, **178**, 477-504.

Bower, T. G. R. Discrimination of depth in premotor infants. *Psychonomic Science*, 1964, **1**, 368.

Bower, T. G. R. Stimulus variables deterimining space perception in infants. *Science*, 1965, **149**, 88–89.

Bower, T. G. R. The development of object permanence: Some studies of existence constancy. *Perception and Psychophysics*, 1967, **2**, 411–418.

Dayton, G. O., Jr., Jones, M. H., Aiu, P., Rawson, R. A., Steele, B., & Rose, M. Developmental study of coordinated eye movements in the human infant. I. Visual acuity in the newborn human. *Archives of Ophthalmology*, 1964, **71**, 865-870. (a)

Dayton, G. O., Jr., Jones, M. H., Steele, B., & Rose, M. Developmental study of coordinated eye movements in the human infant. II. An electro-oculographic study of the fixation reflex in the newborn. *Archives of Ophthalmology*, 1964, **71**, 871-875. (b)

Dodwell, P. C. A coupling system for coding and learning in shape discrimination. *Psychological Review*, 1964, **71**, 148–159.

Doris, J., Casper, M., & Poresky, R. Differential brightness thresholds in infancy. *Journal of Experimental Child Psychology*, 1967, **5**, 522–535.

Doris, J., & Cooper, L. Brightness discrimination in infancy. *Journal of Experimental Child Psychology*, 1966, **3**, 31–39.

Fantz, R. L. Pattern vision in young infants. *Psychological Record*, 1958, **8**, 43–47.

Fantz, R. L. Pattern vision in newborn infants. *Science*, 1963, **140**, 296-297.

Fantz, R. L. Ontogeny of perception. In A. M. Schrier & H. F. Harlow (Eds.), *Behavior of nonhuman primates. Vol.* 2. New York: Academic Press, 1965. Pp. 365-403.

Fantz, R. L. Pattern discrimination and selective attention as determinants of perceptual development from birth. In A. H. Kidd & J. L. Rivoire (Eds.), *Perceptual development in children*. New York: International University Press, 1966. Pp. 143-173.

Fantz, R. L., & Nevis, S. M. Pattern preferences and perceptual-cognitive development in early infancy. *Merrill-Palmer Quarterly*, 1967, **13**, 77–108.

Fantz, R. L., Ordy, J. M., & Udelf, M. S. Maturation of pattern vision in infants during the first six months. *Journal of Comparative and Physiological Psychology*, 1962, **55**, 907-917.

Freeman, R. B. Perspective determinants of visual size-constancy in binocular and monocular cats. *American Journal of Psychology*, 1968, **81**, 67-73.

Garner, W. R., Hake, H. W., & Eriksen, C. W. Operationism and the concept of perception. *Psychological Review*, 1956, **63**, 149-159.

Gibson, J. J. *The perception of the visual world.* Boston: Houghton, 1950.

Gibson, J. J. *The senses considered as perceptual systems.* Boston: Houghton, 1966.

Gibson, J. J., & Gibson, E. J. Perceptual learning: Differentiation or enrichment? *Psychological Review*, 1955, **62**, 32-41.

Gorman, J. J., Cogan, D. C., & Gellis, S. S. An apparatus for grading the visual acuity of infants on the basis of optico-kinetic nystagmus. *Pediatrics*, 1957, **19**, 1088-1092.

Gunter, R. Visual size constancy in the cat. *British Journal of Psychology*, 1951, **42**, 288-293.

Haynes, H., White, B. L., & Held, R. Visual accommodation in the human infant. *Science*, 1965, **148**, 528-530.

Hebb, D. O. *The organization of behavior.* New York: Wiley, 1949.

Heller, D. P. Absence of size constancy in visually deprived rats. *Journal of Comparative and Physiological Psychology*, 1968, **65**, 336-339.

Helson, H. *Adaptation level theory.* New York: Harper, 1964.

Hershenson, M. Visual discrimination in the human newborn. *Journal of Comparative and Physiological Psychology*, 1964, **58**, 270-276.

Hershenson, M. Development of the perception of form. *Psychological Bulletin*, 1967, **67**, 326-336.

Hershenson, M., Kessen, W., & Munsinger, H. Pattern perception in the human newborn: A close look at some positive and negative results. In W. Wathen-Dunn (Ed.), *Models for the Perception of Speech and Visual Form.* Cambridge, Mass.: MIT Press, 1967. Pp. 282-290.

Hubel, D. H., & Wiesel, T. N. Receptive fields, binocular interaction and functional architecture in the cat's visual cortex. *Journal of Physiology (London)*, 1962, **160**, 106-154.

Hubel, D. H., & Wiesel, T. N. Receptive fields and functional architecture in two nonstriate visual areas (18 & 19) of the cat. *Journal of Neurophysiology*, 1965, **28**, 229-289.

Hubel, D. H., & Wiesel, T. N. Receptive and functional architecture of monkey striate cortex. *Journal of Physiology (London)*, 1968, **195**, 215-243.

Ittelson, W. H. *Visual space perception.* Berlin: Springer, 1960.

Kessen, W., Salapatek, P., & Haith, M. M. The visual response of the human newborn to horizontal and vertical linear contour. Paper presented at the meeting of the American Psychological Association, Chicago, September, 1965.

Kohler, I. The formation and transformation of the perceptual world. *Psychological Issues*, 1964, **3**, No. 4.

Kolers, P. A. Clues to a letter's recognition: Implications for the design of characters. *Journal of Typographic Research*, 1969, **3**, 145-168.

Lettvin, J. Y., Maturana, H. R., Pitts, W. H., & McCulloch, W. S. Two remarks on the visual system of the frog. In W. A. Rosenblith (Ed.), *Sensory communication.* New York: Wiley, 1961. Pp. 757-776.

Lewis, M. Infants' responses to facial stimuli during the first year of life. *Developmental Psychology*, 1969, **1**, 75-85.

Locke, N. A. Comparative study of size constancy. *Journal of Genetic Psychology*, 1937, **51**, 255-265.

Neisser, U. *Cognitive psychology*. New York: Appleton, 1967.

Nelson, K., & Kessen, W. Visual scanning by human newborns: Responses to complete triangle, to sides only, and to corners only. Paper presented at the meeting of the American Psychological Association, Washington, D.C., September, 1969.

Pastore, N. Form perception and size constancy in the duckling. *Journal of Psychology*, 1958, **45**, 259-262.

Posner, M. Abstraction and the process of recognition. In G. Bower & J. T. Spence (Eds.), *The psychology of learning and motivation*. Vol. 3, New York: Academic Press, 1969. Pp. 43-100.

Riesen, A. H. Reception functions. In P. H. Mussen (Ed.), *Handbook of research methods in child development*. New York: Wiley, 1960. Pp. 284-307.

Sackett, G. P. A neural mechanism underlying unlearned, critical period, developmental aspects of visually controlled behavior. *Psychological Review*, 1963, **70**, 40-50.

Sackett, G. P. Monkeys reared in isolation with pictures as visual input: Evidence for an innate releasing mechanism. *Science*, 1966, **154**, 1468-1473.

Salapatek, P. Visual scanning of geometric figures by the human newborn. *Journal of Comparative and Physiological Psychology*, 1968, **66**, 247-258.

Salapatek, P. The visual investigation of geometric patterns by the one- and two-month old infant. Paper presented at the meeting of the American Association for the Advancement of Science, Boston, December, 1969.

Salapatek, P., & Kessen, W. Visual scanning of triangles by the human newborn. *Journal of Experimental Child Psychology*, 1966, **3**, 155-167.

Stechler, G., Bradford, S., & Levy, H. Attention in the newborn: Effect on motility and skin potential. *Science*, 1966, **151**, 1246-1248.

Stechler, G., & Latz, E. Some observations on attention and arousal in the human infant. *Journal of the American Academy of Child Psychiatry*, 1966, **5**, 517-525.

Walls, G. L. *The vertebrate eye and its adaptive radiation*. Bloomfield Hills, Michigan: Cranbrook Institute of Science, 1942.

Watson, J. S. Perception of object orientation in infants. *Merrill-Palmer Quarterly*, 1966, **12**, 73-94.

Weisstein, N. What the frog's eye tells the human brain: Single cell analyzers in the human visual system. *Psychological Bulletin*, 1969, **72**, 157-176.

Wickelgren, L. W. Convergence in the human newborn. *Journal of Experimental Child Psychology*, 1967, **5**, 75-85.

Wiesel, T. N., & Hubel, D. H. Spatial and chromatic interactions in the lateral geniculate body of the rhesus monkey. *Journal of Neurophysiology*, 1966, **29**, 1115-1156.

CHAPTER 3

EARLY EXPERIENCE AND PROBLEM–SOLVING BEHAVIOR

Barton Meyers

I. Introduction

During the past few decades, a great deal of research has been performed to uncover the effects of various types of early experience on later behavior. It is *au courant* to attribute this high level of interest to several sources, among them the psychoanalytic and ethological literatures (Beach & Jaynes, 1954; Haywood & Tapp, 1966), although universal admission could probably be obtained for the notion that the *Zeitgeist* was the real agent. However, when the behavior in question is problem-solving, it seems clear that major responsibility for the study of its modulation by early experience belongs to D. O. Hebb. Although limited in number, his experiments (Clarke, Heron, Fetherstonhaugh, Forgays, & Hebb, 1951; Hebb, 1947, 1949) were the first to explore this area; and his provocative monograph (Hebb, 1949) provided a theoretical framework that encouraged further research.

At the present time, the manner in which early experience influences adult problem-solving behavior, whether by enhancing or by degrading it, is widely regarded as a critical question. It plays a central role in the controversy as to whether problem-solving ability is fixed at birth or is significantly mutable by experience (Haywood, 1967; Hunt, 1961). The results of experimentation with animals in this area have penetrated the popular press (Perlman, 1968) and have been considered as providing clues for the abolition of differences between the intelligence quotients of black and white children (Pettigrew, 1964) and of children from different socioeconomic classes (Haywood, 1967). Thus, research in this area has not only theoretical interest but also profound social significance.

A. Approach

The present chapter adopts a developmental approach in relation to the effects of early experience on later problem-solving ability. Its scope is restricted to nonhuman animals. Although the presence of genetic influences on problem-solving behaviors has been well demonstrated (e.g., Cooper & Zubek, 1958; Rosenzweig, 1964), implicit throughout the present treatment is the view that such behaviors develop dynamically, that they are capable of formation, or transformation, as a result of the cofunction of the genetic substrate and of the experiences which the organism encounters. In the same way, the literature covered indicates that those experiential manipulations which affect problem-solving behaviors also induce reliable changes in the anatomy, chemistry, and physiology of the brain. Hopefully, efforts to correlate these two sets of data will aid in the attempt to understand at least some aspects of the behavioral literature.

B. Tests of Problem-Solving Ability

The specification of a behavioral task which measures problem-solving ability, a term for which even a nebulous definition is usually lacking, is a difficult matter. It would appear that most investigators have regarded a task which demands a relatively novel solution based upon past experience to satisfy such a specification.* Conversely, it seems that tasks such as simple discrimination problems or multiple-unit mazes, which require the animal to traverse repeatedly an unchanging apparatus in order to eliminate errors gradually, have been considered to be analogous to rote learning in humans and, therefore, have been little used to assess problem-solving ability. Thus, perhaps by instant tradition, the apparatus most frequently used has been the Rabinovitch and Rosvold (1951) modification of the Hebb-Williams (1946) closed-field test. In this test, after appropriate pretraining to eliminate the animal's exploratory behavior, to adapt him to handling, and to train him to procure food quickly, he is tested on a sequence of problems defined by the altered patterns of barriers within the field (Fig. 1). Each day the animal encounters two new problems, and his error score over a number of trials is recorded. Despite its widespread use, however, Warren (1965b) has suggested that performance on the Hebb-Williams maze is probably a poor measure of problem-solving ability since such performance neither improves with age in cats and monkeys nor differs between them. Reversal learning, which does show orderly improvement over a segment of the phyletic scale at least (Bitterman, 1965; Warren, 1965a, 1965b), has been advanced as a more appropriate measure of problem-solving than the Hebb-Williams test. The problems involved in devising a suitable test can best be illustrated by the words of Lashley (1929):

> The concept of intelligence is becoming essentially a statistical one; it is the correlation between certain of the activities of the organism which are closely related among themselves and relatively independent of other activities. Among such groupings of activities, what correlations constitute intelligence as opposed to other capacities? There is no accepted statement which really defines the concept (pp. 11–12).

C. Environmental Manipulations

The environmental manipulations used in the studies to be discussed have been such as to either reduce or increase the complexity of the perceptual and/or motor experiences which the organism re-

*Such a definition has led many investigators to discuss these behaviors as measures of intelligence. The suggestion that problem-solving ability and intelligence, as typically used, are identical is probably correct. However, the latter term inhibits attempts to treat the dependent variable as a behavior, rather than as a state of mind, which can be affected significantly by experience and which can be either coextensive with or in conflict with other behaviors such as exploration or emotionality. Such considerations undoubtedly have equal relevance at the human level.

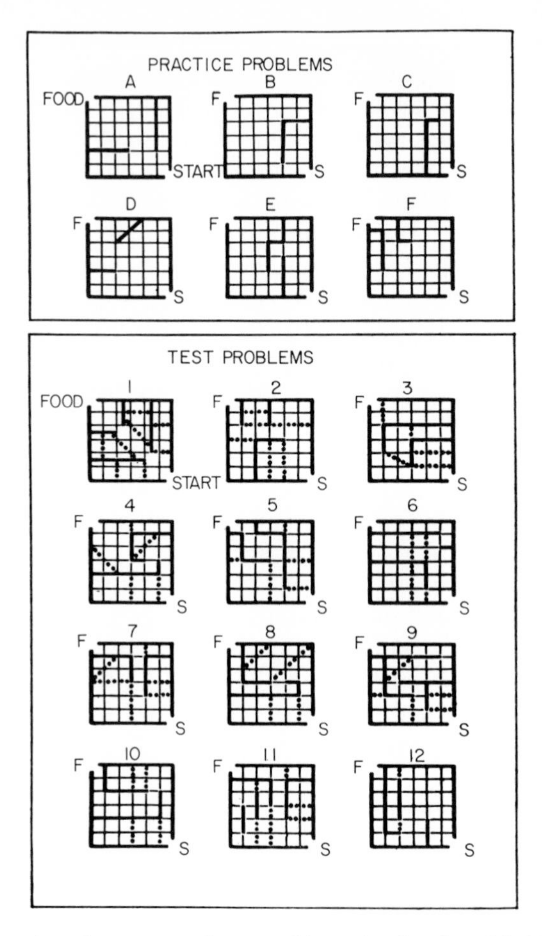

Fig. 1. Floor plan of training and test problems for the closed-field test. The solid lines indicate the maze barriers, while the dotted lines indicate error zones. (From Rabinovitch & Rosvold, 1951. Reprinted by permission of the authors and *Canadian Journal of Psychology*.)

ceives. Those designed to reduce complexity have been referred to as isolation (Krech, Rosenzweig, & Bennett, 1960), a poor (Schwartz, 1964), or restricted environments (Schweikert & Collins, 1966); typically, they have consisted of a small cage, often devoid of objects and with limited and relatively uniform visual and auditory stimulation available to the one or two occupants. In contrast, the environments designated as free (Forgays & Forgays, 1952), unrestricted (Hoffman, 1959), enriched (Cooper & Zubek, 1958), or complex (Brown, 1968) have consisted characteristically of a large living space which contains such objects as ramps, tunnels, barriers, and swings and which is occupied by as many as 10 to 12 animals (Fig. 2).

For the purpose of uniformity, this paper will adopt the terms "restricted environment" (RE) and "complex environment" (CE), respectively. While far from perfect, these choices do not have some of the obvious faults of other terms, such as the often inapplicable suggestion of individual housing by the label "isolation" or the interpretive bias introduced by the label "enriched." Moreover, as Haywood and Tapp (1966) have observed, one must often rely on the experimenter's description in order to distinguish CE from RE since what has been designated as the "normal" control environment in some studies has been virtually identical to the restricted environment in other studies. This confusion results, in large part, from the difficulty of defining what is a "normal" environment for the typical experimental subject which is domesticated and laboratory-reared. In effect, it is the purpose of this paper to arrive at an understanding of which environmental factors contribute to various problem-solving behaviors and through which avenues their contributions are made. Thus, while it may be germane at the outset to indicate baldly the

Fig. 2. Rats in one type of CE, the environmental complexity and training cage (ECT). (From Bennett *et al.*, 1964a. Reprinted by permission of the authors and *Science*.)

effects of RE and CE on such behaviors, ultimately it is necessary to analyze the factors—sensory, motor, social, etc.—which are determinant.

D. Temporal Factors

In discussing the effects of *early* experience on later behavior, the period of primary interest is that extending from birth to puberty, the period, presumably, during which behavioral and physiological organization and growth are proceeding at the fastest rate (King, 1958; Thompson & Schaefer, 1961). Of course, whether early and later experience or, for that matter, whether experience within various periods during early development differentially influences problem-solving behavior remain empirical questions (Section II, D). Most of the experiments included in this paper have tested the behavior of subjects immediately or soon after the environmental manipulation. Occasionally this has meant that the subjects have been tested while still immature, so that some caution concerning the meaning of "later" behavior is advisable. It sometimes does not mean adult.

E. Empirical Overview

The large majority of the experiments performed to determine the effects of early experience on problem-solving behavior have used the rat, although some studies have been conducted with the cat, dog, and monkey. One of the most reliable findings to emerge is that performance on the Hebb-Williams maze is enhanced in rats reared in CE (Brown, 1968; Cooper & Zubek, 1958; Dawson & Hoffman, 1958; Denenberg & Morton, 1962b; Denenberg, Woodcock, & Rosenberg, 1968; Eingold, 1956; Forgays & Forgays, 1952; Forgays & Read, 1962; Hebb, 1947, 1949; Hymovitch, 1952; Mochidome & Fukumoto, 1966; Nyman, 1967; Rabinovitch & Rosvold, 1951; Schwartz, 1964; Smith, 1959; Woods, 1959; Woods, Ruckelshaus, & Bowling, 1960) and impaired in rats reared in RE (Brown, 1968; Cooper & Zubek, 1958; Forgays & Forgays, 1952; Hymovitch, 1952; Mochidome & Fukumoto, 1966; Woods, 1959; Woods *et al.*, 1960). A single exception to these results, in which CE and RE groups did not differ on the Hebb-Williams maze, was provided by a study that markedly altered the usual pretraining and testing conditions (Reid, Gill, & Porter, 1968).

The behavior of rats on tasks other than the Hebb-Williams maze has been affected less reliably following environmental manipulation. Reared in CE, rats have been reported to display superior performance on various mazes including the Warner-Warden and inclined plane mazes (Bingham & Griffiths, 1952), the Lashley III and Dashiell checkerboard mazes (Bennett, Rosenzweig, & Diamond, 1971), a single pattern of the Hebb-Williams maze (Schweikert & Collins, 1966), and the Einstellung apparatus (Hoffman, 1959; Luchins & Forgus, 1955) and on repeated brightness discrimination reversals (Krech, Rosenzweig, & Bennett, 1962; Rosenzweig, 1964). However, rats tended to show no beneficial effects of CE rearing on brightness (Dawson & Hoffman, 1958) and on black versus white discrimination tasks (Bingham & Griffiths, 1952; Woods *et al.*, 1960) or on performance in a five-arm elevated maze (Schwartz, 1961, 1964). In addition, form discrimination has been reported to be both improved (Forgus, 1954) and unchanged (Gill, Ried, & Porter, 1966). In the case of RE rearing, rats have been reported to display deficient performance on repeated brightness-discrimination reversals (Krech *et al.*, 1962; Rosenzweig, 1964) and on Lashley III and Dashiell checkerboard mazes (Bennett *et al.*, 1971). However, no effect of RE rearing was demonstrated when the behavioral task was a black-white discrimination (Bingham & Griffiths, 1952; Woods *et al.*, 1960) or various mazes including the Warner-Warden and inclined plane mazes (Bingham & Griffiths, 1952), the Einstellung apparatus (Hoffman, 1959), or a single pattern of the Hebb-Williams maze (Schweikert & Collins, 1966). Form generalization has also yielded conflicting results (Forgus, 1954; Gill *et al.*, 1966).

Wilson, Warren, and Abbott (1965) have reported the only pertinent experiment with the cat. Their cats were reared in CE and then compared with normal controls. On the Hebb-Williams maze, the experimental animals proved superior to the controls, making significantly fewer errors. No significant differences, however, were obtained on successive reversals of a brightness spatial-discrimination task.

A number of studies have used dogs, but none used a rearing condition defined as CE by the investigator. With the sole exception of performance on a brightness-discrimination (Melzack, 1962), the behavior of dogs reared in RE has been uniformly impaired on a wide variety of tasks: Hebb-Williams maze (Clarke *et al.*, 1951; Thompson & Heron, 1954b), Umweg (or detour) behavior (Clarke *et al.*, 1951; Fox & Stelzner, 1966; Lessac, 1965), delayed response (Thompson & Heron, 1954b), black-white discrimination and its reversal (Melzack, 1962), and spatial discrimination reversals (Fuller, 1966).

Few experiments have been performed on monkeys, and the results have been uniformly negative. Thus, early rearing in RE has produced no effect on discrimination learning (Angermeier, Phelps, & Reynolds, 1967; Griffin & Harlow, 1966; Mason & Fitz-Gerald, 1962; Rowland, 1964), on delayed-response or shock avoidance learning (Rowland, 1964), or on acquisition of learning sets or reversal learning (Mason & Fitz-Gerald, 1962). Rearing in CE was similarly without effect on discrimination learning in monkeys (Angermeier *et al.*, 1967).

The present overview was intended to provide an orientation to research in this area. It indicated that manipulations early in life can affect various problem-solving behaviors of at least some species. But as an overview, it has several limitations. First, important temporal factors concerning the early environment and subsequent testing (see Section I, D) were completely ignored. Second, experimental manipulations were often collapsed to make them fit under the rubric of either CE or RE. This fact, and the difficulty of specifying precisely what is meant by a "normal environment" (see Section I, C), do not allow for easy analysis of conflicting results. For example, if one had a clear understanding of the "normal," it might be evident that in an experiment in which enhanced performance failed to occur the CE omitted some critical dimension of stimulation and, thus, was not actually a CE. And finally, the frequent use of CE and RE groups in experiments that do not include a group even resembling a normal control makes it difficult to decide to which experimental condition to attribute either the presence of an observed behavioral effect (e.g., Krech *et al.*, 1962; Woods, 1959) or its absence (e.g., Reid *et al.*, 1968). A related problem is the case in which a group designated by the investigator to be a normal control actually appears to be a CE group (e.g., Clarke *et al.*, 1951; Melzack, 1962) so that once again attribution of the observed effect is difficult.

From what has just been said, it should be clear that a closer analysis of the data along various of its dimensions is required to reveal its pattern and to suggest its deficiencies.

II. Critical Experiential Factors

A. Visual Perception

It has been suggested that early perceptual experience—particularly early visual experience—crucially determines the quality of the later problem-solving behavior (Hebb, 1949). Several types of data support this suggestion. First, rats blinded early in life tend to make more errors on a Hebb-Williams maze than rats blinded later in life (Hebb, 1947, 1949). However, on a task which demanded rote solution and, therefore, presumably did not measure problem-solving, no behavioral differences obtained between early- and late-blinded animals (Hebb, 1947, 1949). Second, rats which commit relatively few errors on the Hebb-Williams maze by virtue of CE rearing subsequently show a significantly greater increment of errors than control- or RE-reared rats after the maze is rotated through an angle of 45 degrees or more (Brown, 1968; Forgays & Forgays, 1952; Hymovitch, 1952). Such results indicate that the superior performance of the CE groups depended, at least in part, upon locomoting to the goal box by orienting to distant visual cues. Since it was these cues that were altered by rotation of the maze, it was the performance of the CE rats that was significantly impaired. On this basis, it has been suggested that CE rearing produces enhanced perceptual learning which in turn facilitates problem-solving by the increased attention devoted to extrafield visual cues. Finally, Hymovitch (1952) found that while the degree of motor experience available to rats in early life did not correlate highly with their error scores on the Hebb-Williams maze, a "wide" visual environment was necessary for good performance. He concluded that "the differential opportunities presented the various groups for *perceptual learning* were responsible for the results" (Hymovitch, 1952, p. 319).

At this point it is necessary to cite data which indicate the need for examining more closely the nature as well as the effects of early visual experience. For example, exposure of rats to CE or RE has been shown to produce, respectively, superior or inferior problem-solving behavior on a brightness-discrimination reversal task (Krech *et al.*, 1962) which minimized extramaze visual cues (Krech, Rosenzweig, & Bennett, 1956). Thus, the importance of early CE experience in producing greater attention to distant visual cues becomes questionable, although such experience might still be important for attention to visual intramaze and test cues. Furthermore, although some investigators (Bingham & Griffiths, 1952; Hymovitch, 1952) have characterized as "wide" the critical quality of the early visual environment, the available evidence does not unequivocally support this contention. Hymovitch (1952), for example, found no differences on Hebb-Williams maze performance between a group of rats confined in mesh cages within (and occasionally outside of) a CE and a group allowed to move about within the same CE; Forgays and Forgays (1952), on the other hand, found that groups which were allowed to move about a CE were superior to those which were confined in mesh cages within the same CE. The two studies are not strictly comparable, however, since Hymovitch's (1952) caged rats received more extensive and variable experience. Nonetheless, the fact remains that, at least within certain limits, simple exposure to a wide environment is not sufficient to facilitate problem-solving behavior. This statement is further supported by the recent report that a group of rats reared in a large but empty enclosure did not perform as well on the Hebb-Williams apparatus as did groups of rats reared in an enclosure of identical size but supplied with objects and/or barriers (Brown, 1968). On the basis of these and related data, Brown (1968) suggested that the superior performance of CE groups must depend upon their ability to assess visually the relative position of barriers in the Hebb-Williams maze, as well as upon their use of distant visual cues.

B. Locomotion and Proprioception

The importance of early motor and proprioceptive experiences has usually been evaluated relative to the operation of visual factors and, therefore, has already been alluded to in the above section. Perhaps the clearest and most elegant experiment is that of Hymovitch (1952). He used four groups defined according to the condition of their early rearing: (a) CE, (b) mesh cage rotated to various positions within (and without) a CE, (c) solid-walled cage of the same dimensions as the mesh cage, and (d) solid-walled activity wheel. When run subsequently in the Hebb-Williams maze, the first two groups made fewer errors than the second two groups. The interpretation of these results seems obvious. The complex visual experience of the CE and mesh-cage groups was necessary for good performance. The amount of motor experience, in contrast, was irrelevant since the mesh-cage group, with little opportunity for motor experience, performed well, while the solid-walled cage and activity wheel groups, with quite different opportunities for locomotion, both made many errors. Similar conclusions were reached by Schweikert and Collins (1966) who reported that performance on a Hebb-Williams maze tended to vary as a function of the visual complexity of the rearing environment rather than as a function of the amount of physical space available for locomotion.

Although these studies appear to indicate the irrelevance of motor experience and its consequent proprioception, a series of studies by Forgus (1954, 1955a, 1955b) suggest the conditions under which such experiences might contribute to problem solving. In an initial study, he found that rats reared in a glass enclosure placed within a larger CE so that they could view but not contact the objects located in it performed better on a form discrimination test (but not on a form generalization test) than did a group allowed to roam freely throughout the CE (Forgus, 1954). The explanation offered was that the confined group, having been motorically restricted within a complex visual environment, found visual cues particularly salient during subsequent testing and consequently did not respond unduly to irrelevant motor cues, which on this task would have interfered presumably with the solution of the problem. In a *spatial* problem-solving test, however, in which visual cues are evidently of less importance, the confined group did not differ significantly from the free-roaming group, although both performed better than an RE group. From these data it should be clear that the claim for the importance of motor and proprioceptive experiences is highly particular to the nature of the test. In subsequent experiments, once again using groups confined within or allowed free access to a CE but now tested on an 11-unit T-maze, Forgus (1955a, 1955b) found that, after an initial phase of learning based primarily on visual cues, the elimination of these visual cues conferred superiority in performance on the group that had been allowed to traverse the entire CE. Presumably, the motor skills acquired by the latter group during its rearing were beneficial in the absence of visual cues. Again, Forgus (1955b) argued that "the relative influence of early experience on adult cognitive abilities depends largely on the relationship between the kind of early experience and the requirements of the problem task" (p. 213). Unfortunately, in an apparently close replication of one of Forgus' experiments, Walk (1958) did not obtain similar results.

C. Handling

Abundant evidence exists to indicate that early handling of rats can materially alter their later behaviors with particular reference to emotionality (e.g., Denenberg, 1964, 1967). The numerous demonstrations that such handling can facilitate the learning of a shock-avoidance habit (e.g., Denenberg, 1962; Levine, Chevalier, & Korchin, 1956) raise the question of whether the learning effect derives from the impact of this handling on the emotion- or on the problem-solving aspect of the task. Several lines of data intersect to suggest that problem-solving ability is in fact unchanged by handling.

Although Bernstein (1952, 1957) found that postweaning handling improved the performance of juvenile rats on a brightness-discrimination task, it should be noted that, since no pretraining was administered, the possibility exists that the nonhandled control subjects were more emotional during training and, therefore, inferior in performance. It should be noted also that neither CE- nor RE-rearing modifies behavior on this kind of task (see Section I, E). Preweaning handling appears to be ineffective in altering the behavior of rats on a black-white discrimination task (Wong, 1966).

More persuasive perhaps of the fact that preweaning handling does not affect problem-solving ability are those data obtained using the type of correlative analysis recommended by Lashley (1929). When groups of rats, for example, were similarly handled but reared either in CE or in RE, the CE group committed fewer errors on the Hebb-Williams maze (Hymovitch, 1952; Woods, 1959). It is obvious that unequal handling is not a necessary condition for differential performance. Even more direct is the finding that, compared to nonhandled rats, identically housed but handled rats perform no better on the Hebb-Williams maze (Denenberg & Morton, 1962b; Schaefer, 1963, 1968). Once again, early handling does not appear to be a manipulation capable of producing significant modifications in problem-solving behavior.

For the cat, the only available experiment (Wilson *et al.*, 1965) provides data similar to those for the rat. Although handled cats displayed changed (impaired) avoidance learning, they performed no differently than nonhandled cats either on the Hebb-Williams maze or on successive discrimination reversals.

D. Time of Exposure

Having demonstrated that exposure to selected environments during a circumscribed period of development is particularly effective in influencing the organization of various behaviors (e.g., Beach & Jaynes, 1954; Hebb, 1949), many investigators emphasized the possible importance of temporal variables for problem-solving. In this connection, the idea of a critical period was advanced, the idea that during development there is a circumscribed period of time when the substrate for problem-solving behavior can be most affected by appropriate experiences. It should be pointed out that a critical period hypothesis not only demands a major effect of a specific stimulus at one point during the development but also no effect, or at least only a minor effect, at other times (Denenberg, 1968). By contrast, in the situation where stimulation has a maximal effect at one point in development and significant effects at other times as well, one can speak only of an optimal period.

Perhaps, the major concern of studies investigating the differential effects of early experience is just what kind of environment the animal should be subjected to when not undergoing the experimental treatment. As Fuller and Waller put it: "The impossibility of putting animals on a shelf and maintaining them without any experience makes critical tests of the critical-period hypothesis extremely difficult . . ." (1962, p. 238). This is particularly relevant in the case of problem-solving behavior since it is the rearing environment itself that is being manipulated. Theoretically, one would like to rear the subjects in a "normal" environment and then expose them to either the CE or RE only at the desired time. Since at present the definition of a normal environment remains arbitrary, one can only counsel caution in the interpretation of results from studies designed to investigate temporal variables. The type of CE or RE, the time of its presentation, and the nature of the baseline rearing environment must all be considered. Additionally, it must be admitted that although the analysis of the data depends importantly on an understanding of such terms as juvenile and mature, no commonly accepted definition of these terms can be found in the literature, nor are their correct designations readily apparent.

Hebb (1947, 1949) was the first to suggest that the time of presentation of an environmental manipulation might be critical with respect to its effect on problem-solving behavior. Hebb blinded rats by optic enucleation either at the time of eye-opening or at maturity and subsequently tested them on the Hebb-Williams maze. The fact that the late-blinded rats performed better—a finding that Hymovitch (1952) could not confirm—led Hebb to conclude that early visual experience was particularly important. Although it should be clear that only conclusions regarding the relative importance of early versus no visual experience, and not of early versus later visual experience, can be drawn from this experiment, it served nonetheless to provoke interest in the latter question.

Hymovitch (1952) was able to demonstrate persuasively that it is *early* rather than later experience which is particularly effective in influencing problem-solving behavior. He showed that a group of rats exposed to a CE for a long period before maturity and then to a RE for a long period after maturity performed as well on the Hebb-Williams maze as a group that was exposed to a CE both before and after maturity. Moreover, both these groups performed better than groups which received RE exposure prior to maturity and CE exposure after maturity or which were reared "normally" throughout. It appears, then, that early CE experience is as beneficial as a combination of both early and late CE and that late CE experience alone is ineffective (Forgays & Read, 1962). Of course, these data might also be interpreted to suggest that late RE experience has no effect and that early RE experience, if it has an effect, can be compensated for by late CE rearing. Although Woods (1959) showed that the deleterious effects in rats on Hebb-Williams maze performance of early RE rearing can be compensated for by subsequent CE rearing, his results are equivocal in regard to the possible effect of late CE rearing since the pertinent rearing period spanned the latter part of juvenile development and early maturity (66 to 95 days of age).

If it is true that early experience modifies later problem-solving behavior, the next issue that arises concerns the possibility of delimiting a yet more circumscribed period of time responsible for the effect. Several pertinent studies have been conducted but no consensus has emerged (Brown, 1968; Eingold, 1956; Forgays & Read, 1962; Nyman, 1967). Brown (1968), for example, using rather long periods, found no differential effects in rats of RE or CE rearing from 20- to 60-, 60- to 100-, or 20- to 100-days of age. However, the fact that his subjects lived in restricted environments when not receiving the experimental treatment may well have affected the results. Similarly, the data of Eingold (1956) and Nyman (1967) are difficult to interpret, being also afflicted with several methodological weaknesses. But here again, if one ignored such weaknesses, it could be concluded that exposing rats to a CE during a period with a mean of 55 days of age and ranging from approximately 46 to 65 days of age is particularly beneficial for later problem-solving behavior. Such conclusions, the methodological difficulties notwithstanding, are supported by the data of Forgays and Read (1962). In a well-designed study, these investigators found that rats living in a CE from 23 to 43 days of age tended to perform better on a problem-solving task than other groups exposed to the same CE but at other ages.

It should be noted that in the three studies just mentioned (Eingold, 1956; Forgays & Read, 1962; Nyman, 1967) the Hebb-Williams maze was used and the rats, when not in the CE, were housed in small groups in what were evidently normal laboratory cages. However, it must be emphasized that while the various experimental groups exposed to the CE differed from their control counterparts, many of them did not differ among themselves (Table I).

In summary, it seems reasonable to conclude that the period prior to puberty is probably critical for the development of problem-solving behavior (e.g., Forgays & Read, 1962; Hymovitch, 1952). In other words, when the environmental manipulation is in the direction of greater complexity and when this manipulation is carried out during the first 50 or 60 days of life, a significant effect on problem

Table I
MEAN ERRORS AND STANDARD DEVIATIONS OF SIX GROUPS OF Ss ON THE HEBB-WILLIAMS TEST AND *t*s OF THE VARIOUS MEAN COMPARISONS[a]

Group	N	Time of enriched experience (days of age)	Hebb-Williams Test: Mean error score	Hebb-Williams Test: Standard deviation	t^b: 2	3	4	5	6
1	8	0-21	174.25	15.56	2.53*	1.30	1.08	0.21	2.15*
2	17	22-43	153.29	25.29		0.64	0.56	2.07*	3.70**
3	9	44-65	160.33	25.34			0.02	1.22	2.70*
4	9	66-87	160.56	31.94				1.07	2.37*
5	9	88-109	176.67	28.00					1.38
6	8	No enriched experience	195.75	25.41					

[a]Adapted from Forgays and Read (1962). Reprinted by permission of the authors and *Journal of Comparative and Physiological Psychology*.
[b]$^{*}p < .05$, $^{**}p < .01$.

solving is evident; when, however, the same manipulation is applied postpuberally, no effect of environmental complexity is evident. Whether such a conclusion is warranted for a period more precisely delimited than the first "50 or 60" days of age remains to be established.

E. MISCELLANEOUS FACTORS

Only one experiment has attempted to determine whether discrete and intermittent kinds of stimulation other than handling affect problem-solving behavior. Griffiths and Stringer (1952) found no effects of intense auditory stimulation, rapid rotation, extremes of environmental temperature, or shock to the feet on the behavior of rats on the Warner-Warden maze or on a discrimination problem on a modified Lashley jumping stand. Here the use of more standard types of problem-solving tasks should probably be explored.

Although it seems obvious that whether animals are reared in isolation or with conspecifics might be an important condition modifying problem-solving behavior, this social variable has been essentially unexplored. Often it has been embedded among other variables, the CE having animals housed collectively and the RE having them individually caged (e.g., Krech *et al.*, 1962). The only study to investigate this problem systematically, although imperfectly for the purposes of this chapter, uncovered no difference in error scores on a multiple-U maze in rats reared from 21 to 240 days of age either individually or in groups of eight (Myers & Fox, 1963).

But even here, conclusions are difficult by virtue of the fact that the experimental treatment was extended across developmental periods and also by the absence of any behavioral "norms" for the particular apparatus employed.

Another variable richly deserving experimental attention but concerning which only a single investigation (Denenberg *et al.*, 1968) has been conducted is the duration of the treatment-test interval. Typically, the subjects are tested while still in their experimental environment (e.g., Dawson & Hoffman, 1958; Woods *et al.*, 1960) or within a few days after removal from that environment (e.g, Cooper & Zubek, 1958; Griffin & Harlow, 1966). Obviously enough, it is important to know whether or not manipulating the early environment produces a permanent or merely a transient change in problem-solving ability. If the change is of only short-term duration, the equation of problem-solving ability with intelligence must certainly be incorrect since intelligence is typically regarded as an enduring characteristic of the organism. Thus, the need is for studies in which the subjects, subsequent to the experimental treatment, are housed under a control condition for varying periods of time, prior to being tested. The critical datum here would be the relationship between the test score and the treatment-test interval.

Experiments performed to investigate critical periods have employed various treatment-test intervals, but, of course, the time of treatment and the duration of the interval until testing were confounded (Brown, 1968; Eingold, 1956; Forgays & Read, 1962; Nyman, 1967). With this methodological reservation in mind, however, it can be observed that significant performance differences on the Hebb-Williams maze have been found between CE-reared and control rats after treatment-test intervals of as long as 109 days (Forgays & Read, 1962). Deficient performance was obtained from RE-reared dogs as long as 7 months after treatment (Thompson & Heron, 1954b). In the only study designed explicitly to investigate treatment-test interval, a period of almost 1 year was employed, during which time all rats lived in the control environment (Denenberg *et al.*, 1968). It was found that a relatively short time spent in a CE either before or after weaning facilitated performance on the Hebb-Williams maze. Since at the time of testing the animals had already lived one-third to one-half of their total life span, Denenberg *et al.* concluded that the effect of the CE experience is of long-term duration. Further research on this question is needed, but it does appear that the rearing effect endures.

A final issue to be covered under the present heading concerns the question of whether the behavioral changes produced by experimental manipulations result from the complexity (enrichment) or the restriction (impoverishment) of the environment. This question is not a simple one; its answer depends in large part upon the reliable specification of the "normal" environment (see Section I, C). Thus, the report (Bennett *et al.*, 1971) that rats reared in a CE performed no better than those reared in the control condition but that both groups performed better than those reared in RE should probably not be taken as evidence that it is impoverishment rather than enrichment of the environment that modifies problem-solving behavior. Indeed, it is quite possible that, in this experiment, the CE and the control environment shared a critical "enriching" dimension such as an extended visual environment; the more apparent conclusion that only the RE affected behavior relative to the "norm" is not forced. On the other hand, an elegant experiment by Cooper and Zubek (1958) does suggest that the effectiveness of early environmental manipulation depends upon the genetic substrate of the organism upon which it acts. These investigators found that for maze-bright and maze-dull rats of the McGill strains CE rearing enhanced the Hebb-Williams maze performance of the latter group but not of the former, while RE rearing depressed the performance of the former but not that of the latter. The question, then, of enrichment *vs.* impoverishment is a complicated one, the answer depending upon both the specification of the "normal" environment and the interaction with the genetic substrate.

III. Cerebral Effects of Environmental Manipulation

The assumption is commonplace that experimental treatments which produce behavioral changes should also induce alterations in the structure and function of the brain. Less obvious is the notion that such cerebral effects might be detectable following CE or RE rearing. It is the case, however, that these cerebral modifications have been demonstrated in great detail. Since the focus of the present chapter is on experiments with behavioral end points, it is not possible to describe exhaustively all the studies appropriate to the above heading. Instead, emphasis will be given to reviewing the cerebral effects produced by environmental manipulations similar to those which have been shown to affect problem-solving behavior. The pertinent research has been performed on the rat—almost entirely by a single group of investigators including Bennett, Diamond, Krech, and Rosenzweig from the University of California at Berkeley—and, in only small measure, on the dog.

A. Research on the Rat

1. *Procedures*

Investigations by the Berkeley groups have used subjects typically reared in one of three environments: (a) a CE, which they refer to as environmental complexity and training (ECT), in which the rats live in groups of 10 to 12 and encounter various objects and from which they are removed at frequent intervals to be given trials in mazes (see Fig. 2), (b) a control condition, which they refer to as the social control (SC), in which the rats live in groups of three in a standard laboratory cage, and (c) an RE, which they refer to as the isolated control (IC), in which the rats live individually in a cage with three opaque walls (Krech *et al.*, 1960). The animals, typically male descendants of Tryon's maze-bright rats (strain S), are characteristically reared in one or another of the environments just mentioned for a period of 80 days from the time of weaning (25 days of age), although in recent studies some modifications in this basic procedure have been made (Bennett *et al.*, 1971; Rosenzweig, Bennett, & Diamond, 1967c; Rosenzweig, Love, & Bennett, 1968b). In any event, following the experiental phase, or after behavioral testing when such testing is employed (Krech *et al.*, 1962), the rats are decapitated and their brains removed and dissected. The dissection usually yields five parts, sampling, respectively, visual cortex, somesthetic cortex, remaining dorsal cortex, ventral cortex, and the remaining brain or subcortex (Rosenzweig, Krech, Bennett, & Diamond, 1962).

2. *Anatomical Results*

Although it was not expected initially, the research of the Berkeley group soon indicated that the manipulation of the early environment produced small but reliable changes in cerebral anatomy (Bennett, Diamond, Krech, & Rosenzweig, 1964a; Bennett, Krech, & Rosenzweig, 1964b; Rosenzweig, 1966; Rosenzweig *et al.*, 1962). More specifically, their ECT rats had a significantly heavier total cortex than their IC rats, with the difference being much more marked in the visual than in the somesthetic area. With respect to subcortex, however, just the reverse was true—rats reared in the ECT condition had a lighter subcortex than those reared in the IC condition. SC rats displayed a weight of intermediate value for the total cortex that was significantly different from the weight values obtained from both the ECT and IC subjects. With respect once again to the weight of the subcortex, the ECT and SC rats were closely similar and were each significantly different from the IC rats. The increased cortical weight of the ECT rats relative to the IC rats was reflected in their greater depth of cortical tissue (Diamond, Krech, & Rosenzweig, 1964), although it is possible that this brain weight change might also have been due to cerebral elongation which has been found to result from rearing in CE (Altman, Wallace, Anderson, & Das, 1968). The fact that rats reared under the ECT condition had lower body weights than the IC animals precludes the alternative that their increased cortical weight simply reflected increased body weight (Diamond, 1967; Krech *et al.*, 1960; Zolman & Morimoto, 1962).

Further experiments have demonstrated that similar patterns of change in brain anatomy can also be produced by exposure to ECT or IC for as little as 2 hours per day over a period of days (Rosenzweig *et al.*, 1968b). Although this appeared to depend on the age of the subject when the rearing was begun and upon the duration of that rearing (Bennett *et al.*, 1971; Rosenzweig *et al.*, 1967c; Rosenzweig, Bennett, & Krech, 1964; Zolman & Morimoto, 1962, 1965), it was shown that modifications of cerebral anatomy could be obtained even with the adult rat (Bennett *et al.*, 1971; Rosenzweig *et al.*, 1964). This latter finding has led to the conclusion that the usual cerebral effects of the ECT and IC conditions cannot be viewed merely as "alterations in the rate of early cerebral growth" (Rosenzweig *et al.*, 1964, p. 429).

3. *Chemical Results*

It was with the expectation that neurochemical differences would result from differential early rearing that the Berkeley group initiated their research. They predicted that rats living in an environment of enhanced complexity would have increased central neural activity; that this increased neural activity, in turn, would be reflected at some central synapses by the greater release of acetylcholine (ACh), the chemical agent most clearly implicated in synaptic transmission (Koelle, 1965); and that increment in the level of ACh would then lead to a greater concentration of acetylcholinesterase (AChE),* the enzyme primarily responsible for hydrolyzing ACh (Rosenzweig,

*In their earlier work (e.g., Krech *et al.*, 1960), this agent is referred to as the less specific enzyme or enzymes, cholinesterase (ChE). Subsequently, they determined that although their measure was of all the enzymic activity that hydrolyzed ACh, AChE was responsible for over 95% of this activity in the rat brain and, therefore, that the overall activity could reasonably be attributed to AChE (Bennett, *et al.*, 1964b). Thus, in this paper the term AChE is used often to refer to results originally reported as being for ChE. Where the term ChE is used, it is employed intentionally and pertains to the activity of those relatively nonspecific enzymes which also hydrolyze ACh.

1964; Rosenzweig, Krech, Bennett, & Diamond, 1968a). Most of their chemical data concern the level of AChE in their various brain samples.

Among those rats receiving one or another of the experimental treatments for 80 days beginning at the time of weaning (Bennett *et al.*, 1964a, 1964b; Krech *et al.*, 1960; Rosenzweig, 1966; Rosenzweig *et al.*, 1962), the ECT animals showed the highest total level of AChE activity in the cortex because of the greater weight of their cortices; in fact, they exhibited a decrease in cortical specific ACh activity (i.e., activity per unit weight). Correlatively, although IC rats had a smaller amount of total AChE activity relative to their ECT counterparts, they showed a higher level of cortical specific AChE activity due to the decreased weight of their cortices. The SC animals had an intermediate level of cortical specific AChE activity. Interestingly enough, for the subcortical specific AChE activity, the reverse order obtained among the three experimental groups. In the case of ChE, however, total and specific activity increased in the cortex but remained unchanged in the subcortex of ECT animals. Again, the activity of these chemical substances could be altered by administering the experimental treatments for a period of less than 80 days (Rosenzweig *et al.*, 1968b; Zolman & Morimoto, 1962, 1965) or by administering these same treatments during adulthood (Rosenzweig *et al.*, 1964, 1967c).

Finally, the levels of specific activity of a number of other chemical substances were found not to change consequent to the experimental treatments, suggesting that the neurochemical effect is relatively specific. These substances include hexokinase (Bennett *et al.*, 1964a; Rosenzweig, 1966), serotonin (Bennett *et al.*, 1964a), ribonucleic acid (Rosenzweig, 1966), and protein (Altman & Das, 1966; Bennett *et al.*, 1964a; Das & Altman, 1966; Rosenzweig, 1966).

4. *Histological Results*

The greater activity levels of ChE in the cortex of the ECT animals suggested the possibility that the increased cortical weight might be due to a proliferation of glial cells, the cells in which the greatest concentration of ChE occurs (Koelle, 1965). This speculation has recently been confirmed (Altman & Das, 1964; Diamond, 1967; Diamond, Law, Rhodes, Linder, Rosenzweig, Krech, & Bennett, 1966) and is of marked interest since an important role in brain function has been suggested for the neuroglia (Galambos, 1961).

Similarly, it has been suggested that the increase of ChE and AChE in the cortex of ECT animals might reflect, at least in part, a multiplication of dendritic branching (Rosenzweig *et al.*, 1968a). Holloway (1966) has tentatively concluded that this does occur, at least in the visual cortex. Diamond (1967) has found an increase in the size of the cells of the visual cortex and their nuclei, a change that would be demanded by increased dendritic branching.

5. Results of Other Environmental Manipulations

Various control experiments have eliminated from consideration a number of factors which might possibly have been implicated in the production of the above cerebral changes. Apparently, neither the additional handling nor the greater opportunity for motor activity afforded the ECT animals relative to those in IC is critical for the establishment of the cerebral differences (Krech *et al.*, 1960; Rosenzweig *et al.*, 1968a; Zolman & Morimoto, 1965). On the other hand, preweaning handling did induce changes in brain weight and brain chemistry which were significant but which, in turn, were of a pattern different from that produced by ECT (Kling, Finer, & Nair, 1965; Tapp & Markowitz, 1963). Although the extra-cage environment and the daily trials on mazes appeared to have had no effect on the cerebral measures for ECT rats, both the social grouping and the complexity of the cage environment were involved (Rosenzweig, Bennett, & Diamond, 1967a). Finally, a factorial experiment with ECT and IC environments and with blinded and sighted rats demonstrated that the usual cerebral effects, even those in the visual cortex itself, occurred in the absence of visual experience (Krech, Rosenzweig, & Bennett, 1963).

6. Cerebro-Behavioral Correlations

Krech *et al.* (1962) reared rats of the S_1 strain for 30 days from time of weaning either in ECT (with the maze training eliminated) or IC situations and then housed them in SC for the next 30 days while behavioral testing on brightness discrimination and reversal problems proceeded. After this testing, the rats were killed and the chemical and anatomical effects assessed. Although the ECT subjects made fewer errors on the reversal problems, *no* significant differences emerged between the groups on any of the cerebral indices measured. However, the error scores of the ECT group did correlate significantly with the cortical-subcortical ratio both of brain weight and of specific AChE activity; a significant correlation between the error score and the cortical-subcortical ratio of specific AChE activity also obtained for the IC groups. These data on cerebral measures, as well as the data of Zolman and Morimoto (1962), suggest that SC housing and behavioral testing do not produce further cerebral changes among the ECT subjects but do change the brains of IC animals in the direction of their ECT littermates. This conclusion presents a paradox, however, since it indicates that while the behavior of the ECT and IC groups on successive reversals was becoming progressively divergent, their brains were becoming progressively similar. In other words, the behavioral and cerebral events appear not to have covaried over time.

There are other data, recently reported in brief detail, which cast further doubt that the performance of these experimentally reared rats on reversal problems is related reliably to the cortical-subcortical ratio of brain weight (Bennett *et al.*, 1971). First, animals reared in ECT or SC for 30 or 60 days from the time of weaning performed similarly on reversal problems, although they had significantly different cortical-subcortical ratios of brain weight. Second, rats reared in ECT from 60 to 90 or from 90 to 120 days of age showed higher brain weight ratios than rats reared in IC at the same ages, but the performance of the IC rats on reversal problems was superior to that of the ECT rats at the younger age and not different at the later age. Furthermore, the level of cortical AChE activity was found to be unrelated to problem-solving performance across six strains of rats (Rosenzweig, Krech, & Bennett, 1960). However, the ACh/AChE ratio for the cortex did appear to be related to the behavioral performance of these strains.

It seems obvious that at the present time the relationship between cerebral measures and problem-solving behavior is largely unknown. It is possible, of course, that determination of cortical-subcortical ratios is too gross an approach (Rosenzweig, Bennett, & Diamond, 1967b) and that smaller parts of the brain must be sampled. For example, it has been suggested that the hippocampus functions in memory but that other limbic structures do not (Penfield & Milner, 1958). Yet these structures are combined in the ventral-cortex sample (Rosenzweig *et al.*, 1962) so that critical changes in the hippocampus might be masked by random events in adjacent structures. Another reason for the difficulty in correlating behavioral and cerebral events may be the fact that although some of the anatomical and chemical changes produced by the experimental environments reflect active processes instrumental in establishing memory traces, these changes regress once the subject is removed from the environment and memory consolidation is completed (Rosenzweig *et al.*, 1967c). Such an hypothesis would explain the absence of the usual ECT-IC cerebral differences after a period of removal from the experimental environments (Krech *et al.*, 1962; Zolman & Morimoto, 1962) while still allowing for the prediction of behavioral differences. Of course, the two hypotheses outlined here are not mutually exclusive.

B. Research on the Dog

1. *Chemical Results*

Agrawal, Fox, and Himwich (1967) have reported findings on mongrel puppies placed in an RE for 7 days beginning at 4 weeks of age. Compared with normal controls, these dogs showed no significant changes in the neocortex but they did exhibit a number of chemical changes in subcortical structures. Specifically, they had increased levels of glutamic acid and γ-aminobutyric acid (GABA) in the diencephalon, increased levels of glutamine in the superior colliculus and caudate nucleus, and decreased levels of glutamic acid and GABA in the caudate nucleus.

2. *Physiological Results*

Beagle and cross-bred puppies have been implanted with chronic neocortical electrodes and raised in a RE for 7 days beginning at 4 weeks of age (Fox, 1967). Upon emergence from the RE, their electroencephalograms (EEG) showed a pattern of arousal activation not seen characteristically in the records of control animals. Similarly Melzack and Burns (1965) have shown that beagle puppies reared in RE for 1½ to 2 years beginning at 3 weeks of age display a persistently activated EEG from cortical and brainstem reticular formation when presented with a novel environment. When presented with the same novel environment, auditory and visual evoked-potentials, however, were persistently diminished.

IV. Explanatory Suggestions

None of the hypotheses reviewed here encompass all the data, either behavioral or neurological. Some appear to be designed to accommodate the behavioral results only of animals raised in RE. Still others have an inituitive cogency but no data in their support. In a few instances, much of the evidence to which these hypotheses address themselves has already been reviewed and will be referred to only briefly.

A. HEBB'S PERCEPTUAL THEORY

In essence, Hebb stated that it is early perceptual experience—particularly, the early visual experience—that forms whatever organization within the nervous system is critical for the determination of the organism's intelligence (Hebb, 1949; Hebb & Thompson, 1954). "Cell assemblies" and their combinations, "phase sequences," constitute the hypothetical neurological constructs employed. These are conceived as being formed at an early age and as then functioning to subserve the various problem-solving behaviors. Thus, Hebb suggested that early exposure to a CE functions to establish and integrate a particular kind of neurological substrate, while early exposure to an RE is insufficient to initiate or support development of the same substrate. Since it was hypothesized that both efficient learning and efficient problem solving are dependent upon this neurological organization, the critical role indicated for *early experience* in determining intelligence is apparent.

Unquestionably, the organism's visual experiences can be of great importance for subsequent problem-solving behavior, although just which characteristics these experiences must possess has yet to be determined (see Section II, A). Furthermore, the evidence indicates that visual experience early in life is critical (see Section II, D). Nonetheless, it must be emphasized that the importance of such experience probably depends, in large part, upon the behavioral test used (Forgus, 1954, 1955a, 1955b) and had tests radically different from the Hebb-Williams been employed, very different conclusions may well have been reached.

Perhaps the crucial question for Hebb's position is whether or not the behavioral tests that have been used do, in fact, measure intelligence (see Section I, B). Indeed, since first asserting that late-blinded were superior to early-blinded rats on the Hebb-Williams maze—an instrument, as already mentioned, for measuring animal intelligence, not rote learning—Hebb (1947, 1949) has asserted that it is intelligence that is manipulated by certain relevant early experiences and that it is intelligence that is measured by the problem-solving tasks. Since then (see Section I, E), the data have appeared to indicate, although with obvious exceptions, that modifications of early experience do not alter performance on rote tasks, such as various discrimination problems, but that such modifications do significantly change behaviors on tasks requiring a flexible and/or complex response sequence, such as the Hebb-Williams maze, the Einstellung apparatus, and Umweg problems. This is essentially correct even if one accepts the suggestion that successive reversal tests are a better index of intelligence than is the Hebb-Williams maze (Warren, 1965a, 1965b). Thus, it would appear that Hebb's construction of the data, at least in its broad aspects, has empirical support. Still, the question remains far from settled due to the imprecision of our specifications of behavioral tests of intelligence.

B. Exploration Dysfunction

Zimbardo and Montgomery (1957) reported that, compared to rats reared in CE for varying periods subsequent to weaning, control rats showed more exploration when tested in a Y-maze. Despite this finding, the investigators suggested that exploratory activity in rats is enhanced by a testing environment which is "richer" or "more complex" than the rearing environment. They proceeded to speculate that the advantage which CE-reared rats display in problem solving "may arise as a result of the relatively decreased novelty of the test situation for the . . . [CE] Ss. Therefore, they should be less likely to respond as intensely and indiscriminately to all the cues in the environment (the irrelevant as well as the relevant ones)" (Zimbardo & Montgomery, 1957, p. 593).

Advancing this same speculation, Woods (1959) noticed that, unlike rats reared in CE, those reared in RE explored the Hebb-Williams maze excessively, even retracing after having reached the goal box, thereby accumulating additional errors. Although discounting such errors did *not* eliminate the inferiority of the RE rats, Woods tentatively attributed their poor performance to their enhanced tendency to explore. A later experiment determined that two tests of exploration for CE-reared and RE-reared rats yielded significant positive correlations (.48 and .67) with errors scores on the Hebb-Williams maze—a result partially replicated by Brown (1968)—although the differences present between groups for error scores were not always paralleled by differences in exploration measures (Woods *et al.*, 1960). Furthermore, adaptation and preliminary training for the Hebb-Williams maze were found to increase the exploratory behavior of RE rats and decrease the same behavior for CE rats. Again, the effects of modified environments on the problem-solving behavior were ascribed to a primary effect on exploratory tendencies.

In the final experiment in this series, Woods, Fiske, and Ruckelshaus (1961) ran RE- and CE-reared rats on the Hebb-Williams maze either under a low-drive condition or under a high-drive condition, the latter having been produced by increasing the duration of food deprivation or by superimposing upon the deprivation shocks for errors. The purpose of these manipulations was to reduce exploration by establishing a strong conflicting drive and in this way to presumably abolish the differences in problem solving between groups reared in the two environments. The results indicated that the behavior of the CE and RE high-drive groups was indistinguishable. However, in one of the comparisons, the CE and RE low-drive groups were similar in performance, which of course contradicts the basic notion. These, then, are essentially the data upon which Woods and his colleagues built their "exploration hypothesis."

Two other lines of empirical evidence seem relevant to the present hypothesis, but each presents difficulties of interpretation. The first is experiments in which CE and RE rats performed similarly on the Hebb-Williams maze after the animals were given extensive pretraining and then run under relatively severe deprivation conditions. The pretraining was designed to reduce emotionality and exploration, with the deprivation added to reduce exploration even further. Nonetheless, it is not clear that a reduction of "emotionality" in the RE rats accounted for the results. During the 18 days of pretraining, all animals were housed in RE, which may have counteracted the beneficial effect previously accrued by the CE rats, while the 90 to 162 runs in the maze may well have constituted an enriching experience for the RE rats, collapsing group differences even further (Reid *et al.*, 1968).

Of course, the major support for the exploratory hypothesis derives from the fact that, in some studies at least, RE animals have been found to be more exploratory than CE animals. Although these studies, to be sure, have reported a direct relationship in rats between the degree of restriction present in the early environment and the amount of subsequent locomotory exploration (Mochidome & Fukumoto, 1966; Sackett, 1967; Woods *et al.*, 1960; Zimbardo & Montgomery, 1957), others have noted either no such relationship (Brown, 1968; Dawson & Hoffman, 1958; Denenberg & Morton, 1962a; Ehrlich, 1959; Forgays & Read, 1962; Lore & Levowitz, 1966; Montgomery & Zimbardo, 1957; Patrick & Laughlin, 1934) or even an inverse relationship (Forgus, 1954; Gill *et al.*, 1966; Hoffman, 1959; Luchins & Forgus, 1955; McCall, Lester, & Dolan, 1969; Snowdon, Bell, & Henderson, 1964). Clearly, then, the basic data with which Woods and his colleagues have attempted to support their hypothesis are unreliable, so that the hypothesis itself is questionable.

Although having performed an experiment largely irrelevant to the matter under discussion, Rajalakshmi and Jeeves (1968) have correctly pointed out that the excessive locomotion sometimes observed with RE animals could just as well be a secondary condition of poor problem solving. Something of this order must be true since there seems to be no reliable relationship between locomotory exploration and problem-solving behavior.

Finally, the superior performance by CE-reared rats on the Einstellung apparatus (Hoffman, 1959; Luchins & Forgus, 1955), a task which presumably demands exploration and flexibility for its solution, appears to discount further the exploration hypothesis.

C. Melzack's Information-Processing Model

Dealing explicitly with the consequences of RE rearing (and almost solely with dogs), Melzack constructed an explanatory model (Melzack, 1962, 1965, 1968; Melzack & Burns, 1965) which makes interlocking statements on behavioral and physiological levels, many of which seem clearly to have been influenced by earlier work from the McGill laboratories (Hebb, 1949, 1955; Thompson, 1955; Thompson & Heron, 1954a, 1954b).

Two main behavioral facts were noted. First, dogs with early RE experience display a high level of excited activity (Melzack, 1954, 1962, 1965; Thompson & Heron, 1954b; Thompson, Melzack, & Scott, 1956). Later experiments established also that as little as 7 days of RE rearing between the fourth and fifth weeks of life can produce this hyperexcitability in dogs upon their emergence from the environment (Agrawal *et al.*, 1967; Fox, 1967; Fox & Stelzner, 1966) and that the same phenomenon can occur in cats, monkeys, and chimpanzees (Riesen, 1961). Second, restricted dogs are unable to attend selectively to significant environmental stimuli, often dashing from object to object in a room without investigating effectively any one of them and frequently failing to perceive or respond to important or intrusive stimulation (Melzack & Scott, 1957; Thompson & Heron, 1954b). Obviously, such behavior may well be characteristic of the RE-reared rats as well (Lore & Levowitz, 1966; Sackett, 1967).

On the level of physiological explanation, Melzack and Burns (1965) suggest that

> The neural substrates of memories of earlier experience are able to exert control over information selection (via centrifugal fibers) at the earliest synaptic levels of sensory pathways. As a result of severe restriction of early sensory experi-

> ence, most stimuli in a totally new environment have no meaning (prior associations) to provide a basis for selective filtering of the sensory input. Consequently, all inputs, irrelevant as well as relevant, would reach the brain where they could bombard the neural systems that produce sensory and affective arousal. The excessive arousal, in turn, would disrupt central nervous system mechanisms, both innate and acquired, that underlie perceptual discrimination and adaptive response . . . (p. 164).

The fact that RE-reared dogs with chronically implanted neocortical and brainstem-reticular electrodes display "arousal" in the presence of a novel environment of moderate complexity and, in response to clicks or flashes, evoked potentials reduced in amplitude, was adduced as positive evidence for this hypothesis.

Melzack's hypothesis is attractive first, because it attempts to interrelate both behavioral and physiological data, and second, because it can incorporate both Hebb's perceptual hypothesis (Section IV, A) and Woods' exploratory hypothesis (Section IV, B). Yet while the data reliably indicate hyperactivity for RE-reared dogs, as Woods' hypothesis would demand, the fact that such a relationship has not been found reliably in rats (Section IV, B) limits its phylogenetic generality. Furthermore, the seeming consistency between the perceptual and behavioral data, on the one hand, and the electrophysiological data, on the other hand, is difficult to evaluate in light of their uncertain correlation under some circumstances (e.g., Meyers, Roberts, Riciputi, & Domino, 1964; Wikler, 1952) and also because of the crudeness of the analogy between reduced attention to stimuli and evoked potentials reduced in amplitude. Indeed, the reduced amplitude of the evoked potentials could have been fairly well predicted from the activated EEG alone. Finally, the hypothesis does not address itself explicitly to the results of CE-rearing, although it might be argued that these results would be the reciprocal of those reported for RE-rearing.

D. Degradation of Preformed Intellectual Capacity

Posed as an alternative to developmental models (such as Hebb's) that RE rearing prevents the normal acquisition and organization of perceptual, motivational, and/or behavioral substrates, this hypothesis suggests that RE rearing might instead produce a deterioration of an intellectual capacity already established and organized prior to the rearing period (Lessac, 1965; Solomon, 1949; Solomon & Lessac, 1968). In essence, empirical support for this hypothesis demands four groups arranged in a 2 × 2 matrix according to whether or not each group is reared in RE and whether or not the necessary behavioral comparisons to normal controls are made before and after treatment or only after treatment. Such a design permits the resolution of the question of whether the poor performance of the RE subjects represents a failure to acquire habits or a deterioration of habits already present.

Lessac (1965) has performed a study using this design, but only one of his behavioral tests, an Umweg problem, is directly pertinent to problem solving. When tested prior to isolation at 12 weeks of age, beagles were capable of learning the problem; and pretesting had no effect on post-test performance. However, isolation significantly impaired the Umweg behavior such that the isolates then took longer than controls to solve the problem and showed no improvement over trials. It was disturbing, however, that the isolated dogs were generally sluggish and tended to nibble at rather than eat the food rewards. Thus, this experiment provides only qualified support for the deterioration hypothesis.

With only a single experiment performed on a single species, it is not possible at this time to evaluate the validity of the present hypothesis. However, the demonstration that in cats the presumed neural substrate for visual pattern perception is organized neonatally and in fact deteriorates with patterned-light deprivation (Hubel & Wiesel, 1963; Wiesel & Hubel, 1963) suggests that it must be seriously regarded. Again, this hypothesis does not treat the CE data. If it is to treat these data, it must assume that CE rearing enhances problem solving by enhancing an already organized neural substrate.

E. Miscellaneous Hypotheses

Under the present rubric can be included a number of hypotheses, each of which, although possessing some face validity, has not been developed sufficiently and consequently cannot be translated readily into concrete experimental operations. They are mentioned here only in the interest of completeness.

Fuller (1966, 1967) has argued that the behavioral deficits consequent to RE rearing result from competing emotional responses occasioned by emergence from an RE into a varied environment. He advanced this position as an alternative to the failure of behavioral or perceptual organization to occur, as Hebb (1949) would have it, or to the deterioration of such organization, as Lessac and Solomon (Lessac, 1965; Solomon & Lessac, 1968) would propose.

It has been suggested also that the dramatic effects of early experience are due to the high degree of plasticity of the young organism (Fuller & Waller, 1962; Thompson, 1955; Thompson & Schaefer, 1961). Here it is presumed that various functions are particularly amenable to change while undergoing development. Obviously, this suggestion is very similar to the critical period hypothesis. In parallel also is the proposal that young organisms are relatively undifferentiated so that their reaction to stimulation is more general and, consequently, that the effect of such stimulation is more pervasive at a younger than at a later age (Thompson, 1955; Thompson & Schaefer, 1961). Finally, under the concept of primacy, it has been asserted that changes which occur early in life not only endure but influence the acquisition of behaviors in situations occurring later in life (Beach & Jaynes, 1954; Thompson, 1955; Thompson & Schaefer, 1961).

V. Conclusions

It should be clear now that none of the available hypotheses provides a justification, *a priori,* for the assertion that early experience is in fact critically important for the development of problem-solving ability. Indeed, many areas of concern permit only vague and uncertain suggestions. Perhaps this should not be surprising in an area in which there is so much difficulty in specifying accurately both the point of departure and the terminal point, that is, in specifying what constitutes a "normal" environment (and, by contrast, restricted and complex environments) and what behavioral measures qualify as tests of intelligence. It is within this uncertain framework that one must operate.

Nevertheless, as this review has documented, some progress has been made. Using tests that demand novel solutions based upon past experience and, therefore, having presumably the imprimatur of problem solving or intelligence, investigators have reliably demonstrated that the performance of rats can be enhanced or degraded by raising them in a CE or RE, respectively. It is clear, moreover, that it is the manipulation of experience early in the organism's life that is the effective strategy. A wide variety of anatomical, chemical, and physiological effects also result from these modulations of experience, although their relationship to specific behavioral effects is largely unrevealed. Furthermore, it seems that the behaviors under investigation are fairly discrete and therefore distinguishable from what has been designated as emotionality and exploration. And finally, although uncertainty remains concerning the nature of the early experience requisite for the production of enhanced intelligence, Forgus (1955a) was very likely correct in asserting that "the important theoretical issue . . . seems to be the relationship between the quality of early experience and the nature of the task to be solved" (p. 220).

Two aspects of the present area are least understood and, therefore, most imperatively in need of investigation. The first involves the nature of the test used to measure problem solving. Although the Hebb-Williams maze continues to be the instrument of choice, its dependence on visual cues (e.g., Brown, 1968; Forgays & Forgays, 1952) makes it perhaps a highly biased test. Warren (1965a, 1965b), in fact, has specifically suggested that it is a poor measure of problem-solving behavior and that tests of reversal learning would be better. As a beginning, then, studies should be performed in which RE- and CE-reared rats are tested, in a counterbalanced design on both the Hebb-Williams maze and on reversal problems in order to determine their relationship to each other. Similar studies should also be performed for other frequently used tests such as the Umweg problem.

Second, research on the effects of early experience on later problem-solving behavior has concentrated disproportionately on the rat. Only one study has been performed on the cat (Wilson *et al.*, 1965). Of the six investigations using dogs (Clarke *et al.*, 1951; Fox & Stelzner, 1966; Fuller, 1966; Lessac, 1965; Melzack, 1962; Thompson & Heron, 1954b), none has employed an environment labeled as CE. With monkeys there have only been four studies (Angermeier *et al.*, 1967; Griffin & Harlow, 1966; Mason & Fitz-Gerald, 1962; Rowland, 1964). Of these, only one has provided a task which qualifies as a test of problem solving or intelligence, and this one used only a single subject (Mason & Fitz-Gerald, 1962). Moreover, for the one study to rear monkeys in a CE, such a poverty of detail is reported that it is of little value (Angermeier *et al.*, 1967). Clearly, then, to the extent that one desires to generalize the results of this review to humans (Haywood, 1967; Haywood & Tapp, 1966; Hunt, 1961; Pettigrew, 1964), increased use of a variety of species is necessary.

Nonetheless, the possible relation of the animal literature to the human condition should not be passed over lightly, since under the proper circumstances certain human experiences can either elevate or depress the I.Q. of children (Haywood & Tapp, 1966). In fact, Haywood and Tapp have asserted that

> this view would see a large percentage of mild and moderate mental retardates as victims of relatively unstimulating early environments, constituting what has come recently to be called 'cultural deprivation.' While lower-class status does not necessarily involve a condition of cultural deprivation, the sad fact is that the two most often go hand in hand. By the same token, middle-class status does not insure rearing in a highly stimulating environment, but it does carry a higher probability of heightened stimulation. A disproportionately high percentage of mild and moderate retardates come from lower-class homes . . . (p. 144).

From this one might reasonably conclude that our increasing knowledge concerning the effects of early experience on intelligence is matched by a tragically unjust inequity in the way in which our society distributes opportunities necessary for enrichment.

References

Agrawal, H. C., Fox, M. W., & Himwich, W. A. Neurochemical and behavioral effects of isolation-rearing in the dog. *Life Science,* 1967, **6,** 71-78.

Altman, J., & Das, G. D. Autoradiographic examination of the effects of enriched environment on the rate of glial multiplication in the adult rat brain. *Nature* (*London*), 1964, **204,** 1161-1163.

Altman, J., & Das, G. D. Behavioral manipulations and protein metabolism of the brain: Effects of motor exercise on the utilization of leucine-H^3. *Physiology & Behavior,* 1966, **1,** 105-108.

Altman, J., Wallace, R. B., Anderson, W. J., & Das, G. D. Behaviorally induced changes in length of cerebrum in rats. *Developmental Psychology,* 1968, **1,** 112-117.

Angermeier, W. F., Phelps, J. B., & Reynolds, H. H. The effects of differential early rearing upon discrimination learning in monkeys. *Psychonomic Science,* 1967, **8,** 379-380.

Beach, F. A., & Jaynes, J. Effects of early experience upon the behavior of animals. *Psychological Bulletin,* 1954, **51,** 239-263.

Bennett, E. L., Diamond, M. C., Krech, D., & Rosenzweig, M. R. Chemical und anatomical plasticity of brain. *Science,* 1964, **146,** 610-619. (a)

Bennett, E. L., Krech, D., & Rosenzweig, M. R. Reliability and regional specificity of cerebral effects of environmental complexity and training. *Journal of Comparative and Physiological Psychology,* 1964, **57,** 440-441. (b)

Bennett, E. L., Rosenzweig, M. R., & Diamond, M. C. Time courses of effects of differential experience on brain measures and behavior of rats. In W. Byrne (Ed.), *Molecular approaches to learning and memory.* New York: Academic Press, 1971.

Bernstein, L. A note on Christie's: Experimental naïveté and experiential naïveté. *Psychological Bulletin,* 1952, **49,** 38-40.

Bernstein, L. The effects of variations in handling upon learning and retention. *Journal of Comparative and Physiological Psychology,* 1957, **50,** 162-167.

Bingham, W. E., & Griffiths, W. J., Jr. The effect of different environments during infancy on adult behavior in the rat. *Journal of Comparative and Physiological Psychology*, 1952, **45**, 307-312.

Bitterman, M. E. Phyletic differences in learning. *American Psychologist*, 1965, **20**, 396-410.

Brown, R. T. Early experience and problem-solving ability. *Journal of Comparative and Physiological Psychology*, 1968, **65**, 433-440.

Clarke, R. S., Heron, W., Fetherstonhaugh, M. L., Forgays, D. G., & Hebb, D. O. Individual differences in dogs: Preliminary report on the effects of early experience. *Canadian Journal of Psychology*, 1951, **5**, 150-156.

Cooper, R. M., & Zubek, J. P. Effects of enriched and restricted early environments on the learning ability of bright and dull rats. *Canadian Journal of Psychology*, 1958, **12**, 159-164.

Das, G. D., & Altman, J. Behavioral manipulations and protein metabolism of the brain: Effects of restricted and enriched environments on the utilization of leucine-H^3. *Physiology & Behavior*, 1966, **1**, 109-110.

Dawson, W. W., & Hoffman, C. S. The effects of early differential environments on certain behavior patterns in the albino rat. *Psychological Record*, 1958, **8**, 87-92.

Denenberg, V. H. An attempt to isolate critical periods of development in the rat. *Journal of Comparative and Physiological Psychology*, 1962, **55**, 813-815.

Denenberg, V. H. Critical periods, stimulus input, and emotional reactivity: A theory of infantile stimulation. *Psychological Review*, 1964, **71**, 335-351.

Denenberg, V. H. Stimulation in infancy, emotional reactivity, and exploratory behavior. In D. H. Glass (Ed.), *Neurophysiology and emotion*. New York: Russell Sage Found. & Rockefeller Univ. Press, 1967. Pp. 161-203.

Denenberg, V. H. A consideration of the usefulness of the critical period hypothesis as applied to the stimulation of rodents in infancy. In G. Newton & S. Levine (Eds.), *Early experience and behavior: The psychobiology of development*. Springfield, Ill.: Thomas, 1968. Pp. 142-167.

Denenberg, V. H., & Morton, J. R. C. Effects of environmental complexity and social groupings upon modification of emotional behavior. *Journal of Comparative and Physiological Psychology*, 1962, **55**, 242-246. (a)

Denenberg, V. H., & Morton, J. R. C. Effects of preweaning and postweaning manipulations upon problem-solving behavior. *Journal of Comparative and Physiological Psychology*, 1962, **55**, 1096-1098. (b)

Denenberg, V. H., Woodcock, J. M., & Rosenberg, K. M. Long-term effects of preweaning and postweaning free-environment experience on rats' problem-solving behavior. *Journal of Comparative and Physiological Psychology*, 1968, **66**, 533-535.

Diamond, M. C. Extensive cortical depth measurements and neuron size increases in the cortex of environmentally enriched rats. *Journal of Comparative Neurology*, 1967, **123**, 111-120.

Diamond, M. C., Krech, D., & Rosenzweig, M. R. The effects of an enriched environment on the histology of the rat cerebral cortex. *Journal of Comparative Neurology*, 1964, **123**, 111-120.

Diamond, M. C., Law, F., Rhodes, H., Linder, B., Rosenzweig, M. R., Krech, D., & Bennett, E. L. Increases in cortical depth and glia numbers in rats subjected to enriched environment. *Journal of Comparative Neurology*, 1966, **128**, 117-125.

Ehrlich, A. Effects of past experience on exploratory behaviour in rats. *Canadian Journal of Psychology*, 1959, **13**, 248-254.

Eingold, B. Problem-solving by mature rats as conditioned by the length, and age at imposition, of earlier free-environmental experience. Unpublished doctoral dissertation, University of Florida, 1956.

Forgays, D. G., & Forgays, J. W. The nature of the effect of free-environmental experience in the rat. *Journal of Comparative and Physiological Psychology*, 1952, **45**, 322-328.

Forgays, D. G., & Read, J. M. Crucial periods for free-environmental experience in the rat. *Journal of Comparative and Physiological Psychology*, 1962, **55**, 816-818.

Forgus, R. H. The effect of early perceptual learning on the behavioral organization of adult rats. *Journal of Comparative and Physiological Psychology*, 1954, **47**, 331-336.

Forgus, R. H. Early visual and motor experience as determiners of complex maze-learning ability under rich and reduced stimulation. *Journal of Comparative and Physiological Psychology*, 1955, **48**, 215-220. (a)

Forgus, R. H. Influence of early experience on maze-learning with and without visual cues. *Canadian Journal of Psychology*, 1955, **9**, 207-214. (b)

Fox, M. W. The effects of short-term social and sensory isolation upon behavior, EEG, and averaged evoked potentials in puppies. *Physiology & Behavior*, 1967, **2**, 145-151.

Fox, M. W., & Stelzner, D. Behavioural effects of differential early experience in the dog. *Animal Behaviour*, 1966, **14**, 273-281.

Fuller, J. L. Transitory effects of experiential deprivation upon reversal learning in dogs. *Psychonomic Science*, 1966, **4**, 273-281.

Fuller, J. L. Experiential deprivation and later behavior. *Science*, 1967, **158**, 1645-1652.

Fuller, J. L., & Waller, M. B. Is early experience different? In E. L. Bliss (Ed.), *Roots of behavior*. New York: Harper, 1962. Pp. 235-245.

Galambos, R. A glia-neural theory of brain function. *Proceedings of the National Academy of Sciences of the United States*, 1961, **47**, 129-136.

Gill, J. H., Reid, L. D., & Porter, P. B. Effects of restricted rearing on Lashley stand performance. *Psychological Reports*, 1966, **19**, 239-242.

Griffin, G. A., & Harlow, H. F. Effects of three months of total social deprivation on social adjustment and learning in the rhesus monkey. *Child Development*, 1966, **37**, 533-547.

Griffiths, W. J., Jr., & Stringer, W. F. The effects of intense stimulation experienced during infancy on adult behavior in the rat. *Journal of Comparative and Physiological Psychology*, 1952, **45**, 301-306.

Haywood, H. C. Experiential factors in intellectual development: The concept of dynamic intelligence. In J. Zubin & G. A. Jervis (Eds.), *Psychopathology of mental development*. New York: Grune & Stratton, 1967. Pp. 69-104.

Haywood, H. C., & Tapp, J. T. Experience and the development of adaptive behavior. In N. R. Ellis (Ed.), *International review of research in mental retardation*. Vol. 1. New York: Academic Press, 1966. Pp. 109-151.

Hebb, D. O. The effects of early experience on problem-solving at maturity. *American Psychologist*, 1947, **2**, 306-307. (Abstract)

Hebb, D. O. *The organization of behavior: A neuropsychological theory*. New York: Wiley, 1949.

Hebb, D. O. Drives and the C.N.S. (conceptual nervous system). *Psychological Review*, 1955, **62**, 243-254.

Hebb, D. O., & Thompson, W. R. The social significance of animal studies. In G.

Lindzey (Ed.), *Handbook of social psychology.* Vol. 1. Reading, Mass.: Addison-Wesley, 1954. Pp. 532-561.

Hebb, D. O., & Williams, K. A. A method of rating animal intelligence. *Journal of General Psychology,* 1946, **34,** 59-65.

Hoffman, C. S. Effect of early environmental restriction on subsequent behavior in the rat. *Psychological Record,* 1959, **9,** 171-177.

Holloway, R. L., Jr. Dendritic branching: Some preliminary results of training and complexity in rat visual cortex. *Brain Research,* 1966, **2,** 393-396.

Hubel, D. H., & Wiesel, T. N. Receptive fields of cells in striate cortex of very young, visually inexperienced kittens. *Journal of Neurophysiology,* 1963, **26,** 994-1002.

Hunt, J. McV. *Intelligence and experience.* New York: Ronald Press, 1961.

Hymovitch, B. The effects of experimental variations on problem-solving in the rat. *Journal of Comparative and Physiological Psychology,* 1952, **45,** 1313-321.

King, J. A. Parameters relevant to determining the effect of early experience upon the adult behavior of animals. *Psychological Bulletin,* 1958, **55,** 46-58.

Kling, A., Finer, S., & Nair, V. Effects of early handling and light stimulation on the acetylcholinesterase activity of the developing rat brain. *International Journal of Neuropharmacology,* 1965, **4,** 353-357.

Koelle, G. B. Neurohumoral transmission and the autonomic nervous system. In L. S. Goodman & A. Gilman (Eds.), *The pharmacological basis of therapeutics.* (3rd ed.) New York: Macmillan, 1965. Pp. 399-440.

Krech, D., Rosenzweig, M. R., & Bennett, E. L. Dimensions of discrimination and level of cholinesterase activity in the cerebral cortex of the rat. *Journal of Comparative and Physiological Psychology,* 1956, **49,** 261-268.

Krech, D., Rosenzweig, M. R., & Bennett, E. L. Effects of environmental complexity and training on brain chemistry. *Journal of Comparative and Physiological Psychology,* 1960, **53,** 509-519.

Krech, D., Rosenzweig, M. R., & Bennett, E. L. Relations between brain chemistry and problem-solving among rats raised in enriched and impoverished environments. *Journal of Comparative and Physiological Psychology,* 1962, **55,** 801-807.

Krech, D., Rosenzweig, M. R., & Bennett, E. L. Effects of complex environment and blindness on rat brain. *Archives of Neurology (Chicago),* 1963, **8,** 403-412.

Lashley, K. S. *Brain mechanisms and intelligence.* Chicago: Univ. of Chicago Press, 1929.

Lessac, M. S. The effects of early isolation and restriction on the later behavior of beagle puppies. Unpublished doctoral dissertation, University of Pennsylvania, 1965.

Levine, S., Chevalier, J. A., & Korchin, S. J. The effects of shock and handling in infancy on later avoidance learning. *Journal of Personality,* 1956, **24,** 475-493.

Lore, R. K., & Levowitz, A. Differential rearing and free versus forced exploration. *Psychonomic Science,* 1966, **5,** 421-422.

Luchins, A. S., & Forgus, R. H. The effect of differential postweaning environment on the rigidity of an animal's behavior. *Journal of Genetic Psychology,* 1955, **86,** 51-58.

McCall, R. B., Lester, M. L., & Dolan, C. G. Differential rearing and the exploration of stimuli in the open field. *Developmental Psychology,* 1969, **1,** 750-762.

Mason, W. A., & Fitz-Gerald, F. L. Intellectual performance of an isolation-reared rhesus monkey. *Perceptual and Motor Skills,* 1962, **15,** 594.

Melzack, R. The genesis of emotional behavior: An experimental study of the dog. *Journal of Comparative and Physiological Psychology,* 1954, **47,** 166-168.

Melzack, R. Effects of early perceptual restriction on simple visual discrimination. *Science*, 1962, **137**, 978-979.

Melzack, R. Effects of early experience on behavior: Experimental and conceptual considerations. In P. H. Hoch & J. Zubin (Eds.), *Psychopathology of perception.* New York: Grune & Stratton, 1965. Pp. 271-299.

Melzack, R. Early experience: A neuropsychological approach to heredity-environment interactions. In G. Newton & S. Levine (Eds.), *Early experience and behavior: The psychobiology of development.* Springfield, Ill.: Thomas, 1968. Pp. 65-82.

Melzack, R., & Burns, S. K. Neurophysiological effects of early sensory restriction. *Experimental Neurology*, 1965, **13**, 163-175.

Melzack, R., & Scott, T. H. The effects of early experience on the response to pain. *Journal of Comparative and Physiological Psychology*, 1957, **50**, 155-161.

Meyers, B., Roberts, K. H., Riciputi, R. H., & Domino, E. F. Some effects of muscarinic cholinergic blocking drugs on behavior and the electrocorticogram. *Psychopharmacologia*, 1964, **5**, 289-300.

Mochidome, H., & Fukumoto, F. Shikakuteki shoki keiken ga mondai kaiketsu ni oyobosu koka. *Annual of Animal Psychology* (*Tokyo*), 1966, **16**, 11-19. (*Psychological Abstracts*, 1967, **41**, No. 10100)

Montgomery, K. C., & Zimbardo, P. G. Effect of sensory and behavioral deprivation upon exploratory behavior in the rat. *Perceptual and Motor Skills*, 1957, **7**, 223-229.

Myers, R. D., & Fox, J. Differences in maze performance of group- vs. isolation-reared rats. *Psychological Reports*, 1963, **12**, 199-202.

Nyman, A. J. Problem solving in rats as a function of experience at different ages. *Journal of Genetic Psychology*, 1967, **110**, 31-39.

Patrick, J. R., & Laughlin, R. M. Is the wall-seeking tendency in the white rat an instinct? *Journal of Genetic Psychology*, 1934, **44**, 378-389.

Penfield, W., & Milner, B. Memory deficit produced by bilateral lesions in the hippocampal zone. *AMA Archives of Neurology and Psychiatry*, 1958, **79**, 475-497.

Perlman, D. The search for the memory molecule. *New York Times Magazine*, July 7, 1968. Pp. 8-37.

Pettigrew, T. F. *A profile of the Negro American.* Princeton, N.J.: Van Nostrand, 1964.

Rabinovitch, M. S., & Rosvold, H. E. A closed-field intelligence test for rats. *Canadian Journal of Psychology*, 1951, **5**, 122-128.

Rajalakshmi, R., & Jeeves, M. A. Performance on the Hebb-Williams maze as related to discrimination and reversal learning in rats. *Animal Behaviour*, 1968, **16**, 114-116.

Reid, L. D., Gill, J. H., & Porter, P. B. Isolated rearing and Hebb-Williams maze performance. *Psychological Reports*, 1968, **22**, 1073-1077.

Riesen, A. H. Excessive arousal effects of stimulation after early sensory deprivation. In P. Solomon, P. E. Kubzansky, P. H. Leiderman, J. H. Mendelson, R. Trumbull, & D. Wexler (Eds.), *Sensory deprivation.* Cambridge, Mass.: Harvard Univ. Press, 1961. Pp. 34-40.

Rosenzweig, M. R. Effects of heredity and environment on brain chemistry, brain anatomy and learning ability in the rat. *Kansas Study on Education*, 1964, **14**, 3-34.

Rosenzweig, M. R. Environmental complexity, cerebral change, and behavior. *American Psychologist*, 1966, **21**, 321-332.

Rosenzweig, M. R., Bennett, E. L., & Diamond, M. C. Cerebral effects of differential experience. Paper presented at the meeting of the American Psychological Association, Washington, D.C., September, 1967. (a)

Rosenzweig, M. R., Bennett, E. L., & Diamond, M. C. Effects of differential environments on brain anatomy and brain chemistry. In J. Zubin & G. A. Jervis (Eds.), *Psychopathology of mental development.* New York: Grune & Stratton, 1967. Pp. 45-56. (b)

Rosenzweig, M. R., Bennett, E. L., & Diamond, M. C. Transitory components of cerebral changes induced by experience. Paper presented at the meeting of the American Psychological Association, Washington, D. C., September, 1967. (c)

Rosenzweig, M. R., Bennett, E. L., & Krech, D. Cerebral effects of environmental complexity and training among adult rats. *Journal of Comparative and Physiological Psychology,* 1964, **57**, 438-439.

Rosenzweig, M. R., Krech, D., & Bennett, E. L. A search for relations between brain chemistry and behavior. *Psychological Bulletin,* 1960, **57**, 476-492.

Rosenzweig, M. R., Krech, D., Bennett, E. L., & Diamond, M. C. Effects of environmental complexity and training on brain chemistry and anatomy: A replication. *Journal of Comparative and Physiological Psychology,* 1962, **55**, 429-437.

Rosenzweig, M. R., Krech, D., Bennett, E. L., & Diamond, M. C. Modifying brain chemistry and anatomy by enrichment or impoverishment of experience. In G. Newton & S. Levine (Eds.), *Early experience and behavior: The psychobiology of development.* Springfield, Ill.: Thomas, 1968. Pp. 102-141. (a)

Rosenzweig, M. R., Love, W., & Bennett, E. L. Enriched experience for two hours a day alters brain chemistry and anatomy in rats. *Physiology & Behavior,* 1968, **3**, 819-825. (b)

Rowland, G. L. The effects of total social isolation upon learning and social behavior of rhesus monkeys. Unpublished doctoral dissertation, University of Wisconsin, 1964.

Sackett, G. P. Response to stimulus novelty and complexity as a function of rats' early rearing experiences. *Journal of Comparative and Physiological Psychology,* 1967, **63**, 369-375.

Schaefer, T., Jr. Early "experience" and its effects on later behavioral processes in rats. II. A critical factor in the early handling phenomenon. *Transactions of The New York Academy of Sciences,* 1963, **25**, 871-889.

Schaefer, T., Jr. Some methodological implications of the research on "early handling" in the rat. In G. Newton & S. Levine (Eds.), *Early experience and behavior: The psychobiology of development.* Springfield, Ill.: Thomas, 1968. Pp. 102-141.

Schwartz, S. The effect of neo-natal brain damage and early environment on adult behavior in the hooded rat. Unpublished doctoral dissertation, University of Michigan, 1961.

Schwartz, S. Effect of neonatal cortical lesions and early environmental factors on adult rat behavior. *Journal of Comparative and Physiological Psychology,* 1964, **57**, 72-77.

Schweikert, G. E., III, & Collins, G. The effects of differential postweaning environments on later behavior in the rat. *Journal of Genetic Psychology,* 1966, **109**, 255-263.

Smith, C. J. Mass action and early environment in the rat. *Journal of Comparative and Physiological Psychology,* 1959, **52**, 154-156.

Snowdon, C. T., Bell, D. D., & Henderson, N. D. Relationships between heart rate and open-field behavior. *Journal of Comparative and Physiological Psychology,* 1964, **58**, 423-426.

Solomon, R. L. An extension of control group design. *Psychological Bulletin,* 1949, **46**, 137-150.

Solomon, R. L., & Lessac, M. S. A control group design for experimental studies of developmental processes. *Psychological Bulletin,* 1968, **70**, 145-150.

Tapp, J. T., & Markowitz, H. Infant handling: Effects on avoidance learning, brain weight and cholinesterase activity. *Science,* 1963, **140,** 486-487.

Thompson, W. R. Early environment—its importance for later behavior. In P. H. Hoch & J. Zubin (Eds.), *Psychopathology of childhood.* New York: Grune & Stratton, 1955. Pp. 120-139.

Thompson, W. R., & Heron, W. The effects of early restriction on activity in dogs. *Journal of Comparative and Physiological Psychology,* 1954, **47,** 77-82. (a)

Thompson, W. R., & Heron, W. The effects of restricting early experience on the problem-solving capacity of dogs. *Canadian Journal of Psychology,* 1954, **8,** 17-31. (b)

Thompson, W. R., Melzack, R., & Scott, T. H. "Whirling behavior" in dogs as related to early experience. *Science,* 1956, **123,** 939.

Thompson, W. R., & Schaefer, T., Jr. Early environmental stimulation. In D. W. Fiske & S. R. Maddi (Eds.), *Functions of varied experience.* Homewood, Ill.: Dorsey, 1961. Pp. 81-105.

Walk, R. D. "Visual" and "visual-motor" experience: A replication. *Journal of Comparative and Physiological Psychology,* 1958, **51,** 785-787.

Warren, J. M. The comparative psychology of learning. *Annual Review of Psychology,* 1965, **16,** 95-118. (a)

Warren, J. M. Primate learning in comparative perspective. In A. M. Schrier, H. F. Harlow, & F. Stollnitz (Eds.), *Behavior of nonhuman primates.* Vol. 1. New York: Academic Press, 1965. Pp. 249-281. (b)

Wiesel, T. N., & Hubel, D. H. Single-cell responses in striate cortex of kittens deprived of vision in one eye. *Journal of Neurophysiology,* 1963, **26,** 1003-1017.

Wikler, A. Pharmacologic dissociation of behavior and EEG "sleep patterns" in dogs: Morphine, N-allylnormorphine, and atropine. *Proceedings of the Society for Experimental Biology and Medicine,* 1952, **79,** 261-265.

Wilson, M., Warren, J. M., & Abbott, L. Infantile stimulation, activity, and learning by cats. *Child Development,* 1965, **36,** 843-853.

Wong, R. Infantile handling and performance in the T-maze. *Psychonomic Science,* 1966, **5,** 203-204.

Woods, P. J. The effects of free and restricted environmental experience on problem-solving behavior in the rat. *Journal of Comparative and Physiological Psychology,* 1959, **52,** 399-402.

Woods, P. J., Fiske, A. S., & Ruckelshaus, S. I. The effects of drives conflicting with exploration on the problem-solving behavior of rats reared in free and restricted environments. *Journal of Comparative and Physiological Psychology,* 1961, **54,** 167-169.

Woods, P. J., Ruckelshaus, S. I., & Bowling, D. M. Some effects of "free" and "restricted" environmental rearing conditions upon adult behavior in the rat. *Psychological Reports,* 1960, **6,** 191-200.

Zimbardo, P. G., & Montgomery, K. C. Effects of "free-environment" rearing upon exploratory behavior. *Psychological Reports,* 1957, **3,** 589-594.

Zolman, J. F., & Morimoto, H. Effects of age of training on cholinesterase activity in the brains of maze-bright rats. *Journal of Comparative and Physiological Psychology,* 1962, **55,** 794-800.

Zolman, J. R., & Morimoto, H. Cerebral changes related to duration of environmental complexity and locomotor activity. *Journal of Comparative and Physiological Psychology,* 1965, **60,** 382-387.

CHAPTER 4

THE ONTOGENY OF EMOTIONAL BEHAVIOR*

Douglas K. Candland

I. Problems of Experimental Design and Interpretation

The fact that there is no inclusive and commonly accepted definition of the term "emotion" has not deterred students of behavior from investigating the topic, editors from assigning chapters on it, or authors from writing about it. The view guiding the preparation of

*The preparation of this chapter was aided especially by a postdoctoral fellowship from the National Science Foundation taken at the Delta Regional Primate Research Center of Tulane University and by a special research fellowship from the National Institute of Mental Health taken at the Pennsylvania State University. I am extremely grateful to a number of investigators who graciously provided me with unpublished papers and preprints and who patiently answered my questions about their work and opinions. A number of persons active in the field kindly supplied me with thoughtful replies to an inquiry regarding the composition of the chapter. I thank Steven DeKosky and Jay Poliner for their aid in locating sources. Drs. J. Ernest Keen, William A. Mason, Emil Menzel, Jr., and Howard Moltz provided detailed comments on this chapter for which I am very grateful.

this chapter is that "emotion," like "learning," "motivation," and even "threshold," refers to the relationships between selected measurable physiological changes and selected measurable external events. Both a bar press and a face flushed to a measurable degree qualify.

The major problem confronting the student of behavior is how to describe the affective state (e.g., sadness) that is believed to parallel or to be reflected in the behavior he is measuring (e.g., crying). Unfortunately, communication about the nature of affective states is no more exact than descriptions of the sensations accompanying stimulation of the interoceptors, and for that reason contemporary investigators, impressed with the need for precise definition, have not studied them often. Avoiding a subject, however, does not aid enlightenment. In intellectual history, it is noteworthy that within the past century and a quarter, both Charles Bell and Charles Darwin thought the subject of emotion to be of sufficient importance to devote separate, detailed treatises to it. Both wrote books concerned specifically with the role of facial expressions and posture in displays of emotion (Bell, 1806, 1844; Darwin, 1872). In fact, it has been only during the past 50 years that the study of emotion has been neglected as a result, undoubtedly, of the influence of behaviorism and the radical forms of logical positivism. Although the leaders of the behaviorist and radical positivist schools were most certainly concerned with emotion (Skinner, 1938; Watson, 1929; Watson & Rayner, 1920), their adherents often strove for operational purity by the Procrustean method of eliminating or ignoring difficult concepts on the faulty assumption that "If I do not see it, it does not exist." But "it" does exist, and often those who prefer to believe "it" does not exist use "emotion" as an explanatory concept. Those who investigate emotion directly are aware of the richness and complexity of emotional reactivity. Many kinds of emotional reactivity have been labeled (love, fear), but attempts to separate one emotion from another have not been entirely successful. This does not deny the presence of discrete kinds of emotion (Plutchik, 1966), but it does restrict our discussion in the first part of this chapter to the more general phenomena which we call "emotional reactivity."

Experimental investigations of the ontogeny of emotion are of three types, each having special problems. The first is normative observation intended to determine the time at which certain behaviors (e.g., smiling in the human child) appear initially, whether spontaneously or in response to identifiable stimuli. The difficulties inherent in observation, most notably the tendency to anthropomorphize and to report spuriously high reliabilities as a result of selective attention, are well known. A potentially more serious problem in such studies is that a stimulus is often presented periodically until a response is noted. With this procedure, however, one cannot clearly distinguish the effects of developmental level and age on the one hand from the effects of repeated presentations of the stimulus on the other. The solution is to test each animal at one age only—but this is often impractical because of the great number of subjects needed.

The second basic experimental approach involves application of a procedure at a certain age (e.g., isolation from species-mates, introduction of fear-provoking stimuli) and determination of the effects of that procedure at a later age. This design is noteworthy because it has been used frequently in the past two decades to assess the role of "early experience" on emotional responses in later life. As an experimental design, it has three major independent variables: (a) age at time of initial stimulation, (b) age at later test, and (c) time elapsing between initial stimulation and the later test. If any parameter is varied (e.g., two or more ages at the time of initial stimulation), one of the remaining parameters is indeterminate. As an illustration, if animals are stimulated at either 10 or 20 days of age and examined at 100 days of age, then 90 days have elapsed for the first group and 80 for the second. Similarly, if these same animals are tested 100 days following stimulation, then one group will be 110 days of age and the other 120 days of age at the time of the remaining test. Unfortunately, few experiments have been designed with clear awareness of the problems involved here and consequently many of the findings reported are inconclusive and often misleading.

A third type of study, restricted to human subjects, involves the attempt to create specific emotions either by giving emotion-arousing directions to the subjects or by instituting situations designed to arouse emotion. For example, a subject whose heart rate (HR) or galvanic skin reflex (GSR) is being measured is told that the apparatus is sometimes charged with an electric current and may deliver a shock. The change in HR or GSR in response to the information is then measured. Most studies of this kind serve merely an illustrative function, even though it is doubtful whether new illustrations of the presence of emotional reactivity have been necessary since the time of Hippocrates. Some few studies manipulate variables or parameters of the stimulus in the attempt to establish functional relationships. These studies are of special value.

In the present chapter, we shall consider only experimentally derived data, but experimentally derived data based on a variety of animal species. Observational data of one kind or another will be largely ignored. Wherever feasible, interspecific differences and similarities in emotional reactivity will be discussed and theoretical conclusions drawn regarding their development and organization.

II. Major Areas of Investigation

A. Behavioral Critical Periods: The Importance of Experience during Infancy

The concept of the behavioral critical period is patterned after the observation in embryology that a specific stimulus may affect the fetus in a nonreversible way and that the stimulus is effective in this regard only for a limited period during prenatal development. The concept has been used in connection with imprinting because the period of imprinting, or the period of sensitization to an imprinting stimulus, seems comparable to the critical period in embryonic development (Lorenz, 1935; McGraw, 1946; J. P. Scott, 1962). The idea that there may be critical periods in which emotional responses develop is common, and since about 1950 has been generally accepted (Hess, 1959a, 1959b; Freedman, King, & Eliot, 1961; J. P. Scott & Marston, 1950; the major reviews by Moltz, 1960; J. P. Scott, 1962; and a discussion of this position by Denenberg, 1964).

The first attempts to determine the presence of behavioral critical periods used "socialization" as the dependent variable and isolation from species-mates as the independent variable. Socialization was measured by willingness to approach members of the same or different species. Isolation involved separation from conspecifics—although not necessarily from people—during a particular period of development. Among dogs, at least, isolation yields disparate results (e.g., willingness or unwillingness to approach other animals or people) depending upon the period of development during which the isolation is imposed* (Freedman *et al.*, 1961; Scott, Fredericson,

*Although "isolation," "deprivation," and "privation" are often used as if they were interchangeable, hereafter I shall use "isolation" or "restricted" to refer to the absence of *specific* forms of stimulation, "deprivation" to refer to the removal of once-present stimuli, and "privation" to refer to an environment in which the theoretical number of effective stimuli is lower than the number available to the "normal" animal. In some cases, the labels are different from those used by the investigator; however, the number of different situational definitions of "isolation" makes these semantic distinctions necessary.

& Fuller, 1951). Specifically, the effect may be either to increase the degree of emotional reactivity or to prevent the occurrence of an ontogenetically normal decrease in emotional reactivity. That the degree of emotional reactivity modifies the nature of any acquired response has been at least tacitly accepted by students of behavior, and several laws have been proposed to relate the amount of learning, motivation, and performance to the degree of emotion [e.g., the Yerkes-Dodson law of motivation (1908); Broadhurst's application to emotion (1957b); and Schlosberg's (1954) and Hebb's statements (1955a, 1955b)]. But isolation, as an experimental technique, is crude, for the environment of the isolated animal includes numerous stimuli that the investigator, but not necessarily the animal, disregards. Moreover, it invites confusion with the question whether a behavior pattern is "learned" or "innate." The appropriate experimental questions are (a) whether changes in the degree of emotional reactivity result in relatively permanent changes in behavior and (b) whether alterations in behavior occur more readily at one age than at another.

Scott points out that the existence of critical periods has been examined thoroughly in only a few species (dogs, some birds and fowl, and human beings). What limited data are available suggest that postnatal periods such as those devoted primarily to nutrition, transition to mature forms of eating, locomotion, and the development of social bonds, differ among vertebrates in both order, timing, and importance for survival. Scott suggests that a comparison of human beings and dogs indicate that certain postnatal developmental periods differ not only in function, but also in the time and order of their appearance. Consider socialization, for example. From several studies in which puppies were subjected to a variety of treatments, it appears that punishment and in fact ". . . any strong emotion, whether hunger, fear, pain, or loneliness, will speed up the process of socialization" (J. P. Scott, 1962, p. 950). (For sample data, see Elliot & Scott, 1961.) Scott advances his interpretation by suggesting (a) that the state of emotional reactivity that presumably accompanies physical contact with a conspecific animal, usually the mother, is essential for socialization, and (b) that the onset of a behavioral critical period for socialization is related to the possibility and degree of such contact.

Is there a relationship between physical contact, emotional reactivity, and fear? Based on his observations of chick behavior, Hess (1959b) suggested that the termination of the critical period for imprinting coincides with the onset of fear responses. It is possible that this relationship is coincidental, just as the concurrence of fear of strangers and of smiling in the human child may be unrelated. On the other hand, it is possible that the onset of fear does in fact play a causative role, promoting the end of the critical period for socialization. Scott suggests that the form of emotional reactivity called "fear" operates to encourage physical contact and socialization at the onset of the critical period, but later acts to discourage such behavior, thereby bringing the critical period for socialization to an end.

Denenberg, commenting on sensitive and critical periods from his own data on rats and mice, suggests that, although experiences during infancy markedly affect emotional reactivity, acceptance of the critical period hypothesis may mask the importance of the nature of the stimuli experienced. He suggests that the critical period for the development of emotional reactivity is a result of both the amount and kind of stimulation occurring during infancy. Thus, "the amount of the stimulus input in infancy acts to reduce emotional reactivity in a monotonic fashion. . . . An inverted U function should be obtained between amount of infantile stimulation and adult performance for tasks involving some form of noxious elements and which are of 'moderate' difficulty. For tasks which are 'easy' or 'difficult' the relationship between performance in adulthood and infantile stimulation should be monotonic, though opposite in slope" (Denenberg, 1964, p. 335). These suggested relationships are shown in Figs. 1 and 2. Denenberg does not deny the presence of critical periods for the establishment of emotional reactivity in behavioral development. Rather, he refines the concept by calling attention to such parameters as stage of development, the nature and kind of stimulation, and the interaction of the two.

The idea of a U-shaped function comparing levels of emotionality with a degree of learning, amount of motivation, or effectiveness of stimulation has been common to activation theory (Duffy, 1962; Schlosberg, 1954). Using this idea, Denenberg interprets the results of certain studies of his own and of his colleagues. Most of these studies varied both the amount of stimulation and the age at which that stimulation was presented. For example, Fig. 3 shows the mean number of avoidance responses by rats at 60 to 69 days as a function of stimulus intensity and age of application in infancy (2 or 4 days). Perhaps the function is U-shaped, but it is certainly not monotonic. Evidently, the relationship between stimulus intensity and age at the time of treatment is complex, as Denenberg himself suggests.

Both Scott and Denenberg have found emotional reactivity to be a critical parameter of the effect on later behavior of stimulation during

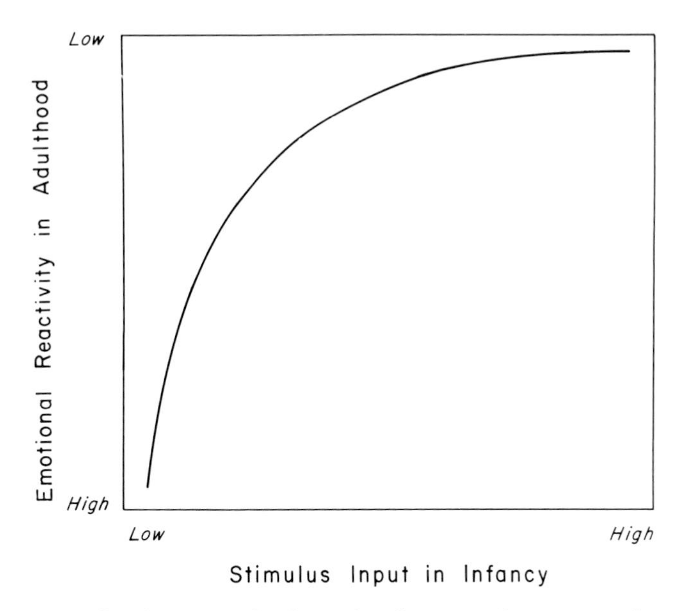

Fig. 1. Denenberg's proposed relationship between the amount of stimulation ("stimulus input") in infancy and the amount of emotional reactivity in adulthood. (From Denenberg, 1964, p. 341.)

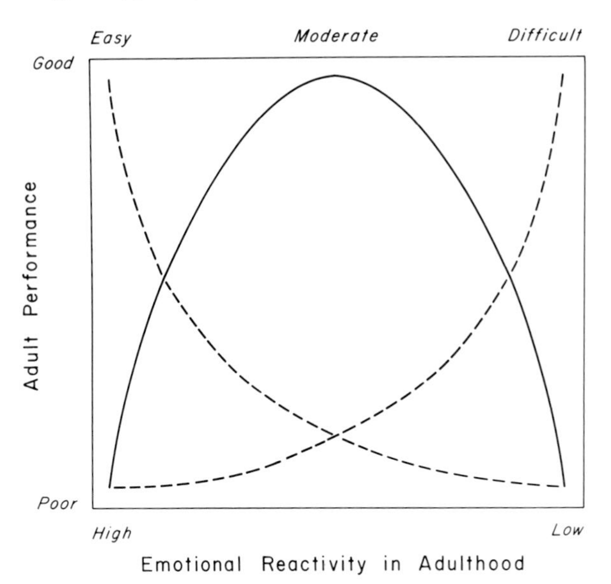

Fig. 2. Denenberg's proposed relationship between the amount of emotional reactivity in adulthood (see Fig. 1) and the quality of adult performance. (From Denenberg, 1964, p. 342.)

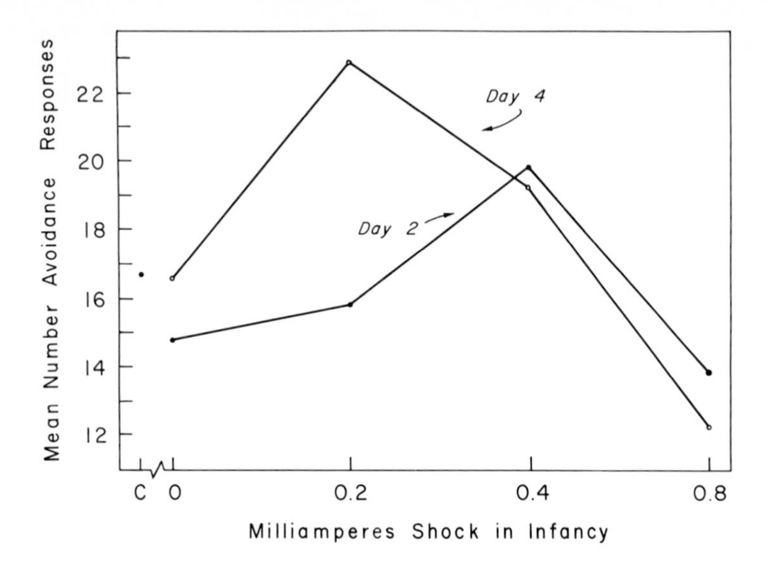

Fig. 3. An example of data in which a factorial design has been used to determine the relationship between experiences during infancy and adult performance. The design is a 2 × 4 with four degrees of stimulation and two ages at the time of application of the stimulation. (From Denenberg, 1964, p. 344.)

infancy. Denenberg makes this clear when he writes that "one would expect little or no relationship between infantile stimulation and later behavior . . . on a task which minimizes the emotional component" (1964, p. 346). Although Scott and Denenberg differ on the importance each gives the critical period suggestion, both assume emotional reactivity to be a primary effect of stimulation during infancy.

One other point must be mentioned here. The abscissa of Fig. 3 reportedly tells us how much "shock" each animal received. Actually, it tells us only how much electric current was applied. The difference between "shock" and "current" leads one to ask: How intense are various kinds of stimulation *to the animal?* Does the perceived intensity of stimulation change as the animal develops? Is 0.5 mA, for example, as intense, painful, or frightening to a 2-day-old rat as to a 4-day-old rat? We simply do not know. To find out, we must attempt to scale intensity, using selected behavioral measures. The scales used may be based either on some later behavior (such as "adult performance" represented in Fig. 3) or on reactions to the stimulus at the time it is presented. For example, do various intensities of current affect HR, GSR, escape, or the startle pattern differently? If so, intensity may be scaled along these dimensions. It is not surprising to find that in most studies, the "scale" is based on "later behavior," since the results are simpler to obtain. Nevertheless, the significance of such findings must remain uncertain until the effectiveness of the stimuli at the time of presentation is calculated.

It is apparent that attention should be given to the nature of the stimulation applied during infancy. Unfortunately, electric current is often used, a form of stimulation not common in the natural development of any species. An important paper by Henderson (1966) examined the relationship between intensity and spacing of stimulation on the one hand and subsequent emotional reactivity on the other. This paper is notable for its appreciation of the need for well-designed parametric research. In his first study, Henderson examined the influence of both rearing condition and amount of stimulus input on emotionality in rats. He varied rearing condition (restricted or normal), treatment (shocked, handled, or undisturbed), and sex in a 2 × 3 × 2 factorial design. An open-field test (Section II, C) was administered at 70 days of age. Henderson's data are plotted in Fig. 4. The

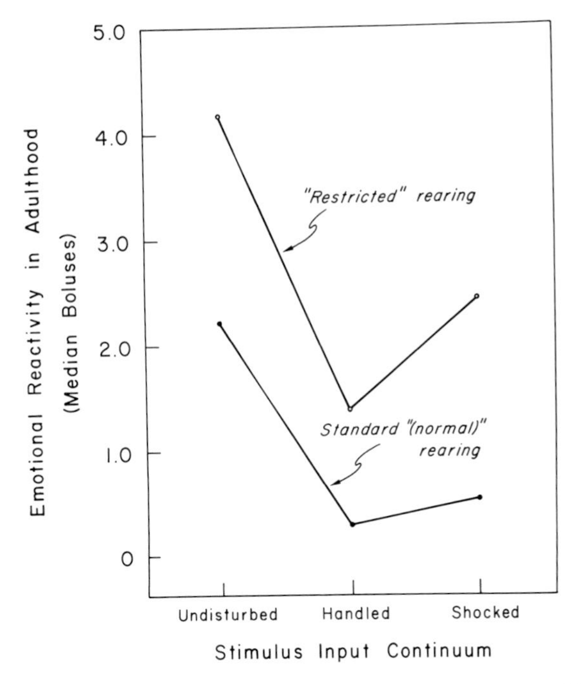

Fig. 4. The relationship between adult emotional reactivity and kind (undoubtedly degree as well) of stimulation during infancy. (The plot was drawn from data given by Henderson, 1966, p. 442.)

U-shaped function obtained differs appreciably from Denenberg's suggestion of a monotonic relationship (Fig. 1). Henderson properly points out that "... until studies quantitatively varying amount of stimulation on a single dimension are carried out, descriptions of functional relationships between stimulus input and emotionality are at best tenuous" (1966, p. 447).

Melzack (1965) has suggested a conceptual framework for interpreting data on socialization and emotional reactivity based on the behavior of several breeds of dogs raised in restricted environments (and, it is assumed, under conditions of reduced stimulation). Among Scottish terriers, sensory and social restriction subsequently produces high excitability, especially in a novel environment. Usually innocuous objects result in the animals showing high, but diffuse, emotional reactivity. [In Section II, C, I shall note that species show different, but characteristic, forms of emotional reactivity (Candland & Nagy, 1969).] The "normal" dog examines new objects leisurely. Restricted dogs, in contrast, dash to new objects but fail to show sustained interest. [This suggests the possibility that emotional level and exploratory activity are related, either monotonically, as I think Halliday's data indicate (1966), or by the U-shaped function which is often evident between learning and emotional and motivational variables. The U-shaped function seems to be the more probable relationship, at least for higher mammals where both extremely high and extremely low levels of activity represent heightened emotional reactivity.]

Scottish terriers raised in restricted environments show unusual responses to noxious stimuli as well; that is, they do not exhibit the prompt avoidance behavior that "normal" dogs show when presented with certain stimuli. Moreover, the restricted dogs are submissive socially to normally raised dogs, even permitting "normal" dogs to steal food from them without quarrel. Finally, restriction has deleterious effects on such commonly measured psychological abilities as delayed reaction, the Hebb-Williams maze, the Umweg problem, and various perceptual tasks.

Behavioral studies (Melzack, 1962, 1965) of beagles raised in restricted environments present a picture similar to that described for Scottish terriers. Like the terriers, the restricted-reared beagles were highly active in novel environments and showed unusual responses to many kinds of noxious stimuli (Melzack, 1952, 1954, 1961; Melzack & Scott, 1957; Melzack & Thompson, 1956; Thompson & Heron, 1954a, 1954b; Thompson, Melzack, & Scott, 1956).

Do these data provide clues as to how the behavioral consequences of early restriction might be mediated? After pointing out deficiencies in two possible interpretations—"that restriction leads to degeneration of nerve fibers or afferent neurons and that experience with patterned stimulation may be necessary for adult stimulation"—Melzack proposes:

> A two-part process in which (a) there is inadequate filtering of inputs on the basis of memories (phase sequences) of the significances of stimuli normally acquired in early experience, so that (b) the total input bombarding the central nervous system produces an excessive central nervous system arousal which, as Hebb (1955a) has suggested, could be responsible for the correspondingly low cue properties necessary for discrimination and adequate responses (1965, pp. 276-277).

One implication of this approach is that, for an animal raised in a restricted environment, all stimuli would be equally effective, since ". . . all inputs, 'irrelevant' as well as 'relevant' would reach the brain where they could bombard the neural systems that produce sensory and affective arousal" (Melzack, 1965, p. 278). Of appreciable importance to our investigation of emotional factors is a second implication of the model, namely, the existence of ". . . a 'vicious circle' in which failure to filter out irrelevant information (on the basis of prior experience) leads to excessive arousal which, in turn, intervenes with mechanisms (perhaps innately determined) that would normally act in the selection of cues for adaptive responses" (Melzack, 1965, p. 278).

Of relevance here are the data reported by Melzack and Burns (1965) on EEG patterns in restricted and normally raised dogs. These data suggest that restricted dogs show not only a shift from low to high EEG frequencies when presented with a view of a novel environment but continue to show higher than normal frequencies (implying a higher CNS arousal level) after restriction is terminated.

It is difficult to summarize our understanding of the relationship between critical periods occurring in infancy and adult emotional reactivity without using so many qualifiers as to imply that the role of critical periods is slight. What we can say is that critical periods or periods of supranormal sensitivity probably determine the level of adult emotional reactivity in some, and possibly in all, mammals. Of course, just how selected stimulation applied during such a period functions to affect the general level of adult emotional reactivity is the crucial and, as yet, largely unanswered question.

B. Prenatal and Maternal Influences on Emotional Reactivity

The idea that selected stimulation of a mother during pregnancy may affect the behavior of the offspring is prevalent in folklore, but until recently it elicited widespread derision in scientific circles. During the past decade, however, experimental evidence has been advanced to suggest that the level of emotional reactivity of offspring is indeed affected by treatment of the mother during pregnancy—at least in the laboratory rat.

Since Thompson's observations (1957a) appear to have launched the current emphasis on prenatal stimulation, a review of his procedure is instructive. Using a Miller-Mowrer box, Thompson trained five female hooded rats to avoid electric current by opening the door between the chambers. These five, in addition to five other females which had not been trained in the double-compartment chamber, were placed in a large cage with five males and bred. When the females which had received the avoidance training became pregnant (as determined by vaginal smears), they were given additional treatment in the avoidance apparatus. This treatment consisted of the following: On each day until the litter was born, the pregnant female was placed on the "shock" side of the apparatus, with the door locked to prevent avoidance, and three times each day the buzzer, which had previously signaled the onset of the current, was sounded. The aim of the procedure was to establish in the pregnant subjects a state of emotional reactivity similar to that described clinically as "free-floating anxiety." The emotional reactivity of the offspring was determined at 30–40 and again at 130–140 days of age by recording activity in the open field and by taking several other measures of "timidity." Pups born of experimental mothers differed from the control pups in showing less activity and greater timidity at 30–40 days of age. These differences were also evident when the pups were tested again at 130–140 days of age, but at a lower (and in one case an insignificant) level of confidence. In the context of open-field studies, the reactions of the experimental pups were interpreted as evidence of heightened emotional reactivity. Thompson provides a reasoned and appropriately cautious discussion that includes several possible interpretations of the results. These interpretations are worthy of examination because they illustrate the kind of alternative explanations we should look for in considering later experiments.

1. The buzzer acted on the fetus by altering the hormonal balance of the mother.
2. The differences in behavior between pups of experimental and

control mothers may reflect genetic differences through inadvertent mate selection. For example, the "emotional" mothers may have been mated more or less easily and thus the high reactivity of their offspring may not have been caused by the stimulation as such, but by choice of mates.

3. The stress of the training period itself may have affected the mothers and may thus have been responsible for the degree of emotional reactivity in the offspring (Kaplan, 1957; Thompson, 1957b). That is, handling of the mothers may have created a level of reactivity different from that of the nonhandled control animals. (Section II, A notes that handling may be a traumatic form of stimulation to rodents.)

4. And finally, the training or stress period may have affected, not the hormonal balance of the mother, but her behavior toward the pups. As a result, the reactivity level of the pups may have been raised.

Joffe (1969b) reminds us that the procedure characteristically employed in studies on the effects of prenatal stimulation involves treatment of pregnant females and subsequent comparison of the offspring of females with those of untreated controls. If the behavior of the offspring of the two groups differs, it is often concluded that the treatment was the effective cause. Unfortunately, this experimental design fails to distinguish the effects of prenatal influence from the effects of variables in the postnatal situation. For example, the treatment may have had no prenatal effect, but it may have affected the mother in such a way that her reaction to her young was atypical. Also, it is possible that the treatment applied to pregnant animals may influence the efficiency or time of parturition and this, in turn, may affect the behavior of the offspring. [Data from Ader and Plaut (1967) support this likelihood. Werboff, Anderson, and Haggett (1968), however, did find this effect in mice.] As we shall see, all attempts to determine just what it is that mediates the effects of prenatal stimulation have yielded equivocal interpretations.

Both Hockman (1961) and Ader and Conklin (1963) have extended Thompson's findings. Hockman, using hooded females, subjected half of them during pregnancy to a stress-training situation similar to, but probably more intense than, that used by Thomson. Open-field tests were administered to the offspring at 30–45 and then again at 180–210 days of age. At the earlier age, the offspring of stressed mothers showed lower open-field activity than the offspring of unstressed mothers, but by the time the second test was administered this difference had disappeared. Moreover, a higher frequency of defecation was noted from stressed offspring, but only on the first day of the first test. Hockman used cross-fostering as a control procedure. This was fortunate, for the results showed that the decreased activity (or heightened emotionality, as decreased activity is generally considered to reflect) of the experimental offspring was the result of an interaction between the prenatal treatment (stressful electric current and an escape/avoidance training paradigm) and postnatal experience (separation from the mother/placement with a foster mother). These findings support, at least in principle, those of Thompson, while pointing out the usefulness of designs that permit assessment of the above-mentioned interaction. A more elaborate design used by Ader and Conklin (1963) on Sprague-Dawley rats, however, yielded results which show little relation to those of Hockman. In this design, handling the pregnant mother was the major independent variable, not shock. The investigators handled half these mothers for 10 minutes, three times each day, throughout gestation. Cross-fostering was carried out 48 hours after birth, with half the offspring tested in the open field at 45 days of age, and half at 100 days. No differences in open-field activity were found in relation to cross fostering, sex, number of field tests, or the interactions of these variables. Similar negative findings resulted from comparisons of defecation frequency. Some differences, however, did appear in timidity, as measured by willingness to leave the home cage, indicating that offspring of handled mothers were less emotional, at least by this measure.

Shall we now conclude that increasing the emotional reactivity of the pregnant female through shock and handling, respectively, *increases* or *decreases* emotionality in the offspring? Perhaps the most likely explanation is that handling leads to lowered emotional reactivity of the offspring, while shock leads to heightened reactivity of the offspring. Superficially, this suggests that handling is not a stressor to the rodent, although it has been shown that handling is indeed noxious to the rat (Candland, Faulds, Thomas, & Candland, 1960). It is evident that we do not know the answer to the question just raised, and perhaps we shall not until stimuli other than handling and shock are studied and until several aspects of emotional reactivity are specified.

C. The Open Field

The open-field test deserves our special attention because most quantitative data relating to the ontogeny of emotional reactivity have been acquired through the use of the open field in one form or another (Section II, A,B). When this test is used with rodents, the animal is removed from its cage and placed in an arena having a diameter six to thirty times the length of the animal. The amount of defecation and the level of activity are the most frequently used measures of emotionality.

At the onset, we should recognize that species react differently to novel environments. The domestic chicken (and probably other fowl) "freezes" in such situations, whereas many mammalian forms react by flight (Candland & Nagy, 1969). Thus, the behaviors selected for observation in the open field will vary from one species to another. In addition to defecation, urination, and level of activity, the correlation between these measures and certain autonomic responses (such as heart rate) has been reported.

There are numerous problems accompanying the use of the open field. Among these are (a) unreliable correlations between activity and elimination (Anderson, 1938; Billingslea, 1942; Bindra & Thompson, 1953; Evans & Hunt, 1942; O'Kelly, 1940); (b) unreliable correlations between "emotionality," "timidity," and similar measures obtained in the open field (Billingslea, 1941, 1942; H. F. Hunt & Otis, 1953; Parker, 1939); (c) unreliable correlations with other activities, such as grooming and freezing (Doyle & Yule, 1959a, 1959b); (d) lack of correlation with such physiological indices of emotional reactivity as heart rate (Candland, Pack, & Matthews, 1967) and mononuclear count (Caldwell, 1960); (e) the probability that the utility of the most commonly used measures is restricted to certain rodent species and that within rodents these measures differ from strain to strain (Candland & Nagy, 1969; Farris & Yeakel, 1945; Henderson, 1964).

Accepting the fact that we do not yet know the relationships among various behavioral measures in the open field or between these measures and autonomic indicators and, moreover, that indices of emotional reactivity are species-specific, we may still use the open field to study the ontogenetic development of emotional reactivity provided we use appropriate control animals and show caution in generalizing from strain to strain and certainly from species to species.*

*Readers accustomed to laboratory work will think of the "home cage" as the natural environment, and those accustomed to field work will think of the open field as confinement. To put a laboratory-raised rat in the open field is to place him in a territory larger than his accustomed living space; to place a rat accustomed to roaming in the open field is to confine him. Differences in reaction to the open field should depend upon the animal's past experience with living space. The concept of life space, in a somewhat different form, was applied to human behavior by psychologist Kurt Lewin five decades ago (1928). For the application of a similar concept to the behavior of captive animals, see Hediger (1942, translated 1950). For an attempt at quantification, see Hull's Chapter 8 (1952).

The results of an illustrative study are shown in Figure 5. Wistar albino rats ($n = 160$) were assigned to one of eight groups differing in age. "Home-cage" defecation was recorded for a 24-hour period, following which each animal was placed alone in an open field measuring 7 feet in diameter. The results indicated that, as animals of this strain mature, defecation frequency in the open field increased rapidly, reaching an asymptotic level at about 54 days of age. In contrast, frequency of defecation in the home cage was found to remain stable for the ages tested (Candland & Campbell, 1962). (See Section IV, for some indication of the lack of generality of this finding.) A similar increase in emotionality (if that is the term for the increase in open-field defecation frequency over home-cage frequency) is found in the domestic chicken (Candland, Nagy, & Conklyn, 1963). Within the rodent family, the C3H mouse also shows an increase in defecation frequency, reaching asymptote at 30 days of age. Unlike the CFE rat (Candland, Culbertson, & Moyer, 1968), the C3H mouse shows an *increase* in defecation frequency upon repeated testing (e.g., daily from 30 to 33 days of age; Candland & Nagy, 1969).

Do these data imply that the mouse becomes more emotional and the rat less emotional with increased experience in the open field? Or could it be that bolus deposition, while accompanying fearfulness

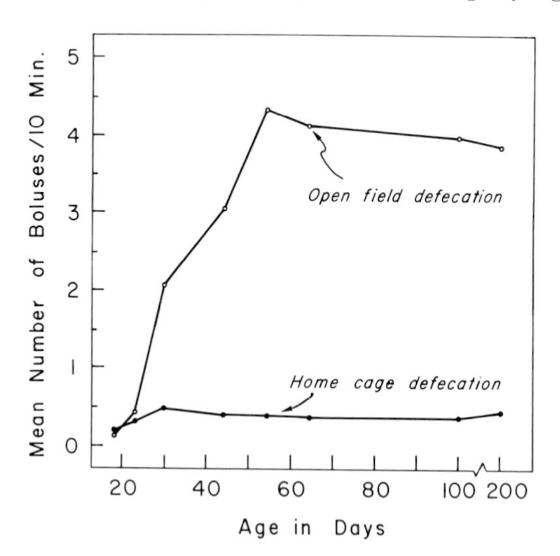

Fig. 5. Frequency of home-cage and open-field elicited emotionality as a function of age. The function, in this example for the Wistar albino rat, is thought to vary ontogenetically with species and strains. (From Candland and Campbell, 1962, p. 594.)

in certain strains, has an additional function—that of territorial marking? We shall discuss an ontogenetic model for specific species in Section III, but we may summarize here the data available on changes in emotional reactivity in the open field. Emotional reactivity increases with age for some rat strains, but possibly decreases in others. Similar data on mice are evidently specific to the strain. The guinea pig, unlike other rodents, remains in the center of the open field and inhibits defecation (Tobach & Gold, 1966). The open field appears to have little effect on swine (Beilharz & Cox, 1967). Emotional reactivity increases with age in the domestic chicken, although freezing (lack of activity) and vocalization frequency are also useful measures.

D. Learning and Motivation

Any experimental study of animal or human learning that includes a deprivation condition deals obviously with motivational factors. Similarly, it is evident that placing an animal in a novel environment, such as that usually provided by an experimental apparatus, alters emotional reactivity. Accordingly, any investigation into the ontogeny of learning must perforce study changes in motivational and emotional levels.

When we consider the question posed by Campbell (1967, p. 43), namely, "Is it true that the young can learn new behavior more readily and retain it longer than adults?", we ask what appears to be a relatively simple question regarding learning capacity. But when we attempt to answer the question experimentally, we see the difficulties involved in equating emotional and motivational levels for conspecifics of different ages. For example, Campbell and his coworkers have shown that age produces differences in shock-aversion thresholds and differences in activity as a function of current intensity and of food deprivation. (For a resumé of these studies, see Campbell, 1967.) Nevertheless, we know so little about the role of emotionality in learning and about the ontogeny of learning as it is affected by emotional reactivity, that at the present time it is almost hopeless to attempt to evaluate the relationships between these two factors.

In the 1920's and 1930's, a number of studies on human subjects attempted specifically to relate emotion to learning and motivation. These were harsh, demonstrative studies. Landis (1926), for example, required of his subject 46 hours of fasting, followed by 36 hours of enforced insomnia, followed by the presentation of electric shock. Duffy (1930) produced a formidable study relating the tremor (tension) of the unused hand to excitability. Washburn and her colleagues reported a series of studies (e.g., Washburn, Giang, Ives, & Pollack, 1925; Washburn, Rowley, & Winter, 1926) attempting to relate GSR with college women's perception of their own emotional reactivity and that of friends. Baker (1934) reported that learning ability could be increased temporarily by dropping a subject's chair 15 inches and thus presumably inducing fear. When pulse and respiration rate increased after the chair-dropping episode, so did performance on a task of learning; conversely, when pulse and respiration decreased, so did performance. Various unspecified strains of rats received similar treatments. Higginson (1930) pinched tails (to produce anger) and generated the odor of cat (to produce fear) and observed the behavior of rats in a circular maze. The "emotional" animals (i.e., pinched or exposed to the odor) showed poorer performance than "untreated" animals. Patrick (1931) discovered that the abrupt presentation of a loud buzzer disrupted maze activity in the rat. As artificial as these studies appear today, they prompted a question that still occupies the student of emotion, namely, is emotion to be considered an "organized" or a "disorganized" response? Several answers have been offered (cf. Duffy, 1948; Leeper, 1948, 1965; Webb, 1948; P. T. Young, 1949, and a clarifying afterword by Bindra, 1955): (a) Emotion disorganizes the normal pattern of behavior and therefore is detrimental to a successful adaptation to whatever motivated or initiated the emotional state. (b) Emotion only *appears* to disorganize; actually, it is reorganizing the pattern of normal behavior to respond to the unusual situation. (c) Emotion is first a disorganized response, but later it becomes organized; that is, general excitement (disorganization) develops into adaptive behavior. (d) "Organized" and "disorganized" are the ends of a continuum—the proper way to look at the matter is to consider the *degree* of organization. (e) Emotion is motivation, *a posteriori*.

If emotion is totally disruptive—if it is, as Leeper so aptly put it, "something to grow out of," like tantrums or fear of strangers, then it is clearly maladaptive. On the other hand, if it is organizing, emotion could be a useful adaptive mechanism. Thus, the evolutionary success of a species would depend, in some degree at least, on its emotional responses. This leads us to ask whether some species are more adaptive emotionally than others, and this question has all the difficulties we have encountered over the past 70 years in asking whether different species have different capacities to learn (Bitterman, 1965; Mayr, 1966; Thorndike, 1898).*

One happy outcome of the present controversy has been the empirical research that has emerged. Thompson and Higgins (1958), for example, arranged a situation in which "disorganization" could be measured. Specifically, they placed rats in one of two compartments, either black or white, until evidence of habituation was noted. The experimental animals were placed in a separate, but adjoining, compartment, and electric current was applied. These subjects jumped to the compartment to which they had been previously habituated. Control animals, in contrast, given equal habituation to either the black or white compartment but then not shocked, chose the compartment to which they had not been previously habituated. The authors comment ". . . the initial phase of emotional stress may be followed by or even preclude behavior that is organized (systematic), adaptive and stable, or temporarily predictable."

In an important parametric study, Broadhurst (1957b) manipulated motivation and emotion and then measured the effect of such manipulation on performance in a discrimination task. Motivation was varied by forcing rats to remain under water for different periods of time, thus depriving them of air; emotion was varied by the use of selectively bred strains showing high and low emotional reactivity, respectively. The learning task was presented in one of three levels of difficulty: low, moderate, and high. The essential results, in regard to motivation, are shown in Fig. 6 on a three-dimensional surface. It is evident that the optimal level of motivation, at least for the performance of a discrimination task, decreases with the increasing difficulty of the task. The relationship between task difficulty and degree of emotional reactivity, however, is ambiguous. On both the easy and difficult discrimination tasks, or under low or intense motivation, the emotional animals learn more rapidly than the less emotional animals. Nevertheless, these differences did not reach statistical significance and they do not form a consistent pattern. This suggests that, under the conditions used in this experiment, the

*The reader may need reminding that we are discussing "emotion" as a unitary concept. If it becomes possible to separate the emotions in some meaningful experimental way, it may be that we shall find some kinds of emotion to be adaptive and others not to be. Moreover, if one considers "emotion" as a homeostatic mechanism, an assumption more in line with the traditional assumptions regarding motivation in theories of learning, an important question is whether homeostasis is an adaptive response for some species and not for others. I present an alternative in Section III.

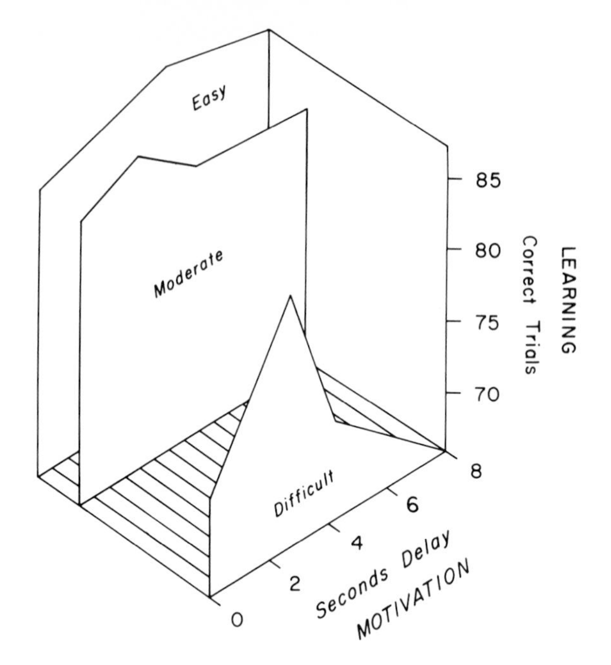

Fig. 6. The relationship between emotion, motivation, and level of difficulty of the task to be learned. (From Broadhurst. 1957b, p. 348.)

Yerkes-Dodson law holds for motivation (as it was originally intended to do), but not for emotion, and, accordingly, that motivation and emotion do not show the same function when related to learning. Recently, Fantino, Kasdon, and Stringer (1970) have shown that the Yerkes-Dodson law does not hold when appetitive reward is used. This finding suggests that the law is valid under aversive, but not appetitive conditions. One searches in vain for parametrically derived data on other species.

E. Emotional Behavior in Individual and Social Situations

Two papers appeared during the early 1930's that deserve special mention under the present rubric. The first was that of Ray (1932) who studied a number of physiological variables, including pulse rate, respiration, and blood pressure, from children reacting to a sudden loss of physical support. His findings indicated that in such an apparently emotion-provoking situation, some children showed an increase in the responses measured and others a decrease. The significance of this, I think, has yet to be completely appreciated (Lacey,

Kagan, Lacey, & Moss, 1964). It is that there is no reason to suppose that emotional reactivity takes the same physiological form in each member of a given species. For example, it is likely that emotional reactivity is indicated by a sudden drop in blood pressure in one person, yet by a sudden rise in another. Although it may be possible to find characteristic patterns of reactivity for an individual, it is unlikely that the same pattern will appear in every other person. Ray's data support the viewpoint that emotion is idiosyncratic, which is an essential assumption of the model to be described in Section III.

In the second important paper, Bridges (1932) reported the development of different emotions in a sample of 62 infants. Bridges' observations are similar in purpose and design to many recent field studies of nonhuman primates, since her observations were made with the intention of providing a catalog of frequency and time of occurrence of specific behavioral patterns. One suggestive finding from her observations was that certain "emotions" appear suddenly, and these, in turn, become differentiated, forming other emotions. For example, her infants showed general excitement to all novel stimuli, but within a few weeks the generalized response became differentiated into "distress" and "delight." This sequence, it may be noted, is what one would expect if separate emotions were under the control of conditioning or training, for it is reasonable to suggest that from "general excitement" one learns, through discrimination, to identify specific emotional stimuli and their responses. Bridges' observations are shown in Fig. 7.

The studies of Ray and Bridges, respectively, suggested (a) that "emotion" is idiosyncratic and (b) that, ontogenetically simple "emotions" become differentiated, either through maturation or learning.

Two other experimental questions occupied investigators in the early 1930's. The first was whether one could judge emotions accurately by examining photographs of faces. It appears that unless one knows the cause of a facial expression (e.g., whether the person was pricked or angered), accurate naming of the emotion reflected in a photograph is a chance affair (Goodenough, 1931; Sherman, 1927). Further, Goodenough (1932) reported the facial display of a congenitally blind and deaf child who exhibited all the expressions noted in normal children. Obviously enough, this suggests that facial expressions in response to emotional situations are not necessarily learned responses.

The second question of concern to experimenters during the 1930's was how emotional maturity was to be assessed. Pressey and Pressey (1933a, 1933b, 1933c) developed a test of emotional ma-

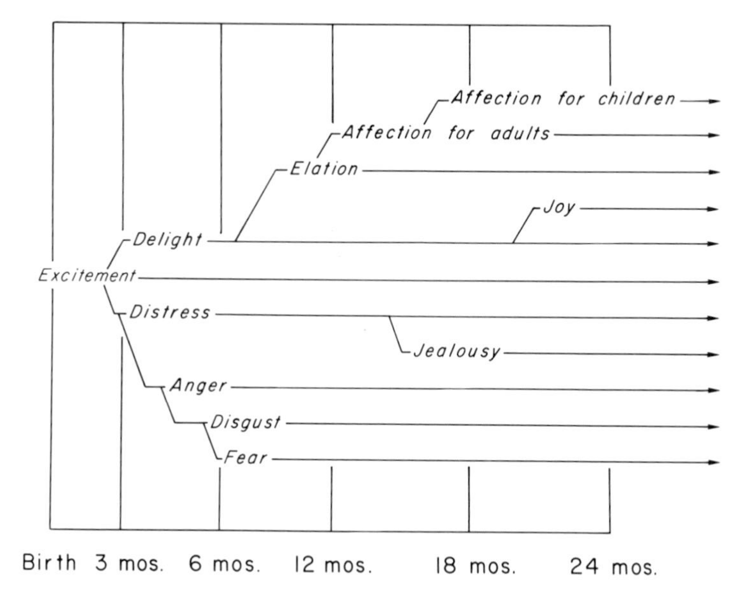

Fig. 7. Description of the ontogeny of human emotional reactivity during the first months of life. Note that specific forms of emotional reactivity appear to develop from more general forms. (Redrawn from Bridges, 1932, p. 340.)

turity—the Pressey X-O test—that they used in a study of American Indians. The test, essentially, is a forced-choice word association test. It is important primarily because so few measures of emotional reactivity, dependable or otherwise, are available. Interestingly enough, these investigators finally concluded (1933c) that what they were actually measuring was the degree of contact their Indian subjects had had with white culture; that is, the test pointed out the obvious, but often overlooked, fact that emotional maturity is what a given culture chooses to call it.

Many hypotheses have been proposed concerning the effects of the social environment on emotion. Some of these hypotheses were mentioned in Sections II, A and B. One commonly asked question on the subject is what is the effect of being raised (a) with other animals and (b) with animals of a different age and sex? Although a number of suggestive and highly refined studies are available here, it would be misleading to use them as a basis for any general statement regarding the development of vertebrate emotional behavior. Few ontogenetic variables have received the systematic attention necessary to support more than vague statements which, in the main, are applicable only to a very limited population. We shall consider selected normative and parametric studies illustrative of the procedures used.

a. Normative Studies. J. A. King has reported observations on the behavior of several species. For example, in two subspecies of the deer mouse *Peromyscus,* King (1958) found periods of emotional development to be parallel, but to mature at different rates. For human beings, Gesell (1934) has published "norms" for the appearance of certain forms of behavior as a function of age. Unfortunately, these norms provide only indirect information on emotional development as a function of social influences. These illustrations indicate clearly the need for basic developmental data on an adequate number of representative species.

b. The Parametric Approach. S. Bernstein and Mason (1962) recorded the behavior of 47 rhesus monkeys (*Macaca mulatta*) from one to 25 months of age in response to 12 stimulus objects: four "complex" objects (e.g., a plastic snake or beetle), four "intermediate," and four "simple" (e.g., a cube) objects. The behavioral responses to the presence of these stimuli indicated that emotional reactivity during the first 3 months of age is characterized in laboratory-raised rhesus by vocalization, rocking, crouching, and sucking. From 3 months to 25 months of age, other signs of emotional reactivity also appear, such as lip-smacking and fear-grimacing.

Studies of population density (Calhoun, 1962a; Christian, 1963) have added valuable data to our understanding of the relationship between social stimulation and emotional development. In particular, they show that when density surpasses a certain ratio of animals to space, selected behavioral and physiological changes ensue. The physiological changes are similar to those which follow stressful situations, the most obvious of these being enlargement of the adrenals. It is evident that emotional reactivity may be altered permanently by population density, and it seems appropriate to consider population density as one form of social stimulation.

A descriptive term of importance in the study of emotional reactivity is "social facilitation." Crawford (1939) defined this term as "any increment in activity resulting from the presence of another individual." For example, a chick eats more grain if other chicks are present and also drinks more water. Indeed, such increases in eating and in drinking have been documented for numerous forms, e.g., dogs (Adolph, 1943), pigs (Jonsson, 1955), ducks (Weidman, 1956). In addition, the presence of other individuals has been shown to affect physical development as well as learning and grooming, and, in fowl, imprinting (Allee, 1938). Although social facilitation almost certainly varies with the period of development and although it must be related to emotional reactivity, these behavioral parameters have received only scant experimental attention.

One of the very few parametric analyses of the role of *communicated* emotion or affect has been reported by R. E. Miller and his associates (Miller, 1967; Miller, Murphy, & Mirsky, 1959a, 1959b). The basic procedure in this series of studies involves yoking two rhesus monkeys (*Macaca mulatta*) in either a conditioned avoidance or a reward situation. One animal, the "stimulus" monkey, receives the stimulation while the other, by pressing a bar is able to terminate an electric current or produce a food reward. The purpose of this "cooperative conditioning," as the investigators called the procedure, was to determine the extent to which one animal could perceive the physiognomic changes occurring in another and react to these changes. It was found that the rhesus monkey is indeed able to use physiognomic cues, although more successfully under conditions of avoidance than under conditions of reward. Apparently, the rhesus' ability to discriminate or to form facial expressions is not innate, for R. E. Miller, Caul, and Mirsky (1967) report that animals reared in social isolation during the first year of life are incapable of learning an avoidance task in response to the facial expressions of another. Correlatively, normally reared rhesus could not utilize the facial expressions (or lack of them) of animals reared in isolation. It is possible, although not likely, that the isolated animals had some deficiency in learning rather than in discrimination. The idea that emotional expression has a learned component is developed in the ontogenetic model presented in Section III.

F. Physiological and Anatomical Correlates

At least since the time of William James,* and certainly since the time of Cannon, attempts have been made to show that there are correlations between physiological and emotional states and that emotional states are, or are not, merely perceptions of particular physio-

*It is time to fell the current myth, promulgated by text writers, including the present author, that the idea that emotion is our perception of physiological change is attributable to either James, Lange, or both. Titchener (1914) pointed out the long history of the concept many years ago and Ruckmick (1934) showed very neatly that McCosh (1880) reported the idea, with the same type of examples used by James, prior to James's statement. Although it is rewarding to know who said what first, it is evident that the so-called James theory was more or less common knowledge in James's lifetime and that the famous statement, "I see a bear, I run, and I am afraid,"

logical states. It is far easier to describe what such studies have failed to do rather than what they have done. The problem is that they have rarely passed from the demonstrative stage to the parametric stage.

Studies of the role of autonomic functions in emotion are separable, topically and chronologically, into the following areas: (a) investigations of the possibility that the amount of epinephrine secreted by the adrenal cortex is directly related to the degree and kind of emotion experienced—ca. 1920-1935; (b) investigations showing that numerous organs and numerous physiological phenomena can be altered by induced emotion—ca. 1935-1945; (c) studies isolating the structures of the CNS necessary for emotional experiences—ca. 1920-1950; and (d) investigations of the role of chemical agents, such as depressants, stimulants, and hormones, on emotional behavior—1955-present. Each of these periods resulted in an appreciable increased in our understanding of emotional processes. We shall consider their contributions separately, although in fact they should not be demarcated sharply.

Marañon (1922, 1924) reported that when he injected persons with adrenaline (epinephrine) many reported nonspecific arousal (the feeling that the bodily processes are working too rapidly, which we often associate with excitement). Those subjects who did report emotional sensations said that they were not duplicates of common emotional feelings, but that they felt "as if" they were afraid, or "as if" they were excited. The failure to experience sensations uncontaminated by this "as if" feeling suggested that the belief that emotion was merely the perception of physiological, particularly, autonomic change, was untenable. Sierra (1921) reported similar results. Cantril and Hunt (1932) and Cantril (1934) attempted to refine Marañon's findings. (In retrospect, it is questionable whether injected epinephrine would have an effect identical to endogenous epinephrine.) Cantril (1934) administered either adrenaline chloride or a placebo and presented stimuli intended to arouse specific emotions, such as fear and disgust. The results were equivocal, yet in the conclusion of the report Cantril notes a distinction to be remembered: when per-

---

was original only because of the inclusion of the bear (see McCosh, 1880, pp. 102-103). Further, it is difficult to believe that James did not know of McCosh's work. McCosh was a famous metaphysician whom James quoted widely (but not in this case) and educator (president of Princeton). Ruckmick (1934) falls only a little short of finding it implausible that James did not know of the McCosh statement. At the very least, we should note that McCosh posited the James, or James-Lange, theory 20 years before James. The next step in ascertaining the history of the concept would be for someone to take the trouble to ascertain which books McCosh read.

sons reported the experience of "as if fear" (my legs trembled, my voice was uncontrollable, *as if* I were afraid), they attempted to "organize the stimulus intellectually." Evidently, they selected some stimulus (e.g., the experimenter) and believed that their sensations of fear were caused by that stimulus. The distinction, like Freud's distinction between "fear" and "free-floating anxiety," stresses the significance of cognitive factors in emotion.

Landis and Hunt (1932) argued that Marañon and others had failed to realize that the meaningful question was not whether adrenaline created the sensation of an "as if" emotion, but whether "genuine emotion" could be created with the injection technique. Cantril and Hunt (1932) injected 1.5 ml of adrenaline chloride into adult human subjects and found that only 10 of 22 persons reported an "as if" emotion; the remainder reported no unusual sensations. When the autonomic nervous system is hyperactive, human beings certainly experience sensations similar to those experienced when frightened or excited. The similarity in experience, however, does not imply identical causal determinants. Moreover, as both Freud and Cantril discerned, sensations accompanying fright are intellectualized (I fear public speaking; I fear thunderstorms) but the sensations of hyperactivity are not.

The experimental data do not permit an unequivocal statement as to whether or not the sensation of physiological change is the emotion. Marañon's technique cannot provide a basis for a decision, even though his experimental data and those of Cantril, Hunt, and Landis document the presence of "as if" emotion and, moreover, establish the experience of emotion as something more than just peripheral change. But whether emotion is always a combination of central and peripheral sensations, we cannot say. Nor do we know whether epinephrine is the major agent of emotion. Rogoff (1945), in a reasoned criticism of the emphasis given epinephrine, cites these objections: (a) Epinephrine is released continuously, and consequently its release as such could not signal the great variety of emotional reactivity known to exist. (b) The human, at least, must be subjected to very intense external stimulation before the rate of release of epinephrine is affected. Moreover, the rate of release is not sufficiently variable such that changes in release could signify differences in the intensity or kind of emotional reactivity. (c) Adrenalectomized animals evidence behavior of an apparently emotional nature. Other investigators have also suggested that emotional reactivity was not caused solely by epinephrine, but was most likely the result of a consortium of causes.

Three other phenomena that have interested students of emotion might be conveniently discussed under the present rubric. These phenomena are gastrointestinal disorders, cardiac change, and GSR.

A number of clinical reports have concluded that peptic ulcer patients often have histories of intense and long-term anxiety, often describe feelings of guilt and frustration, and are characteristically independent and perfectionistic (e.g., Mittelman, Wolff, & Scharf, 1942). It has been noted also by clinical investigators that unrelenting tension and anxiety produce abnormal increases in the hydrochloric acid and pepsin secretions in the stomach cavity, and that those with a high nonspecific level of emotional reactivity are those most likely to develop ulcerations. In this regard, H. G. Wolff (1943) had the good fortune to have access to an employee who, because of an occluded esophagus, had a 3.5-cm gastric fistula inserted for placing food in the stomach. Observation of the stomach lining through the fistula suggested that emotional stress gave rise to the hypersecretion of hydrochloric acid. For example, when the employee thought he would be blamed and, perhaps, dismissed because of the loss of a needed paper, the rate of HCl secretion increased.

Studies on the factors underlying the occurrence of ulceration in animals have also been performed. Brady (1964), for example, used a yoked avoidance procedure in which two rhesus monkeys (*Macaca mulatta*) were placed in restraining chairs. Electric current was applied to the feet of both monkeys until one of them pressed a lever. When he did, the current was delayed for 20 seconds. In other words, one animal could delay or eliminate the painful current; the other animal had no way of controlling it. Both monkeys received the same number of shock applications. After 3 to 4 weeks, all experimental animals, that is, all animals who were in control of the shock, were reported to have developed gastrointestinal lesions with accompanying ulceration; in the four yoked controls no evidence of ulceration was found.

Additional experiments refined these findings by showing that, during the administration of the procedure, the average daily concentration of free hydrochloric acid increased from the tenth to the fourteenth day in experimental animals, although individual differences were numerous. A likely relationship between the development of ulceration and the base level of acid secretion was shown by the finding that the two rhesus with the highest initial acid concentration were the two who developed peptic ulcers during the experimental procedures. These data are consistent with those clinical data (cited above) suggesting that a high rate of secretion or a high concentration of hydrochloric acid produces gastric intestinal lesions more rapidly than lower rates or lower concentrations under the same schedules. This finding suggests that ulceration depends on two factors: a predisposition as indicated by individual differences in normal acid output and the nature and schedule of appearance of emotion-provoking external events. This model, involving both inborn and acquired factors, most likely accounts for the idiosyncratic nature of emotional reactivity.

The resistance of the skin to the passage of electric current has been used also as a variable in some studies of emotion. Although the exact physiological source of the resistance is uncertain, the amount of resistance varies in response to emotional stimulation. Older studies were burdened with the need to constrain and, probably, frighten subjects with cumbersome leads and electrodes. The increasing sophistication in electronic equipment that has made telemetric measurements possible should prompt psychophysiological studies that are able to measure the many subtle physiological changes that probably accompany emotional reactivity (Caceres, 1965; Candland, Taylor, Dresdale, Leiphart, & Solow, 1969; H. S. Wolff, 1967).

James's statement that emotional sensations are the perception of physiological change emphasized the role of the peripheral autonomic nervous system. Cannon's work, in contrast, gave special emphasis to the CNS, particularly the lower centers of the brain. Prompted by this work (e.g., Cannon, 1929) and that of Bard (1942), investigators came to believe that Papez's (1937) suggestion of a "visceral brain" was well founded (MacLean, 1949). This "brain," centered in the rhinencephalon and phylogenetically older than the cerebral cortex, has been considered the center of behavioral patterns essential for survival (e.g., eating, breathing, flight); it also has been held responsible for many behavioral patterns of an emotional nature. Later physiological findings delineated the reticular activating system. The system, shown schematically in Fig. 8, is essentially a feedback system that permits the cerebral cortex to process information from sensory events. Anatomically, two systems are evident, one located in the brainstem and one composed of fibers projecting from the thalamus to the cerebral cortex. This double system provides numerous channels for relating sensation and emotional patterns.

The development of techniques for the placement of chronic implants in the CNS led to important advances in our knowledge of the role of the CNS in emotion (Brady, 1961). For example, it was dem-

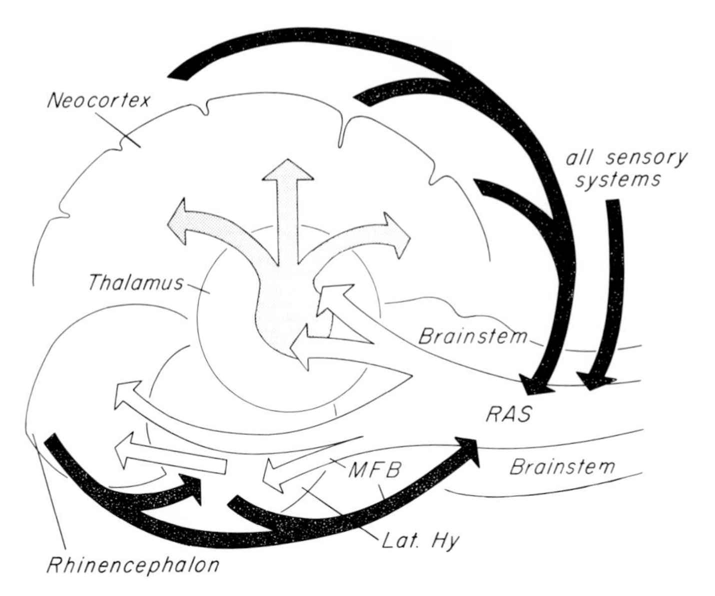

Fig. 8. Highly schematic diagram of the major connections to and from the reticular activating system. The reticular activating system ascends in the core of the brainstem and makes connections with cell bodies in the thalamus and the lateral hypothalamus and sends some fibers through the lateral hypothalamus directly into various structures in the rhinencephalon. As shown by the stippled arrows, some parts of the thalamus diffuse the influence of the system over the neocortex, and some cells in the lateral hypothalamus send fibers into the rhinencephalon for the same purpose. Shown schematically by the arrows are the three major sources of sensory input to the system: all the sensory systems of the body, the neocortex, and the rhinencephalon. The fibers of the ventral portion of this system are carried in the medial forebrain bundle. RAS, reticular activating system; Lat. Hy., lateral hypothalamus; MFB, medial forebrain bundle. (From McCleary and Moore, 1965, p. 27.)

onstrated that electrical stimulation of specific CNS areas results in highly stereotyped emotional behavior, such as aggression in monkeys (Delgado, 1966), threat and flight in the cat (Hunsperger, Brown, & Rosvold, 1964), and rage and flight in the chicken (von Holst & von St. Paul, 1962, 1963). The emotional reactions that have been observed under conditions of selected brain stimulation have been rage and threat, both expressions of aggression, or flight. A lesion in or ablation of the septal area is particularly interesting, since the typical result, at least in rats, is hyperemotionality, generally shown by constant rage, seemingly against any available stimulus. This hyperemotionality continues, however, for only a limited period, subsiding in 20 to 60 days (Brady & Nauta, 1955, Yutzey, Meyer, & Meyer, 1964).

It is apparent that rage and flight are governed by CNS structures. Yet we must ask why, with all the technical advances that have been made during the past 40 years for exploring the CNS, have variations in only flight and rage been produced?

No one doubts that cardiac rate changes in response to emotional stimulation; it is the exact form that the change takes that is open to question. From clinical reports, it would appear that the change may involve either an increase or a decrease, and that it may occur rapidly or slowly. Nevertheless, heart rate is used frequently as a dependent variable in studies both of emotion and of autonomic correlates of conditioning. In contrast to the measurement of heart rate, the measurement of blood pressure is tedious and inexact and this undoubtedly accounts for the fact that it has been used rarely as a dependent variable. The measurement of systolic pressure alone is less difficult. Landis and Gullette (1925) found systolic *variation* to be high among normal subjects (from 18 to 20 mm/min). The presentation of a situation calculated to produce surprise resulted in a sharp increase in systolic pressure, followed by a rapid decrease to the baseline level. J. C. Scott (1932) reported that he was unable to find any relationship between the degree or kind of emotion as determined introspectively, and by measurement of systolic pressure. Morris (1941) confirmed the relatively high variability of blood pressure and also failed to find blood pressure to be predictive of emotional state.

Certain emotional responses are short-term and transient (e.g., GSR and cardiac changes) and others are long-term and irreversible without medical alteration (e.g., gastric intestinal lesions). Once ulceration has appeared, it is clear that the reaction is of the second type. It is not evident, however, that the gastric intestinal changes found in the studies are necessarily of this type; that is, that they necessarily produce lesions. It is possible that the changes found in the yoked design are transient in the same sense that GSR is transient. Among long-term changes are the changes in 17-hydroxycorticosteroid (17-OH-CS) levels. Data on 17-OH-CS levels from parents undergoing severe emotional trauma over the terminal illness of a child have been reported. Measures were made over an 8-month period, starting from the first hospital admission of the child to its death. Note in Fig. 9 that the levels of 17-OH-CS from the mother are extremely regular until the seventh month, when the child's hemorrhage apparently convinced her that the illness was terminal. The father shows both a higher mean level (7.1 mg/24 hours) and greater variability during the 8 months. The investigators note that

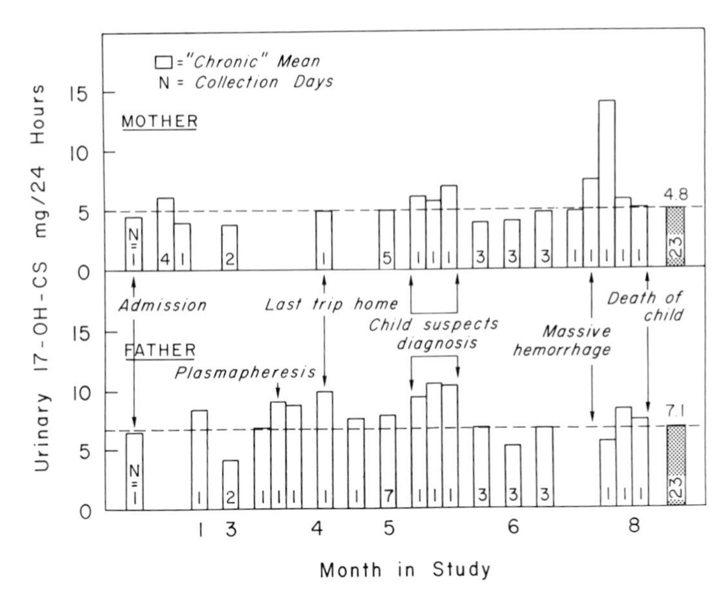

Fig. 9. Urinary 17-OH-CS excretion from a couple during several stages in the terminal illness of their child. (From Friedman *et al.*, 1963.)

the level did not rise in situations in which the parents "appeared to have effectively operating coping or defense behavior" (Friedman, Mason, & Hamburg, 1963), suggesting, again, that cognitive factors overlay emotional reactivity and may, in fact, change our perceptions of the sensations accompanying reactivity.

The studies just reviewed suggest two points. First, external events, and especially the scheduling of these events, lead to emotional responses that can be measured by gastrointestinal change. Although the schedules sometimes generate ulceration and gastrointestinal lesions, it is evident that the basal rate of acid discharge is an important variable determining the probability of ulceration. Second, the studies provide additional evidence that cognitive factors determine how emotion is experienced and, particularly, the way in which it affects bodily organs. The observation in the study by Friedman *et al.*, namely, that the mother did not show divergent 17-OH-CS levels until she found that she could not cope with the emotion-provoking conditions, indicates that perceptual patterns mediate, and perhaps govern, emotional responses.

At this point it would be well to consider how emotional responses can be conditioned. The Russian literature, although often oblivious to fundamental experimental controls, appears to make a strong case for the likelihood of interoceptive conditioning (Razran, 1961). Some of the reported findings must certainly be interpreted with caution, but it does appear that interoceptive responses can be altered by classical conditioning and operant training. The significance of this to the study of emotion is clear, although the evidence, again, is almost entirely of a descriptive nature. However, employing more stringent controls, N. E. Miller and DiCara (1967) have shown that heart rate may be altered, using intracranial electrical stimulation as a reward. The heart rate was shaped, just as response rates may be shaped, under differential low and differential high schedules. The possibility that the observed changes in heart rate were merely reflections of instrumentally trained skeletal responses was decreased by curarization. The findings of Miller and DiCara extend our understanding of emotional response. If visceral "learning" were limited to classical conditioning, the quality of the responses would be severely limited by the structure and restriction of the classical conditioning paradigm. Now it is likely that instrumental training, with its potential to alter states without requiring an available unconditioned response, may account for the obvious wealth and variety of emotional responses, the assortment of individual differences, and the apparent vicissitudes of psychosomatic states (Section III).

What do we know concerning the role of central and peripheral functions in the ontogeny of emotion? Virtually nothing, other than the indirect evidence from studies cited in Sections II, A and B and the significance of early or previous experience on a later level of reactivity discussed in this section. The study of physiological and anatomical functions in emotional reactivity remains mostly at the descriptive level. Until we know which factors can be investigated parametrically, ontogenetic studies are likely to prove unprofitable. Given the hypothesis that emotional reactivity does differ as a function of age and given also the prediction that our understanding of emotional processes is most likely to be enhanced through an understanding of physiological and anatomical mechanisms, we must keep the literature on these mechanisms firmly in mind, however unrelated they seem, knowing that they are apt to become important in the eventual understanding of emotional response.

III. An Ontogenetic Model

The first reactions of the newborn to stimulation are both diffuse and species-specific. In primates, including human beings, one observes a general increase in activity as a function of stimulation. The neonate responds to stimulation, especially tactile stimulation, with a nonspecific increase in activity that can be equated with an increase in arousal level. Heightened activity may have an adaptive function, since it often results in contact with a conspecific, commonly the mother because of her availability. Since such contact serves also to remove the original stimulus, the level of arousal is reduced. This rewarding reduction in tension serves to make the adaptive stimulus (contact with the conspecific, for example) a more powerful reinforcer on future occasions. This suggests that fear is diffuse in the early stages of emotional reactivity and then becomes sufficiently specific to serve as the reinforcing agent in the animal's learning to avoid strange stimuli and unfamiliar objects. No doubt, there are species differences. Dogs, for example, find handling by human beings to be reinforcing (Stanley & Elliot, 1962), whereas the rat will learn a new response in order to avoid contact by human beings (Candland *et al.*, 1960; Candland, Horowitz, & Culbertson, 1962).

The interaction during development of learned rewards and innate rewards (such as different kinds of food) determines an animal's preferences and provides it with a reserve of effective reinforcers of different intensities. Although an increase in general activity may have an adaptive value in the very young, as the animal ages and at the same time acquires different nutritional needs, physical contact alone fails to satisfy these needs and can become maladaptive. Through the processes of discrimination and generalization, the individual associates changes in arousal level, whether determined internally or externally, with more specific stimuli. The primate, for example, clings to objects that tactually are like the mother, runs from strangers, and displays tantrums when handled. These behavior patterns, at first so general that we describe them in broad terms, such as "avoidance" or "approach," become differentiated into a large number of subtle, but discrete, behavioral or physiognomic patterns. These response patterns, in turn, are differentiated in terms of the intensity or severity of the evoking stimulus. Intensity is a function of both the nature of the stimulus (electric shock is probably more severe in general than food deprivation) and the animal's experience with the stimulus. Intensity of the response pattern is almost certainly related proportionately to the degree of arousal. Thus, a reciprocal, feedback relationship connects level of arousal and intensity of the stimulus. The intensity of the response pattern is affected by the animal's perception of the intensity of the stimulus and the intensity of the stimulus is determined, in part, by experience.

What we may call "fixed" patterns, such as facial, vocal, and postural expressions, are adaptive substitutes for the high level of arousal and diffuse activity that were the ontogenetically original, adaptive responses to stimulation. These patterns are adaptive in that they remove the inciting stimulus in a more rapid and less energy-consuming fashion, just as the use of language in the human being and the use of vocalizations by other animals conserves activity. Insofar as responses are patterned to communicate information regarding future action (threat, for example), they reduce the emotional level by ensuring appropriate behavior from other animals (the successful threat wards off attack). It is suggested that when we speak of distinct emotions, we are supplying names to response patterns differentiated from an original increase in the level of arousal and activity and that these processes of differentiation follow the principles of learning and conditioning. Emotion, especially that of the "coarser" (to use James's descriptive phrase), fixed kind, is mainly an unlearned response, unlearned in the sense that irritability to stimulation may be considered unlearned. Emotions, especially those of the finer kind, develop by reinforced discrimination and generalization from the original diffuse irritability. Response patterns that are adaptive are reinforced and persist; those which are nonadaptive are extinguished.

What are the virtues of such an interpretation of the ontogeny of emotion?

1. *It describes how emotion is both general (reflecting changes in physiological functioning) and specific (why we cannot avoid the belief that there are different emotional states).* From the state of arousal and irritability, which is what we measure when we record heart rate, GSR, and other autonomic changes, an infinite number of emotions may be acquired—indeed, they are always changing in degree and intensity as the individual undergoes different experiences. No wonder we cannot agree upon labels or even upon the number and degree of our own emotions at different times.

Some emotions are reasonably fixed by the specific functional limitations of the organism and by its experiences during development. The form that these emotions take is highly specific to the species, for functional limitations permit little variation in the form of the response. For example, in response to heightened emotional reactivity, some species survive by running, others by becoming arboreal, and some by adopting postures, as exemplified to a magnificent degree by the opossum. Because they are immediately critical for survival, we see these responses in extant species as fixed patterns.

The acquired emotions, in contrast, are rarely of immediate survival value to the species and, accordingly, they evidence greater variability of expression. Their variability accounts for the difficulty we experience in nomenclature and the reason that successful experimental studies of emotion are almost always restricted to coarser physiological measures.

2. *The framework provides an explanation of the ontogeny of emotional behavior.* The chief reason that we find it difficult to observe the learned characteristics of emotional behavior is that we are unable to separate maturational from learned effects (Section II, D). Thus, we are apt to complain that at least some forms of emotional behavior appear too rapidly to have been acquired, and we assume, therefore, that the origin of emotional responding is purely maturational and unlearned. Yet available analyses of the relationships between learning and maturation (see Section II, D) are rudimentary. Until we know more about such relationships, it would seem appropriate to consider both experience and maturation as the significant processes in emotional development and concentrate on identifying the sequences and patterns that become differentiated.

3. *The framework is testable.* The privation experiments described earlier are especially well suited to furnish information on the "learnableness" of emotions. Privation does not exclude learning processes. It does, however, provide some control of the number and kind of stimuli available. As already mentioned, data from privation studies on dogs (Section II, A) suggest that the level of arousal is increased under privation conditions and, consequently, that both the form of the emotional response and the provoking stimuli are atypical. This work, as well as that concerned with the effects of different kinds and amounts of stimulation during the infancy of rodents (Section II, A), suggests that even the "fixed" emotions can be altered somewhat by stimulus conditions during maturation. If so, we would expect animals deprived of stimulation to exhibit atypical reactions when stimulation is applied. Melzack's dogs did exactly this. The extrapolation has not been tested directly on rodent species, but among primate species, there is ample evidence that privation produces unusual emotional responding.

If the emotions, as we have suggested, do become differentiated ontogenetically much like other learned responses, they should correspond to the learning ability of the species and to its rate of maturation. Were we to base the phylogenetic scale on learning potential, rather than on morphological similarity, we would probably derive a scale not appreciably different from that which we have now. It follows that the number of emotional responses of which a species is capable should decrease as we go down the scale, and, in addition, that the "fixed" emotions should become proportionately more frequent and the finer states less frequent.

We encounter at least two major difficulties in attempting to use learning processes to explain the ontogeny of certain emotional behaviors:

1. *Some patterns of behavior appear so rapidly and in such complete form that it is difficult to see how they could have been acquired.* Examples are the tantrum in children and in other primate species and the fear grimace common to Macaque and baboon species. It is unlikely, however, that all learning occurs at so slow a rate as we find in the laboratory experiment. Nor does our framework assume that all emotions are acquired. It does, nevertheless, suggest that modification of emotional responses through learning processes is common.

2. *If it is true that the finer emotions are differentiated, then it should be possible to train or condition any response pattern to any specific stimulus.* Training a cat to purr when attacked would be a test of this proposition—as well as a test of the investigator's fervor and fortitude. Nevertheless, it does not seem that it would prove any more difficult than training a housefly to press a bar. The physical and functional limitations of the particular animal limit what it can learn, and certainly the ease with which it can acquire certain responses. Clinical reports provide many indications of the astonishing number and variety of stimuli which can come to provoke emotional behavior and which can be of such intensity as to control a great deal or even all of an individual's behavior. The reason that medical dictionaries have long lists of phobias is that, at least for human beings, any discriminable stimulus can, once learned, come to be an intensely emotion-provoking stimulus.

The framework suggested here represents an attempt to provide a conceptual structure for the ontogenetic study of emotional development based on current knowledge of psychological principles and functions. Such an approach seems desirable because of the sterility of the traditional view, which emphasized study of the phylogeny of emotion. The suggested framework is, nevertheless, appreciably more than a compromise between the idea that emotions are wholly innate and the idea that all emotions are wholly acquired. Like any other behavior we choose to observe and study carefully, emotional behavior has both acquired and evidently nonlearned components.

Our framework makes additional assumptions regarding (a) the influence of arousal level in generating emotion, (b) the development of response patterns, (c) the development of specific responses from those ontogenetically prior, and (d) the feedback relationship between response patterns and level of arousal. Of these, the first and second have been investigated extensively, although under the influence of very different theoretical positions. Schneirla, for example, has developed a highly refined and ingenious model which views early behavior development in terms of approach and withdrawal processes. (See Maier & Schneirla, 1935; Schneirla, 1946, 1951, and especially, 1965.) Although it is impossible to summarize his detailed model succinctly, it can be stated that although Schneirla does not specifically base his data on emotional factors, some of the examples he describes in applying his model can be thought of as emotional behavior. I can see no important differences between the heuristic model stated here and the broad model suggested by Schneirla, although I think it fair to say that I have drawn a sharper distinction between fixed and learned responses than has Schneirla. Nevertheless, his analysis of the development of the facial grimace (1965, p. 64), for example, while not intended to be an exhaustive analysis of that form of behavior, suggests the utility of generalization and discrimination as explanatory concepts. But, to be sure, the analysis presented here is cruder than Schneirla's both because of the need to use data in this chapter only from behavior that can be thought of clearly as emotional in nature and because of the fact that Schneirla's model is of far greater breadth than the topic of emotion alone.

Let us consider data regarding those species that have received systematic experimental attention. Of the abundant literature dealing with some aspect of emotionality, we shall emphasize those data which are concerned especially with ontogeny and are relevant to the conceptual framework that we have developed (see Sources, Section IV, for additional reports).

1. Human Beings

Is there evidence that emotions are refinements of a general state of reactivity? Emotions related to pleasure and pain or those related to satisfaction and disgust appear to have such origins (as both Wundt and James noted). James describes them in this way:

> Disgust is an incipient regurgitation of retching, limiting its expression often to the grimace of the lips and nose; satisfaction goes with a sucking smile, or tasting motion of the lips. The ordinary gesture of negation—among us, moving the head about its axis from side to side—is a reaction originally used by babies to keep disagreeables from getting into their mouths, and may be observed to

> perfection in any nursery. It is now evoked where the stimulus is only an unwelcome idea. Similarly, the nod foreward in affirmation is after the analogy of taking food into the mouth. The connection of the expression of moral or social disdain or dislike, especially in women, with movements having a perfectly definite original olfactory function, is too obvious for comment. Winking is the effect of any threatening surprise, not only of what puts the eyes in danger; and a momentary aversion of the eyes is very apt to be one's first symptom of response to an unexpectedly unwelcome proposition (James, 1890, pp. 390-391).*

Figure 10 describes, in a very general way, some of the emotional changes that occur during the first months of human life. Emotional development and socialization appear to be interdependent. The importance of social experiences during infancy in determining both the degree and nature of adult emotional responding has been emphasized over the past 75 years.** Studies of emotional development of the human infant have been either normative (statistical descriptions of the onset of patterns of behavior) or concerned with abnormal patterns. From the latter, it is evident that certain conditions, especially the removal of the source of primary gratification, result in atypical and maladaptive, emotional reactions. Spitz (1946) describes the syndrome he calls "anaclitic" depression: if the mother is removed when the child is from 6 to 12 months of age, the infant shows a slowness in responding to stimuli, but when it does respond, does so violently, and develops stereotyped behaviors, such as digit-sucking. The observation that the removal of the mother during this period is so important is probably not a matter merely of the age of the child, but of the fact that one object—the mother—has acquired a preponderance of reinforcing properties to the exclusion of other objects. The mother, of course, also provides the food from birth. This assures contact with the mother and makes it virtually certain that the mother is the initial object which the infant approaches. When the infant matures to the time of his seeking food from sources other than the mother, her ability to reinforce emotional reactivity is lessened. By this time, however, the mother has selectively reinforced emotional reactivity in such a way that the more subtle response patterns have been delineated.

*The reader should not assume that this statement represents James's position. In the next paragraph of the original text he maintains that there are many emotions which cannot be explained in this way but which are understandable only in terms of his theory (for example, sweating of the skin and lumping of the throat.) The reader may wish to judge for himself whether or not such behavior can arise out of more gross behavioral patterns or whether James is correct in maintaining that they are of a separate origin.

**Freud's analysis (1953) of 5-year-old Hans's phobia provides rich data on how the nature of emotional responding can be modified.

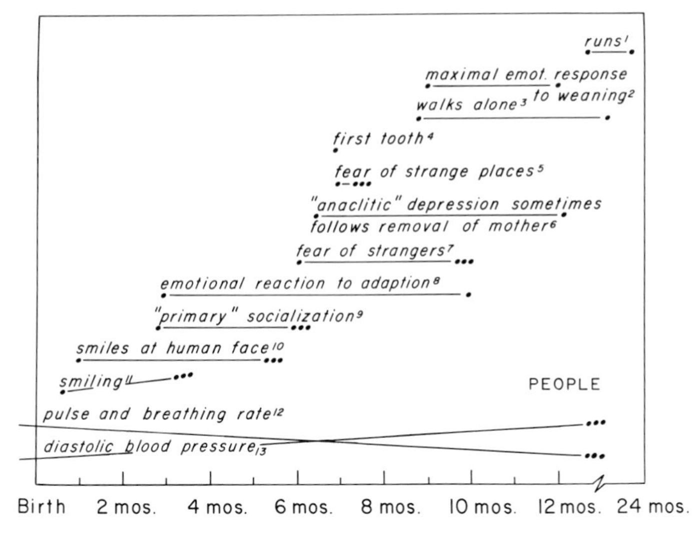

Fig. 10. An ontogenetic plot of the development of emotion in the human infant. Maturational and social points are indicated for comparison with emotional development. The points are numbered to correspond to the following references: 1. Common observation; 2. Doering and Allen (1942); 3. Jones (1926); 4. Doering and Allen (1942); 5. Sherman and Sherman (1929); 6, 7. Spitz and Wolf (1946); 8. Caldwell (1960); 9. J. P. Scott (1963); 10. Spitz and Wolf (1946); 11. Jones (1926); 12, 13. Bayley (1940).

Our framework provides an explanation of the ontogeny of the emotional disruption formed in anaclitic depression and it also suggests several experimental questions. (a) In studies of mother-infant separation in human beings, the infant is removed not only from the mother, but also from other familiar stimulus conditions (crib, room, other persons). It would be helpful to know which of these stimuli or configurations have acquired reinforcing properties and which have not. (b) At birth, before acquired modifications of behavior have had very much of a chance to become established, what kinds of stimulation does the infant respond to—what stimuli initially provoke or change the level of arousal? At the moment, this question can be examined most fruitfully by measuring changes in the level of arousal by autonomic and electrochemical changes. (c) Is there any meaningful relation between either the general level of arousal or arousal lability and the development of emotional states? Bridger (1965), in a neatly interpreted series of studies, finds that, at least in terms of autonomic functioning, human infants demonstrate stable interindividual differences in temperament (lability and soothability, for example). These individual characteristics in level of arousal almost certainly affect the heirarchy of species-specific responses and may be expected to determine, in part, which responses will occur.

2. *Nonhuman Primates; Cats and Dogs*

Studies of nonhuman primates have not been concerned directly with the ontogeny of emotional responding (it is no easier to measure or classify emotion in these species than any other) and this may be the reason that observational data suggest so few, if any, major differences in the *sequence* of development of emotional forms among primate species (see Figure 11). In terms of absolute time, the period of emotional and social development of human beings is longer than that of other primate species, as may be noted by comparing Figs. 10 and 11. The pattern of general activity at birth in nonhuman primates includes mouthing (but probably not sucking), followed by a differentiation of activity into righting responses, responses to unfamiliar

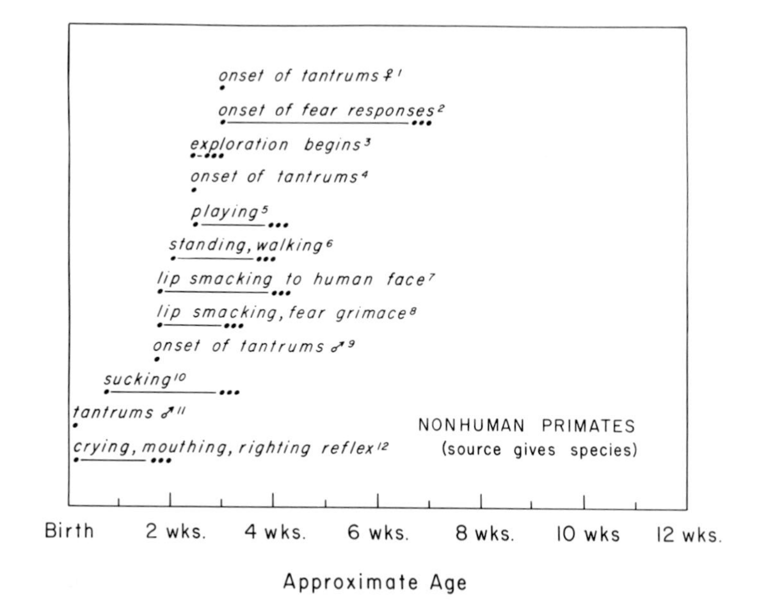

Fig. 11. An ontogenetic plot, like that shown in Fig. 10, for nonhuman primates. The references are: 1. *Theropithecus gelada,* author observation; 2. *Macaca mulatta,* Harlow (1960); 3. *M. irus,* author observation; 4. *M. speciosa,* author observation; 5. *M. mulatta,* Foley (1934); 6. *M. mulatta,* Foley (1934); 7. *T. gelada,* author observation; 8. *M. speciosa,* author observation; 9. *T. gelada,* author observation; 10. *M. mulatta,* Foley (1934); 11. *T. gelada,* author observation; 12. *M. mulatta,* Foley (1934); Mowbray and Caddell (1962).

persons, responses to objects involving the mouth and head region (lip-smacking, smiling), and the development of fear responses and tantrums. Although these behavior patterns appear to the human observer to be maladaptive, or at best nonadaptive, it is likely that they are in fact adaptive for the infant. As they become less so, patterns that are adaptive for the adolescent and adult appear, such as exploration, play, and grooming.

Yerkes and Yerkes (1936) reported the development of avoidance responding in 29 young and adult chimpanzees. They found that the visual characteristics of the test object, such as intensity of movement and rapidity of change, were far more fear-producing than the nature of the stimulus itself (a live snake, a shuttlecock). Haselrud (1938) measured the fearfulness of chimpanzees by placing test objects where they interfered with access to food. Younger chimpanzees adapted to the interposition more rapidly than older animals. In terms of our framework, the findings of both these studies can be only suggestive, for neither attempted to equate motivation for food as a function of age (Section II, D). It is important to the present discussion that the chimpanzees used in these studies were reared with conspecifics, but not in an entirely normal social environment. McCulloch and Haselrud (1939) examined an infant chimpanzee, reared in isolation from conspecifics, and presented it with a variety of test objects. At 7 months, the animal showed itself to be disturbed by moving objects, usually avoiding them but sometimes showing aggression. Eight months later, the animal responded with avoidance and aggression to a wide range of objects, and the authors found it more highly emotional in general (i.e., exhibiting a higher level of arousal) than the socialized animals used in Haselrud's study. Foley (1934, 1935) made a thorough study of the development of the rhesus monkey (*Macaca mulatta*) and recorded the time of onset of many basic responses (Fig. 11). Although the major reflexes appear within a few days of birth, the startle response does not appear clearly until about 1 month of age (Foley, 1934, 1935; Hartman & Tinklepaugh, 1932; Lashley & Watson, 1913). This latter response is often thought to be "fixed," for it appears to be reflexive, but it is not exhibited until some time after the development of other reflexes. It is possible that startle is unlearned; it is also possible, however, that some aspects of startle are acquired through the differentiation of general activity, but that investigators fail to recognize these differentiated aspects until they become a part of the complete startle pattern. Indeed, Tinklepaugh and Hartman (1932) state that they observed a transition from the pinna reflex to the startle response, with movement from different parts of the body added until, after some weeks, all parts were involved in the startle pattern.

If any single activity can be said to be fully developed at birth, it is vocalization, and again in terms of our framework, it is noteworthy that the fear grimace, when it does appear, has all the facial components of intense vocalization, but without sound. An informative difference in vocalization is found between laboratory-raised and feral animals. Among the latter we would expect vocalization to occur in response to emotion-provoking stimuli, since the availability of the mother suggests that vocalization would be reinforced. Conversely, laboratory-reared animals, separated from their mothers soon after birth, are not likely to have had their vocalizations reinforced and, according to our framework, we would not expect them to use vocalization in responding to emotion-provoking situations. Reports by Menzel, Davenport, and Rogers (1963a, 1963b) and by Randolph and Mason (1969) have found this predicted difference in vocalization with chimpanzees.

Foley (1934) describes "contact-seeking" and "seeking of bodily support" as the "most outstanding and persistent behavior pattern" of the young rhesus raised in the laboratory. Clinging remained an impressively frequent behavior pattern through the first 9 months of life. Foley concludes that this clutching reaction to various stimuli, presumably giving rise to what is currently called "contact comfort," was "undoubtedly an undifferentiated continuation of the clutching response present at birth. It probably has its origin in prenatal life, the so-called 'fetal movements' being one such overt manifestation" (1934, p. 70). Although our conceptual framework does not concern prenatal influences, it is evident that the principles of our framework correspond with those suggested by Foley.

Foley also noted that surprisingly few kinds of stimuli could provoke emotional behavior from the neonate. Indeed, he was able to categorize them easily as fear, rage, and love, the three categories that Watson and Rayner (1920) used to describe human infant emotion. These categories are not dissimilar to those proposed as "fixed." Certain environmental conditions imposed by Foley led to rocking (swaying) and digit-sucking. These response patterns were, as we would expect from our assumptions regarding the importance of privation, ". . . *emotional behavior resulting from novel and unfamiliar situations in which the stimuli previously serving a contact or 'support' function were absent*, and adaptation occurred only when other stimulus objects had acquired such instigative functions" (1934, p. 90, italics Foley's). By the end of the animal's second year, "contact seeking" and "seeking of bodily support" occurred only in response to intense emotional stimulation. The "fear reaction," initially shown only in response to loss of support, could now be elicited by intense stimulation, such as a loud noise. Foley suggests that this change is due to "conditioning" (1935, p. 88).

Jacobsen, Jacobsen, and Yoshioka (1932) reported extensive observations on a female chimpanzee during its first year of life. Although the infant responded emotionally and in like manner to the same kinds of stimuli Foley found to provoke emotion in rhesus, one does not find descriptions of the gradation and shading that Foley described. Perhaps inhibition of emotional responding occurs more rapidly in the chimpanzee, although it is possible that the animal differentiated its emotional responding more rapidly than the rhesus. Certainly, descriptions of the chimpanzee's finely differentiated responses to various stimuli suggest that its early, intense, emotional responses quickly became differentiated and generalized. One stimulus situation certain to provoke intense emotional expression was a change in any characteristic of a situation with which the animal was familiar, such as the living quarters. Our framework suggests that the living quarters have the same potential reinforcing power as the mother. Thus, a major change in this configuration should have effects similar to those found when the mother is removed.

Recently, attention has been directed less to descriptions of emotional development than to the effects of early environmental influences on later behavior. Nonhuman primates, usually rhesus, have been raised under different stimulus conditions (most often with stimulus configurations having different degrees of similarity to the mother—at least to the eyes of the human observer) followed by measurement of various responses. These studies, insofar as they involve restriction, attempt to alter not only the general level of arousal, but the number of stimuli potentially available as reinforcers. Because of this they are of special significance to our conceptual framework.

In several studies involving separation, 6- to 7-month-old rhesus showed intense disorientation and vocalization; characteristic of the infants' behavior was an increase in mouthing and sucking (Seay, Hansen, & Harlow, 1962; Seay & Harlow, 1965). Hinde, Spencer-Booth, and Bruce (1965), in their investigation of rhesus, found that separation produces a dramatic decline in activity and manipulation. This pattern is similar phenotypically to anaclitic depression in human infants. On the one hand, the mere presence of the reinforcing stimulus serves to reduce emotional responding; on the other, the stimulus serves to limit the infant's opportunities for manipulation of the environment and, in so doing, restricts its adaptation to other emotion-provoking and novel forms of stimulation. The power of partial reinforcement to control behavior over very long periods is well documented both clinically and experimentally, and it is likely that the partial-reinforcement contingencies that characteristically obtain between infant and mother are a major reason for the persistance of the behavioral effects of separation.

Kaufman and Rosenblum (1967) report separation of infant pigtails (*Macaca nemestrina*) from the mother for 4 weeks. The infants showed agitation and engaged in intense activity and stereotyped sucking. The description of the reaction to separation is of special interest here. Kaufman and Rosenblum write

> that at an undifferentiated level the removal of mother's biotaxic stimulus first produces in the infant a facilitated distress reaction, characterized by a high level of motoric, visceral-autonomic, vocal, and expressive activity, which ordinarily functions to regain mother's comfort-producing stimuli, primarily through the communicative value of the reaction, which has evolved through its selective advantage. We presume that it is accompanied by an undifferentiated unpleasant affect of apprehension (the precursor of anxiety) based on uncertain sensory incongruity. If comfort-producing stimuli do not materialize distress persists but through its catabolic effects suppression occurs. This reaction is characterized by a postural collapse, immobility, withdrawal from the environment and a reduction in visceral-autonomic, vocal and expressive activity (Kaufman & Rosenblum, 1967, pp. 670–671).

To return to the development of emotional reactivity in nonhuman primates, Fig. 12 provides an illustration of expression around the mouth region which could have developed differentially from the general and diffuse activity we have posited as the initial response to

Fig. 12. Facial expressions of *Macaca mulatta*. The left figure is commonly called lip-smacking (although the movement of the lips is not evident in the drawing); the center, low intensity threat; and the right, medium intensity threat. The text suggests that each of these differentiates from the "relatively fixed" pattern at birth of mouthing, rooting, and vocalization. (Drawing by Mary Candland.)

emotional reactivity. Several of the "fixed" responses present at birth involve the mouth, viz., mouthing, rooting, and vocalization. Lip-smacking, which becomes evident in the infant's first month in some species, occurs in the presence of a conspecific, or a human being if he returns the response pattern. Hinde and Rowell (1962) suggest that the occurrence of lip-smacking always involves "positive social advances to another individual: this is often combined with slight fear" (p. 15). The first and second pictures in Fig. 12 show lip-smacking and low intensity threat. The third shows a medium intensity threat or "fear grin." Like lip-smacking, the threat response is often given to conspecifics or to human beings, especially if it is returned. To the human observer the most obvious difference in the three facial expressions and postures is the degree to which the teeth are exposed, but there are other revealing differences. Note the tension of the facial musculature, the flare of the nostrils, and the degree of piloerection. It is not difficult to imagine that the pattern of mouthing, rooting, and vocalization differentiates into lip-smacking (in response to the approach of an object) and into degrees of "threat," the degree being dependent upon the emotion-provoking qualities that the object acquires.

Let us consider several general conclusions taken from various reports on nonhuman primate emotionality:

1. The "fixed" reactions seen at birth are rooting, mouthing, vocalization, and clinging. Since these involve both facial expressions and posture, it may seem obvious to suggest that other emotions develop from them. Nevertheless, it is true that the emotions observed later can be interpreted as differentiations of these basic forms.

2. Laboratory animals raised without mother, conspecific, or other suitable (primarily tactile) objects capable of assuming reinforcing value not only fail to develop the emotional patterns observed among feral animals, but adopt unique patterns of emotional responding. These involve stereotypy of the "fixed" responses, such as digit-sucking, self-clasping, rocking, and swaying. Infants reared with the mother in the laboratory do not develop these stereotyped behaviors (Jensen & Tolman, 1962).

3. Stereotyped patterns, such as the anaclitic depression syndrome, occur only if the mother or mother surrogate is removed at a fairly specific point in the infant's development. Once established, these response patterns persist and tend to recur during later emotion-provoking situations.

4. Some inanimate objects can come to have much the same functions and reinforcing properties as the mother. Evidently, it is not

the mother as such that has an omnipotent influence on the infant, it is certain characteristics which the mother has in excess of other available forms of stimulation that encourage the interdependent relationship between infant and mother. These characteristics are apparently primarily tactile.

Ontogenetic plots, such as those shown for primates in Figs. 10 and 11, are shown for cats and dogs in Figs. 13 and 14, respectively, For the cat, the experimental situation most analogous to the privation studies of the primate are those in which kittens are raised under restricted conditions and examined at maturity. It is evident that restriction and privation do influence the cat's selection of conspecifics and inanimate objects in adulthood, which suggests that socialization in the cat is as plastic as that of other mammalian species (Candland & Milne, 1966). Indeed, Fig. 13 shows that separation from the mother and handling in infancy can affect the cat in much the same way as the infant primate.

When the infant dog is removed from its mother, it shows diffuse activity and intense vocalization. It continues this pattern of behavior, often moving in a circle, until it makes contact with the mother, conspecifics, or some other configuration that provides tactile stimu-

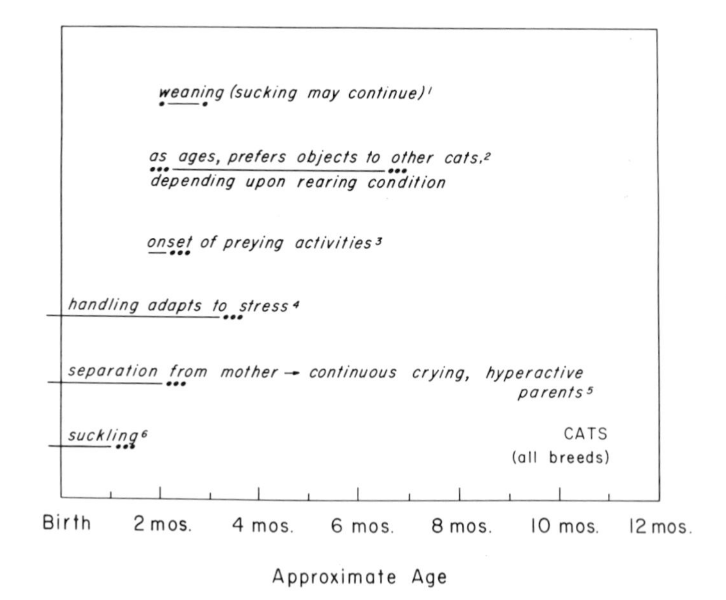

Fig. 13. Ontogenetic plot for cats. The references are: 1. Rosenblatt and Schneirla (1962); 2. Candland and Milne (1966); 3. Rosenblatt and Schneirla (1962); 4. Meier and Stuart (1959); 5. Seitz (1959); 6. Rosenblatt and Schneirla (1962).

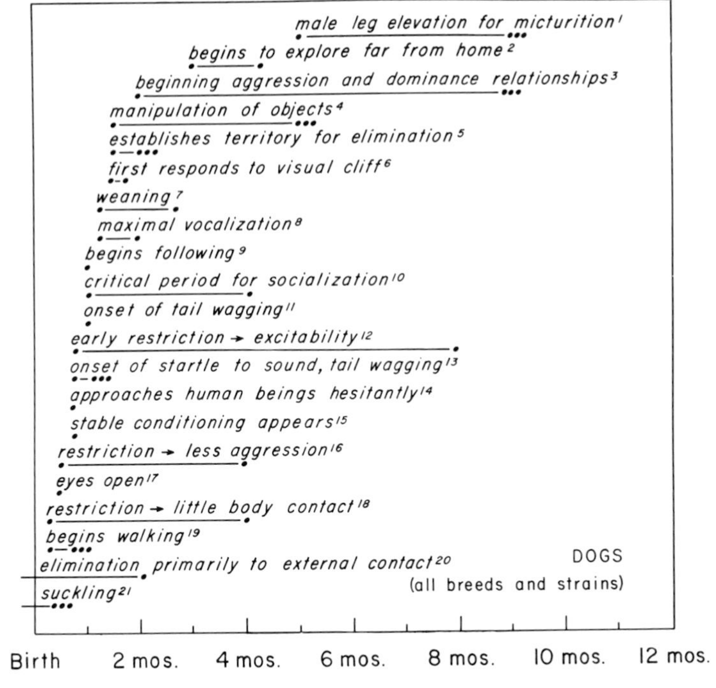

Fig. 14. Ontogenetic plot for dogs. The references are: 1. Fuller and Dubuis (1962); 2. Elliot and Scott (1961); 3. Fuller (1953); 4. Fuller and Dubuis (1962); 5. Ross (1950); 6. J. P. Scott and Fuller (1965); 7. Common observation; 8. Elliot and Scott (1961); 9, 10. J. P. Scott and Fuller (1965); 11. J. P. Scott (1963); 12. Melzack (1965); 13. J. P. Scott and Fuller (1965); 14. Elliot and Scott (1961); 15. Fuller, Easler, and Banks (1950); Cornwall and Fuller (1961); 16. Fuller (1961); 17. J. P. Scott (1963); 18. Fuller (1961); 19, 20, 21. J. P. Scott and Fuller (1965).

lation of an appropriate kind. If other pups are available, as they usually are in the natural situation, contact with them is often sufficient to terminate activity and vocalization (J. P. Scott & Fuller, 1965, p. 85). The pup responds to painful stimulation (temperature, hunger), by yelping. As in primates, vocalization is present at birth and it is the primary response to stimulation. Accompanying this early vocalization is diffuse, general activity. Moreover, as in primates, these behaviors subside in response to appropriate tactile stimulation.

Within a few weeks after birth, sucking in the dog is differentiated into lapping; yelping becomes less frequent in familiar situations, but is still common when the pup is placed in a novel environment. Growling (a form of vocalization) is sometimes heard. The startle response to light and sound, consisting of a backward motion, does not appear until the pup is approximately 3 weeks old. This timing suggests that the startle reaction develops out of more gross responses. Vocalization is differentiated clearly after the first few weeks of life. Several specific sounds become evident and the number of situations that elicit vocalization increases (restraint, injury, and novel situations). That is, finer differentiations appear both in the types of vocalization and in the situations in which these vocalizations occur.

The similarities between the emotional development of the dog and the primate are striking. It is difficult to find any major difference, and the effects of privation and restriction are, on a behavioral level at least, very similar. The primacy of vocalization, the importance of tactile experiences and of configurations that provide such experiences, the development of more precise emotional responses from diffuse general activity, and the interplay between socialization and emotionality are common to primates and domesticated carnivora.

3. *Rodents: Rats and Mice*

The emotional and social development of rats and mice differ from the development of other animals in certain ways. (a) Rats and most mice have underdeveloped sensory capacities of audition and vision until the second week of life. (b) Among domesticated species, their social organization is limited, being restricted primarily to sexual and maternal activities. Dominance relationships, which take a long time to occur, are based on physical size and perhaps previous experience, rather than on "fixed," probably perceptual, cues used by other animals. (c) The emotional responses of these species are limited, so far as we know, and this accounts for the reliance placed by investigators on the open field as a measure of emotionality. If the open field induces anything reflecting emotion, it is gross autonomic activity. Accordingly, the data on the ontogeny of emotionality in the rat are limited to measurements of changes in gross emotional reactivity. We see no facial or postural signs in rats and mice, and little vocalization except squeals in response to pain. When placed in a novel environment, the rat shows an increase in defecation frequency over baseline and this response is maximal and reaches asymptote when the animal is about 54 days old. It is probable that elimination in a novel environment serves some survival function, such as the establishment of territoriality, in addition to the purely emotional component.

The learning ability of the rat and mouse is most certainly inferior to that of primates and domestic carnivora. Our framework suggests that rodents are not likely to show large variations in their emotional responses because of the species-limited potential for acquiring and differentiating new responses.

4. The Domestic Chicken

Upon hatching, the domestic chicken shows no sensory inefficiency comparable with the mammalian species we have discussed, unless the intense need of warmth is included. Separation from conspecifics or familiar objects leads to intense vocalization—"distress calls" which, although occasionally heard in later life, are the primary vocal response of the chick. Our framework would suggest that the distress call becomes differentiated into other vocal responses that the chicken uses in adulthood. As in the other species we have examined, vocalization is present at birth (hatching) in chicks and separation from conspecifics, the mother, or possible objects with

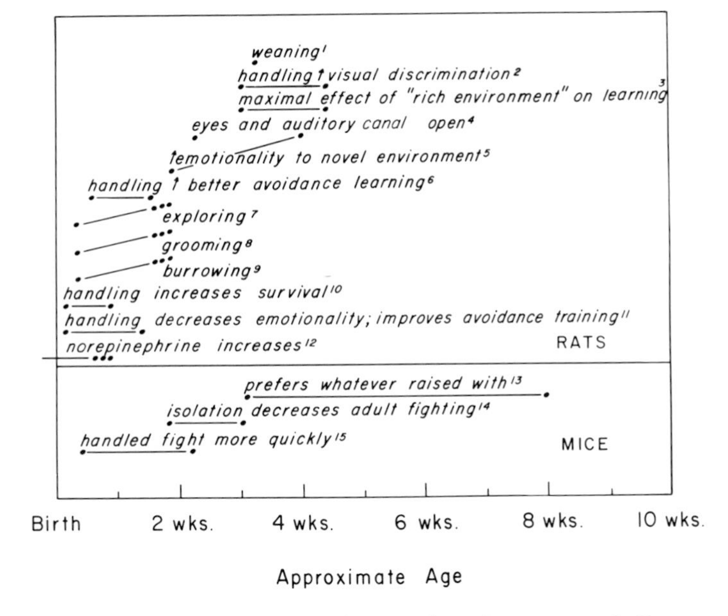

Fig. 15. Ontogenetic plot for rats and mice. The references are: 1. Common observation; 2. L. Bernstein (1957); 3. Forgus (1955); 4. Common observation; 5. Candland and Campbell (1962); 6. Denenberg and Morton (1962); 7, 8, 9. Bolles and Woods (1964); 10. Denenberg and Morton (1962); 11. Levine (1967); 12. R. D. Young (1964a, 1964b); 13. Nagy (1965); 14. J. A. King and Gurney (1954); 15. Levine (1962).

similar tactile (especially warming) attributes, leads to intense, nonspecific agitation until such contact is reestablished. Additionally, socialization and emotionality are interdependent in the chicken, as they are in other species. Imprinting may provide an important avenue for socialization (Moltz, 1960). If it is true that the critical period for imprinting in the chick is terminated by the onset of fear responses, then imprinting is related to emotionality. The "fixed" emotional responses consist of vocalization and activity (Candland *et al.*, 1963).

From Fig. 16 it is evident that in chickens the emotional responses at hatching, separation, and socialization are similar to those of primates and domesticated carnivora. Evidently, such very different forms as these show the same developmental sequence of emotion and socialization.

In addition to imprinting, a singular characteristic of the chicken's behavior is a specialized form of socialization based on aggression—the "pecking," or dominance, order. Although aggression is seen in many vertebrate forms, in the chicken it is highly ritualized. Ontogenetically, it is noteworthy that reasonably stable dominance relation-

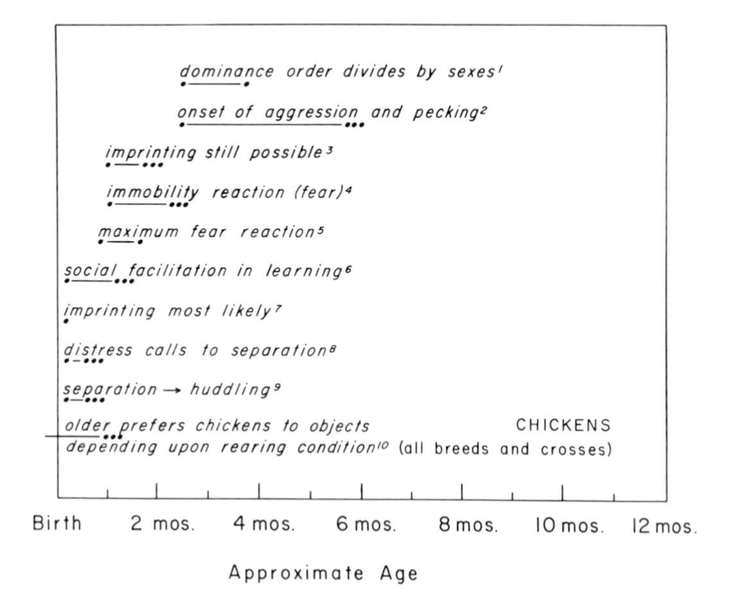

Fig. 16. Ontogenetic plot for chickens. The references are: 1, 2. Guhl (1958); 3. Jaynes (1957); Moltz (1960); Baron and Kish (1960); 4, 5. Ratner and Thompson (1960); 6. W. I. Smith (1957); 7, 8. Common observation; 9. Collias (1950); 10. Candland and Milne (1966).

ships are formed following the onset of fear and at the same time as the onset of aggression and pecking. The establishment of the dominance order serves to reduce aggression, just as do the threat signs of the primates, since position in the dominance order determines which bird can fight which other bird. It serves also in the demarcation of territories. One is tempted to suggest that just as the chicken appears to condense an enormous amount of perceptual-motor learning into the imprinting period, so does it complete the socialization process rapidly. Imprinting and the establishment of dominance orders in the chicken may be regarded as the analog of socialization in mammals and we find, once again, that emotion and socialization are interrelated. It has been shown that the amount of arousal, as measured by heart rate, changes dramatically when two chickens meet for the first time and that the outcome of the aggression which promptly develops can be predicted from the amount of arousal. The function is apparently U-shaped, with chickens high in the dominance order and those low in the order showing the higher heart rates (Candland *et al.*, 1969).

By the time the chicken is 5 months of age, the basic processes of emotion and socialization are present; little is known about development of these patterns after that age, possibly because few changes occur. It is as if those forms of socialization and kinds of emotional reactivity which require a lengthy period of learning in primates and domesticated carnivora are condensed into two brief periods in the chick's life, the one during which imprinting occurs and the other during which dominance orders are established. The brevity of these periods notwithstanding, it is evident that the chicken follows the same general pattern of emotional development as the other groups we have reviewed with respect to vocalization, reactions to separation, and the diffuse character of the early emotional behavior.

The chicken never develops, so far as one can tell, the range of emotional responses that become available to primates or even to cats and dogs. Its emotional responding grows in intensity as it develops physically, yet the nature of the response is the same—intense vocalization, readily identifiable by the human being as a distress call, and flight. Since our framework suggests that the less the learning capacity of the animal, the less the range of emotional responding, we should expect to find a limited range of emotional responding in the chicken, for the species is not notorious for its ability to acquire new responses.

One of the most efficient ways of evoking emotion in the behaviorist is to suggest a "new" theory of emotion. The framework presented here was not designed with this intention. Rather, it grew from consideration of what is known regarding the ontogeny of emotion in several disparate species, with the attempt being made to place these findings within the framework of current explanatory principles of behavior. For some of the facts, the principles appear suitable, for others they are less so. In some places, the data are sparse; in still others, the data are contradictory. I offer the present approach in the hope that it may provide the necessary framework for understanding the ontogeny of emotional behavior. I have attempted to show that (1) the "finer" emotions develop, through generalization and differentiation, from gross levels of emotional reactivity present at birth, (2) the ontogeny of emotion in several species differs in rate rather than in quality, and (3) some species have developed unique ontogenetic periods when highly complex emotions, such as those involved in socialization with conspecifics, are acquired.

IV. Sources

This section describes the sources cited in the chapter. In addition, it lists a number of valuable sources that were not cited.

I. The history of the concept of "emotion" has been admirably reviewed by Arnold (1960). For the reader who prefers to uncover history himself, I recommend sampling the following works in the order indicated: Bain (1859), Darwin (1872), McCosh (1880), James (1890), Ribot (1903), McDougall (1926), Murchison (1928), Watson (1929), Lund (1930, 1939), Ruckmick (1936), and Reymert (1950). The Murchison, Ruckmick, Reymert papers and Arnold (1968) contain reports of the three international symposia held during this century. Each of the other volumes is valuable for its interpretation of contemporary thought. The Tobach symposium (1969) indicates the state of present-day experimental studies.

The literature before 1930 has been reported in a classified bibliography in Raines (1929a, 1929b, 1930a, 1930b, 1931a, 1931b) and prior to 1941, in W. A. Hunt (1941).

II. A. The literature on critical periods continues to amass. It is difficult to discuss critical periods without discussing imprinting, but imprinting requires a discussion of learning, learning requires a separation of "innate" behavior patterns, and soon we have discussed the universe. I have attempted to consider the contemporary view of the topic by showing the arguments developed by several investigators. The ontogeny of emotion is assuredly not evident in anything so strict as day-to-day changes, for nothing is known about such changes. It is recognized, however, that emotional reactivity is vital to socialization and probably to the acquisition of responses beyond infancy. In addition to writers cited in the text of this chapter, Salzen (1967) compares imprinting in primates and chickens, and Scott *et al.* (1951) provide an early statement of the concept applied to socialization of the dog; readers especially interested in imprinting (but not necessarily in the emotional component) will find that Bateson (1966) has made some choice comments. For the history of imprinting, see P. H. Gray (1963).

II. B. One commendable attribute of work in the area of prenatal and maternal influences is that it shows a progressive improvement in experiment methodology and a willingness of investigators to learn from and correct one another. Investigators have not been reluctant to point out problems in their designs and interpretations, and procedures have been improved steadily. This is most obvious if the works are read in chronological order.

A number of important references are not cited in the text; on differences in behavior as a function of cross-rearing, see Ressler (1966) and Joffe (1965a, 1965b); on discrimination by rats between handled and unhandled animals, see R. D. Young (1965); on differential effects of litter size, see Seitz (1954), Ottinger (1961), and Ottinger, Denenberg, and Stephens (1963); on the effects of Caesarean or natural birth, see Meier (1964), Moltz, Robbins, and Parks (1966), and Grota, Denenberg, and Zarrow (1966); on X-irradiation and hyperoxia, consult Furchtgott and Echols (1958), Greenbaum and Gunberg (1962), Fowler, Hicks, D'Amato, and Beach (1962), Werboff, Havlena, and Sikov (1962), and Vierck, King, and Ferm (1966); on hormones or drugs, see Havlena and Werboff (1963), R. D. Young (1964a, 1964b), J. A. Gray, Levine, and Broadhurst (1965), and Joffe (1969a, 1969b); on the result of stress agents, consult Doyle and Yule (1959b), Ader and Belfer (1962), McQuiston (1963), Morra (1965a, 1965b), and Joffe (1965a); for analysis of findings, see Werboff, Gottlieb, Havlena, and Word (1961); and for general discussions, see Broadhurst (1961), R. D. Young (1967), and Joffe (1969a, 1969b).

II. C. I think that we are just beginning to recognize the function of the open field. We have tended to see it as another apparatus, forgetting that apparatus is only a method for forcing the behavior that we wish to observe into measurable form. Sometimes our enthusiasm with technique makes us forget the question that prompted the technique. In this case the question is: What does the open field measure? I have found the comments of Halliday (1966) and the research reports of Bruell (1969) especially stimulating. See also Denenberg (1969b) and Henderson (1969). One advantage of considering the open field a territorial situation is that such a perspective permits a comparison of open-field data and field studies.

The complexity of interpretation of open-field studies is demonstrated by our illustration of the increase in emotional reactivity with age shown by Wistar albino rats. The use of control animals makes these data more readily interpretable. Consider, however, two additional reports on this topic. Furchtgott, Wechkin, and Dees (1961) recorded open-field activity in albino rats at 30, 142, and 390 days of age. They found a decrease in activity after 30 days of age, but essentially identical activity at 142 and 390 days. Werboff and Havlena (1962) report that both activity and defecation continued to decline with each age tested (90, 180, 360, and 540 days). The last two ages are appreciably greater than those used by Candland and Campbell (1962) who found an increase through 54 days of age, at which time asymptote was attained. What is one to conclude? Actually, these studies contain only minor contradictions; nevertheless, the use of different species, different ages, different controls, and different procedures makes it impossible to identify the source of the contradictions.

The development of the open-field method may be followed by consulting Hall (1934, 1936a, 1936b, 1937), Martin and Hall (1941), Hall and Klein (1941), and Broadhurst (1957a, 1958).

II. D. The paucity of source material raises the question: Why include a section on the relationships among learning, motivation and emotion? The answer is that although the paucity of material cannot be denied, neither can the importance of the relationships. P. T. Young (1943, 1961) has commented extensively on this problem. An excellent article by Roberts (1966) makes an impressive experimental attempt to establish the relationships. (See the experimental work of Broadhurst, Sinha and Singh, 1959.)

There is, I think, a common feeling that the issues involved here have been discussed for so long and at such length that nothing remains to be said. Just the reverse is true. This is a very difficult problem; it cannot be ignored in any research on the ontogeny of emotion (or learning, or motivation) and the lengthy discussions only point out how difficult it is. An investigator who is aware of the problem may appear pedantic; one who is not aware is apt to overlook important questions.

II. E. In the section on behavior in social situations, I have found it difficult to present the available data in such a way as to provide anything better than an awkward description of some representative studies and some major experimental problems. The difficulty, I think, is that no one has attempted to evaluate "social" stimuli in terms of their reinforcing properties at different ages, nor has anyone yet explained how social stimuli come to acquire reinforcing powers. Until such systematic knowledge is available, studies can be, at best, of only a demonstrative nature and their value vanishes when we wish to generalize the findings to different ages or conditions. I have not included sexual stimulation in this section, although sexual behavior is of a social nature for the most part and obviously includes some degree of emotional reactivity. The reason for the omission is that the topic touches on emotionality only incidentally and requires lengthy discussion of hormonal and other affects that are beyond the scope of this chapter.

Among the works on behavior in social situations, Guhl (1962a) presents a fine statement of the experimental problems and of the available data on animal interaction. See also J. A. King and Connon (1955), Church (1959), Southwick (1967), Baron, Kish, and Antonitis (1962), McBride, King, and James (1965), Kaufman and Hinde (1961), Denenberg and Morton (1962), Jewell and Loizos (1966), Igel and Calvin (1960), Elliot and Scott (1961), and W. I. Smith and Ross (1952).

For data on population density, see Calhoun (1962a, 1962b), Christian (1963), Christian and Davis (1964), and Davis (1958, 1964). A variation of social facilitation includes studies in which one species is exposed to another under any of numerous conditions. One cannot overemphasize the fact that we study species which, in their nonlaboratory state, coexist in space and interact constantly. Unfortunately, there are not enough reports on this interesting ecological topic to warrant discussion, and it is possible that we do not know enough about any one species at this time to make such a discussion profitable. For illustrative reports, however, see Stanley (1965), Cairns (1966), and Cairns and Werboff (1967).

II. F. The early investigations of the role of the adrenal cortex in emotion provide a casebook in scientific method. For early experimental reports, see Cannon and Britton (1927) and Britton (1928). Bekhterev (1928) contributed a summary of the position emphasizing the adrenal medulla. Further experimental reports are by Harris and Ingle (1937) and Allee and Collias (1938), who investigated the role of epinephrine on the chicken's pecking order. An instructive contemporary report is by Ader, Friedman, and Grota (1967).

During the 1920's and 1930's many papers were published on the effect of emotion on various organs. Typically, the dependent variable (white cell count, blood viscosity) was measured, the independent variable applied (a supposedly fear-producing event, usually), and the dependent variable remeasured. The deficiencies in this design are evident to contemporary workers and may explain why this rich literature is rarely noted in contemporary accounts of emotion. I cite here a few sources in which additional references may be found. On arthritis, see Patterson, Craig, Waggoner, and Feyberg (1943). On blood composition, see Whitehorn (1934), Gellhorn (1964), Nice and Simons (1931), Katz and Nice (1934a, 1934b), Nice and Fishman (1934, 1936), Nice, Katz, Fishman, and Friedman (1935), Nice and Katz (1934, 1935), Gildea, Mailhouse, and Morris (1935), Himwich and Fulton (1931), and Diethelm, Doty, and Milhorat (1945). On diaphragm, see Faulkner (1941). On renal circulation, consult Blomstrand and Löfgren (1956). On oxygen supply and respiratory infection, see Totten (1925) and Despert (1945). On the spleen, see Barcroft (1930); and on the thyroid, Gibson (1962).

Early reports of the utility of GSR (Féré, 1888; Tarchanoff, 1890) remained unappreciated for many decades. Since the 1930's, however, GSR has been used extensively in studies of emotion. Two reviews that emphasize the theory and technique of measuring skin resistance are Edelberg (1967) and Venables and Martin (1967).

On heart rate as a measure of emotion, see Brener (1967) for a very careful discussion of procedures and technique. There are many difficulties in translating cardiac information into useful data. Bridger has contributed an excellent collection of papers discussing this problem (Black & Bridger, 1962; Bridger, 1965; Bridger & Reese, 1959). A very good account of the more general problem of emotional measurement is found in Edwards and Hill (1967). See also Steinschneider and Lipton (1965) and Caceres (1965).

The concise text by McCleary and Moore (1965) offers an excellent discussion of the contemporary viewpoint on the CNS and emotion. For historical review, I suggest reading (in order) Bard (1934), Masserman (1941), and Dunbar (1935); the Dunbar book is outdated, but is still an excellent review of psychosomatic problems and emotion. For a sampling of notable experimental reports, see Biel and O'Kelly (1940) and Wortis and Maurer (1942), who report on "sham rage" in human beings. Also see Murphy and Gellhorn (1945), Spiegel, Wyris, Freed, and Orchinik (1951), Olds and Olds (1961), Hohmann (1966), Plutchik, McFarland, and Robinson (1966), and Brady and Conrad (1960).

In using chemical agents, one can vary dosage level, type of drug, and the dependent variable in many ways, and there are ample studies to support this statement. Brady (1956) discusses the use of operant methodology in such studies. For representative studies on hormonal affects, see J. A. Gray *et al.* (1965) and Mason, Brady, Polish, Bauers, Robinson, Rose, and Taylor (1961).

For general reviews, see Darrow (1935), Gellhorn, Cortell, and Feldman (1940), Wilson, Rupp, and Bartle (1943), Gellhorn (1964), and Steinschneider and Lipton (1965).

The history of research on the relationship between hormonal discharge and gastrointestinal activity is long and exciting. For a short review, see Smith, Boren, and McHugh (1964). Mason (1964) has provided an excellent review and discussion of the role of the endocrine responses in emotional reactivity. His comments regarding what is known and what needs to be known are a reasoned description of the state of this field.

III. Forms of life other than those discussed have received only sporadic attention, and although the studies are informative and suggestive in terms of emotional development and our framework, the work is too scattered to permit generalization and interpretation. On the pigeon, see Craig (1909); on fish, see Aronson (1957); on domestic cattle, consult Stratton (1923), Hersha, Moore, and Richmond (1958), and Collias (1953); and on the roach, see Turner (1913).

For additional work on human anaclitic depression, see Pinneau (1955), Goldfarb (1945), Heinicke (1956), and Lowrey (1940). On privation, see Gewirtz (1959, 1967).

On nonhuman primates, see Andrew (1963a, 1963b, 1963c), Chance (1962), Sackett (1967), and Van Hoff (1962). On dogs, see Fredericson (1952), Mahut (1958), Royce (1955, 1966), Solomon and Wynne (1953), and Thompson and Heron (1954a, 1954b). On chicks, see Phillips and Siegel (1966) and Guhl (1962b). On the rat, see Rogers and Richter (1948), Small (1899), Weininger (1956), and Weininger, McClelland, and Arima (1954).

The idea that emotionality is first displayed as gross activity and then becomes differentiated is not novel, although I do not find that anyone else has analyzed the idea or attempted to see if it fits phylogenetic data as I have done. Bridges' (1932) paper certainly pointed out the possibility of differentiation, but I believe that she thought the process was primarily maturational. This assumption makes the processes too "innate" to be useful. Comments of other authors have often made me think that their interpretations of their data, or their theoretical outlook, were not inconsistent with my framework. See, for example, the papers by D. L. King (1966), Salzen (1967), and Bousfield and Orbison (1952). Salzen (1967) specifically compares the behavior of nonhuman primates and domestic fowl in regard to early experience; his discussion, however, is in terms of imprinting phenomena. The observations of Mowbray and Caddell (1962) on development of the neonate rhesus are important, for they represent one of the few attempts to quantify such data ontogenetically. Bronson (1968a, 1968b) has examined the ontogeny of fear, both in human beings, and in other animals. He, too, suggests that some characteristics of emotional responses are unlearned, but clearly influenced by ontogenetic factors.

References

Ader, R., & Belfer, M. L. Prenatal maternal anxiety and offspring emotionality in the rat. *Psychological Reports,* 1962, **10,** 711-718.

Ader, R., & Conklin, R. M. Handling of pregnant rats: Effects on emotionality of their offspring. *Science,* 1963, **142,** 411-412.

Ader, R., Friedman, S. B., & Grota, L. J. 'Emotionality' and adrenal cortical function: Effects of strain, test, and the 24-hour corticosterone rhythm. *Animal Behaviour,* 1967, **15,** 37-44.

Ader, R., & Plaut, S. M. Prenatal maternal stimulation and offspring behavior and susceptibility to gastric erosions. *Psychosomatic Medicine,* 1967, **29,** 541.

Adolph, E. F. *Physiological regulation.* Lancaster, Pa.: Cattell, 1943.

Allee, W. C. *The social life of animals.* London: Heinemann, 1938. (Rev. ed.: *Cooperation among animals with human implications.* New York, Abelard-Schuman, 1939). (Reprinted: *The social life of animals.* Boston, Beacon Press, 1951)

Allee, W. C., & Collias, N. E. Effects of injections of epinephrine on the social order in small flocks of hens. *Anatomical Record,* 1938, **72,** 119. (Suppl.)

Anderson, E. E. The interrelationship of drives in the male albino rat. III. Interrelations among measures of emotional, sexual, and exploratory behavior. *Journal of Genetic Psychology,* 1938, **53,** 335-352.

Andrew, R. J. Evolution of facial expression. (Papio hamadryas, P. theropithecus.) *Science,* 1963, **142,** 1034-1041. (a)

Andrew, R. J. The origin and evolution of the calls and facial expressions of the primates. (Mandrillus sphinx, Papio hamadryas, P. mandrillus, P. papio, P. porcaricis, P. sphinx.) *Behaviour,* 1963, **20,** 1-109. (b)

Andrew, R. J. Trends apparent in the evolution of vocalization in the old world monkeys and apes. (Papio doguera, P. hamadryas, P. papio, P. porcaricis, Theropithecus gelada.) *Symposia of the Zoological Society of London,* 1963, **10,** 89-101. (c)

Arnold, M. B. *Emotion and personality.* Vol. I. *Psychological aspects.* Vol. II. *Neurological and physiological aspects.* New York: Columbia Univ. Press, 1960.

Arnold, M. B. (Ed.) *The nature of emotion.* Baltimore, Md.: Penguin, 1968.

Aronson, L. R. Reproduction and parental behavior. In Margaret E. Brown (Ed.), *The physiology of fishes.* Vol. 2. *Behavior.* New York: Academic Press, 1957. Pp. 271-304.

Bain, A. *The emotions and the will.* London: Parker, 1859.

Baker, L. M. A study of the relationship between changes in breathing and pulse rate, and the amount learned following supposed emotional and supposed non-emotional stimuli. *Journal of Genetic Psychology,* 1934, **11,** 348-368.

Barcroft, J. Some effects of emotion on the volume of the spleen. *Journal of Physiology (London),* 1930, **68,** 375-382.

Bard, P. On emotional expression after decortication with some remarks on certain theoretical views. Part II. *Psychological Review,* 1934, **41,** 429-449.

Bard, P. Neural mechanisms in emotional and sexual behavior. *Psychosomatic Medicine,* 1942, **4,** 171-172.

Baron, A., & Kish, G. B. Early social isolation as a determinant of aggregative behavior in the domestic chicken. *Journal of Comparative and Physiological Psychology,* 1960, **53,** 459-463.

Baron, A., Kish, G. B., & Antonitis, J. J. Effects of early and late social isolation on aggregative behavior in the domestic chicken. *Journal of Genetic Psychology,* 1962, **100,** 355-360.

Bateson, P. P. G. The characteristics and context of imprinting. *Biological Review,* 1966, **41,** 177-220.

Bayley, N. *Studies in the development of young children.* Berkeley: Univ. of California Press, 1940.

Beilharz, R. G., & Cox, D. F. Genetic analysis of open field behavior in swine. *Journal of Animal Science,* 1967, **26,** 988-990.

Bekhterev, V. M. Emotions as somanto-mimetic reflexes. *In* C. Murchison (Ed.), *Feelings and emotions: The Wittenberg symposium.* Worcester, Mass.: Clark Univ. Press, 1928. Pp. 270-283.

Bell, C. *Anatomy and philosophy of expression.* (1st ed.) London: Longmans Green, 1806. (N. S.)

Bell, C. *Anatomy and philosophy of expression.* (3rd ed.) London, Longmans Green, 1844. (N. S.)

Bernstein, L. The effects of variations in handling upon learning and retention. *Journal of Comparative and Physiological Psychology,* 1957, **50,** 162-167.

Bernstein, S., & Mason, W. A. The effects of age and stimulus conditions on the emotional responses of rhesus monkeys: Responses to complex stimuli. *Journal of Genetic Psychology,* 1962, **101,** 279-298.

Biel, W. C., & O'Kelly, L. I. The effect of cortical lesions on emotional and regressive behavior in the rat. I. Emotional behavior. *Journal of Comparative Psychology,* 1940, **30,** 221-240.

Billingslea, F. Y. The relationship between emotionality and various other salients of behavior in the rat. *Journal of Comparative Psychology,* 1941, **31,** 69-77.

Billingslea, F. Y. Intercorrelational analysis of certain behavior salients in the rat. *Journal of Comparative Psychology,* 1942, **34,** 203-211.

Bindra, D. Organization in emotional and motivated behavior. *Canadian Journal of Psychology,* 1955, **9,** 161-167.

Bindra, D., & Thompson, W. R. An evaluation of defecation and urination as measures of fearfulness. *Journal of Comparative and Physiological Psychology, 1953,* **46,** 43-45.

Bitterman, M. E. Phyletic differences in learning. *American Psychologist,* 1965, **20,** 396-410.

Black, J. D., & Bridger, W. H. The law of initial value in psychophysiology: A reformulation in terms of experimental and theoretical considerations. *Annals of the New York Academy of Sciences,* 1962, **98,** 1229-1241.

Blomstrand, R., & Löfgren, F. Influence of emotional stress on the renal circulation. *Psychosomatic Medicine,* 1956, **18,** 420-426.

Bolles, R. C., & Woods, P. J. The ontogeny of behavior in the albino rat. *Animal Behaviour,* 1964, **12,** 427-441.

Bousfield, W. A., & Orbison, W. D. Ontogenesis of emotional behavior. *Psychological Review,* 1952, **59,** 1-7.

Brady, J. V. Assessment of drug effects on emotional behavior. *Science,* 1956, **123,** 1033-1034.

Brady, J. V. Motivational-emotional factors and intracranial self-stimulation. In D. E. Sheer (Ed.), *Electrical stimulation of the brain.* Austin: Univ. of Texas Press, 1961. Pp. 413-430.

Brady, J. V. Experimental studies of psychophysiological responses to stressful situations. In *Symposium on medical aspects of stress in the military climate.* Washington, D. C.: Walter Reed Army Medical Center, 1964. Pp. 271-289.

Brady, J. V., & Conrad, D. G. Some effects of limbic system: Self-stimulation upon conditioned emotional behavior. *Journal of Comparative and Physiological Psychology,* 1960, **53,** 128-137.

Brady, J. V., & Nauta, W. J. H. Subcortical mechanisms in emotional behavior: The duration of affective changes following septal and habenular lesions in the albino rat. *Journal of Comparative and Physiological Psychology, 1955,* **48,** 412-420.

Brener, J. Heart rate. In P. H. Venables and I. Martin (Eds.), *Manual of psychophysiological methods.* New York: Wiley, 1967. Pp. 103-131.

Bridger, W. H. Individual differences in behavior and autonomic activity in newborn infants. *American Journal of Public Health,* 1965, **55,** 1899-1901.

Bridger, W. H., & Reese, M. F. Psychophysiological studies of the neonate: An approach toward the methodological and theoretical problems involved. *Psychosomatic Medicine,* 1959, **21,** 265-276.

Bridges, K. M. B. Emotional development in early infancy. *Child Development,* 1932, **3,** 324-341.

Britton, S. W. Neural and hormonal factors in bodily activity. The prepotency of medulliadrenal influence in emotional hyperglycemia. *American Journal of Physiology,* 1928, **86,** 340-352.

Broadhurst, P. L. Determinants of emotionality in the rat. I. Situational factors. *British Journal of Psychology,* 1957, **48,** 1-12. (a)

Broadhurst, P. L. Emotionality and the Yerkes-Dodson law. *Journal of Experimental Psychology,* 1957, **54,** 345-352. (b)

Broadhurst, P. L. Determinants of emotionality in the rat. II. Antecedent factors. *British Journal of Psychology,* 1958, **49,** 12-20.

Broadhurst, P. L. Analysis of maternal effects in the inheritance of behavior. *Animal Behaviour,* 1961, **9,** 129-141.

Broadhurst, P. L., Sinha, S. N., & Singh, S. D. The effect of stimulant and depressant drugs on a measure of emotional reactivity in the rat. *Journal of Genetic Psychology,* 1959, **95,** 217-226.

Bronson, G. W. The development of fear in man and other animals. *Child Development,* 1968, **39,** 409-431. (a)

Bronson, G. W. The fear of novelty. *Psychological Bulletin,* 1968, **69,** 350-358. (b)

Bruell, J. Genetics and adaptive significance of emotional defecation in mice. *Annals of the New York Academy of Sciences,* 1969, **159,** 825-830.

Caceres, C. A. (Ed.) *Biomedical telemetry.* New York: Academic Press, 1965.

Cairns, R. B. Attachment behavior of mammals. *Psychological Review,* 1966, **73,** 409-426.

Cairns, R. B., & Werboff, J. Behavior development in the dog: An interspecific analysis. *Science,* 1967, **158,** 1070-1072.

Caldwell, L. S. The mononuclear count as an index of emotionality of the rat. *Dissertation Abstracts,* 1960, **20,** 3852.

Calhoun, J. B. Population density and social pathology. *Scientific American,* 1962, **206,** 139-146. (a)

Calhoun, J. B. A "behavioral sink." In E. L. Bliss (Ed.), *Roots of behavior.* New York: Harper, 1962. Pp. 295-315. (b)

Campbell, B. A. Developmental studies of learning and motivation in infraprimate mammals. In H. W. Stevenson, E. H. Hess, and H. L. Rheingold (Eds.), *Early behavior: Comparative and developmental approaches.* New York: Wiley, 1967. Pp. 43-71.

Candland, D. K., & Campbell, B. A. Development of fear in the rat as measured by behavior in the open field. *Journal of Comparative and Physiological Psychology,* 1962, **55,** 593-596.

Candland, D. K., Culbertson, J. L., & Moyer, R. S. Parameters affecting adaptation to and retention of open-field elimination in the rat. *Animal Behaviour,* 1968, **132,** 46-51.

Candland, D. K., Faulds, B., Thomas, D. B., & Candland, M. H. The reinforcing value of gentling. *Journal of Comparative and Physiological Psychology,* 1960, **53,** 55-58.

Candland, D. K., Horowitz, S. H., & Culbertson, J. L. Acquisition and retention of acquired avoidance with gentling as reinforcement. *Journal of Comparative and Physiological Psychology,* 1962, **55,** 1062-1064.

Candland, D. K., & Milne, D. W. Species differences in approach-behavior as a function of developmental environment. *Animal Behaviour,* 1966, **14,** 539-545.

Candland, D. K., & Nagy, Z. M. The open field: Some comparative data. *Annals of the New York Academy of Sciences,* 1969, **159,** 831-851.

Candland, D. K., Nagy, Z. M., & Conklyn, D. H. Emotional behavior in the domestic chicken (White leghorn) as a function of age and developmental environment. *Journal of Comparative and Physiological Psychology,* 1963, **56,** 1069-1073.

Candland, D. K., Pack, K., & Matthews, T. J. Heart rate and defecation frequency as measures of rodent emotionality. *Journal of Comparative and Physiological Psychology,* 1967, **64,** 146-150.

Candland, D. K., Taylor, D. B., Dresdale, L., Leiphart, J. L., & Solow, S. P. Heart rate and aggression in the domestic chicken (White leghorn). *Journal of Comparative and Physiological Psychology,* 1969, **67,** 70-76.

Cannon, W. B. *Bodily changes in pain, hunger, fear, and rage.* (2nd ed.) New York: Appleton, 1929.

Cannon, W. B., & Britton, S. W. Studies on the conditions of activity in endocrine glands. XX. The influence of motion and emotion on medulli-adrenal secretion. *American Journal of Physiology,* 1927, **79,** 433-465.

Cantril, H. The roles of the situation and adrenalin in the induction of emotion. *American Journal of Psychology,* 1934, **46,** 568-579.

Cantril, H., & Hunt, W. A. Emotional effects produced by the injection of adrenalin. *American Journal of Psychology,* 1932, **44,** 300-307.

Chance, M. R. A. An interpretation of some agonistic postures: The role of "cut-off" acts and postures. *Symposia of the Zoological Society of London,* 1962, **8,** 71-89.

Christian, J. J. The pathology of overpopulation. *Military Medicine*, 1963, **128**, 571-603.

Christian, J. J., & Davis, D. E. Endocrines, behavior, and population: Social and endocrine factors are integrated in the regulation of growth of mammalian populations. *Science*, 1964, **146**, 1550-1560.

Church, R. M. Emotional reactions of rats to the pain of others. *Journal of Comparative and Physiological Psychology*, 1959, **52**, 132-134.

Collias, N. E. The socialization of chicks. *Anatomical Record*, 1950, **108**, 553.

Collias, N. E. Some factors in maternal rejection of sheep and goats. *Bulletin of the Ecological Society of America*, 1953, **34**, 78.

Cornwall, A. C., & Fuller, J. L. Conditioned responses in young puppies. *Journal of Comparative and Physiological Psychology*, 1961, **54**, 13-15.

Craig, W. The expression of emotion in the pigeon. I. The blond ring dove. *Journal of Comparative Neurology*, 1909, **19**, 29-80.

Crawford, M. P. The social psychology of vertebrates. *Psychological Bulletin*, 1939, **36**, 407-446.

Darrow, C. W. Emotion as relative functional decortication: The role of conflict. *Psychological Review*, 1935, **42**, 566-578.

Darwin, C. *The expression of the emotions in man and animals*. London: Murray, 1872. (Reprinted: Chicago, Univ. of Chicago Press, 1965.)

Davis, D. E. The role of density in aggressive behaviour of house mice. *Animal Behaviour*, 1958, **6**, 207-210.

Davis, D. E. Physiological analysis of aggressive behaviour. In W. Etkin (Ed.), *Social behavior and organization among vertebrates*. Chicago, Ill.: Univ. of Chicago Press, 1964. Pp. 53-74.

Delgado, J. M. R. Aggressive behavior evoked by radio stimulation in monkey colonies. *American Zoologist*, 1966, **6**, 669-681.

Denenberg, V. H. Critical periods, stimulus input, and emotional reactivity: A theory of infantile stimulation. *Psychological Review*, 1964, **71**, 335-351.

Denenberg, V. H. The effects of early experience. In E. S. E. Hafez (Ed.), *The behaviour of domestic animals*. (2nd ed.) Baltimore, Md.: Williams & Wilkins, 1969. Pp. 95-130. (a)

Denenberg, V. H. Open-field behavior in the rat: What does it mean? *Annals of the New York Academy of Sciences*, 1969, **159**, 852-859. (b)

Denenberg, V. H., & Morton, J. R. C. Effects of environmental complexity and social groupings upon modification of emotional behavior. *Journal of Comparative and Physiological Psychology*, 1962, **55**, 242-246.

Despert, J. L. The incidence of upper respiratory infections as related to emotional reactions and adjustment. *Journal of Nervous and Mental Disease*, 1945, **101**, 600-602.

Diethelm, O., Doty, E. J., & Milhorat, A. T. Emotions and adrenergic and cholinergic changes in the blood. *Archives of Neurology and Psychiatry*, 1945, **54**, 110-115.

Doering, C. R., & Allen, M. F. Data on eruption and caries of deciduous teeth. *Child Development*, 1942, **13**, 113-129.

Doyle, G. A., & Yule, E. P. Grooming activities and freezing behavior in relation to emotionality in albino rats. *Animal Behaviour*, 1959, **7**, 12-18. (a)

Doyle, G. A., & Yule, E. P. Early experience and emotionality. I. The effects of prenatal maternal anxiety on the emotionality of albino rats. *Journal of Social Responses, Pretoria*, 1959, **10**, 57-66. (b)

Duffy, E. Tensions and emotional factors in reaction. *Genetic Psychology Monographs,* 1930, **7,** 1-79.

Duffy, E. Leeper's "Motivational theory of emotion." *Psychological Review,* 1948, **55,** 324-328.

Duffy, E. *Activation and behavior.* New York: Wiley, 1962.

Dunbar, H. F. *Emotions and bodily changes. A survey of literature on psychosomatic interrelationships, 1910-1933.* New York: Columbia Univ. Press, 1935.

Edelberg, R. Electrical properties of the skin. In C. C. Brown (Ed.), *Methods in psychophysiology.* Baltimore, Md.: Williams & Wilkins, 1967. Pp. 1-53.

Edwards, A. E., & Hill, R. A. The effect of data characteristics on theoretical conclusions concerning the physiology of emotions. *Psychosomatic Medicine,* 1967, **29,** 303-311.

Elliot, O., & Scott, J. P. The development of emotional distress reactions to separation in puppies. *Journal of Genetic Psychology,* 1961, **99,** 3-22.

Evans, J. T., & Hunt, J. McV. The "emotionality" of rats. *American Journal of Psychology,* 1942, **55,** 528-545.

Fantino, E., Kasdon, D., & Stringer, N. The Yerkes-Dodson law and alimentary motivation. *Canadian Journal of Psychology,* 1970, **24,** 77-84.

Farris, E. J., & Yeakel, E. H. Emotional behavior of gray Norway and Wistar albino rats. *Journal of Comparative and Physiological Psychology,* 1945, **38,** 109-118.

Faulkner, W. B. The effect of the emotions upon diaphragmatic function. *Psychosomatic Medicine,* 1941, **3,** 187-189.

Féré, C. Note sur les modifications de le résistance électrique sous l'influence des exatations sensorielles et des émotions. *Comptes Rendus des Seances de la Societe de Biologie,* 1888, **40,** 217-219. (N. S.)

Foley, J. P., Jr. First year development of a rhesus monkey (*Macaca mulatta*) reared in isolation. *Journal of Genetic Psychology,* 1934, **45,** 39-105.

Foley, J. P., Jr. Second year development of a rhesus monkey (*Macaca mulatta*) reared in isolation during the first eighteen months. *Journal of Genetic Psychology,* 1935, **47,** 73-97.

Forgus, R. H. Early visual and motor experience as determiners of complex maze-learning ability under rich and reduced stimulation. *Journal of Comparative and Physiological Psychology,* 1955, **48,** 215-220.

Fowler, H., Hicks, S. P., D'Amato, C. J., & Beach, F. A. Effects of fetal irradiation on behavior in the albino rat. *Journal of Comparative and Physiological Psychology,* 1962, **55,** 309-314.

Fredericson, E. Perceptual homeostasis and distress vocalization in puppies. *Journal of Personality,* 1952, **20,** 472-477.

Freedman, D. G., King, J. A., & Elliot, E. Critical period in the social development of dogs. *Science,* 1961, **133,** 1016-1017.

Freud, S. Three essays on the theory of sexuality. In J. Strachey (Ed.), *The complete works of Sigmund Freud.* Vol. VII. London: Hogarth and Institute of Psycho-analysis, 1953. Pp. 135-245.

Friedman, S. B., Mason, J. W., & Hamburg, D. A. Urinary 17-Hydroxycorticosteroid levels in parents of children with neoplastic disease: A study of chronic psychological stress. *Psychosomatic Medicine,* 1963, **25,** 364-376.

Fuller, J. L. Cross-sectional and longitudinal studies of adjustive behavior in the dog. *Annals of the New York Academy of Sciences,* 1953, **56,** 214-224.

Fuller, J. L. Effects of experiential deprivation upon behavior in animals. *Proceedings of the Institute of Psychiatric Congress, 3rd, 1961.*

Fuller, J. L., & Dubuis, E. M. The behavior of dogs. In E. S. E. Hafez (Ed.), *Behaviour of domestic animals*. Baltimore, Md.: Williams & Wilkins, 1962. Pp. 415-452.

Fuller, J. L., Easler, C., & Banks, E. M. Formation of conditioned avoidance responses in young puppies. *American Journal of Physiology*, 1950, **160**, 462-466.

Furchtgott, E., & Echols, M. Activity and emotionality in pre- and neonatally X-irradiated rats. *Journal of Comparative and Physiological Psychology*, 1958, **51**, 541-545.

Furchtgott, E., Wechkin, S., & Dees, J. W. Open-field exploration as a function of age. *Journal of Comparative and Physiological Psychology*, 1961, **54**, 386-388.

Gellhorn, E. Motion and emotion: The role of proprioception in the physiology and pathology of the emotions. *Psychological Review*, 1964, **71**, 457-472.

Gellhorn, E., Cortell, R., & Feldman, J. The autonomic basis of emotion. *Science*, 1940, **92**, 288-289.

Gesell, A. L. *An atlas of infant behavior: A systematic delineation of the forms and early growth of human behavior patterns*. Vol. I. *Normative series*. Vol. II. *Naturalistic series*. New Haven, Conn.: Yale Univ. Press, 1934.

Gewirtz, J. L. A learning analysis of the effects of normal stimulation, privation and deprivation on the acquisition of social motivation and attachment. In B. M. Foss (Ed.), *Determinants of infant behaviour*. New York: Wiley, 1959. Pp. 213-299.

Gewirtz, J. L. Deprivation and satiation of social stimuli as determinants of their reinforcing efficacy. In J. P. Hill (Ed.), *Minnesota symposia on child psychology*. Vol. I. Minneapolis: Univ. of Minnesota Press, 1967. Pp. 3-56.

Gibson, J. G. Emotions and the thyroid gland: A critical study. *Journal of Psychosomatic Research*, 1962, **6**, 93-116.

Gildea, E. F., Mailhouse, V. L., & Morris, D. P. The relationship between various emotional disturbances and the sugar content of the blood. *American Journal of Psychiatry*, 1935, **92**, 115-130.

Goldfarb, W. Psychological deprivation in infancy and subsequent stimulation. *American Journal of Psychiatry*, 1945, **102**, 18-33.

Goodenough, F. L. The expression of the emotions in infancy. *Child Development*, 1931, **2**, 96-101.

Goodenough, F. L. Expression of the emotions in a blind-deaf child. *Journal of Abnormal and Social Psychology*, 1932, **27**, 328-333.

Gray, J. A., Levine, S., & Broadhurst, P. L. Gonadal hormone injections in infancy and adult emotional behavior. *Animal Behaviour*, 1965, **13**, 33-45.

Gray, P. H. A checklist of papers since 1951 dealing with imprinting in birds. *Psychological Record*, 1963, **13**, 445-454.

Greenbaum, M., & Gunberg, D. L. The effect of neonatal hyperoxia on sexual arousal and emotionality in the male rat. *Animal Behaviour*, 1962, **10**, 28-33.

Grota, L. J., Denenberg, V. H., & Zarrow, M. X. Normal versus Caesarian delivery: Effects upon survival probability, weaning weight, and open field activity. *Journal of Comparative and Physiological Psychology*, 1966, **61**, 159-160.

Guhl, A. M. The development of social organization in the domestic chick. *Animal Behaviour*, 1958, **6**, 92-111.

Guhl, A. M. The social environment and behaviour. In E. S. E. Hafez (Ed.), *The behaviour of domestic animals*. Baltimore, Md.: Williams & Wilkins, 1962. Pp. 96-108. (a)

Guhl, A. M. The behaviour of chickens. In E. S. E. Hafez (Ed.), *The behaviour of domestic animals*. Baltimore, Md.: Williams & Wilkins, 1962. Pp. 491-530. (b)

Hall, C. S. Emotional behavior in the rat. I. Defecation and urination as measures of individual differences in emotionality. *Journal of Comparative Psychology,* 1934, **18,** 385-403.

Hall, C. S. Emotional behavior in the rat. II. The relationship between need and emotionality. *Journal of Comparative Psychology,* 1936, **22,** 61-68. (a)

Hall, C. S. Emotional behavior in the rat. III. The relationship between emotionality and ambulatory activity. *Journal of Comparative Psychology,* 1936, **22,** 345-352. (b)

Hall, C. S. Emotional behavior in the rat. IV. The relationship between emotionality and stereotyping of behavior. *Journal of Comparative Psychology,* 1937, **24,** 369-375.

Hall, C. S., & Klein, L. L. Emotional behavior in the rat. VI. The relationship between persistence and emotionality. *Journal of Comparative Psychology,* 1941, **32,** 503-506.

Halliday, M. S. Exploration and fear in the rat. *Symposia of the Zoological Society of London,* 1966, **18,** 45-49.

Harlow, H. F. Primary affectional patterns in primates. *American Journal of Orthopsychiatry,* 1960, **30,** 676-684.

Harris, R. E., & Ingle, D. J. The influence of destruction of the adrenal medulla on emotional hyperglycemia. *American Journal of Physiology,* 1937, **120,** 420-422.

Hartman, C. G., & Tinklepaugh, O. L. Weitere Beobachtungen über die Geburt beim Affen *Macacus rhesus. Archive fuer Gynaekologie,* 1932, **149,** 21-37. (N. S., from Foley, 1934; see also Tinklepaugh & Hartman, 1932)

Haselrud, G. M. The effect of movement of stimulus objects upon avoidance reactions in chimpanzees. *Journal of Comparative Psychology,* 1938, **25,** 507-528.

Havlena, J., & Werboff, J. Postnatal effects of control fluids administered to gravid rats. *Psychological Reports,* 1963, **12,** 127-131.

Hebb, D. O. The mammal and his environment. *American Journal of Psychiatry,* 1955, **111,** 826-831. (a)

Hebb, D. O. Drives and the C. N. S. (Conceptual nervous system). *Psychological Review,* 1955, **62,** 243-254. (b)

Hediger, H. *Wildtiere in Gefangenschaft. Ein Grundriss der Tiergartenbiologie.* Basel: Schwabe, 1942. (*Wild animals in captivity.* London: Butterworth, 1950.)

Heinicke, C. M. Some effects of separating 2-year old children from their parents; a comparative study. *Human Relations,* 1956, **9,** 105-176.

Henderson, N. D. A species difference in conditioned emotional response. *Psychological Reports,* 1964, **15,** 579-585.

Henderson, N. D. Effects of intensity and spacing of prior stimulation on later emotional behavior. *Journal of Comparative and Physiological Psychology,* 1966, **62,** 441-448.

Henderson, N. D. Prior treatment effects on open field emotionality: The need for representative design. *Annals of the New York Academy of Sciences,* 1969, **159,** 860-868.

Hersha, L., Moore, A. U., & Richmond, J. B. Effects of post-partum separation of mother and kid on maternal care in the domestic goat. *Science,* 1958, **128,** 1342-1343.

Hess, E. H. Imprinting. *Science,* 1959, **130,** 133-141. (a)

Hess, E. H. The relationship between imprinting and motivation. In M. R. Jones (Ed.), *Nebraska symposium on motivation.* Lincoln: Univ. of Nebraska Press, 1959. Pp. 44-77. (b)

Higginson, G. D. The after-effects of certain emotional situations upon maze learning among white rats. *Journal of Comparative Psychology*, 1930, **10**, 1-10.

Himwich, H. E., & Fulton, J. F. The effect of emotional stress on blood fat. *American Journal of Physiology*, 1931, **97**, 533-534.

Hinde, R. A., & Rowell, T. E. Communication by postures and facial expressions in the rhesus monkey (*Macaca mulatta*). *Proceedings of the Zoological Society of London*, 1962, **138**, 1-21.

Hinde, R. A., Spencer-Booth, Y., & Bruce, M. Effects of 6-day maternal deprivation on rhesus monkey infants. *Nature* (*London*), 1965, **210**, 1021-1023.

Hockman, C. H. Prenatal maternal stress in the rat: Its effects on emotional behavior in the offspring. *Journal of Comparative and Physiological Psychology*, 1961, **54**, 679-684.

Hohmann, G. W. Some effects of spinal cord lesions on experienced emotional feelings. *Psychophysiology*, 1966, **3**, 143-156.

Hull, C. S. *A behavior system*. New Haven, Conn.: Yale Univ. Press, 1952. (Paperback, New York: Wiley, 1964.)

Hunsperger, R. W., Brown, J. L., & Rosvold, H. E. Combined stimulation in areas governing threat and flight behavior in the brain stem of the cat. In W. Bargmann & J. P. Schadé (Eds.), *Progress in brain research*. Vol. 6. *Topics in basic neurology*. Amsterdam: Elsevier, 1964.

Hunt, H. F., & Otis, L. S. Conditioned and unconditioned emotional defecation in the rat. *Journal of Comparative and Physiological Psychology*, 1953, **46**, 378-382.

Hunt, W. A. Recent developments in the field of emotion. *Psychological Bulletin*, 1941, **38**, 249-276.

Igel, G. J., & Calvin, A. D. The development of affectional responses in infant dogs. *Journal of Comparative and Physiological Psychology*, 1960, **53**, 302-305.

Jacobsen, C. F., Jacobsen, M. M., & Yoshioka, J. G. Development of an infant chimpanzee during her first year. *Comparative Psychology Monographs*, 1932, **9**, No. 1.

James, W. *The principles of psychology*. Vol. II. New York: Holt, 1890.

Jaynes, J. Imprinting: The interaction of learned and innate behavior. II. The critical period. *Journal of Comparative and Physiological Psychology*, 1957, **50**, 6-10.

Jensen, G. D., & Tolman, C. W. Mother-infant relationship in the monkey, *Macaca nemestrina:* The effect of brief separation and mother-infant specificity. *Journal of Comparative and Physiological Psychology*, 1962, **55**, 131-136.

Jewell, P. A., & Loizos, Caroline. (Eds.) *Play, exploration and territory in mammals*. New York: Academic Press, 1966.

Joffe, J. M. Emotionality and intelligence of offspring in relation to prenatal maternal conflict in albino rats. *Journal of General Psychology*, 1965, **73**, 1-11.

Joffe, J. M. Genotype and prenatal and premating stress interact to affect adult behavior in rats. *Science*, 1965, **150**, 1844-1845. (b)

Joffe, J. M. Prenatal determinants of emotionality. *Annals of the New York Academy of Sciences*, 1969, **159**, 668-680. (a)

Joffe, J. M. *Prenatal determinants of behaviour*. Oxford: Pergamon, 1969. (b)

Jones, M. C. The development of early behavior patterns in young children. *Pedagogical Seminary*, 1926, **33**, 537-585.

Jonsson, P. Statistiske undesgelser over grisenes daglige tilvaekst samt joderforbruget pr./kg tilvaekst. *Tidsisker*, 1955, **12**, 405-429. (N. S.)

Kaplan, A. R. Influence of prenatal maternal anxiety on emotionality in young rats. *Science*, 1957, **126**, 73-74.

Katz, H. L., & Nice, L. B. Changes in the chemical elements of the blood of rabbits during emotional excitement. *American Journal of Physiology,* 1934, **107,** 709-716. (a)

Katz, H. L., & Nice, L. B. The number of reticulocytes in the blood of emotionally excited rabbits. *American Journal of Physiology,* 1934, **109,** 60-61. (b)

Kaufman, I. C., & Hinde, R. A. Factors influencing distress calling in chicks with special reference to temperature changes and social isolation. *Animal Behaviour,* 1961, **9,** 197-204.

Kaufman, I. C., & Rosenblum, L. A. The reaction to separation in infant monkeys: Anaclitic depression and conservation withdrawal. *Psychosomatic Medicine,* 1967, **29,** 648-675.

King, D. L. A review and interpretation of some aspects of the infant-mother relationship in mammals and birds. *Psychological Bulletin,* 1966, **65,** 143-155.

King, J. A. Maternal behavior and behavioral development in two subspecies of *Peromyscus maniculatus. Journal of Mammalogy,* 1958, **39,** 177-190.

King, J. A., & Connon, Helen. Effects of social relationships upon mortality in C57BL/10 mice. *Physiological Zoology,* 1955, **28,** 233-239.

King, J. A., & Gurney, N. L. Effects of early social experience on adult aggressive behavior in C57BL/10 mice. *Journal of Comparative and Physiological Psychology,* 1954, **47,** 326-330.

Lacey, J. I., Kagan, J., Lacey, B. C., & Moss, H. A. Situational determinants and behavioral correlates of autonomic response patterns. In P. Knapp (Ed.), *Expression of the emotions in man.* New York: International University Press, 1964. Pp. 161-196.

Landis, C. Studies of emotional reactions. V. Severe emotional upset. *Journal of Comparative Psychology,* 1926, **6,** 221-242.

Landis, C., & Gullette, R. Studies of emotional reactions. III. Systolic blood pressure and inspiration-expiration ratios. *Journal of Comparative Psychology,* 1925, **5,** 221-224.

Landis, C., & Hunt, W. A. Adrenalin and emotion. *Psychological Review,* 1932, **39,** 467-485.

Lashley, K. S., & Watson, J. B. Notes on the development of a young monkey. *Journal of Animal Behaviour,* 1913, **3,** 114-139.

Leeper, R. W. A motivational theory of emotion to replace "emotion as disorganized response." *Psychological Review,* 1948, **55,** 5-21.

Leeper, R. W. Some needed developments in the motivation theory of emotions. In M. R. Jones (Ed.), *Nebraska symposium on motivation.* Lincoln: Univ. of Nebraska Press, 1965. Pp. 25-122.

Lewin, K. Die Entwicklung der experimentellen Willens-und Affektpsychologie und die Psychotherapie. *Archiv fuer Psychiatrie und Nervenkrankheiten,* 1928, **85,** 515-537. (N. S.)

Lorenz, K. Der Kunipar in der Umwelt des Vogels. *Journal of Ornithology,* 1935, **83,** 137-214, 289-413.

Lowrey, L. G. Personality distortion and early institutional care. *American Journal of Orthopsychiatry,* 1940, **10,** 576-585.

Lund, F. H. *Emotions of men.* New York: Whittlesey House, 1930.

Lund, F. H. *Emotions: Their psychological, physiological, and educative implications.* New York: Ronald Press, 1939.

McBride, G., King, M. G., & James, J. W. Social proximity effects on galvanic skin responses in adult humans. *Journal of Psychology,* 1965, **61,** 153-157.

McCleary, R. A., & Moore, R. Y. *Subcortical mechanisms of behavior.* New York: Basic Books, 1965.

McCosh, J. *The emotions.* New York: Scribner's, 1880.

McCulloch, T. L., & Haselrud, G. M. Affective responses of an infant chimpanzee reared in isolation from its kind. *Journal of Comparative Psychology,* 1939, **28,** 437-445.

McDougall, W. The principal instincts and the primary emotions of man. In *An introduction to social psychology.* Boston: Luce, 1926.

McGraw, M. B. Maturation of behavior. In L. C. Carmichael (Ed.), *Manual of child psychology.* New York: Wiley, 1946. Pp. 332-369.

MacLean, P. D. Psychosomatic disease and the "visceral brain"; recent developments bearing on the Papez theory of emotion. *Psychosomatic Medicine,* 1949, **11,** 338-353.

McQuiston, M. D. Effects of maternal restraint during pregnancy upon offspring behavior in rats. *Dissertation Abstracts,* 1963, **24,** 853-854.

Mahut, H. Breed differences in the dog's emotional behaviour. *Canadian Journal of Psychology,* 1958, **12,** 35-44.

Maier, N. R. F., & Schneirla, T. C. *Principles of animal psychology.* New York: McGraw-Hill, 1935. (Republished: New York, Dover, 1964.)

Marañon, G. *Problemas actuales de la doctrina de las secreciones internas.* Madrid: Ruiz Brothers, 1922.

Marañon, G. Contribution à l'étude de l'action émotive de l'adrénaline. *Revue Francaise d'Endocrinologie,* 1924, **2,** 301-325.

Martin, R. F., & Hall, C. S. Emotional behavior in the rat. V. The incidences of behavior derangements resulting from airblast stimulation in emotional and nonemotional strains of rats. *Journal of Comparative Psychology,* 1941, **32,** 191-204.

Mason, J. W. Psychoendocrine approaches in stress research. In *Symposium on medical aspects of stress in the military climate.* Washington, D. C. Walter Reed Army Medical Center, 1964. Pp. 375-417.

Mason, J. W., Brady, J. V., Polish, E., Bauers, J. A., Robinson, J. A., Rose, R. M., & Taylor, E. D. Patterns of corticosteroid and pepsinogen change related to emotional stress in the monkey. *Science,* 1961, **133,** 1596-1598.

Masserman, J. H. Is the hypothalamus a center of emotion? *Psychosomatic Medicine,* 1941, **3,** 3-25.

Mayr, E. *Animal species and evolution.* Cambridge, Mass.: Harvard Univ. Press, 1966.

Meier, G. W. Behavior of infant monkeys: Differences attributable to mode of birth. *Science,* 1964, **143,** 968-970.

Meier, G. W., & Stuart, J. L. Effects of handling on the physical and behavioral development of Siamese kittens. *Psychological Reports,* 1959, **5,** 497-501.

Melzack, R. Irrational fears in the dog. *Canadian Journal of Psychology,* 1952, **6,** 141-147.

Melzack, R. The genesis of emotional behavior: An experimental study of the dog. *Journal of Comparative and Physiological Psychology,* 1954, **47,** 166-168.

Melzack, R. The perception of pain. *Scientific American,* 1961, **204,** 41-49.

Melzack, R. Effects of early perceptual restricton on simple visual discrimination. *Science,* 1962, **137,** 978-979.

Melzack, R. Effects of early experience on behavior: Experimental and conceptual considerations. In *Psychopathology of perception.* New York: Grune & Stratton, 1965. Pp. 271-298.

Melzack, R., & Burns, S. K. Neurophysiological effects of early sensory restriction. *Experimental Neurology*, 1965, **13**, 163-175.

Melzack, R., & Scott, T. H. The effects of early experience on the response to pain. *Journal of Comparative and Physiological Psychology*, 1957, **50**, 155-161.

Melzack, R., & Thompson, W. R. Effects of early experience on social behavior. *Canadian Journal of Psychology*, 1956, **10**, 82-90.

Menzel, E. W., Jr., Davenport, R. K., Jr., & Rogers, C. M. The effects of environmental restriction upon the chimpanzee's responsiveness to objects. *Journal of Comparative and Physiological Psychology*, 1963, **56**, 78-85. (a)

Menzel, E. W., Jr., Davenport, R. K., Jr., & Rogers, C. M. Effects of environmental restriction upon the chimpanzee's responsiveness in novel situations. *Journal of Comparative and Physiological Psychology*, 1963, **56**, 329-334. (b)

Miller, N. E., & DiCara, L. Instrumental learning of heart rate changes in curarized rats: Shaping and specificity to discriminative stimulus. *Journal of Comparative and Physiological Psychology*, 1967, **63**, 12-19.

Miller, R. E. Experimental approaches to the physiological and behavioral concomitants of affective communication in rhesus monkeys. In S. A. Altmann (Ed.), *Social communication among primates*. Chicago, Ill.: Univ. of Chicago Press, 1967. Pp. 125-134.

Miller, R. E., Caul, W. F., & Mirsky, I. A. Communication of affects between feral and socially isolated monkeys. *Journal of Personality and Social Psychology*, 1967, **7**, 231-239.

Miller, R. E., Murphy, J. V., & Mirsky, I. A. Relevance of facial expression and posture as cues in communication of affect between monkeys. *AMA Archives of General Psychiatry*, 1959, **1**, 480-488. (a)

Miller, R. E., Murphy, J. V., & Mirsky, I. A. Nonverbal communication of affect. *Journal of Clinical Psychology*, 1959, **15**, 155-158. (b)

Mittelman, B., Wolff, H. G., & Scharf, M. P. Emotions and gastroduodenal function: Experimental studies on patients with gastritis, duodenitis, and peptic ulcer. *Psychosomatic Medicine*, 1942, **4**, 5-61.

Moltz, H. Imprinting: Empirical basis and theoretical significance. *Psychological Bulletin*, 1960, **57**, 291-314.

Moltz, H., Robbins, D., & Parks, M. Caesarean delivery and maternal behavior of primiparous and multiparous rats. *Journal of Comparative and Physiological Psychology*, 1966, **61**, 455-460.

Morra, M. Prenatal sound stimulation on postnatal rat offspring open field behaviors. *Psychological Record*, 1965, **15**, 571-575. (a)

Morra, M. Level of maternal stress during two pregnancy periods on rat offspring behaviors. *Psychonomic Science*, 1965, **3**, 7-8. (b)

Morris, D. P. Blood pressure and pulse changes in normal individuals under emotional stress; their relationship to emotional instability. *Psychosomatic Medicine*, 1941, **3**, 389-398.

Mowbray, J. B., & Caddell, T. E. Early behavior patterns in rhesus monkeys. *Journal of Comparative and Physiological Psychology*, 1962, **55**, 350-357.

Murchison, C. (Ed.) *Feelings and emotions. The Wittenberg symposium*. Worcester, Mass.: Clark Univ. Press, 1928.

Murphy, J. P., & Gellhorn, E. Further investigations on diencephalic-cortical relations and their significance for the problem of emotion. *Journal of Neurophysiology*, 1945, **8**, 431-447.

Nagy, Z. M. Effect of early environment upon later social preferences in two species of mice. *Journal of Comparative and Physiological Psychology*, 1965, **60**, 98-101.

Nice, L. B., & Fishman, D. Changes in the viscosity of the blood in normal, splenectomized and adrenalectomized animals following emotional excitement. *American Journal of Physiology*, 1934, **107**, 113-119.

Nice, L. B., & Fishman, D. The specific gravity of the blood of pigeons in the quiet state and during emotional excitement. *American Journal of Physiology*, 1936, **117**, 111-112.

Nice, L. B., & Katz, H. L. The distribution of white blood cells in the peripheral circulation of emotionally excited rabbits. *American Journal of Physiology*, 1934, **109**, 80-81.

Nice, L. B., & Katz, H. L. The specific gravity of the blood of normal rabbits and cats and splenectomized rabbits before, during, and after emotional excitement. *American Journal of Physiology*, 1935, **113**, 205-208.

Nice, L. B., Katz, H. L., Fishman, D., & Friedman, D. L. The non-filament and filament neutrophil count during emotional excitement. *American Journal of Physiology*, 1935, **113**, 102.

Nice, L. B., & Simons, A. H. The specific gravity of the blood of emotionally excited rats. *American Journal of Physiology*, 1931, **97**, 548.

O'Kelly, L. I. The validity of defecation as a measure of emotionality in the rat. *Journal of General Psychology*, 1940, **23**, 75-87.

Olds, M. E., & Olds, J. Emotional and associative mechanisms in the rat brain. *Journal of Comparative and Physiological Psychology*, 1961, **54**, 120-126.

Ottinger, D. R. Some effects of maternal inconsistency and emotionality level upon offspring behavior and development. *Dissertation Abstracts*, 1961, **22**, 324-325.

Ottinger, D. R., Denenberg, V. H., & Stephens, M. W. Maternal emotionality, multiple mothering, and emotionality in maturity. *Journal of Comparative and Physiological Psychology*, 1963, **56**, 313-317.

Papez, J. W. A proposed mechanism of emotion. *Archives of Neurology and Psychiatry*, 1937, **38**, 725-743.

Parker, M. M. The interrelationship of six different situations in the measurement of emotionality in the adult albino rat. *Psychological Bulletin*, 1939, **36**, 564-565.

Patrick, J. R. The effect of emotional stimuli on the activity level of the white rat. *Journal of Comparative Psychology*, 1931, **12**, 357-364.

Patterson, R. M., Craig, J. B., Waggoner, R. W., & Feyberg, R. Studies of the relationship between emotional factors and rheumatoid arthritis. *American Journal of Psychiatry*, 1943, **99**, 775-780.

Phillips, R. E., & Siegel, P. B. Development of fear in chicks of two closely related genetic lines. *Animal Behaviour*, 1966, **14**, 84-88.

Pinneau, S. R. The infantile disorders of hospitalism and anaclitic depression. *Psychological Bulletin*, 1955, **52**, 429-452. (includes reply and rejoinder)

Plutchik, R. Psychophysiology of individual differences with special references to emotions. *Annals of the New York Academy of Sciences*, 1966, **134**, 776-781.

Plutchik, R., McFarland, W. L., & Robinson, B. W. Relationships between current intensity, self-stimulation rates, escape latencies, and evoked behavior in rhesus monkeys. *Journal of Comparative and Physiological Psychology*, 1966, **61**, 181-188.

Pressey, S. L., & Pressey, L. C. A comparative study of the emotional attitudes and interests of Indian and white children. *Journal of Applied Psychology*, 1933, **17**, 227-238. (a)

Pressey, S. L., & Pressey, L. C. A study of the emotional attitudes of Indians possessing different degrees of Indian blood. *Journal of Applied Psychology*, 1933, **17**, 410-416. (b)

Pressey, S. L., & Pressey, L. C. A comparison of the emotional development of Indians belonging to different tribes. *Journal of Applied Psychology*, 1933, **17**, 535-541. (c)

Raines, L. Emotion: A classified bibliography. *Bulletin of Bibliography*, 1929, **13**, 153-155. (a)

Raines, L. Emotion: A classified bibliography. *Bulletin of Bibliography*, 1929, **13**, 180-182. (b)

Raines, L. Emotion: A classified bibliography. *Bulletin of Bibliography*, 1930, **14**, 9-11. (a)

Raines, L. Emotion: A classified bibliography. Part IV. *Bulletin of Bibliography*, 1930, **14**, 53. (b)

Raines, L. Emotion: A classified bibliography. Part V. *Bulletin of Bibliography*, 1931, **14**, 82-83. (a)

Raines, L. Emotion: A classified bibliography. Part VI. *Bulletin of Bibliography*, 1931, **14**, 103-108. (b)

Randolph, Mary C., & Mason, W. A. Effects of rearing conditions on distress vocalizations in chimpanzees. *Folia Primatologica*, 1969, **10**, 103-112.

Ratner, S. C., & Thompson, R. W. Immobility reactions (fear) of domestic fowl as a function of age and prior experience. *Animal Behaviour*, 1960, **8**, 186-191.

Ray, W. S. A study of the emotions of children with particular reference to circulatory and respiratory changes. *Journal of Genetic Psychology*, 1932, **40**, 100-117.

Razran, G. The observable unconscious and the inferable conscious in current Soviet psychophysiology: Interoceptive conditioning, semantic conditioning, and the orienting reflex. *Psychological Review*, 1961, **68**, 81-147.

Ressler, R. H. Inherited environmental influences on the operant behavior of mice. *Journal of Comparative and Physiological Psychology*, 1966, **61**, 264-267.

Reymert, M. L. (Ed.) *Feelings and emotions: The Mooseheart symposium*. New York: McGraw-Hill, 1950.

Ribot, T. A. *The psychology of emotions*. New York: Scribner's, 1903.

Roberts, W. W. Learning and motivation in the immature rat. *American Journal of Psychology*, 1966, **79**, 3-23.

Rogers, P. V., & Richter, C. P. Anatomical comparison between the adrenal glands of wild Norway, wild Alexandrine, and domestic Norway rats. *Endocrinology*, 1948, **42**, 46-55.

Rogoff, J. M. A critique on the theory of emergency function of the adrenal glands: Implications for psychology. *Journal of General Psychology*, 1945, **32**, 249-268.

Rosenblatt, J. S. and Schneirla, T. C. The behaviour of cats. In E. S. E. Hafez (Ed.), The behaviour of domestic animals. Baltimore: Williams and Wilkins, 1962. Pp. 453-488.

Ross, S. Some observations on the lair dwelling behavior of dogs. *Behaviour*, 1950, **2**, 144-162.

Royce, J. R. A factorial study of emotionality in the dog. *Psychological Monographs*, 1955, **69**, No. 407.

Royce, J. R. Concepts generated in comparative and physiological psychological observations. In R. B. Cattell (Ed.), *Handbook of multivariate experimental psychology*. Chicago, Ill.: Rand McNally, 1966. Pp. 642-683.

Ruckmick, C. A. McCosh on the emotions. *American Journal of Psychology*, 1934, **46**, 506-508.

Ruckmick, C. A. *The psychology of feeling and emotions*. New York: McGraw-Hill, 1936.

Sackett, G. P. Some persistent effects of different rearing conditions on pre-adult social behavior of monkeys. *Journal of Comparative and Physiological Psychology,* 1967, **64,** 363-365.

Salzen, E. A. Imprinting in birds and primates. *Behaviour,* 1967, **28,** 232-254.

Schlosberg, H. Three dimensions of emotion. *Psychological Review,* 1954, **61,** 81-88.

Schneirla, T. C. Problems in the biopsychology of social organization. *Journal of Abnormal and Social Psychology,* 1946, **41,** 385-402.

Schneirla, T. C. An evolutionary and developmental theory of biphasic processes underlying approach and withdrawal. In M. R. Jones (Ed.), *Nebraska symposium on motivation.* Lincoln: Univ. of Nebraska Press, 1951. Pp. 1-42.

Schneirla, T. C. Aspects of stimulation and organization in approach/withdrawal processes underlying vertebrate behavioral development. In D. S. Lehrman, R. A. Hinde, and Evelyn Shaw (Eds.), *Advances in the study of behavior.* Vol. 1. New York: Academic Press, 1965. Pp. 1-74.

Scott, J. P. Systolic blood pressure fluctuations with sex, anger, and fear. *Journal of Comparative Psychology,* 1932, **10,** 97-114.

Scott, J. P. Critical periods in behavioral development. *Science,* 1962, **138,** 949-958.

Scott, J. P. The process of primary socialization in canine and human infants. *Monographs of the Society for Research in Child Development,* 1963, **28,** No. 1, 1-47.

Scott, J. P., Fredericson, E., & Fuller, J. L. Experimental exploration of the critical period hypothesis. *Personality,* 1951, **1,** 162-183.

Scott, J. P., & Fuller, J. L. *Genetics and the social behavior of the dog. Chicago, Ill.: Univ. of Chicago Press, 1965.*

Scott, J. P., & Marston, M. V. Critical periods affecting the development of normal and mal-adjustive social behavior of puppies. *Journal of Genetic Psychology,* 1950, **77,** 25-60.

Seay, B., Hansen, E. W., & Harlow, H. F. Mother-infant separation in monkeys. *Journal of Child Psychology and Psychiatry,* 1962, **3,** 123-132.

Seay, B., & Harlow, H. F. Maternal separation in the rhesus monkey. *Journal of Nervous and Mental Disease,* 1965, **140,** 434-441.

Seitz, P. F. D. The effects of infantile experiences upon adult behavior in animal subjects. I. Effects of litter size during infancy upon adult behavior in the rat. *American Journal of Psychiatry,* 1954, **110,** 916-927.

Seitz, P. F. D. Infantile experience and adult behavior in animal subjects. II. Age of separation from the mother and adult behavior in the cat. *Psychosomatic Medicine,* 1959, **21,** 353-378.

Sherman, M. The differentiation of emotional responses in infants. I. Judgments of emotional responses from motion picture views and from actual observation. *Journal of Comparative Psychology,* 1927, **7,** 265-284.

Sherman, M., & Sherman, I. C. *The process of human behavior.* New York: Norton, 1929.

Sierra, A. M. Estudio psicólógico acerca de la emoción experimental. *Revista de Criminologia, Psiguiatria y Medicina Legal,* 1921, **8,** 445-461.

Skinner, B. F. *The behavior of organisms: An experimental analysis.* New York: Appleton, 1938.

Small, W. S. Notes on the psychic development of the young white rat. *American Journal of Psychology,* 1899, **11,** 80-100.

Smith, G. P., Boren, J. J., & McHugh, P. R. The gastric secretory response to acute environmental stress. In *Symposium on medical aspects of stress in the military climate.* Washington, D. C.: Walter Reed Army Medical Center, 1964. Pp. 353-365.

Smith, W. I. Social "learning" in domestic chicks. *Behaviour,* 1957, **11,** 40-55.

Smith, W. I., & Ross, S. The social behavior of vertebrates: A review of the literature (1939-1950). *Psychological Bulletin,* 1952, **49,** 598-627.

Solomon, R. L., & Wynne, L. C. Traumatic avoidance learning: Acquisition in normal dogs. *Psychological Monographs,* 1953, **67,** (4, Whole No. 354).

Southwick, C. H. An experimental study of intragroup agonistic behavior in rhesus monkeys (*Macaca mulatta*). *Behaviour,* 1967, **28,** 182-209.

Spiegel, E. A., Wyris, H. T., Freed, H., & Orchinik, C. The central mechanism of the emotions (experiences with circumscribed thalamic lesions). *American Journal of Psychiatry,* 1951, **108,** 426-432.

Spitz, R. A. The smiling response: A contribution to the ontogenesis of social relations. *Genetic Psychology Monographs,* 1946, **34,** 57-125.

Spitz, R. A., & Wolf, K. M. Anaclitic depression; an inquiry into the genesis of psychiatric conditions in early childhood. II. *Psychoanalytic Study of the Child,* 1946, **2,** 313-342.

Stanley, W. C. The passive person as a reinforcer in isolated beagle puppies. *Psychonomic Science,* 1965, **2,** 21-22.

Stanley, W. C., & Elliot, O. Differential human handling as reinforcing events and as treatments influencing later social behavior in Basenji puppies. *Psychological Reports,* 1962, **10,** 775-788.

Steinschneider, A., & Lipton, E. L. Individual differences in autonomic responsivity: Problems of measurement. *Psychosomatic Medicine,* 1965, **27,** 446-456.

Stratton, G. M. The color red, and the anger of cattle. *Psychological Review,* 1923, **30,** 321-325.

Tarchanoff, J. Über die galvanischen erscheinungen an der Haut des Menschen bei Reizung der Sinnesorgane und bie verschiedenen Farmen der psycheschen Tätigkeit. *Pfluegers Archiv fuer die Gesamte Physiologie des Menschen und der Tiere,* 1890, **46,** 46-55.

Thompson, W. R. Influence of prenatal maternal anxiety on emotionality in young rats. *Science,* 1957, **125,** 698-699. (a)

Thompson, W. R. (Comment on Kaplan, 1957). *Science,* 1957, **126,** 74. (b)

Thompson, W. R., & Heron, W. The effects of restricting early experience on the problem-solving capacity of dogs. *Canadian Journal of Psychology,* 1954, **8,** 17-31. (a)

Thompson, W. R., & Heron, W. The effects of early restriction on activity in dogs. *Journal of Comparative and Physiological Psychology,* 1954, **47,** 77-82. (b)

Thompson, W. R., & Higgins, W. H. Emotion and organized behavior: Experimental data bearing on the Leeper-Young controversy. *Canadian Journal of Psychology,* 1958, **12,** 61-68.

Thompson, W. R., Melzack, R., & Scott, T. H. "Whirling behavior" in dogs as related to early experience. *Science,* 1956, **123,** 939.

Thorndike, E. L. Animal intelligence: An experimental study of the associative processes in animals. *Psychological Review,* 1898, **2,** 1-109. (Monogr. Suppl. No. 8)

Tinklepaugh, O. L., & Hartman, C. G. Behavior and maternal care of the newborn monkey (*Macaca mulatta-"Macaca rhesus"*). *Journal of Genetic Psychology,* 1932, **40,** 257-286.

Titchener, E. B. A historical note on the James-Lange theory of emotion. *American Journal of Psychology,* 1914, **25,** 427-447.

Tobach, Ethel (Ed.). Experimental approaches to the study of emotional behavior. *Annals of the New York Academy of Sciences,* 1969, **159,** 621-1121.

Tobach, Ethel, & Gold, P. S. Behavior of the guinea pig in the open-field situation. *Psychological Reports,* 1966, **18,** 415-425.

Totten, E. Oxygen consumption during emotional stimulation. *Comparative Psychology Monographs*, 1925, **3**, 13.
Turner, C. H. Behavior of the common roach in an open-maze. *Biological Bulletin*, 1913, **25**, 348-365.
Van Hoff, J. A. R. A. M. Facial expressions in higher primates. *Symposia of the Zoological Society of London*, 1962, **8**, 97-125.
Venables, P. H., & Martin, I. Skin resistance and skin potential. In P. H. Venables and I. Martin (Eds.), *Manual of psychophysiological methods*. New York: Wiley, 1967. Pp. 53-102.
Vierck, C. J., Jr., King, F. A., & Ferm, V. H. Effects of prenatal hypoxia upon activity and emotionality of the rat. *Psychonomic Science*, 1966, **4**, 87-88.
von Holst, E., & von St. Paul, U. Electrically controlled behavior. *Scientific American*, 1962, **206**, 50-59.
von Holst, E., & von St. Paul, U. On the functional organization of drives. *Animal Behaviour*, 1963, **11**, 1-20.
Washburn, M. F., Giang, F., Ives, M., & Pollack, M. Memory revival of emotions as a test of emotional and phlegmatic temperaments. *American Journal of Psychology*, 1925, **36**, 456-458.
Washburn, M. F., Rowley, J., & Winter, C. A further study of revived emotions as related to emotional and calm temperaments. *American Journal of Psychology*, 1926, **37**, 280-283.
Watson, J. B. *Psychology from the standpoint of a behaviorist*. (3rd ed.) Philadelphia, Pa.: Lippincott, 1929.
Watson, J. B., & Rayner R. Conditioned emotional reactions. *Journal of Experimental Psychology*, 1920, **3**, 1-14.
Webb, W. B. "A motivational theory of emotions . . ." *Psychological Review*, 1948, **55**, 329-335.
Weidman, U. Verhaltensstudien an der Stockente. *Anas platyrhynghos L.* I. Das Aktionsystem. *Zeitschrift fue Tierpsychologie*, 1956, **13**, 271-308. (N. S.)
Weininger, O. The effects of early experience on behavior and growth characteristics. *Journal of Comparative and Physiological Psychology*, 1956, **49**, 1-9.
Weininger, O., McClelland, W. J., & Arima, R. K. Gentling and weight gain in the albino rat. *Canadian Journal of Psychology*, 1954, **8**, 147-151.
Werboff, J., Anderson, A., & Haggett, B. N. Handling of pregnant mice: Gestational and postnatal behavioral effects. *Physiology & Behavior*, 1968, **3**, 35-39.
Werboff, J., Gottleib, J. S., Havlena, J., & Word, T. J. Behavioral effects of prenatal drug administration in the white rat. *Pediatrics*, 1961, **27**, 318-324.
Werboff, J., & Havlena, J. Effects of aging on open field behavior. *Psychological Reports*, 1962, **10**, 395-398.
Werboff, J., Havlena, J., & Sikov, M. R. Effects of prenatal X-radiation on activity, emotionality, and maze-learning ability in the rat. *Radiation Research*, 1962, **16**, 441-452.
Whitehorn, J. C. The blood sugar in relation to emotional reactions. *American Journal of Psychiatry*, 1934, **13**, 987-1005.
Wilson, G., Rupp, C., & Bartle, H., Jr. Emotional factors in organic disease of the central nervous system. *American Journal of Psychiatry*, 1943, **99**, 788-792.
Wolff, H. G. Emotions and gastric function. *Science*, 1943, **98**, 481-484.
Wolff, H. S. Telemetry of psychophysiological variables. In P. H. Venables and I. Martin (Eds.), *Manual of psychophysiological methods*. New York: Wiley, 1967. Pp. 521-545.

Wortis, H., & Maurer, W. S. "Sham rage" in man. *American Journal of Psychiatry,* 1942, **98,** 638-644.

Yerkes, R. M., & Dodson, J. D. The relation of strength of stimulus to rapidity of habit-formation. *Journal of Comparative Neurology and Psychology,* 1908, **18,** 459-482.

Yerkes, R. M., & Yerkes, Ada W. Nature and conditions of avoidance (fear) response in chimpanzees. *Journal of Comparative Psychology,* 1936, **21,** 53-66.

Young, P. T. *Emotion in man and animal: Its nature and relation to attitude and motive.* New York: Wiley, 1943.

Young, P. T. Emotions as disorganized response—a reply to Professor Leeper. *Psychological Review,* 1949, **56,** 184-191.

Young, P. T. *Motivation and emotion: A survey of the determinants of human and animal activity.* New York: Wiley, 1961.

Young, R. D. Drug administration to neonatal rats: Effects on later emotionality and learning. *Science,* 1964, **143,** 1055-1057. (a)

Young, R. D. Effect of prenatal drugs and neonatal stimulation on later behavior. *Journal of Comparative and Physiological Psychology,* 1964, **58,** 309-311. (b)

Young, R. D. Influence of neonatal treatment on maternal behavior: A confounding variable. *Psychonomic Science,* 1965, **3,** 295-296.

Young, R. D. Developmental psychopharmacology: A beginning. *Psychological Bulletin,* 1967, **67,** 73-86.

Yutzey, D. A., Meyer, P. M., & Meyer, D. R. Emotionality changes following septal and neocortical ablations in rats. *Journal of Comparative and Physiological Psychology,* 1964, **58,** 463-465.

CHAPTER 5

ONTOGENY OF PLAY AND EXPLORATORY BEHAVIORS: A DEFINITION OF PROBLEMS AND A SEARCH FOR NEW CONCEPTUAL SOLUTIONS

W. I. Welker

I. Introduction

"Play" and "exploration" are popular words that have been used to refer to certain varieties of vigorously exuberant and delicately attentive activities of young mammals and birds (Beach, 1945; Berlyne, 1960; Groos, 1898, 1901; Loizos, 1966; Millar, 1968; Welker, 1961).

Play and exploration have their origins early in ontogeny and may, under appropriate conditions, develop during infancy and childhood into extraordinarily intricate and interesting sequences of action. These behaviors are intrinsically fascinating and amusing, and for several reasons they are of particular interest and concern to biological scientists. First, they comprise a large part of the behavioral repertoire of young mammals, especially of humans. Second, their motivating determinants and their complex spatiotemporal patterns of organization are as yet only poorly understood. Third, they clearly promote, during infancy and childhood, the development of a complicated behavioral repertoire that is adaptable to a demanding, changing environment. Indeed, if these activities are inhibited or prevented from developing because of the occurrence of certain environmental and physiological conditions, then the behavioral repertoire may become severely distorted. Finally, play and exploration may have permitted exploitation of novel ecological, ethological, and social niches during phylogeny since they appear to have become more prominent in those living mammals with more complicated and extensive behavioral repertoires. It would seem that play and exploratory activities have been important aspects of such advances.

Despite the subjective certainty with which most of us feel that we *understand* these behaviors when we watch our playful and curious children, house pets, or laboratory subjects, there are many reasons to believe that the factual and conceptual foundations supporting such understanding are scientifically imperfect, and are in several fundamental ways inaccurate and inadequate. To begin with, definitions of play and exploration have not been generally acceptable because the definitive stimulus and response characteristics are usually left vague and inconsistent. Moreover, narrow operational or phenomenological definitions of motivational concepts have created confusion since, deriving from divergent viewpoints, they have led to disparate conceptualizations of play and exploration.

The inadequacy of current conceptions of play and exploration is emphasized when these behaviors are viewed as they emerge from the extremely limited reflex and postural repertoire of the infant into the highly complex and differentiated behavioral repertoire of the juvenile and adult. Thus, the concepts that seem to account for behavior at one age do not easily or appropriately apply to either similar, changed, or changing act sequences at later developmental stages. Very few adequate ontogenetic studies have addressed these problems.

Other impediments to parsimonious conceptualization and classification of behavioral phenomena are not only the enormous size, diversity, and complexity of the total behavioral repertoire itself, but also the intricate spatiotemporal interpolation of so-called play and exploratory actions within the context of this larger total repertoire. Moreover, the existence of wide interspecies differences further complicates the search for adequate general classificatory and explanatory concepts.

Finally, a definite gap in communication exists between the several biological disciplines that are concerned with the ontogenetic development of animal structures and functions. Data regarding the development of play and exploration, and the dependency of each upon internal motivational, physiological, maturational, neurological, and early experiential factors, clearly point to the need for explanatory concepts that are pertinent to virtually all biological disciplines. As attempts to comprehend the great variety of behavioral phenomena in mammals have grown more broadly interdisciplinary, it has become increasingly clear that comprehensive explanatory concepts must encompass those neural, muscular, skeletal, somatic, visceral, endocrine, biochemical, physiological, biophysical, and molecular phenomena that increasingly seem to interface with one another and with behavioral phenomena that we wish to explain (Schmitt, 1965, 1967).

Originally, I intended to summarize concepts related to the ontogeny of play and exploratory behaviors in mammals. I viewed the project as one of reevaluation of these and related concepts from the perspective that I took a decade ago (Welker, 1961). However, after reading some of the relevant literature since the beginning of the 1960's, it has become clear that most of the problems and questions raised in my previous review still exist. Moreover, I am more than ever convinced that some of the major concepts that have guided thinking about behavioral organization and about the motivation of play and exploration, as well as of other "classes" of behavior, now require reformulation within an interdisciplinary neurobiological context. My own confusion in attempting to read and comprehend the voluminous output of neurobehavioral biologists during the last decade has prompted me to try to reappraise critically some of the problem areas relating to the study of play and exploration.

Particular attention will be given to a review of some of the major concepts and problems which pertain to the growth and development of the full repertoire of integrated act sequences that characterize the exploring and playing young animal. I shall emphasize not only the gradual augmentation and expansion of the neonatal behavioral repertoire, but also the physiological, anatomical, and experiential factors which initiate and promote these progressive developments.

I find it virtually impossible to continue to think of play and exploration as unique behavioral categories with characteristics distinct from other phenomena within the behavioral repertoire of mammals. To be sure, I believe that there are real and important varieties of behavior to be dealt with. It is with their conceptualization and explanation that I find fault.

I do not pretend that the specific proposals and concepts that I shall attempt to introduce are either original or theoretically concise and comprehensive. Rather, they are conceived of as research strategies designed to extricate current thinking from certain conceptual "dead ends." I shall begin by reviewing briefly some of the literature pertaining to exploration and play that has accumulated during the last decade.

II. Current Status of Concepts Pertaining to Exploration and Play

In contrast to exploratory behavior, vigorous play activities of animals and humans have received scant attention during the past few years. However, even with respect to exploration, essentially no new major concepts or theories have been developed; the considerable amount of new data which has appeared has simply permitted elaboration of concepts and testing of hypotheses proposed earlier (Barnett, 1963; Berlyne, 1960, 1966; Butler, 1965; Glanzer, 1961; Hebb, 1958b; Welker, 1961). I shall briefly summarize these new data as a background for the subsequent discussion. In the next few paragraphs of this section, the italicized words and phrases correspond to the subject headings of Table I wherein are cited the relevant references to the literature.

There has been a much greater interest in stimulus variables that determine exploration than in the response characteristics of exploratory acts. The general thrust of most recent studies is that there is a powerful organismic *need for external stimulation* and, correlatively, that there are optimal intensity ranges of stimulation that are sought. Such optimal ranges are assumed to exist for stimuli within all sensory modalities and submodalities, and the attendant preference levels are believed to be dependent upon both remote and recent past experience as well as upon certain hypothetical internal states such as degree of arousal, boredom, fatigue, etc. Preferences for specific kinds of motor activity also have been documented, action-induced (feedback) sensory stimulation presumably being the dominant determinant of such activity. It is now quite clear that concepts of optimal stimulation do not have absolute meaning since stimulus preferences and aversions may vary widely from day to day and from moment to moment.

The concept of *novelty*, particularly mild novelty, is still the most commonly accepted "determinant" of "curiosity" or exploration. It has been repeatedly emphasized that novelty is not a stimulus variable but a transactional concept referring to an hypothetical centrally registered ratio of past-to-recent experience with a particular stimulus. Consequently, the measurement of degree of novelty can be expressed only in terms relative both to the duration of exposure to a stimulus and to the time elapsing since that exposure last occurred. Relative novelty clearly applies to all aspects of stimulus change in any modality. *Familiarization* (see "habituation" and "learning") is a term now often used to refer to past exposure to, or experience with, a novel stimulus.

Berlyne, in his reviews (1960, 1966) of determinants of "curiosity," has proposed the term *collative variables*, which he uses to refer to a variety of operationally and phenomenally derived concepts pertaining to an organism's multiple interests in specific aspects of the environment. These concepts include complexity, surprisingness, conflict, and uncertainty as well as novelty, all of which have overlapping definitions. There has been some agreement, despite the difficulties involved in measuring most of these variables, that they elicit positive interest or are approached only at optimal experiential levels of "impact" (Fiske, 1961), with greater than optimal "intensities" being avoided and lower levels of intensity eliciting little or no interest.

Habituation or decrement of exploratory responses with increased exposure to, or familiarization with, mildly novel stimuli has been repeatedly demonstrated, as has been *recovery* of interest to stimuli during certain periods of time intervening between successive exposures. The opposite effect, that is, increased exploration of a stimulus following prolonged exposure to it, is thought to be due to the fact that the stimulus was too strongly novel, thereby eliciting greater initial caution or aversion (see "relationships to fear" below). In those cases involving a gradual increase in exploration, habituation or adaptation to aversive stimuli has been thought to occur.

Stimulus dimensions other than those characterizing the "collative variables" are also believed to elicit exploration. The concept of

Table I
CONCEPTS PERTAINING TO EXPLORATORY BEHAVIOR

Concepts	Relevant literature
Need for stimulation	Feather, 1967; Haude & Oakley, 1967; Houston & Mednick, 1963; Sheldon, 1969.
Novelty	Acker & McReynolds, 1967; Berlyne, Koenig, & Hirota, 1966; Bindra & Claus, 1960; Bindra & Spinner, 1958; Clapp & Eichorn, 1965; Collard, 1962; Ehrlich, 1961; Faw & Nunnally, 1968; Fiske & Maddi, 1961a; Glickman & Van Laer, 1964; Gullickson, 1966; Harris, 1965; R. N. Hughes, 1968; Mendel, 1965; Menzel, 1963; Menzel, Davenport, & Rogers, 1961, 1963b; Meyers & Cantor, 1966; Saayman, Ames, & Moffett, 1964; Sheldon, 1969; Singh & Manocha, 1966; Smock & Holt, 1962.
Familiarization	Cantor & Cantor, 1966; Sheldon, 1969.
Collative variables	Berlyne, 1960, 1966; Charlesworth, 1966; Clapp & Eichorn, 1965; Heckhausen, 1964; Karmel, 1966; McCall & Kagan, 1967; Minton, 1963; Munsinger & Weir, 1967; Sackett, 1966.
Habituation	Bagshaw, Kimble, & Pribram, 1965; Bridger, 1961; Clapp & Eichorn, 1965; Collard, 1962; Fantz, 1964; Glickman & Van Laer, 1964; I. P. Howard, Craske, & Templeton, 1965; N. H. Mackworth & Otto, 1970; R. D. Odom, 1964; Walden, 1968.
Recovery	Collard, 1962.
Preferences and aversions	Clapp & Eichorn, 1965; Collard, 1962; Fantz, 1965; Hershenson, Munsinger, & Kessen, 1965; Karmel, 1966; Kavanau, 1969; Lore, Kam, & Newby, 1967; Suchman & Trabasso, 1966; Thomas, 1966.
Sequential variability	Bryden, 1967; Elkind & Weiss, 1967; Fiske, 1961; Lester, 1968a; M. E. Thompson & Thompson, 1964; Timberlake & Birch, 1967.
Varieties of exploration	Berlyne, 1960, 1966; Gibson, 1962; Halliday, 1968; Haude & Oakley, 1967; Lester, 1968b; Lore & Levowitz, 1966; Maslow, 1963.
General activity	Gross, 1968; Pereboom, 1968; Walden, 1968.
Exploratory drive	Berlyne, 1967; Fowler, 1965.
Learning and reinforcement	Angermeier & Hitt, 1964; Berlyne, 1969, 1967; Berlyne *et al.*, 1966; Bridger, 1961; Cantor & Cantor, 1966; Donahoe, 1965, 1967; Eacker, 1967; Friedlander, 1966b; Glickman & Schiff, 1967; Harlow, 1959; Hunt & Quay, 1961; Jeffrey, 1955; Kiernan, 1964; Kish, 1966; Leuba & Friedlander, 1968; Mason, 1967; Meyer, 1968; Paradowski, 1967; Rheingold, 1961, 1963; Rheingold, Stanley, & Doyle, 1964; Sheldon, 1969; Symmes, 1963; Thoman, Wetzel, & Levine, 1968.
Discrimination	Berlyne, 1960, 1967; Bindra, 1959; Bridger, 1961; Cantor & Cantor, 1966; Fantz, 1966; Gibson, 1962; Saayman *et al.*, 1964.

Arousal	Berlyne, 1960, 1967; Berlyne & Lewis, 1963; Lynn, 1966; Meyer, 1968.
Orienting responses	Bagshaw *et al.*, 1965; Beitel, 1969; A. S. Bernstein, 1969; Kavanau, 1969; Lynn, 1966; N. H. Mackworth & Otto, 1970; Sokolov, 1963; Voronim, Leontiev, Luria, Sokolov, & Vinogradora, 1965.
Boredom	Heron, 1957.
Effectance, competence, and self-actualization	Gibson, 1962; Lester, 1968b; White, 1959.
Early experience	D'Amato & Jagoda, 1962; Dawson & Hoffman, 1958; DeNelsky & Denenberg, 1967a, 1967b; Denenberg, 1964, 1967; Denenberg & Grota, 1964; Denenberg & Karas, 1960; Doty & Doty, 1967; Ehrlich, 1961; Green & Gordon, 1964; Haude & Oakley, 1967; Hebb, 1958b; Held & Hein, 1963; Hunt & Quay, 1961; J. A. King & Eleftheriou, 1959; Korner & Grobstein, 1966; Krech, Rosenzweig, & Bennett, 1965; Lore & Levowitz, 1966; Mason, 1968; Menzel, 1963, 1964; Menzel, Davenport, & Rogers, 1963a, 1963b; Meyers, 1962; Morton, 1968; Neiberg, 1964; Riesen, 1961a, 1961b, 1961c, 1966; Rosenzweig, 1966; Wells, Lowe, Sheldon, & Williams, 1969; Whittier & Littman, 1965.
Prenatal effects	Joffe, 1969.
Critical periods	Ambrose, 1961; Denenberg, 1968; Denenberg & Kline, 1964; Meyers, 1962.
Response characteristics	Barsch, 1965; Bindra, 1961; Bindra & Blond, 1958; Bindra & Spinner, 1958; D. Campbell, 1968; Cobb, Goodwin, & Sailens, 1966; A. I. Cohen, 1966; Denenberg, 1964; Glickman & Van Laer, 1964; J. A. Gray, 1965; Halliday, 1968; Hein & Held, 1967; Kessen, 1967; Korner, Chuck, & Dontchos, 1968; Lewis, Kagan, & Kalafat, 1966b; Leyhausen, 1956; Prechtl, 1958; Rheingold, 1961; Scott, 1958b, 1968b.
Comparative studies	Brookshire, & Rieser, 1967; Denenberg, 1965; Ehrlich, 1970; Glickman & Van Laer, 1964; Gray, 1965; R. N. Hughes, 1969a; Jarrard & Bunnell, 1968; Karmel, 1966; J. A. King, 1958, 1968; Singh & Manocha, 1966.
Ethological field studies	Bridger, 1962; Eibl-Eibesfeldt, 1967; Hinde, 1959b; Jay, 1965; Menzel, 1969; Murray & Brown, 1967; Prechtl, 1958.
Instinctive act sequences	Barnett, 1968; Breland & Breland, 1966; Glickman & Van Laer, 1964; Kessen, 1967; Leyhausen, 1956; Prechtl, 1958; Scott, 1958a, 1968a.

Table I (*Cont'd*)

Concepts	Relevant literature
Human studies	Berlyne, 1960; Berlyne & Frommer, 1966; Berlyne & Lewis, 1963; Birns, 1965; Bridger, 1961; D. Campbell, 1968; Charlesworth, 1966; Clapp & Eichorn, 1965; Collard, 1962, 1968; Cox & Campbell, 1968; Elkind & Weiss, 1967; Fantz, 1964, 1965; Faw & Nunnally, 1968; Friedlander, 1966a, 1966b; Gilmore, 1966; Goldberg & Lewis, 1969; Gullickson, 1966; Harris, 1965; Hershenson *et al.*, 1965; Hutt, 1966; Jeffrey, 1955; Kessen, 1967; Korner *et al.*, 1968; S. C. H. Kuo & Marshall, 1968; Leuba & Friedlander, 1968; Lewis *et al.*, 1966b; Lieberman, 1965; Little & Creaser, 1968; McCall & Kagan, 1967; N. H. Mackworth & Otto, 1970; Maw & Maw, 1962a, 1962b; Mendel, 1965; Meyers & Cantor, 1966, 1967; Mittman & Terrell, 1964; Munsinger & Weir, 1967, Murray & Brown, 1967; Peters & Penney, 1966; Prechtl, 1958; Rheingold, 1963; Rheingold *et al.*, 1964; Rubenstein, 1967; Smock & Holt, 1962; Spitz, 1951; Suchman & Trabasso, 1966; Thomas, 1966.
Imitation	Bowers & London, 1965.
Verbal behavior	Berlyne & Frommer, 1966; Berlyne & Peckham, 1966; Brown & Fraser, 1963; DiVesta, 1965.
Thinking	Berlyne, 1960, Berlyne & Peckham, 1966; Lieberman, 1965; Little & Creaser, 1968.
Attention	Fantz, 1964, 1966; Henker, Asano, & Kagan, 1968; Horn, 1965; Lewis, Kagan, Campbell, & Kalafat, 1966a; Lynn, 1966; McCall & Kagan, 1967; Moyer & Gilmor, 1955.
Social studies	Angermeier, 1962; Chamove, Harlow, & Mitchell, 1967; Denenberg & Grota, 1964; Goodrick, 1965; Hinde & Spencer-Booth, 1967b; R. N. Hughes, 1969b; Jensen & Bobbitt, 1968; Mason, 1965, 1967, 1968; Scott, 1967; Simmel, 1962; Simmel & McGee, 1966; Tolman, 1968.
Maternal-infant interactions	Blauvelt & McKenna, 1961; Cairns, 1966; Green & Gordon, 1964; Hansen, 1966; Harlow, 1961, 1963; Hinde & Spencer-Booth, 1967a, 1967b; Jay, 1965; Jensen & Bobbitt, 1965; D. L. King, 1965; J. A. King, 1958; Korner & Grobstein, 1966; Mason, 1967; Mitchell, 1968, 1969; Rosenblatt, 1965; Rosenblum & Kaufman, 1968; Rubenstein, 1967; Rumbaugh, 1965; Scott, 1968a; Scott & Bronson, 1964.
Fear, hunger, sex, dominance, punishment, stereotypy, schizophrenia	Baron, 1964; Berkson & Mason, 1964; S. Bernstein & Mason, 1962; Bolles, 1965; Bolles & de Lorge, 1962; Bronson, 1968a, 1968b; Butler, 1964; B. A. Campbell & Jaynes, 1966; Candland & Campbell, 1962; Chamove *et al.*, 1967; J. S. Cohen & Stettner, 1968; Collard, 1967, 1968; Davenport & Berkson, 1963; Denenberg, 1964, 1967; Gilmore, 1966; Glickman & Jensen, 1961; Goldberg & Lewis, 1969; Goodrick, 1966; Green & Gordon, 1964; Halliday, 1966, 1968; R. N. Hughes, 1965; Lester, 1967b, 1968c; McReynolds, 1962, 1963; Maslow, 1963; Medinnus & Love, 1965; Neuringer, 1969; Poole, 1966; Richards & Leslie, 1962; Riesen, 1961a; Rodgers & Rozin, 1966; Scott, 1957; Timberlake & Birch, 1967; Walden, 1968.

Neural mechanisms	Bagshaw *et al.*, 1965; Barbizet, 1963; Douglas, 1967; Douglas & Isaacson, 1964; Glickman & Schiff, 1967; Goddard, 1964; Hebb, 1959; Jarrard, 1968; Jarrard & Bunnell, 1968; Kamback, 1967; Kimble, Bagshaw, & Pribram, 1965; Kirkby, Stein, Kimble, & Kimble, 1967; Leaton, 1965; Lester, 1967a; Nielson, McIver, & Boswell, 1965; Roberts, Dember, & Brodwick, 1962; Rosenzweig, 1966; Rubel, 1969; Sokolov, 1960; Welker, 1964.
Biochemical and pharmacological determinants	Battig & Wanner, 1963; Carlton, 1966, 1968; Essman, 1968; Leaton, 1968; Leventhal & Killackeg, 1968.
Autonomic correlates	Bagshaw *et al.*, 1965; Ducharme, 1966; M. L. Gray & Crowell, 1968; Kimmel, 1964; Lewis, *et al.*, 1966a; Meyers & Cantor, 1967.
Brain stimulation	Christopher & Butler, 1968; Glickman & Feldman, 1961.
Electrophysiological recording	Bettinger, Davis, Meikle, Birch, Kopp, Smith, & Thompson, 1967; Hubel, Henson, Rupert, & Galambos, 1959; Komisaruk, 1970; R. F. Thompson & Shaw, 1965.
Methodological developments	Acker & McReynolds, 1967; Bindra, 1961; Bindra & Blond, 1958; Bobbitt, Gourevitch, Miller, & Jensen, 1969; Bobbitt, Jensen, & Gordon, 1964a; Bobbitt, Jensen, & Kuehn, 1964b; D. Campbell, 1968; Cane, 1961; Deutsch, 1966; Friedlander, 1966a, 1966b; J. A. Gray, 1965; Haude, Kruper, & Patton, 1966; Haude & Oakley, 1967; Henker *et al.*, 1968; K. I. Howard, 1961; Jeffrey, 1955; Jensen & Bobbitt, 1965, 1968; Kaufman & Rosenblum, 1966; Kavanau, 1969; Lockard, 1964; Lovaas, Freitag, Gold & Kassorla, 1965; Menzel, 1969; Norton, 1968a, 1968b; Odom *et al.*, 1970; Penney & McCann, 1964; Prechtl, 1965; Rheingold, 1963; Rheingold *et al.*, 1964.
General reviews and theoretical formulations	Ames & Ilg, 1964; Boring, 1964; Brody & Oppenheim, 1966; Bullock, 1965; Caspari, 1960; Fiske & Maddi, 1961b; Fuller, 1962; Groves & Thompson, 1971; Hebb, 1949, 1955, 1958a, 1958b, 1959, 1963; Z.-Y. Kuo, 1967; Lester, 1967b, 1968c; Livson, 1967; Loizos, 1966, 1967; J. F. Mackworth, 1969; Millar, 1968; Scott, 1965, 1967; Sokolov, 1963; Stein, 1966; R. F. Thompson & Spencer, 1966.
Play *vs.* exploration	Burgers, 1966; Jewell & Loizos, 1966; Millar, 1968.
Ontogenetic development and age effects	Ambrose, 1961; Berkson, 1966; Berkson & Karrer, 1968; Bolles & Woods, 1964; D. Campbell, 1968; Elkind & Weiss, 1967; Fantz, 1965; Fantz, Ordy, & Udelf, 1962; Fox, 1964, 1965, 1966; Glickman & Van Laer, 1964; Goodrick, 1965, 1966, 1967; R. N. Hughes, 1968; Jensen & Bobbitt, 1968; Leyhausen, 1956; Mirzakarimova, Stel'makh, & Troshikhin, 1958; Rosenblatt, Turkewitz, & Schneirla, 1969; Welt & Welker, 1963; C. D. Williams, Carr, & Peterson, 1966.

stimulus preference is applicable to approach reactions, and that of *stimulus aversion* to avoidance. Little interest has been devoted to determining parametrically the specific types, intensities, durations, etc., of stimuli involved in such preferences or aversions. Although learning and reinforcement (see below) can influence or establish specific preferences or aversions, unlearned or "natural" preferences and aversions are also believed to occur. However, the possible influence of learning and experience in the genesis of such preferences and aversions must be determined in every instance (Lehrman, 1953, 1962; Moltz, 1963; Schneirla, 1965).

Short-term *sequential variability* in stimulus-response transactions, a prominent aspect of both play and exploration (Glanzer, 1958, 1961; Lepley, 1954; Welker, 1961), has received little attention during the past decade.

Several *varieties of exploratory motives* are now clearly indicated. Thus, an animal may explore "purely" to learn about a novel stimulus, he may explore that stimulus in order to dispel the anxiety or fear that it elicits, or he may explore in the "service" of other "needs" that involve hunger, escape, safety, fear, sexual, etc., motives. Berlyne (1960) has described additional varieties of motives (recognizable predominantly in humans) which he associates with artistic expression, humor, search for knowledge, creativity, and certain types of thinking. Evidently, many investigators still assume that exploration or curiosity is a unitary need. Indeed, the assumption is commonly held that such a need is made manifest merely by dropping a rat into an open-field enclosure or placing it onto an elevated maze or into an enclosed maze. Relationships between exploration and *general activity* have also been discussed which indicate both the complexity of the motivation underlying locomotor activity as well as the simplistic inadequacy of locomotion as an index of exploratory behavior (Gross, 1968).

An *exploratory drive* is still postulated, and the associated concepts of *learning, reinforcement,* and *discrimination* have also been discussed in relation to exploratory activities. There has been much controversy over the roles of drive constructs in explaining certain of the phenomena of exploration mentioned above (Bolles, 1958; Fowler, 1965; Hinde, 1959a; Welker, 1961). For one reason or another, the concept of *arousal* has been given a special position in accounts of the motivation of exploration, as well as of most other behaviors, and both specific and general arousal concepts have been postulated. *Orienting responses* and orienting reflexes have also been used to refer to initial responses to sudden or novel stimuli. However, such behaviors are not unique to exploration since they occur in response to almost all types of stimuli. Concepts such as *boredom, effectance, competence,* and *self-actualization* have also been proposed as important to an understanding of certain motivational aspects of exploration and play. The fact that these terms are so phenomenologically abstract makes it difficult to reach agreement regarding their meanings.

The effects on exploration (as well as on most other types of behavior) of the enrichment and impoverishment of *early experience* have been topics of lively interest and intensive experimentation. Many such influences have been identified and studied, including *prenatal* ones. The importance of selected experiences during early *critical periods* in the development of exploration and play behaviors has also been emphasized.

Specific *response characteristics* of exploratory behaviors have been mentioned, but only when they are so obvious, as they occasionally are in *comparative studies,* that they cannot easily be overlooked. Important observations of exploratory and play behaviors have been made in several *ethological field studies* of different mammals. Such studies have emphasized not only the great variety and complexity of the stimulus and response phenomena that characterize play and exploration, but also the *instinctive* nature of many of the act sequences that such behaviors exhibit in young animals.

Experimental study of play and exploration in *human infants and children* has been a major new development and, as a consequence, *imitation, verbal behavior,* and concepts of *thinking* and *attention* are now encountered. Effects of *social experiences* on play and exploration, especially of those involving *maternal-infant interactions,* have received increased attention.

The relationships between certain features of exploratory behavior and other motivational states and conditions such as *fear, hunger, sex, dominance,* and *punishment,* behavioral *stereotypy* and *schizophrenia* have also been discussed.

Speculation has continued regarding the hypothetical *neural mechanisms* of exploratory behavior. Reticular system, hippocampus, neocortex, amygdala, basal ganglia, and hypothalamus have all been postulated as being involved in one or another general aspect of "curiosity" phenomena (e.g., response to optimal novelty, habituation, etc.). The biological validity of these "mechanisms" as they relate to exploration is in most instances questionable. Similarly, *biochemical determinants* of, *pharmacological influences* on, and *autonomic correlates* of, curiosity phenomena have been examined. Exploration elicited or influenced by *brain stimulation* has been reported, and *electrophysiological recordings* from exploring animals have revealed associated neuroelectric phenomena. The limitations of all these data as they relate to a search for neurobiological mechanisms of behavior will be discussed in later sections.

A number of *methodological developments* have occurred. The use of key-punch apparatus, by means of which a spatiotemporal record of specific or general play and exploratory sequences can be recorded, both in the field and in the laboratory, has proved to be a valuable tool. High-speed and time-sampling cinematographic methods have been used to some extent. Operant techniques have become more common in the quantification of some aspects of "stimulus-seeking" behavior. The types of apparatus and experimental design that have been used have become more numerous in a variety of ways, both with animals and humans. All these innovations have allowed examination of more subtle aspects of the several exploratory phenomena mentioned above. Although field, laboratory, and "free-field" studies have become more numerous, T and Y mazes, and open-field apparatuses are still popular for smaller animals, despite their limitations of assessment of only a particular type of exploration under restricted conditions.

As already mentioned, very little new information has appeared in the literature relative to the vigorous motor activities that I shall continue to refer to as play. A very excellent review of play has been published recently by Millar (1968). Her fine critique of classical *theories* of play (i.e., surplus energy, practice, recapitulation, hedonic), added to earlier critiques of these same theories, should suffice to dispel further discussion of such useless and unproductive conceptualizations. Loizos (1966), in an excellent short review, has also examined and rejected such theories. In addition, both authors have suggested some more fruitful and parsimonious ways of conceiving of the development of play and exploratory activities of animal and human neonates and children. Attempts have been made repeatedly to distinguish between *play and exploration,* but there are complex problems of definition and classification involved in such a distinction. These will be taken up in the next section.

In summary, an information explosion has occurred with respect to exploratory behaviors. The number and variety of conceptual issues that have been implicated in the search for biologically comprehensive explanations of these behaviors is great indeed. Yet little progress has been made in creating a conceptual framework capable of both replacing simple drive-oriented theories and encompassing the large number of act sequences that fall into the general categories of play and exploration. All too little attention has been paid to (a) accurate descriptions of the overt behavioral sequences themselves, (b) complete itemization of the great variety of those sequences generally conceded to comprise play and exploration, (c) realistic neurobiological characterization of the necessary explanatory constructs, and (d) delineation of the *ontogenetic development* of all these phenomena.

III. Definitions of Exploration and Play and Problems of Behavioral Conceptualization

A. Definitions

Before proceeding to an evaluation of the ontogenetic development of play and exploration, it is important to attempt a definition of these behaviors and to determine if they in fact are distinguishable from one another and from other types of behavior. Definitions of play and exploration, and attempts to develop conceptual schemes and theoretical models capable of explaining them, have become increasingly common during the past decade. During this period, play and exploration emerged from moderate obscurity to become of genuine concern to behavioral scientists. Although exactly what constituted play and exploration was not generally agreed upon, it gradually became clear that a great portion of a mammal's daily behavior was clearly irrelevant to the maintenance of physiological homeostasis as classical drive-reduction theories had postulated. Current interest in these apparently "nonhomeostatic" behaviors has had several contributing sources and has been the subject of several excellent reviews (Berlyne, 1960; Butler, 1965; Fowler, 1965; Harlow, 1959; Hebb, 1955, 1958a, 1958b; Loizos, 1966, 1967; Millar, 1968; Nissen, 1951). However, it should be pointed out that the kinds of behaviors identified recently as play and exploration by Millar (1968) are essentially the same as those called "plays" by Groos in his books on play published over 70 years ago (1898, 1901).

Let us begin by summarizing those criteria by which exploration and play have been defined and distinguished from other behavioral classes (such as reproductive, aggressive, fearful, maternal, ingestive, eliminative, etc.). These are: (a) The general or specific characteristics of the *stimulus goals and incentives* with respect to which particular behavioral sequences are organized (cf. Bindra, 1959). For example, preferred food objects for ingestive behavior; a receptive, appropriately postured female for male sexual behavior; an intruding, fighting, or retreating antagonist for aggressive behavior; infants suckling or moving out of the nest for maternal behavior; a familar moving object in the case of vigorous play behavior; a novel object in the case of exploratory behavior, etc. (b) The characteristics of the *terminal act sequences* coinciding with goal achievement or consumption. For example, grasping, biting, chewing and swallowing (eating) for ingestive behavior; copulatory responses for sexual behavior; intense frenetic biting and clawing for aggressive behavior; retrieving young to the nest and nursing them for maternal behavior; pouncing upon, manipulating or batting a familiar moving object for play behavior; sniffing and gently touching a novel stimulus object for exploratory behavior, etc. (c) The characteristics of *preterminal act sequences* or appetitive behaviors associated with approaching or seeking out a goal. For example, hunting, attacking, and killing for ingestive behavior; courtship displays and approach to a prospective partner for reproductive or sexual behavior; posturing, charging, and vocalizing for aggressive behavior; nest building for maternal behavior; stalking or chasing a moving object for play behavior; attentive observation of, or careful approach to a novel object for exploratory behavior, etc. (d) The character of the *internal neurobiological states* as measured directly, defined operationally, or as merely hypothesized. For example, time since last eating or drinking for ingestive behavior; presence of specific changed endocrine levels for reproductive behavior; altered endocrine levels and autonomic processes for aggressive or maternal behavior; heightened "unstressed arousal" for play behavior; heightened mildly "stressed arousal" for exploratory behavior, etc. (e) The assumed or documented *adaptive features* of the transactions of the behavior with the stimulus environment. For example, promotion and maintenance of appropriate tissue metabolism and growth for ingestive behavior; procreation for reproductive behavior; protection of self and conspecific partners for aggression; protection and sustenance of young for maternal behavior; learning and development of new, variable, and relevant problem-oriented act sequences or of perceptual cognizance of the environment for play and exploration, etc. Although influenced greatly by learning and experience, the presumed adaptive features are assumed to be genetically programmed and are thus thought to have evolved because the stimulus-responses transactions referred to have promoted selective survival of procreative individuals, and thus the species. Let us look more closely at these criteria for class inclusion as they relate to play and exploratory behaviors, respectively.

1. Exploratory Behavior

a. Stimulus Goals and Incentives. The external (environmental) or internal (bodily) receptor stimuli which elicit exploration or which constitute its goals are of various types. *Mildly* novel stimuli, for example, elicit exploratory orienting responses as do stimuli which are more complex spatiotemporally. It should again be emphasized that novelty is not a stimulus characteristic. It is, as already mentioned, a transactional concept that relates a current stimulus to previous experience with either that stimulus or with one similar to it. Certain types and intensities of stimuli may also elicit exploratory responses. Such S-R relationships are often referred to as stimulus preferences. However, since approach and orienting responses to mildly novel and preferred stimuli may also characterize ingestion, sexual, aggressive, and escape behaviors, this general criterion, by itself, is not distinctive of exploration. Indeed, it is common to speak of exploration for food, a sexual partner, etc., where the goal may not be physically present. Thus, humans as well as other animals may *seek* novelty and change. Part of this search for change may be associated with partial aversion (e.g., boredom) to physically present, but familiar, stimuli. The possible existence of a shifting mosaic of polyvalent goals and motives complicates the search for stimulus characteristics eliciting exploration.

b. General Characteristics of Terminal and Preterminal Act Sequences. Exploratory act sequences themselves are considered to consist of several types, each beginning with arousal and orienting behaviors. Associated with or following such behaviors are the slow, "cautious," "deliberate," or "attentive" approach act-sequences consisting of locomotor and postural adjustments and attitudes which gradually bring about habituation or culminate in close approach to, or selective choice of, certain specific sensory stimuli. The spatiotemporal character of many of these specific response sequences appears to be predominantly species-specific, but varies with age and may be modified by learning and early experience. Since most act sequences (e.g., cautious approach, alert orientation) that are felt to be prominent aspects of exploration may also characterize most other major behavioral categories, such general response criteria by themselves are not distinctive.

c. Internal Concomitants. Exploration of stimuli commonly occurs when animals (especially young ones) are "unstressed" by internal metabolic or endocrine factors known or assumed to be associated with ingestive, sexual, aggressive, and escape behaviors, although mild "fear" states have been proposed to characterize certain forms of exploration (Berlyne, 1960; Lester, 1968c; McReynolds, 1962). The fact that highly organized behavioral patterns can occur under conditions of "physiological" homeostasis has been a major reason behind the felt need to reevaluate classical modes of behavioral classification, definition, and explanation. Moreover, exploration of novel stimuli is occasionally seen to interrupt even eating, sexual, and aggressive behavior; in such instances, the hypothesized homeostatic imbalances presumably coexist with the hypothetical exploratory "motives." Or it may be that the several behavioral types can be differentially switched on and off regardless of physiological and metabolic states. In any event, many such facts have been adduced to justify the identification of exploration as a separate behavioral category rather than as response sequences derived from the so-called "primary drives." Nevertheless, the view that all behavior has underlying it some unitary internal motivating state has resulted in the postulation of an "exploratory" drive, a "need for novelty," or a "need for optimal stimulation" (Berlyne, 1967). In opposition, however, Fowler (1965) has argued that the hypothetical curiosity states that allegedly motivate exploratory behavior to novel stimuli can be conceived more parsimoniously as response tendencies generalized from past experience with similar stimuli under conditions of "classical" drive motivation, and as such can be accommodated within a drive-reduction framework. But these speculations regarding motivating states are logical word games and, indeed, several investigators (Bindra, 1959; Bolles, 1958; Hebb, 1949; Hinde, 1959a; Welker, 1961) have expressed strong doubt regarding the postulation of drives as a useful means of coping with problems concerning the motivation of exploration, or of any other behavior. In this context, the postulation of "arousal" states has been popular as a means of "explaining" the motivation for exploration (Berlyne, 1967), but here as well such concepts are devoid of demonstrated neurobiological validity.

"Mental state" criteria have also been used to characterize "interal motives" for exploratory behavior. For example, "boredom" has been proposed as an internal state that may prompt an animal to explore when no novel or complex stimuli are present. "Surprise" is also a state that has been implicated in initiating exploration. Similarly, "curiosity" has been used to refer to the phenomenological counterpart of exploratory act sequences in animals as well as in humans. This term is often used synonymously with "exploration."

McReynolds (1962) has defined three classes of exploratory behavior on the basis of postulated differences in central state. Thus, *novelty-adjustive* behavior is that in which anxiety prompts the animal to explore a strongly novel stimulus in order to reduce the ambiguity of that stimulus. *Novelty-seeking* behavior involves preferential approach toward, and choice of, mildly novel stimuli. In *goal-oriented novelty-seeking* behavior on the other hand, the animal seeks a novel stimulus as a means to an end rather than as an end in itself. Berlyne (1960) makes somewhat different distinctions in the motive structures that he postulates as underlying curiosity-type behaviors. Thus, his *collative variables* (i.e., surprise, uncertainty, conflict, complexity, as well as novelty) produce *arousal, attention, orienting responses, stimulus selection, locomotor exploration, investigatory responses,* and *epistemic curiosity.* In all these behaviors, internal predisposing processes are postulated. For Berlyne then, "drives," "appetites," "desires," "wishes," "habits," and "thoughts"—as well as "reinforcement" and "learning"—are conceived to determine the character of the responses. Such terms, although possibly referring to real internal, neurobiological states or processes, will not constitute explanations until they can be validated by identification of actual neurobiological phenomena.

d. Adaptive Functions. Exploration of relatively mild novel stimuli, and the new behavioral patterns that have been shown to result therefrom, may lead an animal to cope with an increasingly greater variety of environmental situations, and may thus have obvious adaptive value. It has been suggested that a casually exploring animal can develop creative and inventive solutions to problem situations that it encounters because it is not forced into the kind of ritualized, stereotyped action patterns that stressful stimuli elicit. In this general sense, exploratory behaviors seem to be as functionally relevant to survival as are act sequences of many other behavioral classes. This of course does not imply that they have in fact led to adaptive success. Experiments must be devised that are able to determine whether or not a particular behavior pattern is important for the survival of either the individual organism or of successive populations (G. C. Williams, 1966). There is certainly no *a priori* reason to believe that all behavior is adaptive; there may be some behaviors which occur in an animal's repertoire which may be either vestigial or preadaptive, and consequently may have only minimal adaptive significance in the animal's current ecological habitats.

2. Play Behavior

a. Stimulus Goals. The stimuli often found to elicit vigorous play activities include mild novelty either of the general environment or of specific objects or surfaces. "Social" play is common when young animals (regardless of species differences) are raised together and/or are familiar with one another. The most vigorous play activities occur to moving stimuli, especially to those that are retaliative, interacting, or collaboratively "teasing." Moveable objects are especially attractive nonsocial goals for such rambunctious activities. Many of the stimuli that elicit "play" sequences in the young animal gradually come to initiate hunting, aggressive attack, ingestive, and reproductive behaviors as it matures. The initiating stimuli and the resultant response patterns may even be similar. It is because of such similarity in stimulus-response transactions that many play activities of young animals have been viewed as being preparatory to adaptive adult patterns. It is also because of this similarity that it seems necessary to reevaluate our criteria for classifying behavior.

b. Terminal and Preterminal Act Sequences. Play as a class of behavior has had even less empirical characterization than has exploration (Loizos, 1966; Millar, 1968; Welker, 1961). In all young animals observed, the term play is used typically to refer to such vigorous action sequences as chasing, wrestling, jumping, swinging, throwing, etc. Although they commonly exhibit many features that are similar to those attributed to other behavioral categories (e.g., stalking, chasing, biting, etc.), it has been argued that their fragmented, incomplete, abortive, and disjunctive sequential character is sufficient justification to classify them as play. The occurrence of such activities in either young or adult animals has led some authors to consider them as displacement activities, that is, as activities which may occur out of their "mature," "complete," "functional" contexts (Loizos, 1966; Marler & Hamilton, 1966; Millar, 1968).

There has be no means been general agreement among various observers as to the distinction between play and exploration. Groos (1898, 1901) and others have used "play" as the more generic term, with curiosity or exploration being regarded as a subclass of play. In humans, "play" usually refers to quiet sedate manipulation of stimuli *as well as* to the more vigorous activities. Millar (1968) has suggested that four forms of play may be distinguished: *exploring play,* when the object or experience is relatively new; *manipulative play,* when the object is familiar but what can be done with it is new; *practice play,* wherein further changes occur in the activities employing the object; and finally *repetitive play*, wherein the same actions are repeated again and again with or without variations. The stimulus goals, response characteristics, and motive states could presumably all differ in these four instances.

c. Internal Concomitants. As in the case of exploration, it is claimed that play also occurs commonly when an animal is unstressed by the metabolic and endocrine factors associated with the general "homeostatic" behavioral classes mentioned earlier. However, play too may interrupt sequences thought to belong to such behavior classes. With regard to the postulation of internal motives, the same cautions apply to play as were mentioned earlier for exploration. The use of phenomenologically derived concepts to refer to the inner mental states of playing animals has been even more prominent in definitions of play than in definitions of exploration. Thus, the mood of the playing animal has been described as joyful, carefree, and pleasurable. "Surplus energy" has been persistently invoked as a motivational state for vigorous play despite its repeated critique as a useful concept (Beach, 1945; Loizos, 1966; Millar, 1968). Boredom has been proposed as a motive for play as well as for exploration. With respect to play, or in his terminology, *ludic behavior*, Berlyne includes everything that is classified as *recreation, entertainment*, and *idle curiosity*, as well as perceptual and intellectual activities such as *stimulus seeking, imagery, thought*, and certain forms of motor activity and emotional arousal. He concludes that "Ludic behavior forms such a motley assortment that it is highly unlikely that all of it has just one function" (Berlyne, 1960, p. 5).

d. Adaptive Functions. Vigorous play behavior, by providing the young animal with numerous opportunities to carry itself successfully through novel engagements with other animals as well as with stimulus objects in the environment, has been judged *a priori* as preparation for similar, but more "serious," encounters in adulthood. This so-called practice-function of play, however, has generally not been viewed as an explanation of those gentle, playful, "nonserious" activities often seen in adaptively successful adults. Nevertheless, none of these views regarding the adaptive function of play constitutes an explanation in a scientific sense; rather, they appear as very general hypotheses that either have not been verified or are not testable.

It is not difficult, at such loose levels of conceptualization, to fabricate almost any "explanation" or assumption regarding the adaptive significance of play activities in adults. Thus, adult play may be viewed as providing opportunity for the animal to learn as well as fostering a state of alertness under nonstressful conditions (Millar, 1968). That extensive learning can occur when animals are allowed to play is well documented (see Table I), although exactly in what ways such experiences contribute to later adaptive or successful behaviors has yet to be examined in detail. The ability to learn more and at faster rates, or to perform more appropriately under naturalistic conditions and in formal testing situations, is generally taken as a sign of better adaptation. It has been suggested that play as well as exploration, being comprised of behavioral sequences that are less ritualized and more variable than other types of behavior, foster the development of creative, flexible, and inventive problem-solving abilities. For example, the learning and expression of certain social gestures and postures during play in young animals may serve subsequently to inhibit aggression in maturing conspecific partners, thus making possible the formation of stable group hierarchies (Jay, 1965; Loizos, 1967). Such arguments assume, of course, that certain gestural and postural activities are more appropriately classed as play and not as some other type of behavior.

B. Problems of Classification

As indicated above, much of the confusion regarding definitions of play and exploration has its origin not only in the subtle phenomenological characteristics often attributed to such behaviors (see definitions in Webster's Unabridged Dictionary), but also in the very generalized and often nontestable *ad hoc* hypotheses conceived as somehow "explaining" these behaviors. Such conceptions undoubtedly structure the thinking of scientists and laymen alike with respect to the causes and characteristics of play and exploration. It has not as yet been demonstrated, of course, that these common-sense conceptions can serve as reliable guides to an appropriate choice of hypotheses and concepts. Millar (1968) has correctly pointed out that scientists cannot go to ordinary language for their definitions, and also that it may be unsatisfactory to restrict arbitrarily the meaning of a frequently used word. At any rate, it seems clear that new sets of criteria must be found for the definition, classification, and explanation of play and exploratory activities.

As interest increases in the analysis of the fine structure of behavioral sequences and in their underlying biological mechanisms, the conceptualizations of play and exploration reviewed above have more and more come to seem incomplete and inaccurate. Although the behavioral phenomena that these concepts denote are real enough, neither their adaptive function nor their physiological determinants have been described satisfactorily. Thus, *adaptive value* is usually judged *a priori* without the support of adequate empirical data. Moreover, underlying *physiological states* are rarely assessed directly, or, at present, are even assessable at all. As a result, operationally defined "drives" are often all that are proposed, and these invariably lead only to vague speculative statements purporting to "account for" the differential probability of occurrence of sets of temporally organized behavioral sequences. The difficulty with the use of the *stimulus goal* as a primary class criterion is that the goal can take on so many different physical characteristics as virtually to defy classification on this basis alone. Thus, an animal may attempt copulation with a human foot, eat nonnutritive substances, show severe aggression to parts of its own body, retrieve inert objects to its nest, etc. Similar criticisms apply to the designation of characteristics of the *appetitive* (preterminal) and *consummatory* (terminal) *responses* as class criteria, particularly when a goal is not clearly sought or reached, or when a behavior appears as a "displacement" or "fixed-action" pattern, possibly even unassociated with sensory feedback from specific stimulus objects. Such behaviors may consist of short sequences which appear out of context and thus do not meet the usual criteria for class inclusion. Even greater concern is generated by the fact that classification of a particular behavior with reference to one or two of the criteria listed above may often force the inclusion of that behavior into one category, whereas when judged by other criteria it could well be identified as belonging to another behavioral category (e.g., playful fighting versus harmful fighting, exploration for food versus exploration of novelty, etc.). Classification of all such preterminal behavior sequences as "appetitive" also has little explanatory value, and again avoids search for neurobiological determinants and diverts attention from analysis of the details of the individual response sequences as well as from the organized behavioral episodes of which they may be a part. Moreover, there are many distinct behavioral sequences which do not fit into any of these general classes, such as washing, grooming, and scratching, as well as numerous other items of the behavioral repertoire, especially of young mammals. In addition, the behaviors identified as play and exploration vary considerably from one mammalian form to another, creating further problems for any general attempts at classification and explanation. Recently, there have been attempts to integrate play and exploratory behaviors within the classical drive-reduction framework (see Fowler, 1965). Here, particularly, experimental studies of behavioral mechanisms have stagnated, resulting in concepts whose biological validity is questionable.

With respect to recent attempts to formulate general and comprehensive schemes for the classification of behavior, the basic conceptual categories of both Schneirla and Scott should be mentioned. Schneirla (1965) proposed a very general conceptual framework within which the behavioral repertoire is viewed as developing out of basic sets of *approach* and *withdrawal* responses. His attempt to present a conceptual scheme that might encompass both the phylogeny and ontogeny of hierachically organized behavioral sequences has given rise to a variety of important explanatory constructs, some of which will be discussed below. However, the dichotomy of approach and withdrawal itself will probably prove of little use in classifying the great variety of specific act sequences comprising the behavioral repertoire of mammals.

Scott's attempt (1958a) at behavioral classification is based on the adaptive function which particular response sequences appear to serve when considered as a group (e.g., epimeletic behaviors; i.e., care-giving behaviors). This scheme, as most of the general classificatory schemes, utilizes all five criteria mentioned above in identifying a particular complex of actions as belonging to one or another general adaptive behavioral category. But it, too, is very general and fails to deal with neurobiological phenomena in sufficient detail.

In summary, the concepts of play and exploration, like those referring to other general classes of behavior, suffer from several shortcomings. Thus, criteria for class inclusion (a) are so vaguely described as to refer only to the most general behavioral events, ignoring finer spatiotemporal action sequences, (b) are overlapping and not mutually exclusive of one another, (c) invoke "explanatory" intraorganismic processes and/or adaptive functions commonly of untested or nontestable biological validity, (d) are derived primarily from observation of already organized behavior patterns of mature experienced animals, and consequently are not applicable to the developing patterns characteristic of neonatal and inexperienced animals, and (e) do not encompass a sufficiently large portion of the natural behavioral repertoire of most mammals. These conceptual deficiencies have each been discussed previously in the literature (Hebb, 1949, 1958a, 1959; Hinde, 1959a, 1959b; Lashley, 1938; Marler & Hamilton, 1966; Schneirla, 1965; Scott, 1968b), and, until they have been remedied, it would not be possible to achieve a neurobiologically based, ontogenetically oriented understanding of behavioral phenomena.

IV. Behavioral Phenomena

Before proceeding to discuss the ontogeny of play and exploration, it might be well to attempt a conceptual analysis of what we have called, respectively, the repertoire of overt behavior and the repertoire of covert, intraorganismic states and processes assumed to underlie it.

A. Repertoire of Overt Behavior

The *overt behavioral repertoire* (Table II) is here defined as the total of all operationally distinguishable unit behavioral sequences exhibited by an individual animal during its entire life history. The size, composition, and diversity of the behavioral repertoire will clearly vary with age and experience as the animal passes through different phases of the life cycle. A *unit behavioral sequence* would be a relatively simple, observable, and measurable segment of overt behavior. Individual sequences differ with respect to the locus and extent of their muscular and skeletal contributions, as well as with respect to their movement duration and spatiotemporal complexity. Overt behavioral sequences can be identified as units simply because the muscle contractions and relaxations that produce them necessarily have starting and stopping points in space and time.

A *response-sequence complex* would consist of organized aggregates or assemblies of unit sequences which exhibit more complex, goal-oriented, or environmentally regulated action patterns. Knowledge of the overt repertoire and its development is essential for any comprehensive theory of behavior. Such knowledge provides clues as to the basic capabilities of an animal in its receptive and reactive transactions with the environment. A complete survey and assessment of such capabilities is also essential for any serious attempt to categorize behavior. The difficulty lies, of course, in classifying and conceptualizing the various "units" into natural groupings and hierarchies. A list of the major overt action patterns that comprise the behavioral repertoire of a variety of mammals is presented in Table II. It must be emphasized that most of the sequences included are mature sequences that have already progressed through earlier developmental stages when they manifest different descriptive and organizational features than they exhibit in adults. Although this particular form of categorization of behavior is conceptually crude, it must suffice until more biologically comprehensive theories can be conceived. For the present, however, our purpose will be served if in looking closely at the enormous size and diversity of the mam-

Table II
THE OVERT BEHAVIORAL REPERTOIRE[a]

1. *Simple reflexes* include muscle twitches and quivers, myotactic (stretch) reflexes, reciprocal "inhibition" of antagonists, lengthening reactions (inverse myotactic reflexes), positive supporting (magnet) extension reactions, negative supporting reactions, ipsilateral extensor reflexes, flexion reflexes, crossed extension reflexes, bilateral intersegmental (crawling) reflex "figures," startle reflexes, scratch reflexes, mouth wiping (by tongue, head, hand or foot) reflexes, head and body shaking reflexes, skin-flick reflexes, urination and defecation reflexes, panting, breathing (inspiration, expiration), gasping, choking, coughing, sneezing, and vomiting reflexes, rooting, mouthing and biting reflexes, licking and swallowing reflexes, clasping and grasping (hand or foot) reflexes, pinna, tympanic, eyeblink and squinting reflexes, lip and other facial reflexes, lordosis, pelvic thrust reflexes, nystagmus, saccadic and drift eye movements, narial dilatation reflexes. The simple responses listed above primarily involve striated muscle but the effects of nonstriated muscle responses may also be observed as the following list illustrates: milk let-down reflexes, bladder and anal reflexes (retentive and releasing), piloerection, gastric and intestinal reflexes, pupillary dilatation, lens accommodation, uterine and vaginal reflexes, penile erection and ejaculation reflexes.
2. *Postures and postural changes* include immobile sitting and standing, leaning, lying (supine or prone), sprawling, crouching, curling up, squatting, balancing, hanging, floating, urination and defecation postures, rearing, hopping, placing, stretching, yawning, tonic head-neck-body reflexes, righting (free fall and supine) reflexes (optic, labyrinthine), labyrinthine acceleratory reflexes, labyrinthine positional reflexes, "freezing" immobilization (e.g., of neonatal kittens when carried by head or scruff of neck). The adjustive recovery from a particular assumed posture (such as stretching) must also be considered as a postural change.
3. *Locomotor sequences* change body locus either in space or usually more specifically with respect to certain types of environmental stimuli, and in such cases may also involve orientation sequences (see below). Locomotor patterns include: walking (quadrupedal, bipedal, bimanual), running (trotting, pacing, cantering, galloping), lunging, charging, creeping, crawling, stalking, dragging, struggling, wriggling, climbing (up or down), rolling, somersaulting, sliding, circling, backing up, skipping, flying, gliding, hopping, sideling, swinging, brachiating, dropping, leaping, jumping, rearing, bucking, diving, surfacing, swimming, and floating.
4. *Simple receptor orientation sequences* include head turning, nodding, lifting, and fixation, eye opening, eye fixation (visual "grasp"), ear turning, sniffing (polypnea), tasting, licking, biting, touching, poking, grasping, holding, batting, dropping, releasing, rubbing, tapping, pushing, pulling, slapping, scratching, twisting, throwing (by proboscis, mouth, hands, feet, or tail).
5. *Simple "fixed" action patterns.*

   (a) *Ingestion (eating and drinking)-related components* include mouthing, biting, lip and tooth grasping, grazing, browsing, nibbling, tasting, chewing, cudding, lapping, sipping, drinking, licking, sucking, swallowing, "chop" licking, and lip smacking.

   (b) *Elimination-related components* include site selection, scratching, digging, squatting, leg lifting, immobilized straining postures, urinating, defecating, parturition, postural recovery, urine and fecal covering and "scenting" or gland secretion "marking."

(c) *Sexual-related components* include (1) *Courtship sequences* such as partner orientation and selection, approach, "teasing," chasing and retreating, dancing, vocalization, body and facial displays or gestures, strutting, specific posturing and "presenting," orogenital and oroanal (nose, lip, and mouth) contacts, biting, sniffing, blowing, sucking, kissing, licking, caressing, nuzzling, and embracing; and (2) *Copulatory and orgastic sequences* include partial and complete mounting (male), immobilization and lordosis (female), clasping and grasping of mate, pelvic thrusting, masturbation, orgasm and ejaculation, dismounting, separation, and after-reactions (rolling, running, cavorting, vocalizing).

(d) *Maternal-related components* include licking, nosing, nuzzling, handling, holding, carrying, retrieving and piling, nipple presenting and nursing, huddling and covering, cradling, corralling of offspring, vocalization.

(e) *Escape, -defense -and attack-related components* include immobilization (freezing), cringing, cowering or submission (e.g., rolling over on back), hiding or running away; crouching, stalking, threat gestures and displays, hissing, tail erecting, or flexing, twitching, slapping, abortive charging, full charge and chasing, bunting, biting, tearing, holding and pinioning, grasping and shaking, hitting, clawing, kicking, sneak attacking, leaping upon and wrestling, growling, barking, grunting, squealing, and screaming.

(f) *Other instinct sequences* include grooming, washing, preening, licking, picking and scratching specific body zones, shaking, awakening, going to sleep, sleeping, "dreaming," nest material retrieval and nest construction, burrowing, digging, excavating, retrieving, holding, carrying, packing, shredding, arranging, weaving, piling, territorial marking (with urine, feces, or glandular secretions) chest beating, various vocalization patterns (cooing, babbling), spitting, blowing, bucking, bathing, basking, cooling, painting, building, wrecking and testing, purring, claw protraction and retraction, resting.

(g) *Miscellaneous gestures, displays and "social" components* include grimaces, tail twitching, ear retraction, eye squinting, lip retraction and tooth baring, protective gestures and postures, smiling, kissing, nuzzling, hugging, huddling, tickling, grabbing, grappling, taking, stealing, staring, gaze avoiding, scratching, specific muscle tensions or relaxations, imitating, encouraging, beckoning, begging, avoiding, separating, mouth sounds, vocalizations (whining, purring, hissing, squealing, etc.), pacing, exhilaration, excitement, restlessness and agitation, rocking, clasping, self-biting and autisms.

6. *Complex organized response sequence aggregates* include exploration, searching or hunting (for food, shelter, escape, a mate, novelty, etc.), building, territorial demarcation, identification and maintenance, games (hide and seek), housekeeping, and various social "structuring" and interactive behaviors.

---

[a] Most of these behavioral sequences are seen only in adult animals or at least in those old enough to have developed complete or matured patterns.

malian behavioral repertoire, the necessity for reevaluating previous ways of studying and explaining it comes to be appreciated.

The list of behavioral sequences included in Table II is arranged from top to bottom along a general *hypothetical continuum of low to high complexity* with respect to all of the following parameters: (a) number of simple unit behavioral sequences involved, (b) degree of stimulus-response variability, (c) role of experiential factors in shaping the efficiency, goal directedness, and accuracy of composite unit sequences, (d) temporal distribution and variability in the composition of the unit sequences, and (e) degree of response sensitivity to fine stimulus characteristics or to any changes in them. Different degrees of such complexity are generally considered to distinguish different *hierarchical levels* when viewed from both ontogenetic and phylogenetic points of view (Schneirla, 1965). Yet, the classification of behavior into levels, classes, and hierarchies such as those underlined in Table II has not been validated since there have been no systematic or thorough attempts to use both ontogenetic and neurobiological criteria. Suggestions for the development of a natural classification of behavioral phenomena will be discussed in a later section.

The list of behavioral sequences included in Table II is intended to be illustrative rather than exhaustive. In most instances the full repertoire, denoted in part by the terms mentioned, has not been described in detail for any mammalian form. It is generally agreed that the number of distinguishable behavioral sequences for most mammals is exceedingly large, and although there are some 330 general types of behavioral sequences listed in Table II, additional subtypes could easily be distinguished, even for a single species. With reference to play and exploratory behaviors, both "cautious" exploration as well as "gentle" and "vigorous" play may contain numerous components listed under *all* of the categories included in Table II.

B. Repertoire of Covert Functions, States, and Processes

Simple descriptions of overt act sequences alone can never lead to an understanding or explanation of behavior. Their changeable, complex characteristics require the postulation of a variety of hypothetical constructs. The overwhelming practical and theoretical necessity for such constructs is attested to by the numerous subtle and complex meanings attributed to the internal states of humans as well as animals. A variety of such concepts are listed in Table III. Most of them are defined in popular, phenomenologically derived, commonsense ways, although for experimental purposes some have been defined operationally. However, even among the latter it is common to find different operational definitions assigned to the same construct. Such inconsistencies are due, in part, to the use of different behavioral indices, experimental methods, types of animals, and theoretical orientations. It might be expected that if the various constructs were validated by reference to underlying neurobiological phenomena, then a single construct could suffice in referring to a particular set of phenomena, whereas at present several may be in use. On the other hand, it cannot be taken for granted that what might appear to be similar constructs do, in fact, have similar neurobiological correlates (habituation and satiation, for example). Moreover, it would seem that many of the more general constructs may be found ultimately to subsume a variety of specific processes that are yet to be specified neurobiologically. Indeed, it is more than likely that numerous additional constructs will be demanded by the very complex nature of the known underlying neurobiological mechanisms.

It is clear that explanations of play and exploration, as well as of other behaviors, require all the major types of constructs included in Table III. With respect to play and exploration, particularly in humans, a variety of highly complex central states has been postulated. For example, we test and explore our environment (including other people) to allay anxiety, to seek a class of esthetic, religious, or emotional experience, or to understand, to explain, or master a situation or problem. So too, in various plays, we may nervously doodle, play games of sport to win, or to experience the exhilaration of skilled exercise, to compete, to demonstrate prowess, for "sheer" joy, out of boredom, or with hate and revenge. The central motives underlying these diverse instances must be different, despite the fact that the behaviors may appear overtly similar.

The *problem of multiple motive structure* also exists in animal behavior. We cannot dismiss out of hand diversity in central motives simply because we do not have access to an animal's phenomenological "world." Indeed, there is clear evidence, even in rats, that some such diversity must exist (McReynolds, 1962; Welker, 1961). In cats and chimpanzees, a multiplicity of the knowing and experiencing motives is even more clearly likely (Hebb, 1949; Littman, 1958). Most of our efforts to comprehend the exploratory and play behaviors of animals have resulted in experiments in which motive

Table III
THE COVERT REPERTOIRE[a]

Conceptual category	Specific concepts
Inactive states	Sleeping, unaware, unconscious, inattentive
Arousal processes	Activation, arousal, alerting
Awake states	Aware, conscious, alert, vigilant
Attentive states	Alerting, attending, expectancy, scanning, focusing, detection, vigilance, sensitivity, orientation
Specific-reactivity processes	Mobilization, threshold, set, preference, aversion, differentiation, image, expectancy, fixed action pattern, scanning, focusing, attitude, perception, detection, hallucination, goal orientation, sensitivity, excitation, orientation, discrimination, tendency, illusion, displacement, identification
Cognitive states and processes	Perception, thinking, planning, purpose, judgment, guessing, trying, will, wish, hypothesis, evaluation, imitation, cognitive content, expectancy, set, decision, insight, optimizing, competence, self-actualizing, recognition, reasoning, understanding, concept formation, abstraction, symbol formation, cognitive map, cognitive model, ideation, aim, creativity, innovation, volition, plasticity, confidence, certainty, effectance, choice, purpose, assumption, conception, goal orientation, seeking
Integrative processes	Generalization, consolidation, judgment, introspection, deduction, homeostasis, programming, mediation, repression, inhibition, facilitation, insight, fixation, plasticity, closure, abstraction, assimilation, feedback, planning, ideation, learning, transaction, creativity, conditioning, symbol formation, association, integration, summation, irradiation, re-afference, displacement, incubation, regulation
Experiential processes	Perception, detection, insight, confidence, introspection, perception, discrimination, surprise, confusion, competence, knowing, symbolizing, feeling, empathy

Motivational processes	Energy, attitude, compulsion, interest, homeostasis, optimizing, thirst, love, aspiration, hope, perserverance, craving, disposition, appetite, preference, aversion, fear, hate, joy, will, wish, drive, need, habit strength, volition, urge, curiosity, hunger, anger, anxiety, value
Affective states	Tension, boredom, sensitivity, joy, emotion, anxiety, conflict, surprise, satiation, anxiety, love, preference, aversion, impact, desire, passion, amusement, sentiment, longing
Learning processes	Discrimination, familiarization, symbolization, consolidation, incubation, learning, insight, fixation, abstraction, imprinting, conditioning, generalization, deduction
Reinforcement processes	Impact, meaning, significance, reward, reinstatement, reinforcement, inhibition, suppression, repression, trace, engram; facilitation, closure, feedback
Other change-type processes and states	Adaptation, adaptation level, satiation, suppression, feedback, forgetting, creativity, innovation, plasticity, inhibition, recovery, displacement, habituation, dishabituation, accommodation, disinhibition
Fixation states and processes	Persistence, fixation, instinct, habit, generalization, fixed action pattern, stereotypy, consistency
Memory states	Memory (immediate, delayed), recall, habit, recognition, amnesia, forgetting, retention, trace, engram, storage, retrieval
Ability states	Discrimination, learning set, perception, achievement, adaptability, habit, acuity, capability, capacity
Maturational processes and states	Critical period, growth, differentiation, readiness, histogenesis, neurogenesis, regionalization, induction, morphogenesis, pattern formation, organization

[a] Behaviorally and phenomenologically derived concepts, constructs, and intervening variables referring to hypothesized, internal, central and covert phenomena, functions, states, contents, and processes.

structures have been conceived all too simplistically. Thus, a rat forced into a strange enclosure is judged to be curious. So too is a laboratory monkey isolated from conspecifics.

Comprehensive behavior theory cannot afford conceptual oversimplifications based upon *naive operationalism.* We must reassess the adequacy not only of our concepts and definitions, but also of the types of explanations we seek and the experimental methods we use. The point is that as long as we employ loosely *or* narrowly defined motivational concepts, we shall fail to discover neurobiological mechanisms that underlie the full repertoire of overt and covert behavioral phenomena. In other words, if the biological substrates of motivational constructs are to be found, then these constructs must be defined in such a way as to have, in principle at least, physiological and anatomical characteristics that can be identified by appropriate experimental studies. The traditional concepts of overt and covert behavioral phenomena listed in Tables II and III, respectively, are largely devoid of validation with respect to neurobiological structures and processes.

V. Ontogeny of Play and Exploration within the Total Overt Behavioral Repertoire

We have already pointed out that concepts based entirely on behavioral phenomena of *mature* animals cannot be used in understanding the behavior of neonatal animals. Indeed, the mature behavioral repertoire itself can be understood fully *only* if its ontogenetic origins are known. Developmental behavioral assays thus permit a descriptive portrayal of the transformation of simpler sequences into spatiotemporally more complex ones, and can, therefore, contribute to the conceptual analysis of behavioral phenomena at all phases of the animal's life cycle.

Ontogenetic development of various *overt behaviors* has been most extensively observed in humans, chimpanzees, rhesus monkeys, domestic cats and dogs, and albino rats. However, in most instances such sequences have been identified only in general ways and exact details concerning the origin and development of specific components have been neglected.

It should become clear in the following paragraphs that the great variety of behavioral sequences that we call play and exploration have *multiple simple origins* in the neonatal repertoire. Moreover, this is probably true as well of all other types of behavior.

Close observation of developing rats illustrates the complexity of the forms and functions of play and exploratory behaviors (Anderson and Patrick, 1934; Blanck, Hard, & Larsson, 1967; Bolles & Woods, 1964; Rosenblatt, 1965; Small, 1899; Stelzner, 1969; Tilney, 1934; Welker, 1964). To begin with, we find the stimulus-oriented adjustive and reactive exploratory and playful responses of the weanling presaged at successively earlier ages by simpler, more rudimentary behavioral sequences not usually classified as play or exploration. Thus, such simple action patterns as turning toward a source of body contact, general arousal to a sudden olfactory, visual, tactile, or auditory stimulus, placing reactions to tactile stimuli, and early fragmentary sniffing, snout contact and vibrissae-whisking sequences—all of which are short and incomplete on first appearance—become integrated into the more fully developed and organized activities that we identify subsequently as exploration. Likewise, early forepaw-fending and head-turning to body contacts, nibbling, sudden sporadic running or jumping, and numerous postural adjustments develop into basic components of later vigorous social play.

A notable feature of these developing behavioral events is that they tend to occur in variable and rapidly changing sequences. Thus, as a burst of running stops, the animal may then groom, nibble on shavings, chase a sibling, leap into midair, climb the cage wall, and so on, all in a period of a few seconds. Current modes of conceptualizing behavioral phenomena are not capable of accounting for such dynamic, shifting action sequences.

The emergence of simple behavioral sequences and the subsequent development of these sequences into more complex response configurations may occur either gradually or rather suddenly. Indeed, the first occurrence may not even be seen since rarely is a complete round-the-clock temporal record taken. Moreover, accurate descriptive spatiotemporal details of any action sequence are rarely recorded since time-sampling and category-rating methods have been most commonly used (see Table I, Methodological Developments). Fine features of action sequences can be recorded only by use of high-speed cinematographic and video methods.

It is evident that all unit behavioral sequences and sequence complexes have a time of first appearance as well as a changing history during ontogeny. Some of these behaviors persist as relatively simple actions throughout life; others are transitional, and are either lost, modified, enmeshed, or integrated with still other sequences to form sequence complexes. Viewed ontogenetically, then, the repertoires of play and exploration are conceived to develop out of a shifting spatiotemporal mosaic of progressively expanding assemblies of behavioral sequences of various types, durations, and degrees of complexity.

In view of the several issues reviewed above, it seems necessary to conclude that thorough studies of the developing behavioral repertoire have not justified identifying play and exploration as either valid behavioral classes or scientifically useful explanatory concepts. We urgently need new constructs that will adequately take into account not only the progressive differentiation of the overt behavioral repertoire but also the covert neurobiological structures and processes that accompany the behavioral changes. It is to these neurobiological phenomena that the next sections are devoted.

VI. Neurobiological Phenomena: Ontogeny of Structures and Functions*

The phenomena to which we must attend in any search for neurobiological mechanisms of the maturing behavioral repertoire (overt as well as covert) reside within the structures and functions of the *muscular, somatic, visceral, skeletal,* and *neural* tissues and organs, as well as within the *biochemical, endocrine, physiological, neurochemical,* and *neuroelectric* processes that comprise the activities of these structures. Since the behavioral repertoire at any given age is dependent upon the levels of biological organization and integration reached at successively early ages, we have emphasized the importance of utilizing the ontogenetic approach in understanding neurobiological mechanisms.

In the next few paragraphs we shall discuss briefly those neurobiological phenomena whose understanding would constitute validation of the developing behavioral repertoire. Obviously not all the available neurobiological facts and concepts could be used in classifying and explaining overt and covert behavioral phenomena.

*For convenience, a list of references to some of this literature is presented here: Adolph (1968), Ambrose (1969), Bernhard and Schadé (1967), Bonner (1965), Bullough (1967), DeHaan and Ursprung (1965), Dodgson (1962), Duncan (1967), Ebert (1965), Glass (1968), Goss (1965), Hahn and Koldovsky (1966), Himwich and Himwich (1964), A. F. W. Hughes (1969), Jacobson (1970), Klosovskii (1963), Locke (1969), McElroy and Glass (1958), Milner (1967), Minkowski (1967), Moore (1965), Newton and Levine (1968), Purpura and Schadé (1964), Quarton, Melnechuk, and Schmitt (1967), Richter (1964), Walsher (1971), Weiss (1955, 1968), Willier, Weiss, and Hamburger (1955), and Wolstenholme and O'Connor (1961, 1968).

However, we are sufficiently far away from a thorough understanding of these phenomena that it is not yet possible to set forth firm guidelines which can delineate those sets of neurobiological data that will be useful. These problems will be discussed briefly in Section VII.

A. Neural Systems

Although single *neurons* (elements or units) are the basic units of function at the cellular level, and a *neural population* represents spatially distinct aggregates of neurons, we must turn to more complex levels of neuroanatomical organization in our search for those structural assemblies that constitute the fundamental operating portions of nervous systems (cf. Kandel & Wachtel, 1968). The simplest of such a functional unit could be called a *simple circuit-component*, two or more of which are organized into a *simple circuit*, several of the latter in turn being interconnected to form a *simple system*, and several of these then into a *system complex*.

Because neural structures are so small, so numerous, and possess such elaborate axonal, dendritic, and synaptic processes, it has been difficult to identify specific circuits. In addition, the great intercircuit mingling within the structural matrices of neural tissue is a major obstacle to the exact delineation of components, circuits, and systems.

Perhaps 300–400 anatomically identifiable simple circuit components comprise the central nervous system(s) of mammals. Although all mammalian species possess essentially the same major types of neural populations, they differ considerably with respect to absolute and relative sizes, as well as with respect to number and complexity of components, circuits, and systems. Although neural populations are commonly architectonically distinct and spatially localized, it must be concluded that *circuits and systems are spatially distributed* rather than localized within the forebrain, midbrain, hindbrain, and spinal cord.

We can see at this point why system-, circuit-, and component-analyses constitute fundamental goals for the neurobiologist. Failure in the search to thoroughly characterize neural circuitry can only mean that we will be unable to understand completely the neural mechanisms underlying behavior. The task is formidable indeed since neural systems of vertebrates are the most complicated of biological tissues and the technical difficulties which obstruct appropriate analyses are numerous.

Although it is not possible in the present context to review the literature pertaining to neural mechanisms of behavior, some reference to this vast literature should be made. There are of course a great variety of ablation, brain stimulation, and chronic macro- and microelectrode recording studies wherein neural mechanisms of overt behavior or of motivation, perception, learning, and problem solving have been sought. The limitation of most such studies is that they appear to conceive of discrete neural regions either as centers wherein a particular behavioral or motive phenomenon is localized, or as having some direct functional relationship to the overt or covert behavioral repertoire. However, from a growing body of neuroanatomical and neurophysiological data, it now appears that a particular spatially localized neural component is only one of several components of a circuit, that each circuit is only one within a system, and that most behavioral functions are distributed among numerous interacting systems. Therefore, it is unlikely that the function of a particular component can be diagnosed when, by a restricted lesion, an entire circuit is divested of its intact operational capabilities. Likewise, electrical stimulation with macroelectrodes within a single component, or in a region where several circuits are commingled, sets up ortho- and antidromic neuroelectric activities that are spatiotemporally unnatural.

It is fair to say that circuit identification and analysis has only just begun. Neuroanatomists have labored long to give descriptive accuracy to the structures they see. However, the hope that cytoarchitectural and nerve-tract descriptions alone can provide adequate insights into behavioral and mental functions has faded, and is essentially invalidated by the actual structural and functional complexity of neural components, circuits, and systems.

We are essentially ignorant about the specific timetable of developmental events that are crucial to the formation of functional neural elements, components, circuits, and systems underlying differentiating behavioral patterns. Yet, enough is known to clearly indicate that there is a marvelously timed complexity and precision exhibited in these developments (A. F. W. Hughes, 1969; Jacobson, 1970; Sperry, 1965; Weiss, 1968).

B. Neuroelectric and Neurosecretory Activities

The important functional features of neural structures relate, of course, to their electrical and neurosecretory capabilities. Since most studies of the ontogenetic development of neuroelectric activities have used macroelectrodes, little relevant data have been obtained regarding the ontogeny of the desired fine details of unit, component, circuit, and system function (see Skoglund, 1969).

The existence in neurons of neurosecretory capabilities, other than those related to transsynaptic neuroelectric activities, is well known, but here too, specific information regarding patterns of ontogenetic development of such neurosecretory components is poorly understood in mammals.

C. Biochemical Systems

An increasing number of studies have been concerned with biochemical maturation of neural systems (see footnote on p. 202). Circulating chemicals from endocrine glands are known to control neuroblast, neuron, component, circuit, and system development. The patterned ontogenetic development of numerous specific endocrines, enzymes, proteins, lipids, etc., and their roles in establishing and maintaining the growth and differentiation of cell bodies, axons, dendrites, and synapses as well as their electrical properties are currently under active study. The involvement of glial cells in such developments is also being investigated. In general, it seems that there is multiple determination of developing neural circuits by biochemical systems. And it is likely that it will be among such systems that we shall find answers regarding the major formative factors responsible for the structural and neuroelectric maturation of neural elements, components, circuits, and systems.

D. Muscle Systems

It is primarily the contraction and relaxation patterns of striated muscle groups, and the somatic structures they move, that constitute the direct or immediate overt observable aspects of behavior. Knowledge of certain structural features of muscles—their sizes, shapes, loci, and modes of attachment and insertion, their ranges and strengths of action, their motor-unit size, etc.—can tell us much about movement mechanisms and about the action capabilities and limitations that such features impose upon particular action sequences. Differences in these features of the muscular mosaic among different animal forms constitute one important class of determinants of species-specific behaviors. However, there have been essentially no studies of the development of the above-mentioned features of muscles as they relate to the growth and organization of the behavioral repertoire.

E. Skeletal, Somatic, and Visceral Systems

Muscles move bones, cartilaginous structures, hair and vibrissae, skin, and other specific specialized ectodermal and endodermal structures. Again, there are species differences in the relative prominence of such anatomical and functional associations, and such differences affect the behavioral repertoire by setting limits to action capabilities.

Unfortunately, there have been few ontogenetic studies devoted to specifying the collaborative influence of these tissue types on the emerging behavioral repertoire of the animal. As is the case for muscle systems, full knowledge of the role of these other nonneural structures in behavior is essential.

F. Experiential Aspects

The embryo, fetus, and neonate are inevitably exposed to certain varieties of tactile, rotational, light, vibrational, acoustic, and chemical stimuli during the development of those neural circuits which receive, transmit, and utilize these stimuli. Evidence has been accumulating to show that specific forms of patterned stimulation are capable of influencing some functions of the nervous system at certain *critical periods* during circuit formation. The exact influences of these stimuli, and the associated neuroelectric activities, upon the elements and components of circuits and systems remain to be determined. It is possible that the effects consist only of the most minute biochemical and structural modifications, but a considerable amount of research will be required to determine their exact nature. Electron microscopic and microbiochemical and microneurochemical methods will be required for such determinations. The facts of learning, discrimination, insight, habituation, and the like, clearly imply associated structural-functional alterations in activated neural components, circuits, and systems at all phases of the life cycle. It is likely that once it is known exactly in which specific ways such structures are affected during both maturation and learning, any controversy regarding differences between these hypothetical processes would be resolved.

The several systems mentioned above constitute the neurobiological phenomena whose interaction during ontogeny results in a differentiated *overt behavioral repertoire* that is delicately programmed and regulated in ways that promote an adaptive exploitation of the environment. The integrative and goal-oriented features of the behavioral repertoire constitute the functions of the *repertoire of covert processes* within the brain. These processes are critical for the regulation and programming of the organism's activities. The concepts that we currently use to refer to these covert processes are almost entirely speculative and loosely defined. If we are ever to comprehend their exact features it will be necessary to investigate their spatiotemporal operations in neurobiological terms. Beginnings have been made in this direction, but methodological and analytic problems have been repeatedly encountered. Because of a lack of full collaborative interdisciplinary investigation into the ontogeny of the interacting neurobiological systems and their behavioral expressions, the search for underlying mechanisms has been narrow, fragmented, and slow. In the next section I shall outline briefly some of the methods I believe capable of advancing our understanding of such mechanisms.

VII. A Natural Classification of Behavioral and Neurobiological Phenomena

In previous sections we have emphasized that explanations and classifications of behavior have relied upon concepts variously formulated by different experimenters utilizing different methodologies. Such traditional approaches have left us with many confused and incomplete concepts and theories, "unnatural" in the sense that they do not address themselves adequately to all the fine details of *overt action sequences,* to the determining *environmental factors* which constitute the adequate stimuli, to the associated central or *covert states and processes,* and to the *ontogenetic development* of all these phenomena.

It has long seemed apparent that ontogenetic studies can disentangle the components of the adult behavioral repertoire by providing descriptive and experimental data relevant to the development of complex patterned response sequences out of simpler ones. By starting with the simplest repertoire of the embryo, fetus, or neonate, the successive addition and alteration of unit behavioral sequences can be traced as they are genetically and experientially programmed. Empirically, the problem is that of analyzing not only the changing differential probabilities of occurrence of specific act sequences but also the neurobiological structures and processes that accompany them. Such analyses would permit accurate determination of the basic building blocks out of which the behavioral repertoire is constructed.

The proposal here, then, is to reorient the processes of classification and conceptualization to focus on those finer, naturally occurring behaviors that actually develop ontogenetically. The relation of such overt behavioral developments to biological mechanisms can then be studied directly. If we start with relatively simple action sequences whose motor components are most easily understood (in terms of activation and inhibition patterns of CNS circuits), and trace not only how the behavioral units are modified progressively, but also how the underlying neurobiological structures and functions are altered correspondingly, then the validity of covert concepts or hypothetical constructs can be more directly assessed. In other words, the primary goal of a *natural classification of neurobiological-behavioral phenomena* is to trace the diversification of behavior into various ontogenetic tracks, and hierarchies of these, wherein the nature of maturational and experiential changes (and their internal neurobiological concomitants or causes) are fully delineated. A *behavior track* would probably consist of natural developmental sequences in which specific neurobiological structures and processes are programmed to yield coherent sets of response sequences that relate to each other in specific ways. In this regard, neither play nor exploration would constitute a natural behavioral track. Both consist of many disparate and differentiable response sequences, each linked to its own set of determining conditions. Thus, touching, looking, or listening probably begin as separate S-R transactions, and only with neurobiological maturation and experience do they become merged into higher order transactions of greater hierarchical complexity. Guidelines for the identification of natural behavioral-neurobiological tracks and hierarchies have yet to be established. As mentioned earlier, it is not likely that all of the myriad neurobiological and behavioral data obtainable will be useful in such endeavors. It will be necessary to explore anew problems and concepts in classification and taxonomy (Hennig, 1966; Novikoff, 1943; Simpson, 1961; Whyte, Wilson, & Wilson, 1969).

VIII. Conclusion

In the preceding sections we reviewed those interdisciplinary approaches thought to be essential for the development of comprehensive explanations of both overt and covert behavioral phenomena. The relatively extensive discussion of these issues is as relevant to the development of play and exploration as it is to those that pertain to the development of other behavioral phenomena whose explanations are sought in neurobiological terms.

Although there has been a dramatic increase in neurobiological training and research during the past two decades, we are repeatedly reminded of just how little we know of the finely structured mechanisms that prompt even the simplest behavioral events. For example, we can scarcely envision those *exact* neural mechanisms that enable an animal to follow with its head and eyes a sudden moving visual stimulus. We know, in a general way, some of the neural components and circuits involved, but not all of them, nor how they interact. We just do not know which spatiotemporally high-speed neurophysiological inputs, central transactions, and outputs are involved in such accomplishments; nor do we know exactly which neural components and circuits are essential.

In an epilogue to a recent paper, Paul Weiss has reviewed the information explosion as he has experienced it in the neurobiological sciences. He wrote:

> I believe, though, that progress could be faster, steadier, and on a broader front if the research forces could distribute themselves more evenly and widely over the vast field remaining to be explored. There is need to rekindle target- or goal-consciousness and to make deliberate choice of goals with a deep sense of relevance, as counterpoise to the downdrift of mass movement along lines of least effort that leaves piles of trivial and redundant data in its wake. Much as I admire the progress in the life sciences in this century if measured as a relative increment over the past, it must appear diminutive to anyone who is aware of the enormous distance to be traveled—and traveled the hard way—to reach a true, comprehensive, and consistent understanding of the phenomena of life (Weiss, 1970, p. 163).

This may sound like a dismal and unproductive point of view, but the theme that I have chosen to develop in the present chapter requires a full realization of just how primitive our understanding is of the neurobiological mechanisms of behavior. Yet, there is hope since there are several practical steps toward improvement that can be taken.

First, neurobiological training must begin early in the educational process. As with language training, it is unproductive and inefficient to thrust a foreign topic upon young minds only at the college level. Nursery school-aged children can know of the brain, they can grasp some important rudiments of structure and functions of the interacting organs and tissues of the body.

Second, multidisciplinary training programs must be fostered. It is unfortunate that the administrative and conceptual inertia of traditional departments tend to prevent broad interdisciplinary training. It is at this level that attention must be focused to promote integrated programs in behavioral neurobiology.

Third, conceptual renovation must be accelerated. New inspired conceptions must successively transcend older exhausted ones. Lashley's critiques of classical conceptions (cf. Beach, Hebb, Morgan, & Nissen, 1960) and Hebb's reformulations (Hebb, 1949) constituted powerful forces for new thinking. Many such pioneering efforts will be repeatedly required before we can begin to feel comfortable with the framework that we are trying to build.

And finally, new methodologies must be developed. Techniques must be improved for the analysis, not only of the high-speed spatiotemporal behavioral sequences that constitute the daily routines of most mammals, but also of the organization of such short-term sequences into hierarchical levels involving larger and longer blocks of space and time. In addition, techniques for simultaneous recording from several circuit components, circuits, and systems are required. Similar micromethods would be required in the collection and analysis of biochemical and physiological data. Along with such data must evolve appropriate mathematical and statistical techniques to cross-correlate and analyze the exceedingly large amount of data that such methods would generate (Mesarovic, 1968). In addition, new conceptual modes must be discovered that can develop criteria for the natural classification of such neurobiological data and for their arrangement into appropriate hierarchies that possess explanatory power and theoretical importance.

Acknowledgments

This project has been supported by U. S. Public Health Service Grants M-2786 and NB-6225. I wish to thank Shirley Hunsaker for her valuable assistance in preparing the bibliographical and tabular material; Vanita Hankel, Barbara Adrianopoli, and Judith Marguardt for typing the manuscript; and Terrill Stewart for preparing photographic materials.

References

Acker, M., & McReynolds, P. The "need" for novelty: A comparison of six instruments. *Psychological Record,* 1967, **17,** 177-182.

Adolph, E. F. *Origins of physiological regulations.* New York: Academic Press, 1968.

Ambrose, J. A. The development of the smiling response in early infancy. In B. M. Foss (Ed.), *Determinants of infant behavior.* Vol. 1. New York: Wiley, 1961. Pp. 179-196.

Ambrose, J. A. The concept of a critical period for the development of social responsiveness in early human infancy. In B. M. Foss (Ed.), *Determinants of infant behavior.* Vol. 2. New York: Wiley, 1963. Pp. 201-225.

Ambrose, A. (Ed.) *Stimulation in early infancy.* New York: Academic Press, 1969.
Ames, L. B., & Ilg, F. L. The developmental point of view with special reference to the principle of reciprocal neuromotor interweaving. *Journal of Genetic Psychology,* 1964, **105,** 195-209.
Anderson, A. C., & Patrick, J. R. Some early behavior patterns in the white rat. *Psychological Review,* 1934, **41,** 480-496.
Angermeier, W. F. The effect of a novel and novel-noxious stimulus upon social operant behavior in the rat. *Journal of Genetic Psychology,* 1962, **100,** 151-154.
Angermeier, W. F., & Hitt, J. C. Basic dimensions of light as reinforcement: Changes in patterns, brightness and movement. *Journal of Genetic Psychology,* 1964, **104,** 147-153.
Bagshaw, M. H., Kimble, D. P., & Pribram, K. H. The GSR of monkeys during orienting and habituation and after ablation of the amygdala, hippocampus and inferotemporal cortex. *Neuropsychologia,* 1965, **3,** 111-119.
Barbizet, J. Defect of memorizing of hippocampalmammillary origin: A review. *Journal of Neurology, Neurosurgery and Psychiatry,* 1963, **26,** 127-135.
Barnett, S. A. *A study in behaviour.* London: Methuen, 1963.
Barnett, S. A. The "Instinct to teach." *Nature (London),* 1968, **220,** 747-749.
Baron, A. Suppression of exploratory behavior by aversive stimulation. *Journal of Comparative and Physiological Psychology,* 1964, **57,** 299-301.
Barsch, R. H. The concept of reach-grasp-release as a visual, auditory, and tactual process. *Journal of Genetic Psychology,* 1965, **106,** 237-243.
Battig, K., & Wanner, N. U. Locomotor exploration in rats as a measure of psychopharmacological effects. *Biochemical Pharmacology,* 1963, **12,** 25. (Abstract)
Beach, F. A. Current concepts of play in animals. *American Naturalist,* 1945, **79,** 523-541.
Beach, F. A., Hebb, D. O., Morgan, C. T., & Nissen, H. W. (Eds.) *The neuropsychology of Lashley, selected papers of K. S. Lashley.* New York: McGraw-Hill, 1960.
Beitel, R. E. The head orienting response, sound localization and habituation in the cat. Unpublished doctoral dissertation, University of Wisconsin, 1969.
Berkson, G. Development of an infant in a captive gibbon group. *Journal of Genetic Psychology,* 1966, **108,** 311-325.
Berkson, G., & Karrer, R. Travel vision in infant monkeys: Maturation rate and abnormal stereotyped behaviors. *Developmental Psychobiology,* 1968, **1,** 170-174.
Berkson, G., & Mason, W. A. Stereotyped behavior of chimpanzees: Relation to general arousal and alternative activities. *Perceptual and Motor Skills,* 1964, **19,** 635-652.
Berlyne, D. E. *Conflict, arousal, and curiosity.* New York: McGraw-Hill, 1960.
Berlyne, D. E. Curiosity and exploration. *Science,* 1966, **153,** 25-33.
Berlyne, D. E. Arousal and reinforcement. In D. Levine (Ed.), *Nebraska symposium on motivation.* Lincoln, Neb.: Univ. of Nebraska Press, 1967. Pp. 1-110.
Berlyne, D. E. The reward value of indifferent stimulation. In J. T. Tapp (Ed.), *Reinforcement and behavior.* New York: Academic Press, 1969. Pp. 178-214.
Berlyne, D. E., & Frommer, F. D. Some determinants of the incidence and content of children's questions. *Child Development,* 1966, **37,** 177-189.
Berlyne, D. E., Koenig, I. D., & Hirota, T. Novelty, arousal, and the reinforcement of diversive exploration in the rat. *Journal of Comparative and Physiological Psychology,* 1966, **62,** 222-226.
Berlyne, D. E., & Lewis, J. L. Effects of heightened arousal on human exploratory behavior. *Canadian Journal of Psychology,* 1963, **17,** 398-411.

Berlyne, D. E., & Peckham, S. The semantic differential and other measures of reaction to visual complexity. *Canadian Journal of Psychology,* 1966, **20,** 125-135.

Bernhard, C. G., & Schadé, J. P. (Eds.) *Developmental neurology.* Vol. 26. *Progress in brain research.* Amsterdam: Elsevier, 1967.

Bernstein, A. S. To what does the orienting response respond? *Psychophysiology,* 1969, **6,** 338-350.

Bernstein, S., & Mason, W. A. The effects of age and stimulus conditions on the emotional responses of Rhesus monkeys: Responses to complex stimuli. *Journal of Genetic Psychology,* 1962, **101,** 279-298.

Bettinger, L. A., Davis, J. L., Meikle, M. B., Birch, H., Kopp, R., Smith, H. E., & Thompson, R. F. "Novelty" cells in association cortex of cat. *Psychonomic Science,* 1967, **9,** 421-422.

Bindra, D. *Motivation. A systematic reinterpretation.* New York: Ronald Press, 1959.

Bindra, D. Components of general activity and the analysis of behavior. *Psychological Review,* 1961, **68,** 205-215.

Bindra, D., & Blond, J. A time-sample method for measuring general activity and its components. *Canadian Journal of Psychology,* 1958, **12,** 74-76.

Bindra, D., & Claus, H. J. A test of the novelty-reactions interpretation of the effects of stimulus change. *Journal of Comparative and Physiological Psychology,* 1960, **53,** 270-272.

Bindra, D., & Spinner, N. Response to different degrees of novelty: The incidence of various activities. *Journal of The Experimental Analysis of Behavior,* 1958, **1,** 341-350.

Birns, B. Individual differences in human neonates' responses to stimulation. *Child Development,* 1965, **36,** 249-256.

Blanck, A., Hard, E., & Larsson, K. Ontogenetic development of orienting behavior in the rat. *Journal of Comparative and Physiological Psychology,* 1967, **63,** 327-328.

Blauvelt, H., & McKenna, J. Mother-neonate interaction: Capacity of the human newborn for orientation. In B. M. Foss (Ed.), *Determinants of infant behaviour.* Vol. 1. New York: Wiley, 1961. Pp. 3-29.

Bobbitt, R. A., Gourevitch, V. P., Miller, L. E., & Jensen, G. D. The dynamics of social interactive behavior: A computerized procedure for analyzing trends, patterns, and sequences. *Psychological Bulletin,* 1969, **71,** 110-121.

Bobbitt, R. A. Jensen, G. D., & Gordon, B. N. Behavioral elements (taxonomy) for observing mother-infant-peer interaction in *Macaca nemestrina. Primates,* 1964, **5,** 71-80. (a)

Bobbitt, R. A., Jensen, G. D., & Kuehn, R. E. Development and application of an observational method: A pilot study of the mother-infant relationship in pigtail monkeys. *Journal of Genetic Psychology,* 1964, **105,** 257-274. (b)

Bolles, R. C. The usefulness of the drive concept. In M. R. Jones (Ed.), *Nebraska symposium on motivation.* Lincoln, Neb.: Univ. of Nebraska Press, 1958. Pp. 1-33.

Bolles, R. C. Effects of deprivation conditions upon the rat's home cage behavior. *Journal of Comparative and Physiological Psychology,* 1965, **60,** 244-248.

Bolles, R. C., & de Lorge, J. Exploration in a Dashiell maze as a function of prior deprivation, current deprivation, and sex. *Canadian Journal of Psychology,* 1962, **16,** 221-227.

Bolles, R. C., & Woods, P. J. The ontogeny of behavior in the albino rat. *Animal Behaviour,* 1964, **12,** 427-441.

Bonner, J. *The molecular biology of development.* London and New York: Oxford Univ. Press, 1965.

Boring, E. G. The trend toward mechanism. *Proceedings of The American Philosophical Society*, 1964, **108**, 451-454.

Bowers, P., & London, P. Developmental correlates of role-playing ability. *Child Development*, 1965, **36**, 499-508.

Breland, K., & Breland, M. *Animal behavior*. New York: Macmillan, 1966.

Bridger, W. H. Sensory habituation and discrimination in the human neonate. *American Journal of Psychiatry*, 1961, **117**, 991-996.

Bridger, W. H. Ethological concepts and human development. *Recent Advances in Biological Psychiatry*, 1962, **4**, 95-107.

Brody, N., & Oppenheim, P. Tensions in psychology between the methods of behaviorism and phenomenology. *Psychological Review*, 1966, **73**, 395-405.

Bronson, G. W. The fear of novelty. *Psychological Bulletin*, 1968, **69**, 350-358. (a)

Bronson, G. W. The development of fear in man and other animals. *Child Development*, 1968, **39**, 409-431. (b)

Brookshire, K. H., & Rieser, T. C. Temporal course of exploratory activity in three inbred strains of mice. *Journal of Comparative and Physiological Psychology*, 1967, **63**, 549-551.

Brown, R., & Fraser, C. The acquisition of syntax. In C. N. Cofer & B. Musgrave (Eds.), *Verbal behavior and learning: Problems and processes*. New York: McGraw-Hill, 1963.

Bryden, M. P. A model for the sequential organization of behaviour. *Canadian Journal of Psychology*, 1967, **21**, 37-56.

Bullock, T. H. Physiological bases of behavior. In J. A. Moore (Ed.), *Ideas in modern biology*. Garden City, N.Y.: Natural History Press, 1965. Pp. 451-482.

Bullough, W. S. *The evolution of differentiation*. New York: Academic Press, 1967.

Burgers, J. M. Curiosity and play: Basic factors in the development of life. *Science*, 1966, **154**, 1680-1681.

Butler, R. A. The reactions of rhesus monkeys to fear-provoking stimuli. *Journal of Genetic Psychology*, 1964, **104**, 321-330.

Butler, R. A. Investigative behavior. In A. M. Schrier, H. F. Harlow, & F. Stollnitz (Eds.), *Behavior of nonhuman primates*. Vol. 2. New York: Academic Press, 1965. 463-493.

Cairns, R. B. Attachment behavior of mammals. *Psychological Review*, 1966, **73**, 409-426.

Campbell, B. A., & Jaynes, J. Theoretical note: Reinstatement. *Psychological Review*, 1966, **73**, 478-480.

Campbell, D. Motor activity in a group of newborn babies. *Biologia Neonatorum*, 1968, **13**, 257-270.

Candland, D. K., & Campbell, B. A. Development of fear in the rat as measured by behavior in the open field. *Journal of Comparative and Physiological Psychology*, 1962, **55**, 593-596.

Cane, V. Some ways of describing behaviour. In W. H. Thorpe & O. L. Zangwill (Eds.), *Current problems in animal behaviour*. London and New York: Cambridge Univ. Press, 1961. Pp. 361-388.

Cantor, G. N., & Cantor, J. H. Discriminative reaction time in children as related to amount of stimulus familiarization. *Journal of Experimental Child Psychology*, 1966, **4**, 150-157.

Carlton, P. L. Scopolamine, amphetamine and light-reinforced responding. *Psychonomic Science*, 1966, **5**, 347-348.

Carlton, P. L. Brain-acetylcholine and habituation. In P. B. Bradley & M. Fink (Eds.),

Anticholinergic drugs. Vol. 28. *Progress in brain research.* Amsterdam: Elsevier, 1968.

Caspari, E. Genic control of development. *Perspectives in Biology and Medicine,* 1960, **4,** 26-39.

Chamove, A., Harlow, H. F., & Mitchell, G. Sex differences in the infant-directed behavior of preadolescent rhesus monkeys. *Child Development,* 1967, **28,** 329-335.

Charlesworth, W. R. Persistence of orienting and attending behavior in infants as a function of stimulus-locus uncertainty. *Child Development,* 1966, **37,** 473-491.

Christopher, Sister Mary, & Butler, C. M. Consummatory behaviors and locomotor exploration evoked from self-stimulation sites in rats. *Journal of Comparative and Physiological Psychology,* 1968, **66,** 335-339.

Clapp, W. F., & Eichorn, D. N. Some determinants of perceptual investigatory responses in children. *Journal of Experimental Child Psychology,* 1965, **2,** 371-388.

Cobb, K., Goodwin, R., & Sailens, E. Spontaneous hand positions of newborn infants. *Journal of Genetic Psychology,* 1966, **108,** 225-237.

Cohen, A. I. Hand preference and developmental status of infants. *Journal of Genetic Psychology,* 1966, **108,** 337-345.

Cohen, J. S., & Stettner, L. J. Effect of deprivation level on exploratory behavior in the albino rat. *Journal of Comparative and Physiological Psychology,* 1968, **66,** 514-517.

Collard, R. R. A study of curiosity in infants. Unpublished doctoral dissertation, University of Chicago, 1962.

Collard, R. R. Fear of strangers and play behavior in kittens with varied social experience. *Child Development,* 1967, **38,** 877-891.

Collard, R. R. Social and play responses of first-born and later-born infants in an unfamiliar situation. *Child Development,* 1968, **39,** 325-334.

Cox, F. N., & Campbell, D. Young children in a new situation with and without their mothers. *Child Development,* 1968, **39,** 123-131.

D'Amato, M. R., & Jagoda, H. Effect of early exposure to photic stimulation on brightness discrimination and exploratory behavior. *Journal of Genetic Psychology,* 1962, **101,** 267-271.

Davenport, R. K., & Berkson, G. Stereotyped movements of mental defectives. II. Effects of novel objects. *American Journal of Mental Deficiency,* 1963, **67,** 879-882.

Dawson, W. W., & Hoffman, C. S. The effects of early differential environments on certain behavior patterns in the albino rat. *Psychological Record,* 1958, **8,** 87-92.

DeHaan, R. L., & Ursprung, H. (Eds.) *Organogenesis.* New York: Holt, 1965.

DeNelsky, G. Y., & Denenberg, V. H. Infantile stimulation and adult exploratory behavior: Effects of handling upon tactual variation seeking. *Journal of Comparative and Physiological Psychology,* 1967, **63,** 309-312. (a)

DeNelsky, G. Y., & Denenberg, V. H. Infantile stimulation and adult exploratory behavior in the rat: Effects of handling upon visual variation seeking. *Animal Behaviour,* 1967, **15,** 568-573. (b)

Denenberg, V. H. Critical periods, stimulus input, and emotional reactivity: A theory of infantile stimulation. *Psychological Review,* 1964, **71,** 335-351.

Denenberg, V. H. Behavioral differences in two closely related lines of mice. *Journal of Genetic Psychology,* 1965, **106,** 201-205.

Denenberg, V. H. Stimulation in infancy, emotional reactivity and exploratory behavior. In D. C. Glass (Ed.), *Neurophysiology and emotion.* New York: Russell Sage Found., 1967. Pp. 161-190.

Denenberg, V. H. A consideration of the usefulness of the critical period hypothesis as applied to the stimulation of rodents in infancy. In G. Newton & S. Levine (Eds.), *Early experience and behavior. The psychogiology of development.* Springfield, Ill.: Thomas, 1968. Pp. 142-167.

Denenberg, V. H., & Grota, L. J. Social-seeking and novelty-seeking behavior as a function of differential rearing histories. *Journal of Abnormal and Social Psychology,* 1964, **69**, 453-456.

Denenberg, V. H., & Karas, G. G. Interactive effects of age and duration of infantile experience on adult learning. *Psychological Reports,* 1960, **7**, 313-322.

Denenberg, V. H., & Kline, N. J. Stimulus intensity versus critical periods: A test of two hypothese concerning infantile stimulation. *Canadian Journal of Psychology,* 1964, **18**, 1-5.

Deutsch, J. Analysis of positional behaviors in the rhesus monkey (*Macaca mulatta*). Technical Report No. 66-1, 1966, Office of Naval Research, Tulane University, Delta Regional Primate Research Center.

DiVesta, F. J. Developmental patterns in the use of modifiers as modes of conceptualization. *Child Development,* 1965, **36**, 185-213.

Dodgson, M. C. H. *The growing brain. An essay in developmental neurology.* Baltimore, Md.: Williams & Wilkins, 1962.

Donahoe, J. W. The reinforcing effects of variable visual stimulation. *Journal of Genetic Psychology,* 1965, **107**, 205-218.

Donahoe, J. W. The effects of temporal variability on the reinforcing properties of visual stimulus change. *Journal of Genetic Psychology,* 1967, **111**, 227-232.

Doty, B. A., & Doty, L. A. Effects of handling at various ages on later open-field behavior. *Canadian Journal of Psychology,* 1967, **21**, 463-470.

Douglas, R. J. The hippocampus and behavior. *Psychological Bulletin,* 1967, **67**, 416-442.

Douglas, R. J., & Isaacson, R. L. Hippocampal lesions and activity. *Psychonomic Science,* 1964, **6**, 187-188.

Ducharme, R. Effect on internal and external cues on the heart rate of the rat. *Canadian Journal of Psychology,* 1966, **20**, 97-104.

Duncan, C. J. *The molecular properties and evolution of excitable cells.* Oxford: Pergamon Press, 1967.

Eacker, J. N. Behaviorally produced illumination change: Visual exploration and reinforcement facilitation. *Journal of Comparative and Physiological Psychology,* 1967, **64**, 140-145.

Ebert, J. D. *Interacting systems in development.* New York: Holt, 1965.

Ehrlich, A. The effects of past experience on the rat's response to novelty. *Canadian Journal of Psychology,* 1961, **15**, 15-19.

Ehrlich, A. Response to novel objects in three lower primates: greater galago, slow loris, and owl monkey. *Behaviour,* 1970, **37**, 55-63.

Eibl-Eibesfeldt, I. Concepts of ethology and their significance in the study of human behavior. In H. W. Stevenson, E. H. Hess, & H. L. Rheingold (Eds.), *Early behavior—comparative and developmental approaches.* New York: Wiley, 1967. Pp. 127-146.

Elkind, D., & Weiss, J. Studies in perceptual development. III. Perceptual exploration. *Child Development,* 1967, **38**, 553-561.

Essman, W. B. Differences in locomotor activity and brain-serotonin metabolism in differentially housed mice. *Journal of Comparative and Physiological Psychology*, 1968, **66**, 244-246.

Fantz, R. L. Visual experience in infants: Decreased attention to familiar patterns relative to novel ones. *Science*, 1964, **146**, 668-670.

Fantz, R. L. Ontogeny of perception. In A. M. Schrier, H. F. Harlow & F. Stollinitz (Eds.), *Behavior of nonhuman primates*. Vol. 2. New York: Academic Press, 1965. Pp. 365-403.

Fantz, R. L. Pattern discrimination and selective attention as determinants of perceptual development from birth. In A. H. Kidd & J. L. Rivoire (Eds.), *Perceptual development in children*. New York: International Universities Press, 1966. Pp. 143-173.

Fantz, R. L., Ordy, J. M., & Udelf, M. S. Maturation of pattern vision in infants during the first six months. *Journal of Comparative and Physiological Psychology*, 1962, **55**, 907-917.

Faw, T. T., & Nunnally, J. C. The influence of stimulus complexity, novelty, and affective value on children's visual fixations. *Journal of Experimental Child Psychology*, 1968, **6**, 141-153.

Feather, N. T. An expectancy-value model of information seeking behavior. *Psychological Review*, 1967, **74**, 342-360.

Fiske, D. W. The inherent variability of behavior. In D. W. Fiske & S. R. Maddi (Eds.), *Functions of varied experience*. Homewood, Ill.: Dorsey Press, 1961. Pp. 326-354.

Fiske, D. W., & Maddi, S. R. A conceptual framework. In D. W. Fiske & S. R. Maddi (Eds.), *Functions of varied experience*. Homewood, Ill.: Dorsey Press, 1961. Pp. 11-56. (a)

Fiske, D. W., & Maddi, S. R. (Eds.) *Functions of varied experience*. Homewood, Ill.: Dorsey Press, 1961. (b)

Fowler, H. *Curiosity and exploratory behavior*. New York: Macmillan, 1965.

Fox, M. W. The ontogeny of behavior and neurologic responses in the dog. *Animal Behaviour*, 1964, **12**, 301-311.

Fox, M. W. Reflex-ontogeny and behavioral development of the mouse. *Animal Behaviour*, 1965, **13**, 234-241.

Fox, M. W. Neuro-behavioral ontogeny. A synthesis of ethological and neurophysiological concepts. *Brain Research*, 1966, **2**, 3-20.

Friedlander, B. Z. Three manipulanda for the study of human infants' operant play. *Journal of the Experimental Analysis of Behavior*, 1966, **9**, 47-49. (a)

Friedlander, B. Z. Effects of stimulus variation, ratio contingency, and intermittent extinction on a child's incidental play for perceptual reinforcement. *Journal of Experimental Child Psychology*, 1966, **4**, 257-265. (b)

Fuller, J. L. *Motivation. A biological perspective*. New York: Random House, 1962.

Gibson, J. J. Observations on active touch. *Psychological Review*, 1962, **69**, 477-491.

Gilmore, J. B. The role of anxiety and cognitive factors in children's play behavior. *Child Development*, 1966, **37**, 397-416.

Glanzer, M. Curiosity, exploratory drive, and stimulus satiation. *Psychological Bulletin*, 1958, **55**, 302-315.

Glanzer, M. Changes and interrelations in exploratory behavior. *Journal of Comparative and Physiological Psychology*, 1961, **54**, 433-438.

Glass, D. C. (Ed.) *Environmental influences*. New York: Rockefeller Univ. Press, 1968.

Glickman, S. E., & Feldman, S. M. Habituation of the arousal response to direct stimulation of the brain stem. *Electroencephalography and Clinical Neurophysiology*, 1961, **13**, 703-709.

Glickman, S. E., & Jensen, G. D. The effects of hunger and thirst on Y-maze exploration. *Journal of Comparative and Physiological Psychology*, 1961, **54**, 83-85.

Glickman, S. E., & Schiff, B. B. A biological theory of reinforcement. *Psychological Review*, 1967, **74**, 81-109.

Glickman, S. E., & Van Laer, E. K. The development of curiosity within the genus Panthera. *Zoologica (New York)*, 1964, **49**, 109-114.

Goddard, G. V. Functions of the amygdala. *Psychological Bulletin*, 1964, **62**, 89-109.

Goldberg, S., & Lewis, M. Play behavior in the year-old infant: Early sex differences. *Child Development*, 1969, **40**, 21-31.

Goodrick, C. L. Social interactions and exploration of young, mature, and senescent male albino rats. *Journal of Gerontology*, 1965, **20**, 215-218.

Goodrick, C. L. Activity and exploration as a function of age and deprivation. *Journal of Genetic Psychology*, 1966, **108**, 239-252.

Goodrick, C. L. Exploration of nondeprived male Sprague-Dawley rats as a function of age. *Psychological Reports*, 1967, **20**, 159-163.

Goss, R. J. *Adaptive growth*. New York: Academic Press, 1964.

Gray, J. A. A time-sample study of the components of general activity in selected strains of rats. *Canadian Journal of Psychology*, 1965, **19**, 74-82.

Gray, M. L., & Crowell, D. H. Heart rate changes to sudden peripheral stimuli in the human during early infancy. *Journal of Pediatrics*, 1968, **72**, 807-814.

Green, P. C., & Gordon, M. Maternal deprivation: Its influence on visual exploration in infant monkeys. *Science*, 1964, **145**, 292-294.

Groos, K. *The play of animals*. (transl. by E. L. Baldwin) New York: Appleton, 1898.

Groos, K. *The play of man*. (transl. by E. L. Baldwin) New York: Appleton, 1901.

Gross, C. G. General activity. In L. Weiskrantz (Ed.), *Analysis of behavioral change*. New York: Harper, 1968. Pp. 91-106.

Groves, P. M., & Thompson, R. F. Habituation: A dual-process theory. *Psychological Review*, 1970, **77**, 419-450.

Gullickson, G. R. A note on children's selection of novel auditory stimuli. *Journal of Experimental Child Psychology*, 1966, **4**, 158-162.

Hahn, P., & Koldovsky, O. *Utilization of nutrients during postnatal development*. Oxford: Pergamon Press, 1966.

Halliday, M. S. Exploration and fear in the rat. In P. A. Jewell & C. Loizos (Eds.), *Play, exploration and territory in mammals*. New York: Academic Press, 1966. Pp. 45-59.

Halliday, M. S. Exploratory behavior. In L. Weiskrantz (Ed.), *Analysis of behavioral change*. New York: Harper, 1968. Pp. 107-126.

Hansen, E. W. The development of maternal and infant behavior in the rhesus monkey. *Behaviour*, 1966, **27**, 107-149.

Harlow, H. F. The development of learning in the rhesus monkey. *American Scientist*, 1959, **47**, 459-479.

Harlow, H. F. The development of affectional patterns in infant monkeys. In B. M. Foss (Ed.), *Determinants of infant behaviour*. Vol. 1. New York: Wiley, 1961. Pp. 75-88.

Harlow, H. F. The maternal affectional system. In B. M. Foss (Ed.), *Determinants of infant behaviour*. Vol. 2. New York: Wiley, 1963. Pp. 3-33.

Harris, L. The effects of relative novelty on children's choice behavior. *Journal of Experimental Child Psychology*, 1965, **2**, 297-305.

Haude, R. H., Kruper, D. C., & Patton, R. A. Relationships among measures of visual exploration in monkeys. *Journal of Comparative and Physiological Psychology*, 1966, **62**, 156-159.

Haude, R. H., & Oakley, S. R. Visual exploration in monkeys as a function of visual incentive duration and sensory deprivation. *Journal of Comparative and Physiological Psychology*, 1967, **64**, 332-336.

Hebb, D. O. *The organization of behavior*. New York: Wiley, 1949.

Hebb, D. O. Drives and the C. N. S. (conceptual nervous system). *Psychological Review*, 1955, **62**, 243-254.

Hebb, D. O. Alice in wonderland or psychology among the biological sciences. In H. F. Harlow & C. N. Woolsey (Eds.), *Biological and biochemical bases of behavior*. Madison, Wis.: Univ. of Wisconsin Press, 1958. Pp. 451-467. (a)

Hebb, D. O. The motivating effects of exteroceptive stimulation. *American Psychologist*, 1958, **13**, 109-113. (b)

Hebb, D. O. A neuropsychological theory. In S. Koch (Ed.), *Psychology: A study of a science*. Vol. 1. *Sensory, perceptual and physiological formulations*. New York: McGraw-Hill, 1959. Pp. 622-643.

Hebb, D. O. The semiautonomous process: Its nature and nurture. *American Psychologist*, 1962, **18**, 16-27.

Heckhausen, H. Complexity in perception: Phenomenal criteria and information theoretic calculus—note on D. E. Berlyne's "complexity effects." *Canadian Journal of Psychology*, 1964, **18**, 168-173.

Hein, A., & Held, R. Dissociation of the visual placing response into elicited and guided components. *Science*, 1967, **158**, 390-392.

Held, R., & Hein, A. Movement-produced stimulation in the development of visually guided behavior. *Journal of Comparative and Physiological Psychology*, 1963, **56**, 872-876.

Henker, B. A., Asano, S., & Kagan, J. An observation system for studying visual attention in the infant. *Journal of Experimental Child Psychology*, 1968, **6**, 391-393.

Hennig, W. *Phylogenetic systematics*. (Transl. by D. D. Davis & R. Zangerl) Urbana, Ill.: Univ. of Illinois Press, 1966.

Heron, W. The pathology of boredom. *Scientific American*, 1957, **196**, 52-56.

Hershenson, M., Munsinger, H., & Kessen, W. Preference for shapes on intermediate variability in the newborn human. *Science*, 1965, **147**, 630-631.

Himwich, W. A., & Himwich, H. E. (Eds.) *The developing brain*. Vol. 9. *Progress in brain research*. Amsterdam: Elsevier, 1964.

Hinde, R. A. Unitary drives. *Animal Behaviour*, 1959, **7**, 130-141. (a)

Hinde, R. A. Some recent trends in ethology. In S. Koch (Ed.), *Psychology: A study of a science*. Vol. 2. New York: McGraw-Hill, 1959. Pp. 561-610. (b)

Hinde, R. A., & Spencer-Booth, Y. The effect of social companions on mother-infant relations in rhesus monkeys. In D. Morris (Ed.), *Primate ethology*. Chicago, Ill.: Aldine, 1967. Pp. 267-286. (a)

Hinde, R. A., & Spencer-Booth, Y. The behaviour of socially living rhesus monkeys in their first two and a half years. *Animal Behaviour*, 1967, **15**, 169-196. (b)

Horn, B. Physiological and psychological aspects of selective perception. In D. S. Lehrman, R. A. Hinde, & E. Shaw (Eds.), *Advances in the study of behavior*. Vol. 1. New York: Academic Press, 1965. Pp. 155-215.

Houston, J. P., & Mednick, S. A. Creativity and the need for novelty. *Journal of Abnormal and Social Psychology*, 1963, **66**, 137-141.

Howard, I. P., Craske, B., & Templeton, W. B. Visuomotor adaptation to discordant exafferent stimulation. *Journal of Experimental Psychology*, 1965, **70**, 189-191.
Howard, K. I. A test of stimulus-seeking behavior. *Perceptual and Motor Skills*, 1961, **13**, 416.
Hubel, D. H., Henson, C. O., Rupert, A., & Galambos, R. "Attention" units in the auditory cortex. *Science*, 1959, **129**, 1279-1280.
Hughes, A. F. W. *Aspects of neural ontogeny*. New York: Academic Press, 1968.
Hughes, R. N. Food deprivation and locomotor exploration in the white rat. *Animal Behaviour*, 1965, **13**, 30-32.
Hughes, R. N. A re-examination of the effects of age on novelty reactions and exploration in rats. *Australian Journal of Psychology*, 1968, **20**, 197-201.
Hughes, R. N. Exploration in three laboratory rodents. *Perceptual and Motor Skills*, 1969, **28**, 90. (a)
Hughes, R. N. Social facilitation of locomotion and exploration in rats. *British Journal of Psychology*, 1969, **60**, 385-388. (b)
Hunt, J. McV., & Quay, H. Early vibratory experience and the question of innate reinforcement value of vibration and other stimuli: A limitation on the discrepancy (burnt soup) principle in motivation. *Psychological Review*, 1961, **68**, 149-156.
Hutt, C. Exploration and play in children. In P. A. Jewell & C. Loizos (Eds.), *Play, exploration and territory in mammals*. New York: Academic Press, 1966. Pp. 61-81.
Jacobson, M. *Developmental neurobiology*. New York: Holt, 1970.
Jarrard, L. E. Behavior of hippocampal lesioned rats in home cage and novel situations. *Physiology & Behavior*, 1968, **3**, 65-70.
Jarrard, L. E., & Bunnell, B. M. Open-field behavior of hippocampal-lesioned rats and hamsters. *Journal of Comparative and Physiological Psychology*, 1968, **66**, 500-502.
Jay, P. Field studies. In A. M. Schrier, H. F. Harlow, & F. Stollnitz (Eds.), *Behavior of nonhuman primates*. Vol. 2. New York: Academic Press, 1965. Pp. 525-595.
Jeffrey, W. E. New technique for motivating and reinforcing children. *Science*, 1955, **121**, 371.
Jensen, G. D., & Bobbitt, R. A. An observational methodology and preliminary studies of mother-infant interaction in monkeys. In B. M. Foss (Ed.), *Determinants of infant behaviour*. Vol. 3. New York: Wiley, 1965. Pp. 47-63.
Jensen, G. D., & Bobbitt, R. A. Implications of primate research for understanding infant development. *Science & Psychoanalysis*, 1968, **12**, 55-81.
Jewell, P. A., & Loizos, C. (Eds.) *Play, exploration and territory in mammals*. New York: Academic Press, 1966.
Joffe, J. M. *Prenatal determinants of behaviour*. Oxford: Pergamon Press, 1969.
Kamback, M. Effect of hippocampal lesions and food deprivation on response for stimulus change. *Journal of Comparative and Physiological Psychology*, 1967, **63**, 231-235.
Kandel, E., & Wachtel, H. The functional organization of neural aggregates in aplysia. In F. D. Carlson (Ed.), *Physiological and biochemical aspects of nervous integration*. Englewood Cliffs, N. J.: Prentice-Hall, 1968. Pp. 17-65.
Karmel, B. Z. Randomness, complexity, and visual preference behavior in the hooded rat and domestic chick. *Journal of Comparative and Physiological Psychology*, 1966, **61**, 487-489.
Kaufman, I. C., & Rosenblum, L. A. A behavioral taxonomy for *Macaca nemestrina* and

Macaca radiata: Based on longitudinal observation of family groups in the laboratory. *Primates,* 1966, **7,** 206-258.

Kavanau, J. L. Behavior of captive white-footed mice. In E. P. Willems & H. L. Raush (Eds.), *Naturalistic viewpoints in psychological research.* New York: Holt, 1969. Pp. 221-270.

Kessen, W. Sucking and looking: Two organized congenital patterns of behavior in the human newborn. In H. W. Stevenson, E. H. Hess, & H. L. Rheingold (Eds.), *Early behavior—comparative and developmental approaches.* New York: Wiley, 1967. Pp. 147-179.

Kiernan, C. C. Positive reinforcement by light: Comments on Lockard's article. *Psychological Bulletin,* 1964, **62,** 351-357.

Kimble, D. P., Bagshaw, M. H., & Pribram, K. H. The GSR of monkeys during orienting and habituation after selective partial ablations of the cingulate and frontal cortex. *Neuropsychologia,* 1965, **3,** 121-128.

Kimmel, H. D. Adaptation of the GSR under repeated applications of a visual stimulus. *Journal of Experimental Psychology,* 1964, **68,** 421-422.

King, D. L. A review and interpretation of some aspects of the infant-mother relationship in mammals and birds. *Psychological Bulletin,* 1965, **63,** 143-155.

King, J. A. Maternal behavior and behavioral development in 2 subspecies of *Peromyscus maniculatus. Journal of Mammalogy,* 1958, **39,** 177-190.

King, J. A. Species specificity and early experience. In G. Newton & S. Levine (Eds.), *Early experience and behavior. The psychobiology of development.* Springfield, Ill.: Thomas, 1968. Pp. 42-64.

King, J. A., & Eleftheriou, B. E. Effects of early handling upon adult behavior in 2 subspecies of deermice, *Peromyscus maniculatus. Journal of Comparative and Physiological Psychology,* 1959, **52,** 82-88.

Kirkby, R. J., Stein, D. G., Kimble, R. J., & Kimble, D. P. Effects of hippocampal lesions and duration of sensory input on spontaneous alternation. *Journal of Comparative and Physiological Psychology,* 1967, **64,** 342-345.

Kish, G. B. Studies of sensory reinforcement. In W. K. Konig (Ed.), *Operant behavior: Areas of research and application.* New York: Appleton, 1966. Pp. 109-159.

Klosovskii, B. N. (Ed.) *The development of the brain and its disturbance by harmful factors.* (Transl. by Basil Haigh) New York: Macmillan, 1963.

Komisaruk, B. R. Synchrony between limbic system theta activity and rhythmical behavior in rats. *Journal of Comparative and Physiological Psychology,* 1970, **70,** 482-492.

Korner, A. F., Chuck, B., & Dontchos, S. Organismic determinants of spontaneous oral behavior in neonates. *Child Development,* 1968, **39,** 1145-1157.

Korner, A. F., & Grobstein, R. Visual alertness as related to soothing in neonates: Implications for maternal stimulation and early deprivation. *Child Development,* 1966, **37,** 867-876.

Krech, D., Rosenzweig, M. R., & Bennett, E. L. Environmental impoverishment, social isolation and changes in brain chemistry and anatomy. *Physiology & Behavior,* 1965, **1,** 99-104.

Kuo, S. C. H., & Marshall, H. R. Visual and tactual curiosity of deaf and hearing children. *Journal of Genetic Psychology,* 1968, **113,** 183-193.

Kuo, Z.-Y. *The dynamics of behavior development. An epigenetic view.* New York: Random House, 1967.

Lashley, K. S. Experimental analysis of instinctive behavior. *Psychological Review,* 1938, **45,** 445-471.

Leaton, R. N. Exploratory behavior in rats with hippocampal lesions. *Journal of Comparative and Physiological Psychology*, 1965, **59**, 325-330.

Leaton, R. N. Effects of scopolamine on exploratory motivated behavior. *Journal of Comparative and Physiological Psychology*, 1968, **66**, 524-527.

Lehrman, D. S. A critique of Konrad Lorenz's theory of instinctive behavior. *Quarterly Review of Biology*, 1953, **28**, 337-363.

Lehrman, D. S. Interaction of hormonal and experiential influences on development of behavior. In E. L. Bliss (Ed.), *Roots of behavior*. New York: Harper, 1962. Pp. 142-156.

Lepley, W. M. Variability as a variable. *Journal of Psychology*, 1954, **37**, 19-25.

Lester, D. Exploratory behavior in peripherally blinded rats. *Psychonomic Science*, 1967, **8**, 7-8. (a)

Lester, D. Sex differences in exploration: Toward a theory of exploration. *Psychological Record*, 1967, **17**, 55-62. (b)

Lester, D. Response alternation: A review. *Journal of Psychology*, 1968, **69**, 131-142. (a)

Lester, D. Curiosity and the role of the individual in the imposition of change. *Psychological Reports*, 1968, **22**, 1109-1112. (b)

Lester, D. The effect of fear and anxiety on exploration and curiosity: Toward a theory of exploration. *Journal of Genetic Psychology*, 1968, **79**, 105-120. (c)

Leuba, C., & Friedlander, B. Z. Effects of controlled audiovisual reinforcement on infants' manipulative play in the home. *Journal of Experimental Child Psychology*, 1968, **6**, 87-99.

Leventhal, G. S., & Killackeg, H. Adrenalin, stimulation and preference for familiar stimuli. *Journal of Comparative and Physiological Psychology*, 1968, **65**, 152-155.

Lewis, M., Kagan, J., Campbell, H., & Kalafat, J. The cardiac response as a correlate of attention in infants. *Child Development*, 1966, **37**, 63-71. (a)

Lewis, M., Kagan, J., & Kalafat, J. Patterns of fixation in the young infant. *Child Development*, 1966, **37**, 331-341. (b)

Leyhausen, P. Verhaltensstudien an Katzen. *Zeitschrift fuer Tierpsychologie*, 1956, (Monogr. Suppl.), 1-120.

Lieberman, J. N. Playfulness and divergent thinking: An investigation of their relationship at the kindergarten level. *Journal of Genetic Psychology*, 1965, **107**, 219-224.

Little, E. B., & Creaser, J. W. Epistemic curiosity and man's higher nature. *Psychological Reports*, 1968, **23**, 615-624.

Littman, R. A. Motives, history and causes. In M. R. Jones (Ed.), *Nebraska symposium on motivation*. Lincoln, Neb.: Univ. of Nebraska Press, 1958. Pp. 114-168.

Livson, N. Towards a differentiated construct of curiosity. *Journal of Genetic Psychology*, 1967, **111**, 73-84.

Lockard, R. B. A method of analysis and classification of repetitive response systems. *Psychological Review*, 1964, **71**, 141-147.

Locke, M. (Ed.) *The emergence of order in developing systems*. New York: Academic Press, 1968.

Loizos, C. Play in mammals. In P. A. Jewell & C. Loizos (Eds.), *Play, exploration and territory in mammals*. New York: Academic Press, 1966. Pp. 1-9.

Loizos, C. Play behavior in higher primates: A review. In D. Morris (Ed.), *Primate ethology*. Chicago, Ill.: Aldine, 1967. Pp. 176-218.

Lore, R. K., Kam, B., & Newby, V. Visual and nonvisual depth avoidance in young and adult rats. *Journal of Comparative and Physiological Psychology*, 1967, **64**, 525-528.

Lore, R. K., & Levowitz, A. Differential rearing and free versus forced exploration. *Psychonomic Science,* 1966, **5,** 421-422.

Lovaas, O. I., Freitag, G., Gold, V. J., & Kassorla, I. C. Recording apparatus and procedure for observation of behaviors of children in free play settings. *Journal of Experimental Child Psychology,* 1965, **2,** 108-120.

Lynn, R. *Attention arousal and the orientation reaction.* Oxford: Pergamon Press, 1966.

McCall, R. B., & Kagan, J. Attention in the infant: Effects of complexity, contour, perimeter, and familiarity. *Child Development,* 1967, **38,** 939-952.

McElroy, W. D., & Glass, B. *The chemical bases of development.* Baltimore, Md.: Johns Hopkins Press, 1958.

Mackworth, J. F. *Vigilance and habituation. A neuropsychological approach.* Baltimore, Md.: Penguin Books, 1969.

Mackworth, N. H., & Otto, D. A. Habituation of the visual orienting response (VOR) in young children. *Perception & Psychophysics,* 1970, **7,** 173-178.

McReynolds, P. Exploratory behavior: A theoretical interpretation. *Psychological Reports,* 1962, **11,** 311-318.

McReynolds, P. Reaction to novel and familiar stimuli as a function of schizophrenic withdrawal. *Perceptual and Motor Skills,* 1963, **16,** 847-850.

Marler, P., & Hamilton, W. J. *Mechanisms of animal behavior.* New York: Wiley, 1966.

Maslow, A. H. The need to know and the fear of knowing. *Journal of Genetic Psychology,* 1963, **68,** 111-126.

Mason, W. A. Determinants of social behavior in young chimpanzees. In A. M. Schrier, H. F. Harlow, & F. Stollnitz (Eds.), *Behavior of nonhuman primates.* Vol. 2. New York: Academic Press, 1965. Pp. 335-364.

Mason, W. A. Motivational aspects of social responsiveness in young chimpanzees. In H. W. Stevenson & E. H. Hess (Eds.), *Early behavior—comparative and developmental approaches.* New York: Wiley, 1967. Pp. 103-126.

Mason, W. A. Early social deprivation in the nonhuman primates: Implications for human behavior. In D. C. Glass (Ed.), *Environmental influences.* New York: Rockefeller Univ. Press, 1968. Pp. 70-101.

Maw, W. H., & Maw, E. W. Selection of unbalanced and unusual designs by children high in curiosity. *Child Development,* 1962, **33,** 917-922. (a)

Maw, W. H., & Maw, E. W. Children's curiosity as an aspect of reading comprehension. *Reading Teacher,* 1962, **15,** 236-240. (b)

Medinnus, G. R., & Love, J. M. The relation between curiosity and security in preschool children. *Journal of Genetic Psychology,* 1965, **107,** 91-98.

Mendel, G. Children's preferences for differing degrees of novelty. *Child Development,* 1965, **36,** 453-465.

Menzel, E. W., Jr. The effects of cumulative experience on responses to novel objects in young isolation-reared chimpanzees. *Behaviour,* 1963, **21,** 1-12.

Menzel, E. W., Jr. Patterns of responsiveness in chimpanzees reared through infancy under environmental restriction. *Psychologische Forschung,* 1964, **27,** 337-365.

Menzel, E. W., Jr. Naturalistic and experimental approaches to primate behavior. In E. P. Willems & H. L. Raush (Eds.), *Naturalistic viewpoints in psychological research.* New York: Holt, 1969. Pp. 78-121.

Menzel, E. W., Jr., Davenport, R. K., & Rogers, C. M. Some aspects of behavior toward novelty in young chimpanzees. *Journal of Comparative and Physiological Psychology,* 1961, **54,** 16-19.

Menzel, E. W., Jr., Davenport, R. K., & Rogers, C. M. The effects of environmental restriction upon the chimpanzee's responsiveness to objects. *Journal of Comparative and Physiological Psychology*, 1963, **56**, 78-85. (a)

Menzel, E. W., Jr., Davenport, R. K., & Rogers, C. M. Effects of environmental restriction upon the chimpanzee's responsiveness in novel situations. *Journal of Comparative and Physiological Psychology*, 1963, **56**, 329-334. (b)

Mesarovic, M. D. (Ed.) *Systems theory and biology*. Berlin: Springer, 1968.

Meyer, M. E. Arousal and reinforcing properties of light onset as a function of duration and of visual complexity for chicks. *Psychological Reports*, 1968, **23**, 419-424.

Meyers, W. J. Critical period for facilation of exploratory behavior by infantile experiences. *Journal of Comparative and Physiological Psychology*, 1962, **55**, 1099-1101.

Meyers, W. J., & Cantor, G. N. Infants' observing and heart period responses as related to novelty of visual stimuli. *Psychonomic Science*, 1966, **5**, 239-240.

Meyers, W. J., & Cantor, G. N. Observing and cardiac responses of human infants to visual stimuli. *Journal of Experimental Child Psychology*, 1967, **5**, 16-25.

Millar, S. *The psychology of play*. Baltimore, Md.: Penguin Books, 1968.

Milner, E. *Human neural and behavioral development. A relational inquiry with implications for personality*. Springfield, Ill.: Thomas, 1967.

Minkowski, A. (Ed.) *Regional development of the brain in early life*. Oxford: Blackwell, 1967.

Minton, H. L. A. A replication of perceptual curiosity as a function of stimulus complexity. *Journal of Experimental Psychology*, 1963, **66**, 522-524.

Mirzakarimova, M. G., Stel'makh, L. N., & Troshikhin, V. A. Directed changes in passive-defensive and investigatory reflexes in ontogenesis. *Pavlov Journal of Higher Nervous Activity*, 1958, **8**, 697-702.

Mitchell, G. D. Intercorrelations of maternal and infant behaviors in *Macaca mulatta*. *Primates*, 1968, **9**, 85-92.

Mitchell, G. Paternalistic behavior in primates. *Psychological Bulletin*, 1969, **71**, 399-417.

Mittman, L. R., & Terrel, G. An experimental study of curiosity in children. *Child Development*, 1964, **35**, 851-855.

Moltz, H. Imprinting: An epigenetic approach. *Psychological Review*, 1963, **70**, 123-128.

Moore, J. A. (Ed.) *Ideas in modern biology*. Garden City, N.Y.: Natural History Press, 1965.

Morton, J. R. C. Effects of early experience "handling and gentling" in laboratory animals. In M. W. Fox (Ed.), *Abnormal behavior in Animals*. Philadelphia, Pa.: Saunders, 1968. Pp. 261-292.

Moyer, K. E., & Gilmor, B. H. Attention span for experimentally designed toys. *Journal of Genetic Psychology*, 1955, **87**, 189-201.

Munsinger, H., & Weir, M. W. Infant's and young children's preference for complexity. *Journal of Experimental Child Psychology*, 1967, **5**, 69-73.

Murray, S. K., & Brown, L. T. Human exploratory behavior in a "natural" versus a laboratory setting. *Perception & Psychophysics*, 1967, **2**, 230-232.

Neiberg, A. Effect of rearing conditions on exploratory behavior. *Psychological Reports*, 1964, **15**, 207-210.

Neuringer, A. J. Animals respond for food in the presence of free food. *Science*, 1969, **166**, 399-401.

Newton, G., & Levine, S. (Eds.) *Early experience and behavior. The psychobiology of development*. Springfield, Ill.: Thomas, 1968.

Nielson, H. C., McIver, A. H., & Boswell, R. S. Effect of septal lesions on learning, emotionality, activity, and exploratory behavior in rats. *Experimental Neurology*, 1965, **11**, 147-157.

Nissen, H. W. Phylogenetic comparison. In S. S. Stevens (Ed.), *Handbook of experimental psychology*. New York: Wiley, 1951. Pp. 347-386.

Norton, S. On the discontinuous nature of behavior. *Journal of Theoretical Biology*, 1968, **21**, 229-243. (a)

Norton, S. Time lapse photographic analysis of rat behavior. *Federation Proceedings, Federation of American Societies for Experimental Biology*, 1968, **27**, 437. (Abstract) (b)

Novikoff, A. B. The concept of integrative levels and biology. *Science*, 1943, **101**, 209-215.

Odom, B., Mitchell, G., & Lindburg, D. A device to record primate social behavior. *Primates*, 1970, **11**, 93-96.

Odom, R. D. Effects of auditory and visual stimulus deprivation and satiation on children's performance in an operant task. *Journal of Experimental Child Psychology*, 1964, **1**, 16-25.

Paradowski, W. Effect of curiosity on incidental learning. *Journal of Educational Psychology*, 1967, **58**, 50-55.

Penney, R. K., & McCann, B. The children's reactive curiosity scale. *Psychological Reports*, 1964, **15**, 323-334.

Pereboom, A. C. Systematic-representative study of spontaneous activity in the rat. *Psychological Reports*, 1968, **22**, 717-732.

Peters, R. D., & Penney, R. K. Spontaneous alternation of high and low reactively curious children. *Psychonomic Science*, 1966, **4**, 139-140.

Poole, T. B. Aggressive play in polecats. In P. A. Jewell & C. Loizos (Eds.), *Play, exploration and territory in mammals*. New York: Academic Press, 1966. Pp. 23-44.

Prechtl, H. F. R. The directed head turning response and allied movements of the human baby. *Behaviour*, 1958, **13**, 212-242.

Prechtl, H. F. R. Problems of behavioral studies in the newborn infant. In D. S. Lehrman, R. A. Hinde, & E. Shaw (Eds.), *Advances in the study of behavior*. Vol. 1. New York: Academic Press, 1965. Pp. 75-98.

Purpura, D. P., & Schadé, J. P. (Eds.) *Growth and maturation of the brain*. Vol. 4. *Progress in brain research*. Amsterdam: Elsevier, 1964.

Quarton, G. C., Melnechuk, T., & Schmitt, F. O. (Eds.) *The neurosciences*. New York: Rockefeller Univ. Press, 1967.

Rheingold, H. L. The effect of environmental stimulation upon social and exploratory behaviour in the human infant. In B. M. Foss (Ed.), *Determinants of infant behaviour*. Vol. 1. New York: Wiley, 1961. Pp. 143-171.

Rheingold, H. L. Controlling the infant's exploratory behavior. In B. M. Foss (Ed.), *Determinants of infant behaviour*. Vol. 2. New York: Wiley, 1963. Pp. 171-178.

Rheingold, H. L., Stanley, W. C., & Doyle, G. A. Visual and auditory reinforcement of a manipulatory response in the young child. *Journal of Experimental Child Psychology*, 1964, **1**, 316-326.

Richards, W. J., & Leslie, G. R. Food and water deprivation as influences on exploration. *Journal of Comparative and Physiological Psychology*, 1962, **55**, 834-837.

Richter, D. (Ed.) *Comparative neurochemistry*. Oxford: Pergamon Press, 1964.

Riesen, A. H. Excessive arousal effects of stimulation after early sensory deprivation. In P. Solomon, P. E. Kubzansky, P. H. Leiderman, J. H. Mendelson, R. Turumbull, & D. Wexler (Eds.), *Sensory deprivation*. Cambridge, Mass.: Harvard Univ. Press, 1961. Pp. 34-40. (a)

Riesen, A. H. Stimulation as a requirement for growth and function in behavioral development. In D. W. Fiske & S. R. Maddi (Eds.), *Functions of varied experience*. Homewood, Ill.: Dorsey Press, 1961. Pp. 57-80. (b)

Riesen, A. H. Studying perceptual development using the technique of sensory deprivation. *Journal of Nervous and Mental Disease*, 1961, **132**, 21-25. (c)

Riesen, A. H. Sensory deprivation. In E. Stellar & J. M. Sprague (Eds.), *Progress in physiological psychology*. Vol. 1. New York: Academic Press, 1966. Pp. 117-147.

Roberts, W. W., Dember, W. N., & Brodwick, M. Alternation and exploration in rats with hippocampal lesions. *Journal of Comparative and Physiological Psychology*, 1962, **55**, 697-700.

Rodgers, W., & Rozin, P. Novel food preferences in thiamine-deficient rats. *Journal of Comparative and Physiological Psychology*, 1966, **61**, 1-4.

Rosenblatt, J. S. The basis of synchrony in the behavioral interaction between the mother and her offspring in the laboratory rat. In B. M. Foss (Ed.), *Determinants of infant behaviour*. Vol. 3. New York: Wiley, 1965. Pp. 3-41.

Rosenblatt, J. S., Turkewitz, G., & Schneirla, T. C. Development of home orientation in newly born kittens. *Transactions of the New York Academy of Sciences*, 1969, **31**, 231-250.

Rosenblum, L. A., & Kaufman, I. C. Variations in infant development and response to maternal loss in monkeys. *American Journal of Orthopsychiatry*, 1968, **38**, 418-426.

Rosenzweig, M. R. Environmental complexity, cerebral change, and behavior. *American Psychologist*, 1966, **21**, 321-332.

Rubel, E. W. A comparison of somatotopic organization in sensory neocortex of newborn kittens and adult cats. Unpublished doctoral dissertation, Michigan State University, 1969.

Rubenstein, J. Maternal attentiveness and subsequent exploratory behavior in the infant. *Child Development*, 1967, **38**, 1089-1100.

Rumbaugh, D. M. Maternal care in relation to infant behavior in the squirrel monkey. *Psychological Reports*, 1965, **16**, 171-176.

Saayman, G., Ames, E. W., & Moffett, A. Response to novelty as an indicator of visual discrimination in the human infant. *Journal of Experimental Child Psychology*, 1964, **1**, 189-198.

Sackett, G. P. Development of preference for differentially complex patterns by infant monkeys. *Psychonomic Science*, 1966, **6**, 441-442.

Schmitt, F. O. The physical basis of life and learning. *Science*, 1965, **149**, 931-936.

Schmitt, F. O. Molecular biology among the neurosciences. *Archives of Neurology (Chicago)*, 1967, **17**, 561-572.

Schneirla, T. C. Aspects of stimulation and organization in approach/withdrawal processes underlying vertebrate behavioral development. In D. S. Lehrman, R. A. Hinde, & E. Shaw (Eds.), *Advances in the study of behavior*. Vol. 1. New York: Academic Press, 1965. Pp. 1-74.

Scott, J. P. The genetic and environmental differentiation of behavior. In D. B. Harris (Ed.), *The concept of development, an issue in the study of human behavior*. Minneapolis, Minn.: Univ. of Minnesota Press, 1957. Pp. 59-77.

Scott, J. P. *Animal behavior*. Chicago, Ill.: Univ. of Chicago Press, 1958. (a)

Scott, J. P. Critical periods in the development of social behavior in puppies. *Psychosomatic Medicine*, 1958, **20**, 42-54. (b)

Scott, J. P. Principles of ontogeny of behavior patterns. *Proceedings of the 16th International Congress of Zoology, 1963*, 1965. Pp. 363-366.

Scott, J. P. The development of social motivation. In D. Levine (Ed.), *Nebraska symposium on motivation*. Lincoln, Neb.: Univ. of Nebraska Press, 1967. Pp. 111-132.

Scott, J. P. Social facilitation and allelomimetic behavior. In E. C. Simmel, R. A. Hoppe, & G. A. Milton (Eds.), *Social facilitation and imitative behavior*. Boston, Mass.: Allyn & Bacon, 1968. Pp. 55-72. (a)

Scott, J. P. *Early experience and the organization of behavior*. Belmont, Calif.: Brooks/Cole Publ. Co., 1968. (b)

Scott, J. P., & Bronson, F. H. Experimental exploration of the et-epimeletic or care-soliciting behavioral system. In P. H. Leiderman & D. Shapiro (Eds.), *Psychobiological approaches to social behavior*. Stanford, Calif.: Stanford Univ. Press, 1964. Pp. 174-193.

Sheldon, A. B. Preference for familiar versus novel stimuli as a function of the familiarity of the environment. *Journal of Comparative and Physiological Psycholoy*, 1969, **67**, 516-521.

Simmel, E. C. Social facilitation of exploratory behavior in rats. *Journal of Comparative and Physiological Psychology*, 1962, **55**, 831-833.

Simmel, E. C., & McGee, D. P. Social facilitation of exploratory behavior in rats: Effects of increased exposure to novel stimuli. *Psychological Reports*, 1966, **18**, 587-590.

Simpson, C. G. *Principles of animal taxonomy*. New York: Columbia Univ. Press, 1961.

Singh, S. D., & Manocha, S. N. Reactions of rhesus monkey and langur to novel situations. *Primates*, 1966, **7**, 259-262.

Skoglund, S. Growth and differentiation. *Physiological Review*, 1969, **31**, 19-42.

Small, W. S. Notes on the psychic development of the young white rat. *American Journal of Psychology*, 1899, **11**, 80-100.

Smock, C. D., & Holt, B. G. Children's reaction to novelty: An experimental study of 'curiosity motivation.' *Child Development*, 1962, **33**, 631-642.

Sokolov, E. N. Neuronal models and the orienting reflex. In M. A. B. Brazier (Ed.), *The central nervous system and behavior*. New York: Josiah Macy, Jr. Found., 1960. Pp. 187-276.

Sokolov, E. N. Higher nervous functions: The orienting reflex. *Annual Review of Physiology*, 1963, **25**, 545-580.

Sperry, R. W. Embryogenesis of behavioral nerve nets. In R. L. DeHaan & H. Ursprung (Eds.), *Organogenesis*. New York: Holt, 1965. Pp. 161-186.

Spitz, R. A. Purposive grasping. *Journal of Personality*, 1951, **1**, 141-148.

Stein, L. Habituation and stimulus novelty: A model based on classical conditioning. *Psychological Review*, 1966, **73**, 352-356.

Stelzner, D. J. The postnatal development of synaptic pattern in the lumbosacral spinal cord and of responses in the caudal parts of the body of the albino rat. Unpublished doctoral dissertation, University of Pennsylvania, 1969.

Suchman, R. G., & Trabasso, T. Color and form preference in young children. *Journal of Child Psychology*, 1966, **3**, 177-187.

Symmes, D. Contingent illumination change and operant responding by kittens. *Psychological Reports*, 1963, **12**, 531-534.

Thoman, E., Wetzel, A., & Levine, S. Learning in the neonatal rat. *Animal Behaviour*, 1968, **16**, 54-57.

Thomas, H. Preferences for random shapes: Ages six through nineteen years. *Child Development*, 1966, **37**, 843-859.

Thompson, M. E., & Thompson, J. P. Alternation and repetition in a multiple choice situation. *Psychological Reports*, 1964, **11**, 523-527.

Thompson, R. F., & Shaw, J. A. Behavioral correlates of evoked activity recorded from association areas of the cerebral cortex. *Journal of Comparative and Physiological Psychology*, 1965, **60**, 329-339.

Thompson, R. F., & Spencer, W. A. Habituation: A model phenomenon for the study of neuronal substrates of behavior. *Psychological Review*, 1966, **173**, 16-43.

Tilney, F. Behavior in its relation to the development of the brain. II. Correlation between the development of the brain and behavior in the albino rat from embryonic states to maturity. *Bulletin of the Neurological Institute of New York*, 1934, **3**, 252-358.

Timberlake, W. D., & Birch, D. Complexity, novelty and food deprivation as determinants of speed of shift of behavior. *Journal of Comparative and Physiological Psychology*, 1967, **63**, 545-548.

Tolman, C. W. The role of the companion in social facilitation of animal behavior. In E. C. Simmel, R. A. Hoppe, & G. A. Milton (Eds.), *Social facilitation and imitative behavior*. Boston, Mass.: Allyn & Bacon, 1968. Pp. 33-54.

Voronim, L. G., Leontiev, A. N., Luria, A. R., Sokolov, E. N., & Vinogradora, O. S. (Eds.), *Orienting reflex and exploratory behavior*. (Transl. by V. Shmelov & K. Hanes. English Trans. Ed., D. B. Lindsley) Washington, D.C.: Am. Inst. Biol. Sci., 1965.

Walden, A. M. Studies of exploratory behavior in the albino rat. *Psychological Reports*, 1968, **22**, 483-493.

Walsher, D. N. (Ed.) *Early childhood: The development of self-regulating mechanisms*. New York: Academic Press, 1971, in press.

Weiss, P. A. Nervous system (Neurogenesis). In B. H. Willier, P. A. Weiss, & V. Hamburger (Eds.), *Analysis of development*. Philadelphia, Pa.: Saunders, 1955. Pp. 346-401.

Weiss, P. A. *Dynamics of development: Experiments and inferences*. New York: Academic Press, 1968.

Weiss, P. A. Whither life science?: What are the great unanswered questions that serve as beacons for biological research? *American Scientist*, 1970, **58**, 156-163.

Welker, W. I. An analysis of exploratory and play behavior in animals. In D. W. Fiske & S. R. Maddi (Eds.), *Functions of varied experience*. Homewood, Ill.: Dorsey Press, 1961. Pp. 175-226.

Welker, W. I. Analysis of sniffing of the albino rat. *Behaviour*, 1964, **22**, 223-244.

Wells, P. A., Lowe, G., Sheldon, M. H., & Williams, D. I. Effects of infantile stimulation and environmental familiarity on exploratory behaviour in the rat. *British Journal of Psychology*, 1969, **60**, 389-393.

Welt, C., & Welker, W. I. Postural and locomotor development of a gibbon (*Hylobates lar*). *Physical Anthropologist*, 1963, **21**, 425. (Abstract)

White, R. W. Motivation reconsidered: The concept of competence. *Psychological Review*, 1959, **66**, 297-333.

Whittier, J. L., & Littman, R. A. Very early weaning, weaning stimulation, and the adult exploratory and avoidance behavior in white rats. *Canadian Journal of Psychology*, 1965, **19**, 288-303.

Whyte, L. L., Wilson, A. G., & Wilson, D. *Hierarchical structure*. Amsterdam: Elsevier, 1969.

Williams, C. D., Carr, R. M., & Peterson, H. W. Maze exploration in young rats of four ages. *Journal of Genetic Psychology*, 1966, **109**, 241-247.
Williams, G. C. *Adaptation and natural selection: A critique of some current evolutionary thought.* Princeton, N.J.: Princeton Univ. Press, 1966.
Willier, B. H., Weiss, P. A., & Hamburger, V. (Eds.) *Analysis of development.* Philadelphia, Pa.: Saunders, 1955.
Wolstenholme, G. E. W., & O'Connor, M. (Eds.) *Somatic stability in the newly born.* Boston, Mass.: Little, Brown, 1961.
Wolstenholme, G. E. W., & O'Connor, M. (Eds.) *Growth of the nervous system.* Boston, Mass.: Little, Brown, 1968.

CHAPTER 6

THE ONTOGENY OF SEXUALITY

Richard E. Whalen

I. Introduction

Reproductive behaviors are unlike many behaviors available to organisms in that they do not appear in their complete form until late in development, and, superficially at least, do not seem to be the culmination of a continuous developmental process. In nonprimate mammalian species, mating behavior does not appear until puberty, and when it does appear, it is displayed in a form characteristic of the mature, sexually experienced animal. These characteristics of the mating pattern of lower mammals have led some theorists to conceive of mating behavior as instinctive, that is, as being encoded isomorphically in the genome. Were this the case, there would of course be little point in discussing the ontogeny of mating behavior. As it stands, however, the instinct notion of the determination of sexual behavior is not only conceptually weak (Whalen, 1971), but represents an inadequate understanding of gene action and too narrow a view of the experiential determinants of behavior. An ontogenetic study of mating behavior is possible and can contribute to our understanding of sexuality.

This review will focus upon the ontogeny of sexual behavior in those relatively few mammals which have been studied intensively. Consideration will be given to the role of genes in the expression of mating behavior, the nature and extent to which the behavior is modulated by experience during ontogeny, and the influence of hormones upon the development of those central and peripheral structures which mediate the expression of sexual behavior in adulthood. No attempt will be made here to review all the research relevant to this broad problem. Rather, selected studies will be drawn from the current literature in an attempt to describe various facets of sexual development, areas of understanding and bafflement, and future research orientations.

II. Genes and Sexual Behavior

When discussing the ontogeny of sexual behavior, it is logical to begin with the genome, the starting point of ontogeny. However, it should be pointed out that sexual behavior, in the individual, is always jointly determined by the genome and the environment; thus it is not possible, in the individual, to distinguish the contribution of each (Whalen, 1971). This, in essence, is the epigenetic model of behavioral development (Moltz, 1965). To investigate the genetic contribution to sexual behavior, we must therefore examine population differences in mating.

The most extensive genetic analysis of male mating to date has been carried out by McGill using inbred mice. McGill (1965) has carefully described the mating pattern of the male mouse and has subjected the pattern to quantitative analysis. Table I illustrates this type of analysis for two inbred strains (C57Bl and DBA) and the hybrid derived from these strains. The table is divided into five components: (a) measures on which the hybrids behave quantitatively like the C57Bl/6J males; (b) measures on which the hybrids were similar to the DBA/2J males; (c) measures on which the hybrid scores were intermediate between the scores of the inbred strains; (d) measures on which the hybrids differed from one or both parent strains; and (e) a measure on which the parent strains and the hybrid strain did not differ significantly from one another.

McGill has continued his study of the inherited characteristics in

Table I
SEXUAL RESPONSES OF INBRED AND HYBRID MALE MICE[a]

Measure	Strain		
	Inbred C57Bl/6J	Inbred DBA/2J	Hybrid BDF1
A.			
1. Mount latency (seconds)	42	85	42
2. Number of intromission thrusts	400	129	546
3. Number of intromissions before ejaculation	17	5	18
B.			
4. Ejaculation duration (seconds)	23	17	19
5. Time of intromission	15	20	19
6. Number of head mounts	2	0.5	0
C.			
7. Interintromission interval (seconds)	28	137	42
8. Length of mounts without intromission	2	7	3
9. Preintromission mount duration	1	4	2
D.			
10. Number of mounts without intromission	18	16	7
11. Thrusts per intromission	16	20	25
12. Intromission latency (seconds)	107	179	93
E.			
13. Ejaculation latency (seconds)	1252	1376	1091

[a]From McGill (1965).

the mating patterns of C57Bl/6J and DBA/2J mice. In a recent study (1970), he focused upon measures of intromission latency in parental strains, in hybrids, and in backcross offspring. Through careful quantitative analysis he was able to separate the total variance due to individual differences in intromission latency into two components, genotypic variance (Vg) and environmental variance (Ve). (V genotype and V environment = V population.) In this population Ve = .045 and Vg = .007. After obtaining these values, McGill was then able to establish the degree of genetic determination (°GD = Vg/Vp). The °GD was found to be .135, indicating that 13.5% of the trait variance was due to genotypic variance. A similar analysis of ejaculation latency measures for the same animals yielded a degree of genetic determination of 18%. Thus, between 13% and 18% of the individual differences in intromission and ejaculation latency could be accounted for by individual differences in genome. Although these values may seem low, they are in fact within the range often found for behavioral phenotypes (Hirsch & Boudreau, 1958; Manning, 1961).

The genetic analysis of female sexuality is less complete than that of the male, yet the literature does contain a valuable study carried out by Goy and Jakway (1959) dealing with the inheritance of sexual patterns in female guinea pigs. They examined five major characteristics of receptivity in two strains, as well as in hybrid and backcross offspring. Their data are shown in Table II. These data allowed Goy and Jakway to determine that in the guinea pig three genetic factors influence female sexuality: one governing the latency, probability, and duration of heat, a second governing the duration of the lordosis response, and a third governing the probability of male-type mounting responses.

Tables I and II reveal several important facts concerning the relationship between genome and behavior. First, inbred strains differ in the quantitative aspects of their behavior. Second, the magnitude of strain differences in behavior is a function of the particular behavioral measure taken, indicating of course that mating behavior is not a unitary response dimension. Third, aspects of sexual behavior are heritable. (The behavior of hybrids is systematically related to the behavior of the parent strains.) And fourth, the inheritance of sexual performance may exhibit the genetic characteristics of additivity, heterosis, or the dominance of one parent over another, depending upon the measure of behavior taken. But of course these same data also indicate that some characteristics of sexual activity in the populations studied are only loosely tied to the genome and are strongly influenced by environmental variations.

The reader interested in pursuing the relationship between the genome and the development of sexual behavior should consult the papers and reviews of McGill (1962, 1965), McGill and Ransom (1968), McGill and Tucker (1964), Goy and Jakway (1959), Jakway

Table II
SEXUAL RESPONSES OF INBRED AND HYBRID FEMALE GUINEA PIGS TREATED WITH ESTROGEN AND PROGESTERONE[a]

Strain	% Tests + for estrus	Latency to estrus (hr)	Duration of estrus (hr)	Duration of maximum lordosis (sec)	No. of male-type mounts per estrus
2	94.0	4.2	7.9	13.6	1.0
13	79.3	6.6	4.8	24.5	19.5
F 1	98.6	4.8	7.2	19.4	5.9
F 2	95.4	4.7	7.2	19.5	7.1
Backcross to 2	94.4	4.3	7.9	13.8	5.1
Backcross to 13	86.7	5.7	5.9	22.4	11.5

[a]From Goy and Jakway (1959).

(1959), Manning (1961), Valenstein, Riss, and Young (1954), and Whalen (1961a). These studies, although they provide convincing evidence that sexuality is heritable, do not, however, provide much insight into the multitude of physiological and maturational events intermediate between the genome and behavior. To date, we know almost nothing about *how* individual differences in the genome result in the development of individual differences in sexual behavior. It is indeed unfortunate that this approach to the understanding of sexual development has not been more fully exploited since genetic analysis *can* be a powerful tool for generating hypotheses and for providing greater understanding of the determinants involved in ontogenetic development.

III. Hormones and the Development of Sexual Behavior

Distinct male-female differences exist in the nature of sexual responses. Typically the male mammal displays a set of sexual responses which involve mounting the female, pelvic thrusting, intromission, and ejaculation. These responses are usually characterized as "masculine." Typically the female mammal displays a different, yet equally distinctive, set of sexual responses including hopping and running, lordosis, tail deviation, treading and, in some cases, postcoital after-reactions. These responses are usually characterized as "feminine." Genetic analysis might lead us to believe that these sex differences in behavior reflect, in some simple way, sex differences in genetic characteristics. Such a conclusion is not unreasonable in view of the well-known sex differences in chromosome characteristics. A somewhat more sophisticated hypothesis, however, might suggest that these sex-chromosome differences lead to the development of a testis in the male and an ovary in the female and that their respective hormonal secretions then underlie the behavioral differences observed. Current evidence indicates that neither of these hypotheses provides an adequate model for understanding the development of sex-typical mating behaviors. For example, in the rat administration of the ovarian hormones, estrogen and progesterone, to the male does not mimic the effects of administering these same hormones to the female. Similarly, injection of testosterone in the female does not mimic the effects of testosterone injection in the male, indicating that adult male and female rats possess different potentials for responding to testicular and ovarian hormones, respectively.

The modern study of the development of sex differences in hormonal reactivity has had a profound impact upon our current ideas of the ontogeny of sexuality. The germinal study in the area was published in 1959 by Phoenix, Goy, Gerall, and Young. These workers treated pregnant guinea pigs with the androgen, testosterone propionate (TP). The female offspring of mothers treated with a "low" dose of TP during pregnancy were genitally unmodified at birth, while those females whose mothers had received a "high" TP. dose were pseudohermaphroditic at birth, possessing external genitals indistinguishable from those of normal males. When these androgenized animals matured, they were treated with estrogen and progesterone at dosage levels known to induce sexual receptivity in normal females. They were found to be significantly less responsive to gonadal hormones than control females. In other words, fewer of the androgenized females displayed any signs of heat or sexual receptivity, and those which did, exhibited heat periods of longer latency and shorter duration than control females. Table III illustrates these findings, and shows further that the degree of inhibition of responsiveness to estrogen and progesterone was a function of the dose of androgen administered prenatally. Clearly, prenatal androgen exposure can make females malelike in their response to gonadal hormones. This characteristic was revealed even more strikingly when males and pseudohermaphroditic females were treated with testosterone propionate in adulthood and examined for display of masculine behavior (Table IV). Prenatal exposure made the females malelike in their response not only to androgen but to estrogen and progesterone as well. Here, then, were the first data to suggest that underlying the development of both male and female sexual behavior is a hormone-controlled differentiation of the brain. As Phoenix *et al.* put it:

> The embryonic and fetal periods, when the genital tracts are exposed to the influence of as yet unidentified morphogenic substances, are periods of differentiation. The adult period, when the genital tracts are target organs of the gonadal hormones, is a period of functional response as measured by cyclic growth, secretion and motility. The response depends on whether Mullerian or Wolffian duct derivatives have developed, and although generally specific for hormones of the corresponding sex, it is not completely specific. For the neural tissues mediating mating behavior, corresponding relationships seem to exist. The embryonic and fetal periods are periods of organization or 'differentiation' in the direction of masculinization or feminization. Adulthood, when gonadal hormones are being secreted, is a period of activation; neural tissues are the target organs and mating behavior is brought to expression (Phoenix *et al.*, 1959, p. 379).

Table III
FEMALE BEHAVIOR IN NORMAL AND ANDROGENIZED GUINEA PIGS ADMINISTERED 0.664 μg ESTROGEN AND 0.2 mg PROGESTERONE IN ADULTHOOD[a]

Group	Percent of tests positive for estrus	Latency to heat (hr)	Duration of heat (hr)	Maximum lordosis duration (sec)
Control females	96	3.7	7.2	9.3
Genitally unmodified androgenized females	77	5.8	3.3	6.0
Genitally hermaphroditic androgenized females	22	9.2	2.0	2.0
Castrated males	–	–	–	–

[a]From Phoenix *et al.* (1959).

Table IV
MASCULINE BEHAVIOR IN MALE, FEMALE, AND HERMAPHRODITIC FEMALE GUINEA PIGS TREATED WITH TESTOSTERONE PROPIONATE[a]

Group	Mounts/test	Tests before first mount	mg of TP prior to first mount
Spayed control females	5.8	7.0	30.0
Spayed hermaphrodites	15.4	3.0	10.0
Spayed males	20.5	1.5	3.8

[a]From Phoenix *et al.* (1959).

The second paper in what is now a long series of studies dealing with hormones and sexual development was published by Harris and Levine (1962) who treated female rats with testosterone propionate 5 days after birth. When mature, these females failed to display sexual receptivity even after being treated with exogenous estrogen and progesterone. When administered androgen (TP), however, these same females exhibited masculine responses to a more marked degree than control females similarly treated in adulthood.

Both the studies of Phoenix *et al.* and of Harris and Levine imply that hormone-induced sexual differentiation is limited to a particular period of development. Evidence subsequently obtained confirms this hypothesis. Goy, Bridson, and Young (1964), for example, also administered testosterone propionate to guinea pigs, but during different periods of gestation. Each female received 5 mg TP daily for the first 6 days of treatment and 1 mg TP each day thereafter. When these animals matured, they were tested for the display of sexual receptivity following estrogen and progesterone treatment. The findings were quite clear: hormone treatment between days 30-35 of pregnancy caused a marked suppression of receptivity. Treatment during days 25-30 or days 35-40 caused partial suppression; the same treatment earlier than day 25 or later than day 45 of pregnancy caused little if any suppression. Obviously, in the guinea pig, days 30-35 of pregnancy represent the period of maximum susceptibility to the effects of TP.

In the rat, however, susceptibility to the behavioral effects of TP is postnatal, ranging from birth to approximately 10 days after birth. Goy, Phoenix, and Young (1962) demonstrated this fact when they examined the receptivity of female rats treated with TP beginning, respectively, on the first, tenth, or twentieth day after birth. Only females treated on the first or tenth day showed a significant suppression of response to estrogen and progesterone in adulthood. Similar studies by other workers also support this conclusion (e.g., Beach, Noble, & Orndoff, 1969).

These studies relating to the "sensitive period" for androgen-induced heat suppression may appear inconsistent. The sensitive period in the guinea pig occurs before birth, while in the rat it occurs after birth. It should be kept in mind, however, that the gestation period of the guinea pig is 68 days and that the guinea pig is quite mature at birth. The rat, on the other hand, is immature at birth following a 21-day gestation period. Clearly, the issue is not one of prenatal *vs.* postnatal effects, but one of developmental stage. Presumably, gonadal hormones can influence the central nervous system during a particular period of development, a period which is not determined by the event of birth.

In 1965, Grady, Phoenix, and Young carried the initial studies involving TP administration one step further. Specifically, they argued that if the effects of early androgenization in the female suppress "femininity," opposite effects should result if androgens are removed from the male prior to the differentiation period. Accordingly, they castrated male rats 1, 5, 10, 20, 30, 50, or 90 days after birth and studied their responsiveness, respectively, to male and female hormones in adulthood. Spayed females served as controls. Grady *et al.* found that males castrated either at 1 or at 5 days after birth showed female-type lordosis responses when treated with estrogen and progesterone, while males castrated 10 days or later after birth rarely displayed lordosis, even when mounted by males.

Feder and Whalen (1965) also followed this approach and examined the feminine behavior of male rats which were either castrated or castrated and treated with estrogen shortly after birth. Again it was found that neonatally castrated males responded to estrogen and progesterone with the display of lordosis. Interestingly, estrogen treatment in infancy did not facilitate the development of this responsivity; in fact, the estrogen-treated castrate males behaved like neonatally androgenized females and did not respond to female gonadal hormones in adulthood.

Later investigations confirmed these observations (Beach *et al.*, 1969; Whalen & Edwards, 1966, 1967). For example, Whalen and Edwards (1967) studied elicitation of the lordosis response in male and female rats gonadectomized at birth or gonadectomized and treated with exogenous testosterone propionate or estradiol benzoate at birth. Their results are shown in Fig. 1. As can be seen, castration of the male permits the lordotic display in adulthood, while castration of the female has no influence on the subsequent display of lordosis; the presence of the testes or stimulation by TP or EB inhibits

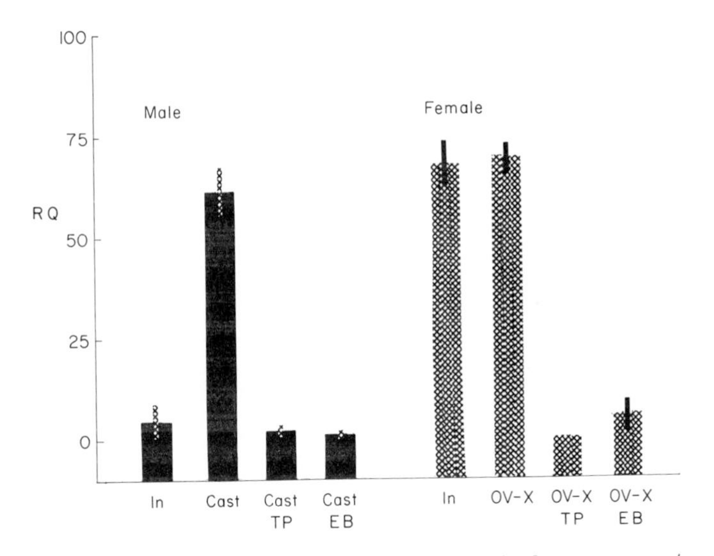

Fig. 1. Relative frequency of lordosis patterns (R.Q. = lordosis responses/mounts × 100) in male and female rats treated with estrogen and progesterone following hormonal manipulation at birth. In, incision at birth; Cast, castration at birth; Cast TP, castration plus testosterone propionate treatment at birth; Cast EB, castration plus estradiol benzoate treatment at birth; OV-X, ovariectomy at birth; OV-X TP, ovariectomy plus testosterone propionate treatment at birth; OV-X EB, ovariectomy plus estradiol benzoate treatment at birth. (Data from Whalen & Edwards, 1967.)

later reactivity to estrogen and progesterone. Thus, testicular secretions, as well as the esters of testosterone and estrogen, can be thought of as "defeminizing" agents.

The defeminization component of hormone-controlled sexual differentiation has been well established. Indeed, many studies have shown that the presence of the testes or the administration of high doses of steroid hormones suppresses responses to estrogen and progesterone in adulthood. Controversy still exists, however, about the "masculinization" component of differentiation. The initial studies of Phoenix *et al.* with the guinea pig and Harris and Levine with the rat indicated that androgenized females exhibit high frequencies of malelike mounting behavior when treated with testosterone propionate in adulthood. While some workers have obtained this facilitation (A. A. Gerall & Ward, 1966), others have not.

Over the past few years several studies have been carried out in the author's laboratory in the attempt to demonstrate that perinatal androgenic stimulation will enhance the frequency of mounting behavior in adulthood and that neonatal castration will inhibit the development of the potential to exhibit mounting behavior. Figure 2 (Whalen & Edwards, 1967) illustrates some of these findings. In this study, male and female rats were gonadectomized within 12 hours after birth. Some of these animals were treated with oil, some with testosterone propionate, and some with estradiol benzoate at the

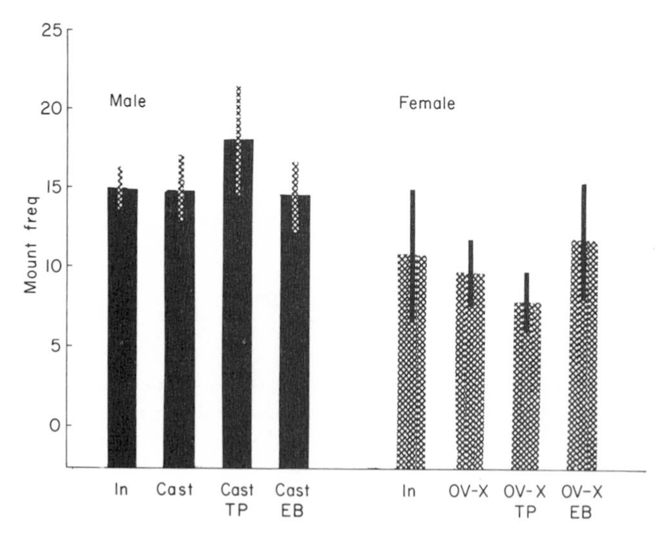

Fig. 2. Mount frequency in male and female rats treated with testosterone propionate following hormonal manipulation at birth. Legend as in Fig. 1. (Data from Whalen & Edwards, 1967.)

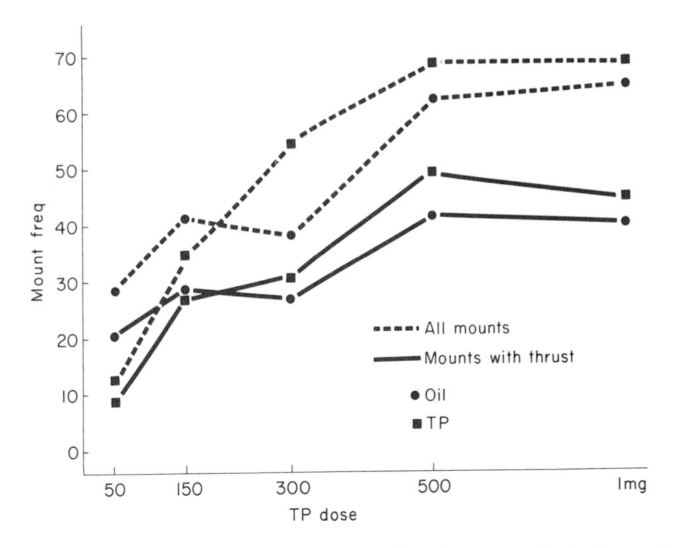

Fig. 3. Mount frequency of female rats treated with progressively increasing doses of testosterone propionate in adulthood. "All mounts" includes mounts with thrusting and mounts without thrusting. TP, females treated with testosterone propionate prenatally and postnatally; Oil, females treated with oil prenatally and postnatally. (Data from Whalen *et al.*, 1969.)

time of castration. When these animals matured, TP was administered, and they were tested for the display of masculine responses. It is clear that postnatal endocrine stimulation neither enhances nor inhibits the potential to display mounting in adulthood. More specifically, males castrated at birth mounted as frequently as males not castrated until adulthood; females which matured either with or without their ovaries mounted as frequently as females given a large dose of testosterone propionate at birth.

Although we found no evidence that early hormone stimulation "masculinized" in the sense of increasing the probability of mounting responses in adulthood, the possibility remained that perinatal androgen might have reduced the threshold of responsiveness to androgen in adulthood. In the Whalen and Edwards study (1967), for example, animals had been observed for masculine behavior only after they had received massive doses of TP in adulthood. This might not have provided a sufficiently sensitive test of the effects of early androgenization. Accordingly, in a second series of experiments (Whalen, Edwards, Luttge, & Robertson, 1969) the attempt was made to determine whether androgenization of the neonatal rat would facilitate responsiveness to low doses of androgen in adulthood. Figure 3 shows some of the findings.

It is clear that regardless of the dose of TP administered in adulthood, control females mounted as frequently as females whose mothers had received 2.0 mg TP daily on days 16–20 of pregnancy and which had themselves received 1.0 mg TP 96 hours after birth. The androgenized females had clearly masculinized genitalia, yet they mounted no more frequently than control females.

Two other studies have provided data consistent with these observations. Grady *et al.* (1965) studied the mounting behavior of male rats castrated, respectively, 1, 5, 10, 30, or 90 days after birth. In contrast, perhaps, to what one might expect, males castrated on day 1 mounted more frequently than males in any of the other groups. Beach *et al.* (1969) found no difference in the rate of mounting between male rats which had been castrated at birth and treated with either oil or with TP on days 1 and 2, days 3 and 4, days 5 and 6, days 9 and 10, or days 13 and 14 after birth.

These findings make it abundantly clear that early androgenic stimulation does not enhance masculine motivation in the sense of increasing the probability that the animal will approach and initiate sexual behavior with a receptive female. But this is not to say that early androgenic stimulation has no influence on adult sexual patterns. Several studies (Beach *et al.*, 1969; Harris & Levine, 1965; Ward, 1969; Whalen & Edwards, 1967; Whalen *et al.*, 1969; Whalen & Robertson, 1968) have demonstrated that neonatal androgenic stimulation enhances the probability that the animal will display intromission and ejaculation-type patterns in adulthood. Our own work, for example (Whalen & Robertson, 1968), has shown that female rats treated with TP both before and after birth are capable of showing intromission- and ejaculation-type responses and that these responses occur with a frequency and timing indistinguishable from that shown by normal males.

In summary, perinatal endocrine stimulation does not alter the probability or rate of mounting behavior, but it does have a profound effect upon the probability that such mounts will result in intromission-type responses and, ultimately, ejaculation-type responses. One might say, therefore, that early androgenization influences sexual performance while leaving sexual motivation unaffected.

IV. The Nature of Sexual Differentiation

Before proceeding with a more analytic examination of the role played by early hormonal stimulation in sexual development, it might be worthwhile to review the observations which have led to what is becoming accepted theory. Table V summarizes the effects typically obtained following manipulation of hormone levels during what we may now call the "critical differentiation period." Three generalizations emerge from these comparisons: (a) Mounting behavior is little influenced by hormonal stimulation during development. This behavior can be displayed by both males and females and it is neither consistently enhanced nor consistently inhibited by variations in hormone level during the critical period. (b) Intromission and ejaculation-type responses are moderately to strongly influenced by the presence or absence of gonadal hormones during development. (c) The probability of lordosis responses is strongly influenced by early hormonal stimulation.

These observations have led to the hypothesis that prior to differentiation the systems which control mating are essentially feminine. That is to say, if the organism continues to develop through the differentiation period without hormonal stimulation, it will later respond in a feminine manner to stimulation with either androgen or estrogen and progesterone. (Conditions 2, 4, and 5 shown in Table V yielded similar behavior.) Stimulation by the testes (condition 1) or by exogenous testosterone propionate (conditions 3, 6, and 7) *masculinize* in the sense of increasing the probability of intromission responses and *defeminize* by decreasing the potential for lordosis responses. These conclusions have obtained substantial experimental support. Theory, however, has not remained closely tied to these observations, but has led rather to three additional generalizations.

Table V

RELATIONSHIPS BETWEEN HORMONAL CONDITION DURING DEVELOPMENT AND SEXUAL RESPONSE IN ADULTHOOD

	Hormone condition during adulthood			
	Androgen			Estrogen and progesterone
Hormone condition during critical period	Mounts	Intromission responses	Ejaculation responses	Lordosis responses
Male				
1. Testes intact	+++	+++	+++	−
2. Castrated at birth	+++	+	−	+++
3. Castration at birth + TP	+++	+++	++	−
Female				
4. Ovaries intact	+++	+	−	+++
5. Ovariectomized at birth	+++	+	−	+++
6. TP at birth	+++	++	+	−
7. TP pre- and postnatally	+++	+++	+++	−

A. Generality

First, the characterization of differentiation presented in Table V is often considered a general model for the mammalian sexual-differentiation process even though virtually every study published to date has dealt with either rat or guinea pig. Although a few studies of the monkey (Goy, 1966) and the mouse (Edwards & Burge, 1971) are consistent with the observations noted in Table V, studies of the hamster are not. Crossley and Swanson (1968), for example, have found that if male hamsters are castrated when adult and administered female sex hormones, they display feminine responses. In the guinea pig and rat, of course, such responses would be unlikely. In addition, these workers found that normal female hamsters administered testosterone propionate in adulthood rarely display malelike mounting responses, again a pattern which is different from that shown by either the rat or guinea pig. We must be cautious, therefore, in the conclusions we draw about the generality of our model. It may well be that the model is appropriate to many species, including man (Ehrhardt & Money, 1967; Stoller, 1967; Whalen, 1967), but we shall not know until we have studied a wider variety of animal forms.

B. The Virilizing Agent

The second major theoretical assumption drawn from the data of sexual differentiation is that sexual differentiation is controlled by the testicular androgen, testosterone. The basis for this assumption is that the action of the male testis can be mimicked by the administration of testosterone propionate to the female or to the neonatally castrated male. The logic of the argument is as follows. The presence of the testes or stimulation by TP during the critical period enhances the probability of intromission responses and diminishes the potential to display lordosis responses. Therefore, either the testis secretes testosterone propionate or testosterone propionate mimics the effects of testicular androgen. The first possible conclusion may be ruled out since the propionic ester of testosterone is a synthetic form not secreted by the testis. The second possibility has face validity and is usually assumed to be correct. In apparent support is the study by Resko, Feder, and Goy (1968) showing that the blood plasma of the newborn rat does contain measurable amounts of testosterone. But this same study also presents a puzzle. According to Resko *et al.* the plasma of the 1-day-old male rat contains .027 μg/100 ml of testosterone. If the newborn rat has 2 ml of plasma, it would have .00054 μg of testosterone present in its circulation at any one time. If circulating levels of testosterone are at such a low level during the critical period, then one would expect that it would be possible to induce lordosis suppression with low doses of exogenous testosterone. But this is not possible. Barraclough and Gorski (1962), for example, treated female rats with either 1250 μg or 10 μg of TP 5 days after birth. The high dose, but not the low dose, resulted in the suppression of lordosis behavior. Similarly, Mullins and Levine (1968) administered different doses of TP to female rats (5 μg, 10 μg, 50 μg, 100 μg, or 500 μg TP) 120 hours after birth and then studied lordosis behavior in adulthood. Only the two highest doses suppressed lordosis and this suppression was not complete—74% of the females receiving either 100 or 500 μg TP showed some lordosis, although in these animals the probability of lordosis was reduced. Finally, Harris and Levine (1965) produced complete inhibition of lordosis by treating female rats with 500 μg TP 96 hours after birth. (Most other studies demonstrating the suppression phenomenon have employed TP doses of 1.0 mg or more.) These studies suggest that the lordosis response is not suppressed by as little as 50 μg TP, it is partially suppressed by 100 μg and is almost completely suppressed by a dosage of 500 μg or more of TP. Thus, although the effective dose in 50% of the animals (ED_{50}) is probably about 100 μg, 100 μg seems extremely high in comparison with the .00054 μg of circulating testosterone in the intact male rat.

There is one possible solution to this problem which is readily apparent. In the studies mentioned, testosterone propionate was administered to female rats no earlier than 96 hours after birth, although in the male rat the testes by that time have already produced their suppressive effects. Feder and Whalen (1965) found that males castrated 16–32 hours after birth displayed moderate levels of lordosis in adulthood (ratio of lordosis responses to mounts by male L/M = 40.0%). If castration was delayed until 96 hours after birth, then lordosis levels were lower (L/M = 19.5%); if delayed until 1 week after birth, then lordosis levels were close to zero (Zucker, 1967). On the other hand, if male rats are castrated less than 12 hours after birth, lordosis levels in adulthood can approach those of females (L/M = 70.0%; Whalen & Edwards, 1967; Fig. 1). A comparison of these and other studies of neonatal castration reveals that the suppression of the potential for lordosis behavior is controlled by the testes and that the testes exert this effect to a large degree between birth and 4 days after birth. (This conclusion rests on the doubtlessly invalid assumption that all rats are born at the same developmental age. Nonetheless, as a first approximation, the conclusion seems reasonable.) On the basis of these findings one might conclude that high doses of TP are needed to suppress lordosis because TP has invariably been administered late in the sensitive period. However, in light of more recent findings, this conclusion is unwarranted.

Clemens, Hiroi, and Gorski (1969) have reported recently on the effects of treating female rats with 10 μg TP on the day of birth or on day 2, 4, 6, or 10, respectively. Treatment on the day of birth or on the day after birth had no suppressive effects. Treatment on day 4 or 6 partially suppressed reactivity to estrogen and progesterone. Treatments on days 2, 4, or 6 even seemed to facilitate responsiveness to estrogen. Thus, it is impossible to conclude that low doses of TP can induce behavioral suppression when administered early in the sensitive period.

These data lead us to the tentative conclusion that lordosis potential is suppressed only by pharmacological doses of testosterone propionate. Unfortunately, no one has carried out the appropriate dose-response and time-response experiments which would allow us to state the minimally effective dose-time unit for the inhibitory effects of TP. Nonetheless, the available data would seem to indicate that the minimally effective dose is probably greater than 10 μg of TP. Of course this dose is still large in comparison with circulating testosterone levels.

Another explanation of the TP effect which might be advanced is that the propionic ester of testosterone does not have ready access to the brain, thus necessitating high doses to inhibit lordosis. This interpretation is also invalid. Long before interest developed in the sexual differentiation of mating behavior, studies were carried out on the differentiation of brain-pituitary-gonad relationships. For example, in 1936 Pfeiffer demonstrated that female rats which had received a testis transplant at birth were sterile in adulthood. The ovaries of these females contained follicles but no corpora lutea, indicating that the ovaries had failed to ovulate. Current research has shown that this anovulatory sterility can be induced in female rats by testosterone propionate treatment after birth. Furthermore, it has been shown that this sterility reflects a change in the neural control of gonadotropin secretion. The work of Barraclough, Gorski, Harris and many others has elucidated this neural differentiation process. In many ways this differentiation process parallels behavioral differentiation. Table VI illustrates these general findings. As can be seen,

Table VI

EFFECTS OF HORMONAL MANIPULATION AT BIRTH IN THE RAT UPON OVARIAN FUNCTION AND VAGINAL CONDITION IN ADULTHOOD

Hormonal condition during critical period	Gonadotropin secretion	Natural or transplanted ovary	Natural or transplanted vagina
Male			
1. Testes intact	Tonic	Follicles; no corpora lutea ("Sterile")	Estrus-type cornification ("constant estrus")
2. Castrated at birth	Cyclic	Follicles, corpora lutea	Diestrus-proestrus-estrus (cyclic changes)
Female			
3. Ovaries intact	Cyclic	Follicles, corpora lutea	Diestrus-proestrus-estrus (cyclic changes)
4. TP at birth	Tonic	Follicles; no corpora lutea	Estrus-type cornification ("constant estrus")

the normal female pattern is cyclic and the male pattern is tonic. The tonic pattern can be induced by the testes themselves or by TP treatment during the sensitive period.

With these findings in mind, we can turn to the question of the minimal sterilizing dose of TP. Gorski (1966) has shown that 70% of female rats administered 10 μg of testosterone propionate 96 hours after birth are sterile in adulthood. The ovaries of the sterile animals contain no corpora lutea, and they weigh less (32.5 mg) than those of cyclic females (65.2 mg); in addition, the vaginal cell pattern of these sterile females is one of persistent cornification.

In another study, Gorski and Barraclough (1963) evaluated the sterilizing effectiveness of TP in doses of 1250 μg, 10 μg, 5 μg, and 1 μg, respectively. The following percentages of animals were sterilized: 99.8%, 70.6%, 44.0%, and 30%. Thus the minimum sterilizing dose of TP can be as low as 1 μg, with ED_{50} between 5 μg and 10 μg of TP. Since the available evidence indicates that anovulatory sterility reflects a central neural effect of TP rather than an effect of TP upon the ovary, we may conclude that even 1 μg of testosterone propionate can influence the brain function of the newborn rat. We cannot conclude, therefore, that 100 μg is needed to produce an inhibition of lordosis since TP is a relatively weak androgen with respect to the brain. Our dilemma remains: testosterone propionate seems to mimic the behavioral action of the testes, but only when administered at doses 10-100 times higher than needed to alter the neural control of gonadotropin secretion.

The simplest and most commonly accepted solution to the problem is to propose that those neurons involved in behavior have a higher threshold of response to androgen than do those neurons involved in gonadotropin control. This argument is somewhat tautological—the threshold determines the effect and the effect defines the threshold—and does not provide an explanation of the observed effect.

Perhaps an alternative explanation is to be found in exploring the fact that the neonatal testis secretes more than one androgen. In other words, while testosterone may be the predominant androgen in the blood of the neonate, it may be a metabolite of testosterone, such as androstenedione, that is the effective inhibiting hormone. Since not all the injected testosterone would be converted to androstenedione, it might be necessary to administer large doses of testosterone propionate to obtain an inhibiting dose of the effective metabolite. We tested this hypothesis in two experiments. In the first (Luttge & Whalen, 1969), we treated female rats four different times after birth (48, 72, 96, and 120 hours) with either oil or with 500 μg of androstenedione. From 65 through 75 days of age, and again from 88 through 94 days of age, daily vaginal smears were taken from each animal to determine whether the vaginal pattern was cyclic or of the "constant estrus" type characteristic of the sterile animal. The animals were then ovariectomized and the ovaries examined for the presence of corpora lutea, the index of ovulation. Finally, these animals were administered estrogen and progesterone and tested for the lordosis response. Our findings are illustrated in Fig. 4. The androstenedione-treated and control groups differed strikingly in their pattern of vaginal smears. Androstenedione animals showed the "constant-estrus" pattern suggesting anovulatory sterility which was confirmed when we examined their ovaries—25 of the 26 had ovaries that lacked corpora lutea. In contrast, the ovaries of the control females all contained corpora lutea, and these animals all showed the normal cyclic pattern of vaginal change. When mated, however, females from both groups displayed a high frequency of lordosis. This massive dose of androstenedione (2 mg) had produced the masculine pattern of gonadotropin secretion, but it did not suppress the animal's ability to display lordosis.

In a second experiment (Luttge and Whalen, 1970b), we treated female rats on days 2, 3, 4, and 5 with oil or with one of five doses of the naturally occurring free alcohol form of testosterone (not TP), androstenedione or dihydrotestosterone, respectively. We again examined the vaginal patterns in adulthood. As indicated in Fig. 5, the proportion of the group showing the "constant-estrus" sterility pattern in-

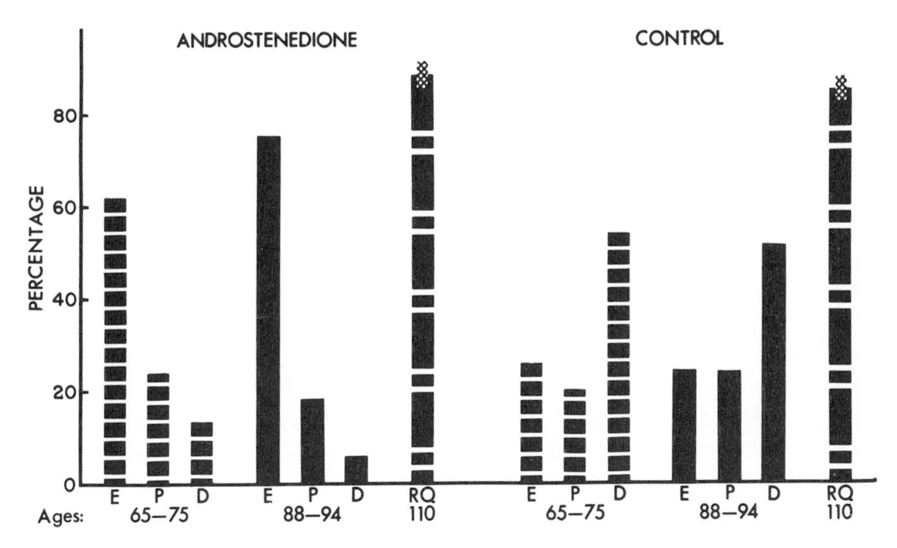

Fig. 4. Effects of androstenedione or vehicle administered to neonatal female rats on vaginal smear pattern and mating behavior in adulthood. Smears were taken at 65-75 and 88-94 days of age and classified as estrus (E), proestrus (P), or diestrus (D). Sexual receptivity following estrogen-progesterone treatment was assessed at 110 days of age and expressed as the ratio of lordosis responses to mounts by the male (R.Q.).

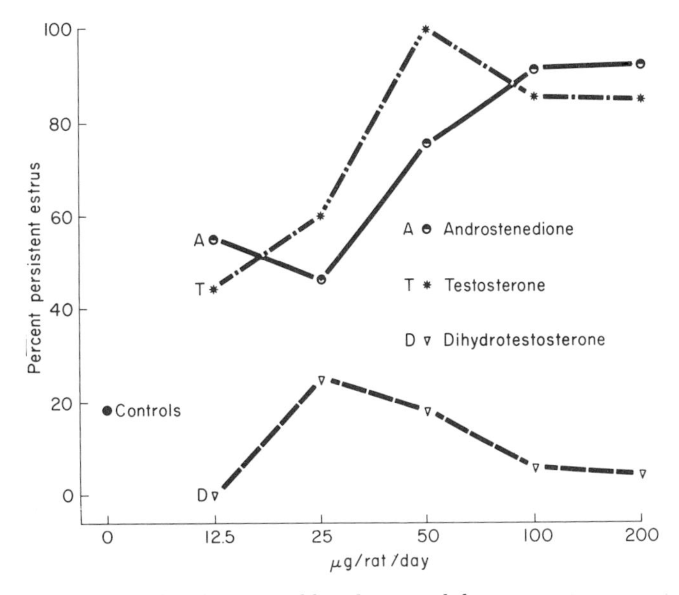

Fig. 5. Percentage of each group of female rats exhibiting persistent vaginal cornification in adulthood following treatment with oil (control) or with one of five doses of androstenedione (A), testosterone (T), or dihydrotestosterone (D) in infancy. Each hormone dose was administered daily on days 2, 3, 4, and 5 after birth.

creased as the dosage of testosterone or androstenedione was increased. Dihydrotestosterone had no effect. Those animals which had received the highest dose of androgen in infancy (800 μg) were administered estrogen and progesterone and tested for the display of lordosis. All exhibited high levels of lordosis behavior—none of these androgens, including testosterone, suppressed lordosis.

The data from these two experiments showed that as little as 50 μg/day/4 days of free testosterone and androstenedione "masculinized" the neuroendocrine control of gonadotropin secretion, while as much as 1000 μg of these same agents failed to "masculinize" the neurobehavior control system. Most striking, of course, was the finding that 200 μg/day/4 days of free testosterone did not mimic the behavior-suppressing effect of a single injection of 100 μg of testosterone propionate. Of course it is possible to argue that the dose of free testosterone was low in relation to TP, the latter, in other words, being considerably more potent than free testosterone. The findings of Goldfoot, Feder, and Goy (1969), however, make this suggestion unlikely. These workers castrated male rats 22-27 hours after birth. Some of the animals were treated with 250 μg of androstenedione every other day from birth until day 19. In adulthood, these animals readily displayed lordosis when administered estrogen and progesterone. It would seem, then, that neither free testosterone nor one of its major androgenic metabolites is capable of suppressing lordosis potential when administered in massive doses during the critical period of differentiation. Only the presence of the testes, or the administration of large doses of testosterone propionate or estradiol benzoate, induces behavioral suppression. It thus becomes difficult to conclude that TP injections mimic the action of the testes.

It would be satisfying indeed if our discussion of behavioral differentiation could be culminated with a review of some simple experiment showing that some "agent X," produced by the body, induces a suppression of the potential to show lordosis behavior when administered in nanogram or picogram quantities during the critical period. Unfortunately, this is not possible. Unless one accepts the idea that massive doses of exogenous testosterone are needed to mimic the effects of testicular testosterone, one is led to the conclusion that testosterone is not *the* virilizing agent. Suppression by the testes of the lordotic potential is clear, and represents a critical characteristic of the sexual differentiation process. However, control of differentiation exercised by the secretion of testicular testosterone has not been proven. If more conclusive evidence can be produced to show that testicular testosterone is the masculinizing agent, or that the testes effect differentiation by some other means, a major advance would be made in our understanding of the ontogeny of sexuality.

C. Neural Differentiation

The third major assumption of sexual differentiation theory is that the enhanced intromission behavior and suppressed lordosis behavior seen in intact males and TP-treated females reflect changes in brain function. In order to consider the issues involved in identifying the locus at which early hormonal stimulation exerts its effects, it is necessary to examine separately the enhancement of masculine intromission responses and the inhibition of feminine lordosis responses.

1. Masculine Responses

As shown in Table V, the presence of testes or TP stimulation during the critical period permits the display of intromission and ejaculation responses in adulthood. It has been proposed that this phenomenon reflects a masculinization of brain tissue. We, on the other hand, have argued that enhanced intromission behavior reflects the effects of androgen on phallic development, and that the degree of phallic development determines the probability of intromission and ejaculation behavior (Whalen, 1968). This latter argument was based upon a variety of observations which indicated that (a) phallic growth and reactivity to androgen in adulthood is positively correlated with hormonal stimulation during development; (b) a positive correlation exists between the size and structural integrity of the phallus on the one hand and intromissive behavior on the other; and (c) treatments which reduce the sensitivity of the phallus (e.g., local anesthesia) reduce the probability of intromission. Recent reports have provided support for this argument. Beach *et al.* (1969) castrated male rats within 24 hours of birth and treated them with oil or with TP for 2 days starting on day 1, 3, 5, 9, or 13. All animals were again treated with androgen and tested for mating. Intromission frequency was high only in those animals treated with TP on day 1 and 2, or 3 and 4. After sacrificing the animals, penis weight was taken. As shown in Fig. 6, a high correlation was found between penis weight and the frequency of intromission. Similarly, Goldfoot *et al.* (1969) studied adult intromission behavior in male rats castrated at birth and treated with androstenedione every other day for 19 days.

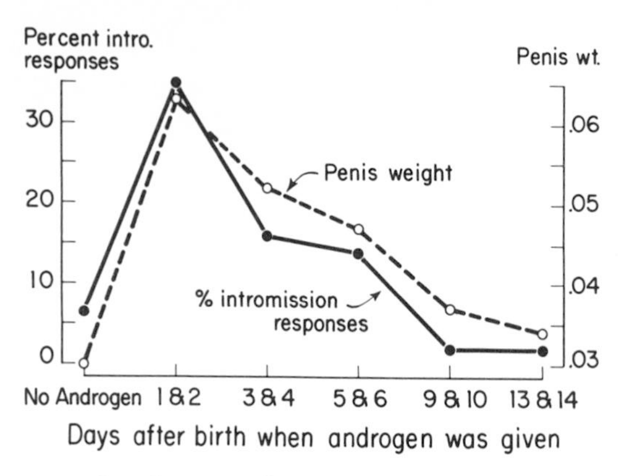

Fig. 6. Percentage of total mounts that involved intromission and the weight of the penis expressed as percentage of body weight in male rats castrated at birth and administered no androgen or administered testosterone propionate for 2 days beginning on day 1, 3, 5, 9, or 13 after birth. (Data from Beach *et al.*, 1969.)

This treatment did not suppress lordosis behavior, but it did facilitate penis growth and intromission behavior. These studies, as well as older studies (Whalen, 1968), show that manipulations enhancing phallic development also enhance the probability of intromission; manipulations which interfere with phallic development reduce the probability that the animal will show intromission. Early hormonal stimulation thus influences sexual performance via its action on peripheral genital development. These findings, of course, do not show that intromission behavior in the male is uninfluenced by the central neural action of hormones during the sensitive period. One can only conclude that phallic development seems to account for a large proportion of the variance in adult intromission behavior.

Possibly the best evidence for a neural effect of early hormone stimulation has been provided by Hart (1968a). Hart studied the genital reflexes of spinal male rats castrated at either 4 or 12 days after birth. The day 4 castrates exhibited fewer erections, quick flips, and long-flip responses than the day 12 castrates. Hart suggested that the reduced frequency of these genital reflexes in rats castrated during the critical period may reflect a failure of sexual differentiation of those systems in the spinal cord normally involved in mating. It should be noted, however, that the reduced reflex activity observed could have reflected inadequate phallic development rather than a nonorganized spinal neuron system. Thus, although these data suggest, they in no way prove that hormonal stimulation during the critical period organizes the neural tissue involved in masculine response patterns.

2. Feminine Responses

The presence of the testes or stimulation by TP during the critical period reduces the probability that the rat will display lordosis following estrogen and progesterone stimulation in adulthood. It is assumed that this phenomenon represents a differentiation of neural tissue.

In our laboratory, we reasoned that since the male and androgenized female appear to be less responsive to estrogen than the normal female, one might find a differential uptake or retention of estradiol in relevant brain tissues. We have explored this possibility in several recent experiments. In our first study (Green, Luttge, & Whalen, 1969), we administered radiolabeled estradiol to gonadectomized adult male, female, and neonatally TP-treated female rats. The animals were sacrificed 5, 30, or 60 minutes after treatment, and several neural and peripheral tissues were examined. Figure 7 illustrates some of our findings. Depicted are radioactivity levels in hypothalamic samples before and after counts from cortex samples were subtracted. Highly reliable male-female differences were found, with radioactivity levels being higher in females than males. Neonatally TP-treated females fell between the males and females. This pattern obtained in samples of hypothalamus, preoptic-diagonal band region, and pituitary, but not in brain cortex.

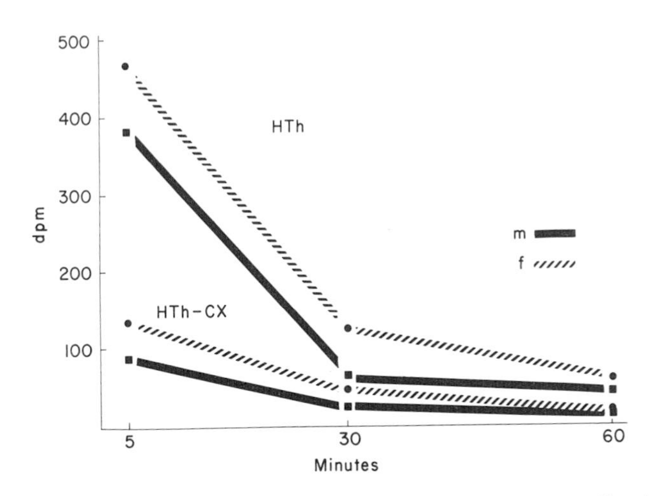

Fig. 7. Radioactivity levels expressed as disintegrations/min/mg (dpm) in hypothalamic tissue of male and female rats before and after counts from cortex samples were subtracted. Males (m) and females (f) were sacrificed 5, 30, or 60 minutes after i.v. administration of 6,7-estradiol-^{3}H. (Data from Green *et al.*, 1969.)

The findings shown in Fig. 7 suggest that the CNS tissues of males are less responsive to estrogen than are these same tissues in females. Similar findings have also been reported by Flerkó and Mess (1968), Flerkó, Mess, and Illei-Donhoffer (1969), and by McGuire and Lisk (1969). Closer examination, however, reveals that such an interpretation is unwarranted. In our study, the groups differed not only in hormonal condition during development, but also in body weight at the time of sacrifice. The males weighed more than the females and the androgenized females fell between these groups. It was possible, therefore, that the males, in being heavier, had received an effectively lower dose of radioactive estrogen than did the females. Accordingly, in our second experiment, we administered radiolabeled estradiol to adult males and females of the same body weight, but of slightly different ages. Under these conditions no male-female differences in brain radioactivity levels were found upon sacrifice 30 minutes after treatment. Further work on the problem has confirmed this observation. In fact, when one measures brain radioactivity 2 hours or 24 hours after treatment with estradiol-^{3}H, radioactivity levels are actually higher in males than in females as is shown in Fig. 8. However, when one uses the ratio of hypothalamic (or preoptic-diagonal band) to plasma levels, or the ratio of hypothalamic to cortex levels as a measure of brain radioactivity, differences are no longer apparent. Thus, under conditions in which rats are matched for body weight, males and females do not differ in either the uptake or the retention of estradiol. Autoradiographic findings are consistent with these results (Pfaff, 1968; Stumpf, 1968). Here again we are faced with a puzzle: the behavioral response of male and female rats to estrogen is dramatically different, yet diencephalic brain tissues of males and females accumulate and retain nearly equivalent amounts of estrogen.

If differentiation of the brain cannot be accounted for in terms of the uptake and retention of hormone, how can behavioral differences be explained? Although the brain is not usually considered capable of steroid metabolism, the brains of males and females may metabolize estrogens differentially. Perhaps brain enzymes differ in males and females or possibly the brain of the male is incapable of producing the appropriate proteins when stimulated by estrogen. Currently we are examining the possibility that, following treatment with estradiol, the major ovarian estrogen, the brains of females and males accumulate and retain different estrogenic metabolites. To test this hypothesis, we administered estradiol-^{3}H to gonadectomized adult male and female rats and sacrificed them 2 hours later. Anterior

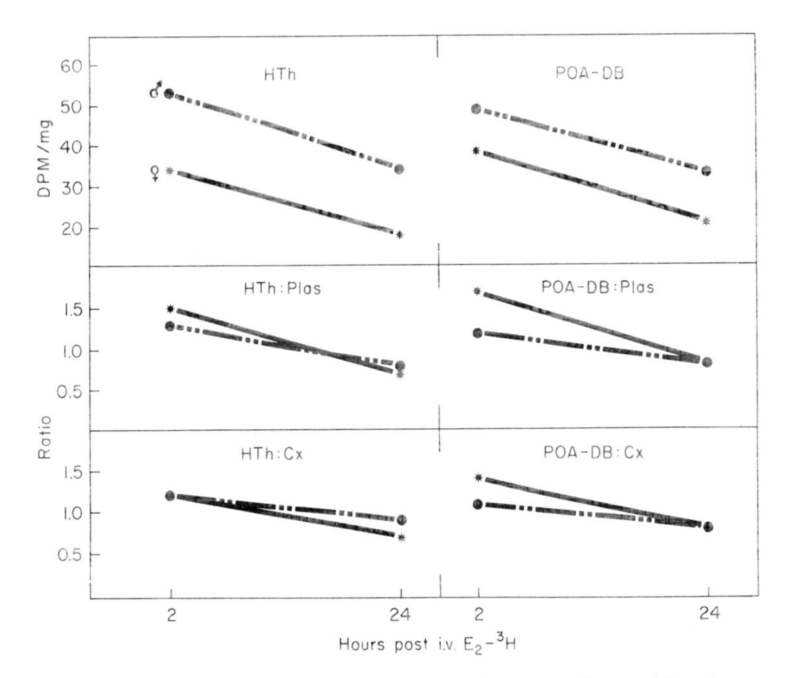

Fig. 8. Radioactivity levels in hypothalamic and preoptic-diagonal band tissues of male and female rats 2 hours or 24 hours after i.v. treatment with 6,7-estradiol-^{3}H. Radioactivity is expressed as disintegrations/min/mg (DPM) and as the ratio of DPM in hypothalamus and preoptic-diagonal band tissue to DPM in plasma or in cortex. (Data from Whalen & Luttge, 1970.)

and posterior hypothalamic samples were taken and subjected to steroid extraction procedures. The extracts were then examined by thin-layer chromatography for the separation of the component metabolites. Figure 9 depicts the resulting chromatograph. The predominant estrogen found in the anterior hypothalamus was estradiol. In this tissue the estradiol/estrone ratio was approximately 7. In posterior hypothalamic samples both estradiol and estrone were found in a ratio of approximately 2. Both absolutely and relatively, significantly more estrone was found in the posterior hypothalamus than in the anterior hypothalamus, but this pattern was almost exactly the same in males and females. Apparently, regional differences exist in the neural metabolism of estrogen or in the uptake and retention of metabolites, but these regional differences are not sex specific. Again, we found no mechanism by which to account for sex differences in the behavioral reactivity to estrogen.

Neither we nor any other workers have provided satisfactory evidence that hormone-controlled behavioral differentiation reflects a

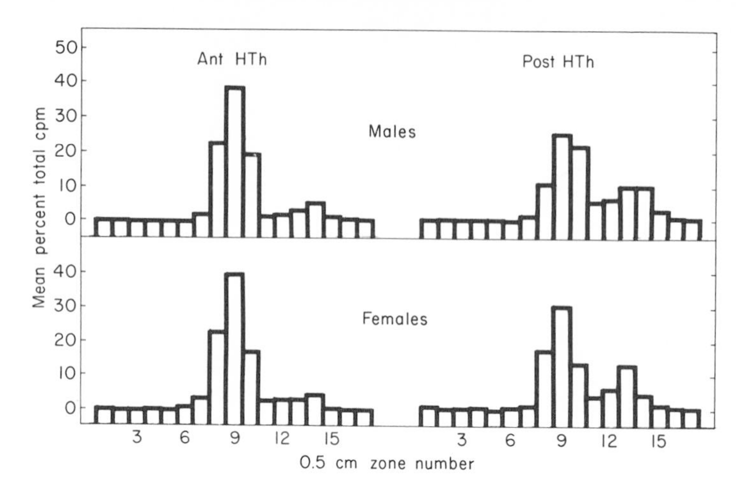

Fig. 9. Thin-layer chromatograph of steroids extracted from the anterior and posterior hypothalamus of male and female rats administered 6,7-estradiol-^{3}H. Data are expressed as percent of total radioactivity in each 0.5-cm zone of the chromatograph. Standards indicated that zones 8–10 correspond to estradiol and zones 13–15 correspond to estrone. (Data from Luttge & Whalen, 1970a.)

neural differentiation process. To be sure, it is not obvious how one can account for behavioral differentiation in terms of peripheral rather than central nervous system mechanisms; nonetheless, there is no evidence which allows us to make the decision. Perhaps it is worth noting that careful consideration of the role which peripheral systems play in mating has proven important in understanding the control and differentiation of masculine sexual response patterns. A similar consideration may be fruitful for understanding feminine behavior patterns.

To summarize: Table V outlines the effects of the testes and of TP stimulation during the critical period upon the display of masculine and feminine sexual responses in adulthood. These findings provide a model for sexual differentiation in some, but not all species. They suggest, but do not prove, that testicular testosterone is the virilizing agent, and they point to, but again do not prove, a neural locus for sexual differentiation. And finally, acceptance of a theory which maintains that sexual differentiation of the brain is effected by the secretion of testicular testosterone during a limited period of development is premature. Many exciting and critical problems still confront us and must be solved before we understand the role of hormones in the ontogeny of sexuality.

V. Experience and the Development of Sexual Behavior

Although we are still puzzled by many aspects of the hormonal control of sexual differentiation, it is nonetheless clear that great progress has been made during the past 10 years toward understanding the variables which influence differentiation. The same, unfortunately, cannot be said for the analysis of the role played by experience in the determination of individual differences in sexuality. Research in this area has been sporadic and there have been no major developments during the past 5 years. Since the study of the experiential determinants involved in sexuality is not at present in a "dynamic phase," and since much of the relevant material has been reviewed recently (Whalen, 1969), a detailed summary of this research seems gratuitous. Rather, some comments will be made which highlight the findings of researchers investigating experiential factors involved in sexuality.

A. Effects of Variations in Rearing Conditions

One class of "experience" studies has focused on the role of rearing conditions in connection with the development of sexual responses. Generally, these studies involve a severe restriction in the social interaction of developing animals, followed by an examination of sexual behavior in adulthood. Using this paradigm, Beach (1942, 1958) found that social isolation of male rats from as early as 14 days after birth caused no major change in the probability, frequency, or timing of sexual responses in adulthood. Others, however, have reported some interference with mating as a result of isolation-rearing in the rat (e.g., Folman and Drori, 1965). In contrast, studies of the guinea pig (A. A. Gerall, 1963; H. D. Gerall, 1965; Goy and Young, 1956; Valenstein, Riss & Young, 1955) have shown repeatedly that early social isolation can have a detrimental effect on adult sexual behavior.

In primates, early social isolation has more profound effects upon sexual behavior than does such treatment with rodents. Mason (1960), for example, isolated three male and three female rhesus monkeys from their mothers 1 month after birth. These animals received almost no intraspecific social contact until they were tested at 2½ years of age. Their mating behavior was seen to be extremely disorganized. While the males attempted to mount the females, their responses were uncoordinated and inappropriately directed. Among the isolated females, sexual presentation responses were found to be less frequent and more stereotyped than were the presentation responses of feral females. Parts of this experiment have been repeated recently by Missakian (1969) and have yielded essentially the same results. Specifically, Missakian found that male monkeys socially deprived from infancy exhibited significantly fewer mounts and mount attempts than wild-reared species mates.

In some species, at least, social interactions during development clearly contribute to the maturation of those response systems which underlie mating. Unfortunately, studies of this type involve extreme manipulation of the animal's normal environment during development and as a result point only to those factors which influence the experientially labile components of sexual development. They tell us little about the critical experiences which lead to normal sexual development.

B. Effects of Postpuberal Experience

A second class of "experience" studies is concerned with effects of adult sexual experience upon the performance of later sexual behavior. Within this class, studies have shown that mating experience tends to reduce response latencies (Larsson, 1959; Whalen, 1961b, 1963), while having relatively little effect upon the basic character of the response itself.

Carr, Loeb, and Dissinger (1965) have shown that in rats copulatory experience determines the male's response to odors of the female, and Bermant and Taylor (1969) have demonstrated that sexual experience partially ameliorates the detrimental effects of olfactory bulb lesions upon mating. These studies suggest that one effect of mating experience may be to reduce the animal's dependence upon any particular sensory cue.

In some species, sexual experience may also reduce the animal's dependence upon high levels of circulating gonadal hormones. Rosenblatt and Aronson (1959), for example, reported that male cats may continue to mate for several years after castration, but only if they had obtained mating experience prior to castration. This effect, however, is not general. Rabedeau and Whalen (1959) and Bloch and Davidson (1968) found no such effect in the rat, and Hart (1968b) has reported that sexually experienced castrated male dogs show a gradual decline in mating following castration—a decline which parallels that shown by sexually naive dogs.

These few studies indicate that no firm generalizations can be drawn about the role of rearing or adult sexual experience in either the development or maintenance of sexual responses. All that can be said at present is that "experience," by altering motivational and performance systems, may influence the ontogeny of sexuality in different ways and in different degrees, depending upon the species. Of course, it is our belief that experience plays a much more pervasive role in the ontogeny of sexuality than has been demonstrated. However, validation of this belief must await future research.

VI. Summary

The present chapter has considered the role of the genome, the effects of perinatal hormonal stimulation, and the action of the environment upon the ontogeny of sexuality in animals. Each, of course, contributes to the development of sexual behavior. Nonetheless, our coverage has been unbalanced. This reflects not only the bias of the author, but also the relative concern of contemporary investigators for the various problem areas. A further analysis of the genetic components in sexual development is important, but little interest has been generated in this line of approach. The same may be said for the analysis dealing with experience in relation to the development of mating patterns. Only the hormonal control of mating behavior has recently received active attention. This should be expected; science never develops in a balanced manner. From time to time a germinal experiment is reported which touches the imagination and sets the direction for an intense study of a problem. It is only hoped that this review has pointed out where the questions remain and the areas in which those germinal experiments are needed.

References

Barraclough, C. A., & Gorski, R. A. Studies on mating behaviour in the androgen-sterilized female rat in relation to the hypothalamic regulation of sexual behaviour. *Journal of Endocrinology*, 1962, **25**, 175-182.

Beach, F. A. Comparison of copulatory behavior of male rats raised in isolation, cohabitation, and segregation. *Journal of Genetic Psychology*, 1942, **60**, 121-136.

Beach, F. A. Normal sexual behavior in male rats isolated at fourteen days of age. *Journal of Comparative and Physiological Psychology*, 1958, **51**, 37-38.

Beach, F. A., Noble, R. G., & Orndoff, R. K. Effects of perinatal androgen treatment on responses of male rats to gonadal hormones in adulthood. *Journal of Comparative and Physiological Psychology*, 1969, **68**, 490-497.

Bermant, G., & Taylor, L. Interactive effects of experience and olfactory bulb lesions in male rat copulation. *Physiology & Behavior*, 1969, **4**, 13-17.

Bloch, G. J., & Davidson, J. M. Effects of adrenalectomy and experience on post-castration sex behavior in the male rat. *Physiology & Behavior*, 1968, **3**, 461-465.

Carr, W. J., Loeb, L. S., & Dissinger, M. L. Responses of rats to sex odors. *Journal of Comparative and Physiological Psychology*, 1965, **59**, 370-377.

Clemens, L. G., Hiroi, M., & Gorski, R. A. Induction and facilitation of female mating behavior in rats treated neonatally with low doses of testosterone propionate. *Endocrinology*, 1969, **84**, 1430-1438.

Crossley, D. A., & Swanson, H. H. Modification of sexual behavior of hamsters by neonatal administration of testosterone propionate. *Journal of Endocrinology*, 1968, **41**, XIII-XIV.

Edwards, D. A., & Burge, K. G. Early androgen treatment and male and female sexual behavior in mice. *Hormones and Behavior* 1971, **2**, 49-58.

Ehrhardt, A. A., & Money, J. Progestin-induced hermaphroditism: IQ and psychosexual identity in a study of ten girls. *Journal of Sex Research*, 1967, **3**, 83-100.

Feder, H. H., & Whalen, R. E. Feminine behavior in neonatally castrated and estrogen-treated male rats. *Science*, 1965, **147**, 306-307.

Flerkó, B., & Mess, B. Reduced estradiol-binding capacity of androgen sterilized rats. *Acta Physiologica Academiae Scientiarum Hungaricae*, 1968, **33**, 111-113.

Flerkó, B., Mess, B., & Illei-Donhoffer, A. On the mechanism of androgen sterilization. *Neuroendocrinology*, 1969, **4**, 164-169.

Folman, Y., & Drori, D. Normal and aberrant copulatory behavior in male rats (R. Norvegicus) reared in isolation. *Animal Behaviour*, 1965, **13**, 427-429.

Gerall, A. A. An exploratory study of the effect of social isolation on the sexual behavior of guinea pigs. *Behaviour*, 1963, **11**, 274-282.

Gerall, A. A., & Ward, I. L. Effects of prenatal exogenous androgen on the sexual behavior of the female albino rat. *Journal of Comparative and Physiological Psychology*, 1966, **62**, 370-375.

Gerall, H. D. Effect of social isolation and physical confinement on motor and sexual behavior of guinea pigs. *Journal of Personality and Social Psychology*, 1965, **2**, 460-464.

Goldfoot, D. A., Feder, H. H., & Goy, R. W. Development of bisexuality in the male rat treated neonatally with androstenedione. *Journal of Comparative and Physiological Psychology*, 1969, **67**, 41-45.

Gorski, R. A. Localization and sexual differentiation of the nervous structures which regulate ovulation. *Journal of Reproduction and Fertility*, 1966, Suppl. 1, 67-88.

Gorski, R. A., & Barraclough, C. A. Effects of low dosages of androgen on the differentiation of hypothalamic regulatory control of ovulation in the rat. *Endocrinology*, 1963, **73**, 210-216.

Goy, R. W. Role of androgens in the establishment and regulation of behavioral sex differences in mammals. *Journal of Animal Science*, 1966, **25**, Suppl., 21-35.

Goy, R. W., Bridson, W. E., & Young, W. C. Period of maximum susceptibility of the prenatal female guinea pig to the masculinizing actions of testosterone propionate. *Journal of Comparative and Physiological Psychology*, 1964, **57**, 166-174.

Goy, R. W., & Jakway, J. S. The inheritance of patterns of sexual behaviour in female guinea pigs. *Animal Behaviour*, 1959, **7**, 142-149.

Goy, R. W., Phoenix, C. H., & Young, W. C. A critical period for the suppression of behavioral receptivity in adult female rats by early treatment with androgen. *Anatomical Record*, 1962, **142**, 307.

Goy, R. W., & Young, W. C. The importance of genetic and experiential factors for the organization of sexual behavior patterns in the female guinea pig. *Anatomical Record*, 1956, **124**, 296.

Grady, K. L., Phoenix, C. H., & Young, W. C. Role of the developing rat testis in differentiation of the neural tissues mediating mating behavior. *Journal of Comparative and Physiological Psychology*, 1965, **59**, 176-182.

Green, R., Luttge, W. G, & Whalen, R. E. Uptake and retention of tritiated estradiol in brain and peripheral tissues of male, female, and neonatally androgenized female rats. *Endocrinology*, 1969, **85**, 373-378.

Harris, G. W., & Levine, S. Sexual differentiation of the brain and its experimental control. *Journal of Physiology* (*London*), 1962, **163**, 42P-43P.

Harris, G. W., & Levine, S. Sexual differentiation of the brain and its experimental control. *Journal of Physiology*, (*London*) 1965, **181**, 379-400.

Hart, B. L. Neonatal castration: Influence on neural organization of sexual reflexes in male rats. *Science*, 1968, **160**, 1135-1136. (a)

Hart, B. L. Role of prior experience in the effects of castration on sexual behavior of male dogs. *Journal of Comparative and Physiological Psychology*, 1968, **66**, 719-725. (b)

Hirsch, J., & Boudreau, J. C. Studies in experimental behavior genetics. 1. The heritability of phototaxis in a population of *Drosophila melanogaster*. *Journal of Comparative and Physiological Psychology*, 1958, **51**, 647-651.

Jakway, J. S. The inheritance of patterns of mating behaviour in the male guinea pig. *Animal Behaviour*, 1959, **7**, 150-162.

Larsson, K. Experience and maturation in the development of sexual behaviour in male puberty rat. *Behaviour*, 1959, **14**, 101-107.

Luttge, W. G., & Whalen, R. E. Partial defeminization by administration of androstenedione to neonatal female rats. *Life Science*, 1969, **8**, 1003-1008.

Luttge, W. G., & Whalen, R. E. Regional localization of estrogenic metabolites in the brain of male and female rats. *Steroids*, 1970, **15**, 605-612. (a)

Luttge, W. G., & Whalen, R. E. Dihydrotestosterone, androstenedione, testosterone: Comparative effectiveness in masculinizing and defeminizing reproductive systems in male and female rats. *Hormones and Behavior* 1970, **1**, 265-281. (b)

McGill, T. E. Sexual behavior in three inbred strains of mice. *Behaviour*, 1962, **19**, 341-350.

McGill, T. E. Studies of the sexual behavior of male laboratory mice: Effects of genotype, recovery of sex drive, and theory. In F. A. Beach (Ed.), *Sex and behavior*. New York: Wiley, 1965. Pp. 76-88.

McGill, T. E. Genetic analysis of male sexual behavior. In G. Lindzey & D. D. Thiessen (Eds.), *Behavior-genetic analysis: The mouse as a prototype*. New York: Meredith, 1970. Pp. 57-88.

McGill, T. E., & Ransom, T. W. Genotypic change affecting conclusions regarding mode of inheritance of elements of behaviour. *Animal Behaviour*, 1968, **16**, 88-91.

McGill, T. E., & Tucker, G. R. Genotype and sex drive in intact and in castrated male mice. *Science*, 1964, **145**, 514-515.

McGuire, J. L., & Lisk, R. D. Oestrogen receptors in androgen or oestrogen sterilized female rats. *Nature* (*London*) 1969, **221**, 1068-1069.

Manning, A. The effects of artificial selection for mating speed in *Drosophila melanogaster*. *Animal Behaviour*, 1961, **9**, 82-92.

Mason, W. A. The effects of social restriction on the behavior of rhesus monkeys. I. Free social behavior. *Journal of Comparative and Physiological Psychology*, 1960, **53**, 582-589.

Missakian, E. A. Reproductive behavior of socially deprived male rhesus monkeys (Macaca mulatta). *Journal of Comparative and Physiological Psychology*, 1969, **69**, 403-407.

Moltz, H. Contemporary instinct theory and the fixed action pattern. *Psychological Review*, 1965, **72**, 27-47.

Mullins, R. F., Jr., & Levine, S. Hormonal determinants during infancy of adult sexual behavior in the female rat. *Physiology & Behavior*, 1968, **3**, 333-338.

Pfaff, D. W. Autoradiographic localization of radioactivity in rat brain after injection of tritiated sex hormones. *Science*, 1968, **161**, 1355-1356.

Pfeiffer, C. A. Sexual differences of the hypophyses and their determination by the gonads. *American Journal of Anatomy*, 1936, **58**, 195-225.

Phoenix, C. H., Goy, R. W., Gerall, A. A., & Young, W. C. Organizing action of prenatally administered testosterone propionate on the tissues mediating mating behavior in the female guinea pig. *Endocrinology*, 1959, **65**, 369-382.

Rabedeau, R. G., & Whalen, R. E. Effects of copulatory experience on mating behavior in the male rat. *Journal of Comparative and Physiological Psychology*, 1959, **52**, 482-484.

Resko, J. A., Feder, H. H., & Goy, R. W. Androgen concentrations in plasma and testis of developing rats. *Journal of Endocrinology*, 1968, **40**, 485-491.

Rosenblatt, J. S., & Aronson, L. R. The influence of experience on the behavioural effects of androgen in prepuberally castrated male cats. *Animal Behaviour*, 1959, **6**, 171-182.

Stoller, R. J. Gender identity and a biological force. *Psychoanalytic Forum*, 1967, **2**, 318-325.

Stumpf, W. E. Estradiol-concentrating neurons: Topography in the hypothalamus by dry-mount autoradiography. *Science*, 1968, **162**, 1001-1003.

Valenstein, E. S., Riss, W., & Young, W. C. Sex drive in genetically hetergeneous and highly inbred strains of male guinea pigs. *Journal of Comparative and Physiological Psychology*, 1954, **47**, 162-165.

Valenstein, E. S., Riss, W., & Young, W. C. Experiential and genetic factors in the organization of sexual behavior in male guinea pigs. *Journal of Comparative and Phsyiological Psychology*, 1955, **48**, 397-403.

Ward, I. L. Differential effect of pre- and post-natal androgen on the sexual behavior of intact and spayed female rats. *Hormones and Behavior*, 1969, **1**, 25-36.

Whalen, R. E. Strain differences in sexual behavior. *Behaviour*, 1961, **18**, 199-204. (a)

Whalen, R. E. Effects of mounting without intromission and intromission without ejaculation on sexual behavior and maze learning. *Journal of Comparative and Physiological Psychology*, 1961, **54**, 409-415. (b)

Whalen, R. E. The initiation of mating in naive female cats. *Animal Behaviour*, 1963, **11**, 461-463.

Whalen, R. E. Discussion of gender identity and a biological force, R. J. Stoller, *Psychoanalytic Forum*, 1967, **2**, 329-331.

Whalen, R. E. Differentiation of the neural mechanisms which control gonadotropin secretion and sexual behavior. In M. Diamond (Ed.), *Perspectives in reproduction and sexual behavior*. Bloomington, Ind.: Indiana Univ. Press, 1968. Pp. 303-340.

Whalen, R. E. The determinants of sexuality in animals. In S. C. Plog & R. B. Edgerton (Eds.), *Changing perspectives in mental illness*. New York: Holt, 1969. Pp. 627-653.

Whalen, R. E. The concept of instinct. In J. L. McGaugh (Ed.), *Psychobiology*. New York: Academic Press, 1971, in press.

Whalen, R. E., & Edwards, D. A. Sexual reversibility in neonatally castrated male rats. *Journal of Comparative and Physiological Psychology*, 1966, **62**, 307-310.

Whalen, R. E., & Edwards, D. A. Hormonal determinants of the development of masculine and feminine behavior in male and female rats. *Anatomical Record*, 1967, **157**, 173-180.

Whalen, R. E., Edwards, D. A., Luttge, W. G., & Robertson, R. T. Early androgen treatment and male sexual behavior in female rats. *Physiology & Behavior*, 1969, **4**, 33-40.

Whalen, R. E., & Robertson, R. T. Sexual exhaustion and recovery of masculine copulatory behavior in virilized female rats. *Psychonomic Science*, 1968, **11**, 319-320.

Whalen, R. E., & Luttge, W. G. Long-term retention of tritiated estradiol in brain and peripheral tissues of male and female rats. *Neuroendocrinology*, 1970, **6**, 255-263.

Zucker, I. Suppression of oestrous behaviour in the immature male rat. *Nature (London)* 1967, **216**, 88-89.

CHAPTER 7

THE ONTOGENY OF MATERNAL BEHAVIOR IN SOME SELECTED MAMMALIAN SPECIES*

Howard Moltz†

For any mammalian species to survive, it must not only reproduce, it must also care for its young. Why mammals exhibit this care, why they come to nurse and otherwise nurture their offspring, is the subject of the present chapter.

However, before approaching that subject, it must first be emphasized that maternal behavior is not expressed uniformly throughout the mammalian series; rather, it is expressed in a diversity of ways, a diversity which seems limited only by the requirements of litter survival. Some mammals, for example, build nests prior to parturition, such as the hamster and mouse; others, such as the rabbit, build nests either before or during parturition, depending on the strain

*The present chapter was written while the author was in receipt of Research Grant GB 23943 from the National Science Foundation.

†The analysis of maternal behavior presented herein originated, for the most part, in discussions with my graduate students, Michael Leon and Michael Lubin. I am deeply grateful. Thanks are due also to Miss Kristyna Hartse for her patient care in the preparation of the manuscript.

(Ross, Sawin, Denenberg, & Zarrow, 1961); and still others (dogs and cats) do not build at any time. Placentophagy is widespread among mammals but there are several families which do not exhibit this behavior—Camilidae is one such family (Pitters, 1954) and Phocidae is another (Bartholomew, 1959). The rat and mouse retrieve their young when they wander from the nest, but under the same circumstances the rabbit remains indifferent. And finally, while all mammals of course nurse their young, the time they actually spend in this activity shows wide variation. At one extreme is the tree shrew (family Tupaiidae) who nurses only at 48-hour intervals and then for only 5 or 10 minutes per session (Martin, 1966); at the other extreme perhaps is the laboratory rat who, at least during the first 3 or 4 days following parturition, spends some 20 hours a day in contact with young (Holland, 1965).

To the extent that the interspecific divergences we have just mentioned reflect fundamental differences in physiological organization and psychological complexity, to that extent no single explanation of maternal behavior is likely to obtain across the entire class of mammals. And of course there is no reason to expect it should. The mammalian series, after all, includes an enormous diversity of species, each representing in some degree a unique level of biological and behavioral integration (Schneirla, 1950). For each, then, to nurture its young in exactly the same way, and for each, in particular, to have its maternal behavior governed by the same causal determinants, would be most improbable indeed.

The natural history of mammals, as Lehrman has pointed out (1961), has not inspired as widespread or as intense an interest as that of birds. As a result, we simply do not have, except for a few familiar mammals, any really precise and complete descriptions of maternal behavior. Rather, the bulk of our information consists of scattered and fragmentary facts about the way in which one species nurses and another retrieves. Such facts, while sufficient to highlight interspecific differences in maternal behavior, are not sufficient to provide the kind of descriptive data needed to begin the search for causal determinants. For this, we must know, for example, whether the female characteristically builds a nest, and, if so, just how it is constructed. We must know also the behavioral details of the birth process and just when, in relation to the time the young emerge, the various items of the maternal complex actually appear. Finally, since the litter situation does not remain static, there must also be included in every maternal profile a detailed description of the way in which the mother-young interaction changes to reflect changes in both the physiological condition of the mother and the stimulus characteristics of the young.

There is still another point that must be emphasized. Before attempting to explain the maternal behavior of any species, we must have available for that species not only the kinds of observational details just mentioned, but some knowledge as well of the intraorganic mechanisms underlying gestation and birth. Unfortunately, it is only for the rat, and to a much more limited extent for the rabbit, hamster, and mouse, that we have both sufficient behavioral data and sufficient physiological evidence to support an integrated analysis. Of necessity, then, it is upon these forms that the present chapter must focus. However, it is hoped that, although limited perforce to only a few species, our discussion will at least raise experimental questions relevant to the development and expression of maternal behavior in a variety of mammals.

I. The Rat

A. Characteristics of Maternal Behavior

Gestation in the rat is characteristically 22 to 23 days in duration. Beginning about 24 hours before term, the female begins to construct a nest. The material employed—usually paper, hay, or excelsior—which heretofore was used to build a flat, matlike structure, is now used to construct what can be designated as the "preparturient nest." To build such a nest, the female pushes and plys the available material, usually with her mouth but often with her paws, until a circular or semicircular mound is formed. Although distinctly different from the matlike structure referred to above, the preparturient nest is only loosely constituted, having sides rarely exceeding 1 or 2 inches in height. Beginning almost immediately after parturition, however, the nest material is reworked to form a mound that is at once higher (four to six inches) and more compact. This the female accomplishes usually by reaching out with her mouth and pulling the material at first toward and then over her. When completed, the postparturient nest often hides from view both mother and pups. In the laboratory, the pre- and postparturient nests are invariably constructed in that corner of the cage which the female had previously selected for sleeping (Eibl-Eibesfeldt, 1961).

The birth of the young takes place in the preparturient nest and is heralded by peristalticlike waves passing posteriorly along the flank musculature. For a period of some 10 hours prior to parturition, the female lies quietly, usually with the ventral surface of her body against the floor and with her legs stretched backwards. As the uterine contractions become more vigorous, however, and as the interval between contractions shortens, she begins to arch her back, pressing her abdomen downward. Suddenly her posture changes, as a result, presumably, of the first fetus entering the lower birth canal; at this time she adopts what Sturman-Hulbe and Stone (1929) appropriately called "the head between heels position." When in such a position, she vigorously licks the vaginal orifice, often biting and tearing the surrounding tissue as well. As the first fetus begins to emerge, it also is licked, and within a short time passes clear of the birth canal. When the placenta is finally expelled, the female takes the pup between her forepaws and devours the fetal membranes; the placenta and umbilicus are eaten thereafter in rapid order. As soon as the pup is free of fetal tissue, the mother begins to lick its body surface, coming gradually to concentrate on its ano-genital area.

The second and subsequent deliveries occur in essentially the same way as the first, except that during these deliveries the female is less likely to adopt the "head-between-heels position." The interval between deliveries is variable and, in the case of a large litter, the delivery process may take several hours. Once the young are born and each has been cleaned, the female proceeds to rebuild the nest, transforming it from the low and loosely structured preparturient nest to the high and compact postparturient nest.

Nursing begins almost immediately after parturition, with the mother crouching over the young in such a way as to expose her mammary region. As the thermotactic pups begin to nuzzle in her fur, she adjusts her posture, allowing attachment first to one nipple then to another. After the pups are attached and sucking has begun, the mother usually remains either entirely passive or engages in no activity other than that of readjusting her position to sustain attachment.

Although simple in appearance, nursing is in fact a complex response. The mother must not only inhibit her own motor activity while adopting a posture that permits sucking, but at the same time must crouch in such a way as to maintain the body temperature of the young at approximately 100°F. Both these functions invariably demand subtle positional changes in response to movement of the pups. Just how subtle these adjustments actually are is exemplified occasionally by the female which to all appearances seems to be nursing but whose pups are discovered to be cold and either dead or moribund.

Retrieving, the last maternal response item to be discussed, can be observed beginning at parturition. The mother grasps the pup, usually with the incisors and usually at the middorsal region, although seizing a pup by its head or paw is not uncommon. Once seized, that pup is rapidly transported back to the nest.

In the laboratory, retrieving is easily elicited by removing the young from the nest and depositing them along the floor of the cage. Once aware that they are thus deployed, the female will grasp each in turn and transport them back to the nest. The fact that the pups are brought to the nest and not simply deposited at random, distinguishes retrieving from what might be called "carrying." Retrieving is directed and serves to collect the young at a specific location; carrying is apparently aimless and ends by leaving the young scattered. Occasionally, a puerperal female that has failed to establish a stable nest site will be seen to carry rather than retrieve. Such a female responds quickly enough to scattered young, but in transporting these young, she will drop some in one and some in another quadrant of the cage. Thus dispersed, at least a few will invariably die of neglect. [The reader is urged to consult also the excellent descriptions of maternal behavior in the rat provided by Rosenblatt and Lehrman (1963) and Wiesner and Sheard (1933).]

B. Immediate Display of Maternal Behavior in the Puerperal Female

Important in understanding the ontogeny of maternal behavior is the fact that the nulliparous female, if kept continuously in the presence of young, will begin eventually to display interest in these young (Rosenblatt, 1967; see also Cosnier, 1963; Wiesner & Sheard, 1933). She will begin, in other words, to build a nest, lick, retrieve, and finally crouch in a nursing posture, all in a manner indistinguishable from the puerperal female. That these responses are not instigated by hormonal changes resulting from enforced association with pups is evidenced by the fact that they occur with the same reliability in similarly confined nulliparae that have been either ovariectomized or hypophysectomized (Rosenblatt, 1967).

But whereas the nulliparous female requires, on the average, some 6 or 7 days of continuous exposure to pups before she will begin attending to them, her puerperal counterpart responds immediately upon the emergence of the pups from the birth canal. What accounts for this alacrity on the part of the puerperal female? Or, to put it another way, why is it that in such a female maternal behavior occurs in almost precise synchrony with the birth of the young? The need for this synchrony is of course obvious and it is to the events responsible we now turn our attention. The correlative question of why the nulliparous female responds to young at all will be considered in a later section.

1. The Experience of Parturition

Perhaps the difference in response latency between the puerperal and the nulliparous female can be explained by reference to an event as obvious as the occurrence of parturition. As is well known, parturition in the rat, as in most other mammalian forms, is characterized by a complex of responses that includes self-licking—particularly of the vaginal area—the ingestion of birth fluids, and the consumption of fetal and placental membranes. These events, paralleled by such endogenous phenomena as oxytocin release, uterine contractions, and the emergence of the fetuses, might by their very nature be instrumental in orienting the puerperal female toward the young and consequently in instigating immediate responsiveness.

This view of parturition as providing experiences important and perhaps even essential for the characteristic behavior of the puerperal female has been advanced by several investigators. Schneirla (1956), for example, speaks of "parturitive behavior" as "facilitating initiation of nursing and other stimulative relations of mother and newborn" (p. 421). Similarly, Rosenblatt and Lehrman (1963) suggest that the immediate responsiveness of the puerperal female might be occasioned, in part at least, by the experience gained during the parturitive process itself.

In an effort to determine whether parturition can be invoked to explain the difference in latency between puerperal and nulliparous rats, Moltz, Robbins, and Parks (1966) undertook to deliver near-term females (21 days, 6 hours *post coitum*) by Caesarean section.

These females, half of whom were breeding for the first time (primiparous), and half of whom had reared two previous litters (multiparous), had fetuses and placentas removed surgically and then, after a period of time sufficient to permit mammary parenchyma to reach a secretory level comparable to that found in normal puerperae, were presented with normally delivered foster pups. The data left no doubt about the question to be answered. Each Caesarean-sectioned animal, whether multiparous or primiparous, responded to foster young with an alacrity indistinguishable from that of the normally delivered female. That the behavior thereafter was also indistinguishable was revealed by systematic observations of the mother as well as by the weight of each litter at the time of weaning. Obviously enough, whatever role the experience of parturition plays in the induction of maternal behavior, it is not of sufficient importance to account for the difference in latency between the puerperal and the nulliparous rat.

2. *Relief from Uterine Distension*

It is possible that it is not the experience of parturition as such but simply the relief from uterine distension that is the critical event underlying puerperal responsivity. If such were in fact the case, then of course whether the near-term conceptus was removed surgically or delivered vaginally would be irrelevant—in each case the female would be expected to show immediate maternal behavior.

Klein (1952) undertook to investigate this possibility by ligating the lower end of each uterine cornu in the midpregnant rat. At the expected time of parturition, but of course while still retaining their own fetuses, such females were offered newborn foster young. Despite the continuance of uterine distension, each quickly showed what the experimenter referred to as "quite normal nestling and maternal behavior" (p. 86).

In a parallel study, Klein extraverted each fetus into the peritoneal cavity, leaving only the placentas inserted in the uterine wall. These females were at midpregnancy at the time of the operation, and it was at this time, rather than at expected parturition, that they were now offered foster young. Although the operation afforded virtually complete relief from uterine distension, they showed no maternal behavior.

It is possible of course that the stimulative events related to the evacuation of the conceptus do play a role in instigating nurtural responsivity in the puerperal female. But like the experience of parturition, this role appears to be in no way critical.

3. *Mammary Engorgement*

It is well known that, in the rat and of course in other mammals as well, the onset of lactation occurs at about the time of parturition, so that when the young emerge the mammae are already distended with milk (Kuhn, 1968, 1969; Kuhn & Lowenstein, 1967; Shinde, Ota, & Yokoyama, 1964). In this condition, the puerperal female promptly begins to nurse and, as we have already mentioned, begins with equal promptness to exhibit other maternal responses. Perhaps it is mammary engorgement or, more precisely, the peripheral stimulation accompanying that engorgement, that can explain the latency difference between puerperal and nulliparous females. Already present during parturition, such stimulation may induce nursing as soon as pups to be nursed become available—and nursing in turn may induce retrieving and nest building in rapid order. The nulliparous female, of course, experiences no intramammary stimulation and thus in contrast to her puerperal counterpart may lack a stimulative dimension critical for inducing immediate attention to young.

One way of assessing the criticality of mammary engorgement in the behavior of the puerperal female is to determine whether such a female will display nursing, nest building, and retrieving with characteristic alacrity even when deprived of all mammary tissue and hence of all mammary stimulation.

Moltz, Geller, and Levin (1967) subjected weanling rats to a total mammectomy and, incidentally, to a total thelectomy (surgical removal of the nipples) as well. When these animals matured and were then impregnated and allowed to give birth normally, they behaved in a manner indistinguishable from intact puerperae. Obviously, intramammary tension is not a critical event in the initiation of maternal behavior nor, of course, can it be enlisted to explain puerperal-nulliparous differences in maternal latency.

4. Self-Licking

The nulliparous rat spends a significant amount of time in grooming or self-licking, focusing this activity primarily on the head and forepaws and on the shoulders and upper back (Roth & Rosenblatt, 1967). Although she licks other areas of her body as well, they receive scant attention relative to the areas just mentioned. With the advent of pregnancy, however, this pattern of self-licking changes. Now the female directs most of her attention to those parts of her body which before she had largely ignored, namely, vagina, pelvis, and nipples (Roth & Rosenblatt, 1967; see also Steinberg & Bindra, 1962).

There undoubtedly are many events underlying this shift in licking. One such class of events, as Roth and Rosenblatt themselves point out, probably arises from pregnancy-induced changes which come to be localized largely in what has been called the "critical body regions." For example, the nipples enlarge during pregnancy, with the base of each nipple becoming bare of fur, the abdomen increases in girth, and the vagina partially evaginates and discharges fluid. In addition to these peripheral changes, self-licking during pregnancy may be regulated as well by an increased need for salt which, as Lehrman (1956) and Bindra (1959) have suggested, could be reduced by the ingestion of the presumably "salty" vaginal exudate. In support of this suggestion, Steinberg and Bindra (1962) found that the addition of sodium chloride to the diet of the pregnant rat significantly reduced genital licking while leaving unaffected the licking of nipples and pelvic region.

But irrespective of just what it is that induces so marked a change in licking focus, the important question for our purposes is what role, if any, this change plays in the induction of maternal behavior? Having had, in other words, the kind of licking experience that pregnancy alone brings, is the puerperal female thereby "prepared" for the appearance of the young in ways that the nulliparous female is not?

Birch (1956) raised female rats with rubber collars around their necks, thereby preventing licking of any body area posterior to the collar. Removing the collars some 2 to 3 hours prior to parturition, he found his experimental subjects markedly deficient in maternal behavior. More specifically, the majority either cannibalized at the time of birth or simply remained indifferent. And even in those cases in which some attention was paid to young, the latencies were, in Birch's words, "abnormally long." On the basis of these data, Birch, and other investigators following Birch (e.g., Lehrman, 1956), hypothesized that ventral and, in particular, vaginal licking during pregnancy is transferred to the pup, as the pup, in emerging from the birth canal, becomes an extension of the mother's body.

This is a simple and attractive hypothesis, one that has been repeatedly invoked to account for the behavioral difference between puerperal and nulliparous females. Unfortunately, however, the data on which it is based have not been confirmed. Coomans (cited in Eibl-Eibesfeldt, 1958), Friedlich (1962), and Kirby and Horvath (1968) have each repeated Birch's study but without obtaining Birch's results. Christophersen and Wagman (1965) have also tried to replicate these results, even depriving their rats of self-licking as early as 11 days of age. They too, however, failed to affect in any way the characteristic behavior of the puerperal female. Although it is difficult to account for Birch's original data, what seems clear is that the unique pattern of self-licking displayed by the pregnant female does not underlie the immediate interest in young she subsequently shows at the time of parturition.

5. *The Hormones of Parturition*

Thus far, neither the experience of parturition, mammary engorgement, relief from uterine distension, nor self-licking during pregnancy has provided a clue to the maternal behavior of the puerperal female. But of course there is still another dimension along which she differs from her nulliparous counterpart.

As long ago as 1925, Stone advanced the hypothesis that the endocrine changes accompanying the termination of pregnancy might underlie as well the onset of maternal behavior. This is an obvious idea of course, readily suggested by the observation that such maternal items as nursing, nest building, and retrieving are each initiated in the puerperal female in close temporal association with the birth process. But just what are the hormonal events which characteristically precipitate parturition and perhaps also maternal behavior? A complete answer, unfortunately, cannot be given, since even now there are just too many gaps in our understanding of the mammalian birth process. Nonetheless, many investigators of maternal behavior have attempted to manipulate the blood concentrations of the gonadal steroids and the pituitary gonadotropins. The aim, of course, was to induce immediate or near-immediate maternal behavior in the nulliparous female.

Stone (1925) joined pairs of females in parabiotic union by means of a small abdominal junction. One of the twins of each pair was impregnated and then allowed to give birth. This female, despite her enforced union, had an "uneventful" parturition and displayed maternal behavior immediately upon the emergence of the young; her nulliparous twin, in contrast, remained altogether indifferent to the foster pups offered her.

Carried out in 1925, Stone's work was indeed pioneering. But considered as an attempt to investigate whether or not maternal behavior in the puerperal female is hormonally facilitated, it raises the critical question of which hormones, if any, passed from the parturient female to her nulliparous twin. Estrogen seemingly did not, as evidenced by the condition of the vaginal epithelium; neither did prolactin apparently, since none of the nulliparae showed signs of mammary change.

Using direct injections of plasma rather than hoping for diffusion across a parabiotic union, Terkel and Rosenblatt (1968) accomplished recently what Stone had set out to do—they demonstrated that substances present in the blood at the time of parturition can induce maternal behavior. Specifically, they compared nulliparous females given plasma taken from puerperal females with those given plasma from proestrus and diestrus females, respectively. Only those nulliparous subjects injected with "maternal plasma" showed a significant reduction in the onset of maternal behavior, responding to foster young, on the average, 48 hours after they were proffered (but with a good deal of variability, however). Each of the remaining groups did not differ significantly from untreated nulliparae, taking, as is characteristic of such females, some 6 or 7 days to respond.

Following this work, Terkel (1970) developed a method for effecting a continuous cross-transfusion of blood between two freely moving rats. Using this technique to transfuse "maternal blood," he was able to reduce not only the latency with which his nulliparae responded to foster young but the high variability they had previously exhibited following discrete injections.

This reduced latency is impressive, although it is counted in hours rather than in minutes. Perhaps the endocrine agents actually involved require time to condition or excite the neuronal substrate mediating the several response items of the maternal complex. In any event, there can be no doubt that substances in the blood of puerperal females can "increase the readiness of virgins to respond maternally to pups" (Terkel & Rosenblatt, 1968, p. 481). Of course, as the experimenters themselves point out, their studies do not offer any clues as to precisely which hormone or hormones were in fact responsible for the results obtained. The studies to be reviewed now have each been addressed to this more specific question.

Wiesner and Sheard (1933) used a battery of hormones, or more precisely a battery of crude hormone extracts, prepared, respectively, from mare and human pregnancy urine, bovine corpora lutea, minced human placenta, and bovine anterior pituitary tissue. Both intact and ovariectomized nulliparae were injected with one or another of these extracts and tested daily for retrieving. In some of the experiments reported, the foster young were housed continuously with the experimental females; in others they were available only during the retrieving test. Neither the luteal, urinary, nor placental products induced either retrieving, or any other response item in the maternal complex. Anterior pituitary extracts, on the other hand, were somewhat more effective, although even here only about 33% of the nulliparae tested responded maternally. A frequency of 33% is of course not an impressive finding and, in the present case, is particularly difficult to interpret because of the crude nature of the pituitary extracts employed. Interestingly enough, of all the agents that must have been contained in these pituitary extracts, prolactin was specifically dismissed as responsible for inducing the limited retrieving that was observed. The basis for this dismissal, simply enough, was that it seemed "exceedingly unlikely that the same factor [would be] responsible for both mammary secretion and maternal behavior quite apart from the mutual independence of the two phenomena" (p. 234).

Riddle and his associates reported a series of studies that have been widely cited in the literature and as widely misinterpreted. Having previously found that broodiness in fowl "seems to be induced only or chiefly by prolactin," they attempted to determine whether prolactin—and a variety of other hormones and chemical agents as well—would induce maternal behavior in the rat (Riddle, 1937; Riddle, Hollander, Miller, Lahr, Smith, & Marvin, 1942a; Riddle, Lahr, & Bates, 1935a, 1935b, 1942b). Nulliparous females of between 60 and 70 days were used, some of which remained intact, while others were ovariectomized. After eliminating what Riddle *et al.* (1942b) called "normal reactors," that is, females which spontaneously showed interest in young during a series of 10-minute tests prior to treatment, each animal was injected daily for 10 days and then tested each day for 10 minutes with a single pup. If the pup was retrieved, the female was considered to have acted maternally.

Riddle *et al.* (1942a) found prolactin, intermedin, and luteinizing hormone to induce retrieving in some 50 to 80% of both their intact and ovariectomized females. In contrast, whole anterior pituitary extract was effective only in the intact females, while follicle stimulating hormone (with thyrotropin), pregnant mare's serum, "adrenotropin," and "Prolan" were each largely without effect. Among the steroid hormones used, progesterone, testosterone, and pellet implants of deoxycorticosterone (but not injections) were followed by retrieving in at least 65% of the animals tested. Phenol and thyroxine were also effective in inducing retrieving, but only in ovariectomized females. Estrone proved to have special properties—it terminated retrieving in animals that had begun to respond after injection of one or another of the above-mentioned substances.

Needless to say, the present results are difficult to interpret. In the first place, one would have wished Riddle to have observed items in addition to retrieving and under conditions that involved more than a single foster pup presented for only 10 minutes. Maternal behavior, obviously enough, includes a complex of responses patterned over time and focused typically on an entire litter. These responses—retrieving, nursing, nest building, and licking—are each essential for litter survival, and to this end must each be integrated by the female into an effective temporal sequence. To study retrieving alone, then, is not to study "maternal behavior"—it is to study retrieving. And under stimulation from just a single pup, it is doubtful whether even retrieving can be tested meaningfully.

In the second place, Riddle used a bewildering variety of compounds, some of which seemed effective in one type of experimental female and some in another. It is instructive that the experimenters themselves do not suggest that each of these agents functions also to induce retrieving in the puerperal female. On the contrary, in such females they suggest the action of a single hormone that (a) is released in increased amounts at the time of parturition, (b) exerts an antigonadal effect, and (c) excites, either directly or indirectly, the "sensorimotor mechanism" mediating maternal behavior. Prolactin, according to Riddle "best fits these requirements." As to progesterone, deoxycorticosterone, and the several other agents claimed effective in inducing retrieving in nulliparae, the experimenters have little to say, except that these agents might have induced retrieving through their respective release of prolactin.

There have been several attempts to confirm the findings of Riddle but each has been without success. Lott (1962), for example, injected .25 mg of progesterone daily into nulliparous females for a period of 10 days. On day 11, 24 hours after the last injection, each experimental and each control (oil-injected) female was tested with a single foster pup for 10 minutes daily for 3 days. There was no significant difference between groups and in fact little evidence of retrieving. Lott and Fuchs (1962) administered 40 IU of prolactin daily for 10 days and once again found little evidence of retrieving. Finally, Beach and Wilson (1963) injected 400 IU of prolactin per day for 5 days, following which not one but six foster pups were proffered daily for three consecutive days. In a second experiment, they administered estrogen for 21 consecutive days followed by concurrent injections of prolactin and progesterone. Neither of these regimens induced retrieving with any appreciable frequency.

Contradictory results are always difficult to explain. Lott, Lott and Fuchs, and Beach and Wilson used strains of rats different from the strain employed by Riddle, they also used pituitary hormones of greater chemical purity. What is certain is that under the conditions of these experiments, neither prolactin, progesterone, nor even prolactin in combination with progesterone was found effective in inducing retrieving.

Convinced that further attempts to replicate the findings of Riddle would be futile, other experimenters adopted a different approach to the problem of hormonal induction. Taylor (1965) and Denenberg, Taylor, and Zarrow (1969), for example, imposed a hormone regimen found effective in evoking maternal nest building in ovariectomized nulliparous rabbits (Zarrow, Denenberg, & Kalberer, 1965b).

This regimen involved injection of both estrogen and progesterone with the subsequent withdrawal of progesterone prior to the termination of estrogen treatment. For use with the ovariectomized nulliparous rat, estradiol (either 0.2 μg or 1 μg) was administered from day 1 to day 19 and progesterone (2 mg) from day 2 to day 15. In a second study, the progesterone treatment was extended to 19 days and the estradiol treatment (now at a dose level of either 2 or 4 μg) to 27 days. The only index of maternal behavior employed was dowel-shredding which, like nest building, has been found to exhibit a peak at or just prior to the time of parturition (Taylor, 1965). Contrary to expectation, there was no increase in the amount of material shredded following either the withdrawal of progesterone (day 15 or day 19) or the termination of estradiol (day 19 or day 27). If as the experimenters claim, an increase in dowel-shredding is "representative of maternal nest building" (Denenberg *et al.*, 1969, p. 14), then in the rat maternal nest building cannot be evoked by an estrogen-progesterone regimen found effective in the rabbit.

Over the past several years a series of studies has been carried out in the author's laboratory addressed also to the problem of hormonal induction. The approach adopted initially involved not the nulliparous but the preparturient female and the question asked was whether, in such females, selective disruption of the endocrine events accompanying parturition would subsequently interfere with the display of maternal behavior. If it did—if an imposed hormonal imbalance were in fact to render the puerperal female indistinguishable from her nulliparous counterpart—then perhaps some insight would be gained into the problem of maternal latency.

In our first study (Moltz & Wiener, 1966), we drew attention to the fact that associated with the birth process and in some part responsible for the occurrence of that process is a dramatic change in ovarian output, one that reverses the prevailing gestational ratio of estrogen to progesterone (Eto, Hsoi, Musudo, & Suzuki, 1962; Zarrow, 1961). Simply enough, the question posed was whether or not manipulation of this apparently focal event would affect the onset of maternal behavior and, in the event it did, whether the influence thus exerted would be manifested differentially in primiparous and multiparous puerperae.

To this end we ovariectomized (and of necessity delivered by Caesarean section) each of our animals shortly before term. In response to the steroid imbalance imposed by our ovariectomy, only 50% of our primiparous females accepted immmediately the foster litters offered them; the remaining 50% delayed acting maternally until, after 3 days, the young died of neglect. In contrast, virtually all of the multiparous animals behaved in characteristic puerperal fashion.

The ovariectomy just referred to resulted, of course, in the reduction of circulating levels of estrogen and, in view of the positive feedback effect of this steroid on prolactin discharge (Deis, 1967; Everett, 1966; Meites & Nicoll, 1965; Nicoll & Meites, 1962; Ramirez & McCann, 1964; Ratner, Talwalker, & Meites, 1963; Rothchild & Schwartz, 1965; Welch, Sar, Clemens, & Meites, 1968) probably in the reduction of circulating prolactin as well. In addition, removing the ovaries also resulted in a decrease in plasma progesterone.

It seemed to us unlikely that the reduced concentration of progesterone effected through ovariectomy was responsible for the observed delay of maternal behavior in our primiparous females. As already mentioned, associated with the birth process in the rat is a shift in ovarian output, resulting, *inter alia,* in a *decrease* in progesterone titer. Both from the time course of this decrease (Fajer & Barraclough, 1967; Grota & Eik-Nes, 1967; Hashimoto, Henricks, Anderson, & Melampy, 1968) and from the fact that maternal behavior is initiated typically at a point of low progesterone concentration, we did not think our ovariectomy could have produced the observed behavioral effect merely as a result of having realized a still further reduction in progesterone. Indeed, not only did we consider this particular consequence of the operation to be irrelevant for the results obtained, but, on the contrary, hypothesized that if the characteristic near-term decrease in blood progesterone were to be prevented from occurring, then the initiation of maternal behavior in turn would effectively be delayed (cf. Richards, 1967; Rothchild, 1965). Accordingly, we attempted to determine whether such manipulation of progesterone would in fact delay the appearance of maternal behavior and, if so, whether the effect thus exerted would once again be manifested differentially in the primiparous and multiparous female (Moltz, Levin, & Leon, 1969a).

Here as well, both primiparous and multiparous females were used. Each, beginning on day 19 of pregnancy, was injected with 2 mg of progesterone. On day 21, Caesarean sections were performed, necessitated by the failure of our rats to give birth normally under the progesterone dosage administered.

The results obtained paralleled those of our ovariectomy study. There, as here, a significant proportion of the primiparous females delayed responding to young until the young died of neglect. At the same time, virtually all the multiparous females responded immediately.

There is no doubt that selected hormonal manipulation can affect the latency to maternal behavior. However, before proceeding to discuss just which endocrine mechanisms are actually involved in governing the onset of maternal behavior in the puerperal female, it might be well at this point to ask how the observed influence of breeding history was mediated, of how, in other words, our multiparous females came to respond in a manner different from our primiparous females.

We can assume that selected hormonal agents function to increase the excitability of all or some of the neuronal mechanisms mediating maternal behavior, with the result that the puerperal female becomes immediately responsive to the sight, sound, and perhaps odor of the young. We can also assume that these same neuronal mechanisms are modified by previous breeding experience in such a way as to increase their responsiveness to endocrine activation.

A plausible model for conceptualizing the effects such an increase might exert would involve the idea of threshold. Specifically, the very fact that the normal primiparous mother is as maternally proficient as the normal multiparous mother (Moltz & Robbins, 1965) indicates that the substrate mediating maternal behavior is already differentiated and organized by the time the first birth is completed. Since a substrate of this kind must surely have threshold characteristics unique, in part at least, to each female, individual differences in tissue excitability would be expected to exist among primiparous animals, with some primiparous animals, of course, having much lower arousal thresholds than others. Previous parity might increase the level of tissue excitability, so that, although individual differences persist, the average level of tissue excitability among multiparous animals would be higher than the average level that already prevails among primiparous animals. The threshold distribution of a multiparous group would then have a lower mean value than that of a primiparous group but the distributions of these two groups would overlap because of individual differences in tissue excitability which are unrelated to the occurrence of parity.

Clearly enough, a primiparous-multiparous difference of this kind, assuming it does typically exist, would not be manifested in behavior under conditions in which stimulating hormones prevail at blood concentrations above that demanded for arousal by even the most elevated tissue thresholds. Presumably, these are the concentrations that characteristically obtain following normal delivery and apparently following caesarean delivery as well. However, if only residual titers of key hormones are made to prevail or if neuronal responsiveness to these key hormones is made to decrease, then only tissue mediators having low thresholds of arousal might respond. As a consequence, immediate maternal behavior would be seen only in animals possessing these particular thresholds.

In brief, we are suggesting that low thresholds to hormonal stimulation prevail among multiparous females but that, because of individual differences in tissue excitability, some primiparous animals have thresholds which fall within the lowered threshold distribution of the multiparous population. This suggestion could explain (a) why a significantly smaller incidence of immediate acceptance was found among the primiparous as compared with the multiparous females of both our ovariectomy and progesterone studies and (b) why as many as half these primiparae nonetheless behaved without delay.

But of course the major problem still remains, namely, the identity of the inductor hormones. Included in the report of our progesterone study was the hypothesis that these hormones are estrogen and prolactin, or more precisely, the synergy of estrogen and prolactin. We now believe that progesterone is also essential. Consider first the titers of these endocrine agents in the normal pregnant female.

Progesterone, as assayed in both peripheral plasma (Grota & Eik-Nes, 1967; Wiest, Kidwell, & Balogh, 1968) and ovarian venous plasma (Eto, Hsoi, Musudo, & Suzuki, 1962; Fajer & Barraclough, 1967; Hashimoto *et al.*, 1968), begins to increase on day 4 of pregnancy, reaching maximal concentrations on day 14. Thereafter, the output of the steroid starts to fall slowly until, on day 20, the decline becomes abrupt. [This apparently is due to the sharp increase of 20 α-hydroxysteroid dehydrogenase, the enzyme largely responsible for the conversion of progesterone to 20α-hydroxypregn-4-en-3-one (Kuhn, 1969; Wiest, 1968; Wiest *et al.*, 1968).]

That progesterone is taken up in rat brain during most of pregnancy can reasonably be inferred from studies showing the ready accumulation of the tritiated preparation in the nonpregnant female (Hamburg, 1966; Laumas & Farooq, 1966; Raisinghani, Dorfman, Forchielli, Gyermek, & Genther, 1968). In such an animal, it has been found to elevate activation thresholds in cortical (Arai, Hiroi, Mitra, & Gorski, 1967; Beyer, Ramirez, Whitmoyer, & Sawyer, 1967; Komisaruk, McDonald, Whitmoyer, & Sawyer, 1967; Ramirez, Komisaruk, Whitmoyer, & Sawyer, 1967), diencephalic (Barraclough &

Cross, 1963), and hippocampal (Kobayashi, Kobayashi, Takezawa, Oshima, & Kawamura, 1962) neurons. Probably many other neurons are similarly affected since, unlike estrogen, progesterone shows no evidence of selective uptake in the central nervous system (Seiki, Higashida, Imanishi, Miyamoto, Kitagawa, & Kotani, 1968; Seiki, Miyamoto, Yamashita, & Kotani, 1969).

Virtually coincident with the onset of progesterone withdrawal is a characteristic increase in plasma estrogen (Yoshinaga, Hawkins, & Stocker, 1969). Prevailing evidently at low concentrations prior to the time of midpregnancy, the secretory rate of estrogen begins increasing about day 15. At first slowly and then more rapidly, estrogen rises to peak value on or shortly before the day of parturition. It is selectively bound in specific brain regions, at least in the nonpregnant female, as the work of Eisenfeld and Axelrod (1965, 1966), Pfaff (1965, 1968), and Stumpf (1968) show.

Prolactin, the third hormone conceived essential for the immediate induction of maternal behavior, is released at the time of coition (Dilley & Adler, 1968), although thereafter it is maintained at low plasma concentrations until about day 20 of pregnancy (Grindeland, McCulloch, & Ellis, 1969; Kwa & Verhofstad, 1967). On day 22, the day of parturition, it is seen to increase sharply (Amenomori, Chen, & Meites, 1970). To the author's knowledge, there is no evidence concerning prolactin uptake in the central nervous system nor, in turn, is there evidence concerning the effect of prolactin on the uptake of other hormones.

At this point it might be well to ask why, of all the endocrine changes occurring at or near the time of parturition, progesterone, estrogen, and prolactin were selected as critical? What of the role of other hormones, particularly other hypophyseal and ovarian hormones?

The pituitary content of follicle-stimulating hormone (FSH) increases dramatically between days 12 and 16 of pregnancy, remaining at "remarkably high levels" until the day of delivery (Greenwald, 1966). These high levels, however, reflect merely increased storage rather than increased synthesis and release as evidenced by the fact that the number of vesicular follicles remains constant throughout pregnancy.

There is a threefold increase in the pituitary content of luteinizing hormone (LH) between day 1 and day 8 of pregnancy, but thereafter the content remains at a fairly steady level (Greenwald, 1966). Apparently it is only following delivery—and several hours after maternal behavior has been established—that a surge of LH occurs. It is this surge that triggers the postpartum ovulation exhibited characteristically by the female rat.

Oxytocin, a polypeptide hormone synthesized in the hypothalamus and stored in the posterior pituitary (e.g., Cross, 1966), has long been suspected of being one of the endocrine agents responsible for the initiation of labor. This suspicion has been strengthened by the detection recently of increased levels of oxytocin during parturition in the jugular blood of sheep, cows, horses, and goats (cf. Fitzpatrick, 1966). In the rat, however, assays of oxytocin have not been carried out, and consequently for this species at least we simply do not know whether concentrations of the peptide in peripheral blood change in any systematic way during labor and prior to the onset of maternal behavior.

Finally, there is the question of 20α-hydroxypregn-4-en-3-one (20α-OH) and its possible effect on the behavior of the puerperal female. As we have already mentioned, 20α-OH is a metabolite of progesterone, synthesized within the ovary by the enzyme 20α-hydroxysteroid dehydrogenase. Assays of peripheral blood have revealed a sharp increase in the concentration of 20α-OH between days 19 and 21 of gestation (Wiest *et al.*, 1968). In itself, however, this near-term increase is probably without functional significance, 20α-OH representing merely a by-product of the "enzymatic process by which progesterone levels in the peripheral plasma and uterus are lowered" (Wiest, 1968, p. 1183). But thus far 20α-OH has been studied only in relation to the initiation of labor (Wiest, 1968); its possible effect on the initiation of maternal behavior has not been investigated directly.

As is evident, we cannot decisively reject the possibility that oxytocin, or 20α-OH, or even FSH is involved in the induction of maternal behavior. Nonetheless, these hormones were not manipulated by the present author in the attempt to identify the endocrine mechanisms underlying responsivity of the puerperal female. It was simply that, for purposes of experimentation, a choice had to be made, and progesterone, estrogen, and prolactin seemed functionally more relevant to the induction of maternal behavior than other ovarian and hypophyseal hormones. Moreover, it was possible to picture just how they might operate to evoke immediate attention to young.

We conceive of the maternal mediating system as being essentially insensitive to both estrogen and prolactin arousal so that, at even relatively high titers, these critical hormones characteristically fail to exert an activational effect. Probably before such an effect can occur, the mediating system must be primed, or in other words must be made to sustain lower-than-normal thresholds to estrogen and prolactin. It is here that progesterone, or rather that progesterone withdrawal, is believed to function.

Existing at high concentrations during the larger part of pregnancy, progesterone probably at first acts to maintain or even to increase activation thresholds within the mediating system. Beginning, however, on day 15, blood levels of progesterone, as we have already mentioned, typically start to decline. It is conceivable that this decline, occurring from the high progesterone titers of pregnancy, brings about within the mediating system an effect functionally akin to what Kawakami and Sawyer (1959) called the "rebound from progesterone dominance." As pictured here, such a rebound would not only decrease selected thresholds within the mediating system but would carry these thresholds to levels lower than normal. It would, in other words, have the effect—at a time more or less coincident with the birth process—of rendering the mediating system sensitive to endocrine influence, particularly to the influence of estrogen and prolactin. It is only then, presumably, that estrogen and prolactin, functioning synergistically and at concentrations increased by near-term secretory rates, would be able to excite the mediating substrate to a point that makes it acutely responsive to the sight, sound, and odor of the young. Thus affected, the puerperal female can respond not in 6 or 7 days but as soon as the young emerge from the birth canal.

With this picture in mind, we began to manipulate progesterone, estrogen, and prolactin in the attempt to reduce the latency to maternal behavior in the nulliparous female. The experimental regimen that we finally adopted can be described briefly (Moltz, Lubin, Leon, & Numan, 1971).

At between 90 and 110 days of age, nulliparous females were ovariectomized and 3 weeks later administered estradiol, progesterone, and prolactin. The estradiol was injected once daily from day 1 through day 11 at a dosage level of 12 μg per injection; the progesterone twice daily on days 6 through 9 at a dosage level of 3 mg per injection; and the prolactin on the evening of day 9 and the morning of day 10, each at a dosage of 50 IU. On the afternoon of day 10—approximately 20 hours after the progesterone had been withdrawn—six foster pups, 10-15 hours old were given each female. The pups remained until the following morning, at which time a fresh litter was substituted. This procedure continued until maternal behavior was displayed or, failing maternal behavior, until 7 days had elapsed. To be scored as maternal, a female was required not only to retrieve, but to build a nest, assume a nursing posture, lick the young, and keep them warm. She was, in other words, obliged to exhibit the full spectrum of nurtural attachment.

Of the ten nulliparous females subjected to our hormone schedule, each, without exception, showed full maternal behavior at between 35 and 40 hours from the time the pups were first proffered. Not only does this represent a significant reduction in latency from the average of 6 to 7 days characteristic of untreated nulliparae, but represents as well a uniformity in time of onset closely approaching that exhibited by the puerperal female. In contrast, control females given, respectively, only two of the three "inductor hormones" (the vehicle in each case having been substituted for the hormone omitted), or simply injected with all three vehicles, showed marked variability in onset and, of course, a significantly higher median latency. Figure 1 represents the data graphically.

The picture that we now have of maternal behavior places in perspective the roles played by both pup-related stimuli and endogenously produced hormones. Clearly, the presence of young, quite apart from any endocrine intervention, can activate the neuronal system mediating the expression of nursing, nest building, retrieving, and licking. But activation by pups alone takes some 6 or 7 days, an interval entirely too long for the survival of a neonatal litter. In the puerperal female then, endocrine intervention becomes necessary to ensure immediate responsiveness. Here our data enable us to conceive of the action of a triad of hormones, involving substrate sensitization by progesterone withdrawal and substrate excitation by estrogen and prolactin. It is this triad that is pictured as inducing the immediate responsivity critical for neonatal survival.

Although the nulliparous females subjected to our triad of hormones showed a significant reduction in latency as well as a marked uniformity in the time of onset of maternal behavior, the question still remains as to why they took as long as 35 to 40 hours to respond. This may indicate that the dosage level of one or another of the hormones administered was too low, or perhaps too high. Or it may indicate simply that progesterone, estrogen, and prolactin are in themselves sufficient to effect only a limited reduction in latency and that to induce immediate maternal behavior additional hormones must be administered. Or finally, it may mean only that the time elapsing between the injection of one or another of our inductor agents was not optimal. With respect to this last alternative we already know how critical is the interval between the last progesterone injection and the first prolactin injection. For example, when progesterone

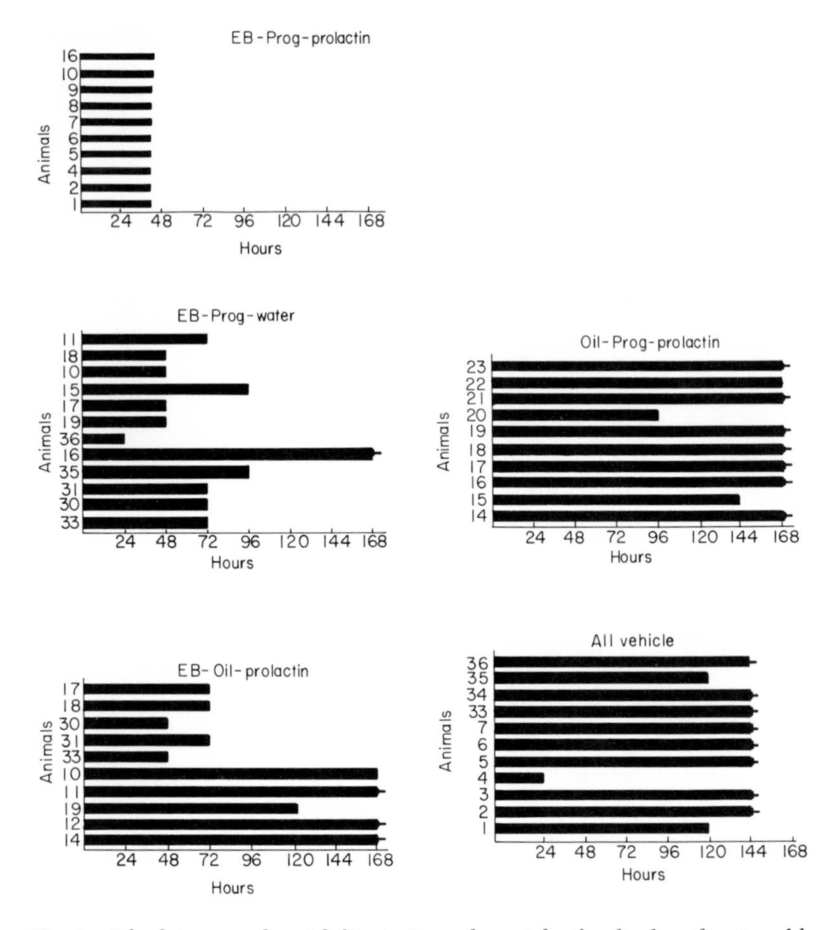

Fig. 1. The latency and variability in time of onset for the display of maternal behavior in experimental and control females. Broken bar indicates that the animal failed to act maternally at the conclusion of the observation period. (From Moltz *et al.*, 1971.)

was administered on days 4 through 7, rather than, as in our present schedule, on days 6 through 9, only 20% of our ovariectomized nulliparous females responded within the 40 hours found to obtain previously. Correlatively, when progesterone was extended to the morning of day 10 rather than terminated on the evening of day 9, 60 as compared with 100% displayed full maternal behavior (Moltz *et al.*, 1971). It is at least conceivable that a different scheduling of the present regimen—one perhaps that allows for complete sensitization by progesterone withdrawal—would reduce our obtained latency of 35 to 40 hours to perhaps 15 to 20 hours (cf. Terkel, 1970). However, whether with discrete injections one can ever duplicate the effects of endogenous release—and thereby induce immediate attention to young—remains to be determined.

C. Maternal Behavior in the Adult Male

Like the nulliparous female, the adult male rat also responds to young. Given the same 6 or 7 days of exposure to foster pups, he too begins to act maternally, building a nest, licking, retrieving, and finally hovering in a nursing posture (Rosenblatt, 1967; Wiesner & Sheard, 1933). That the male can exhibit these nurtural responses and in the appropriate temporal sequence, indicates that he also possesses a fully developed mediating system which, like that of the female, is capable of responding to pup-related stimuli alone. The question that arose is whether this system is also capable of activation by our triad of inductor hormones (Moltz, Leon, & Lubin, unpublished data).

Adult male rats were castrated and subjected to the same endocrine regimen as experimental nulliparous females, namely, estrogen and prolactin imposed on progesterone withdrawal. In contrast to their female counterparts, these males did not respond at 35 to 40 hours, but instead responded as did control animals, taking 6 to 7 days.

These data lead us to suspect that the mediating system of the male is largely insensitive to endocrine facilitation; that while it can respond to pup-related stimuli, it cannot be readily sensitized to these same stimuli through hormone injection. Thus far we have not attempted to vary dosage level systematically, although to some gonadectomized males we did administer twice the amount of estrogen and progesterone found effective in our nulliparous females. It is instructive that these males did show a significant reduction in latency as compared with males receiving standard dosages; they did not, however, approach the average latency, nor did they exhibit the uniformity in latency shown by our nulliparous females.

But whether or not the mediating system of the male is in fact susceptible to hormonal facilitation, it evidently differs in sensitivity from that of the female. It becomes of interest to determine how this difference might have arisen.

During the past decade, there has been much research on the role of androgen in the neonatal male, revealing, among other effects, its influence in suppressing the reactivity of the neuronal system mediating lordotic behavior (see Chapter 6). Neonatal androgen may well have a parallel effect on the neuronal system mediating maternal behavior. This system, more specifically, might begin to develop in both the male and female rat at or perhaps even before the time of birth. In the male, however, androgen may come early in ontogeny to exert an effect on threshold, making his maternal mediating system subsequently insensitive to hormonal excitation. If this is correct, the result might well be that of a system which, although capable of supporting full maternal behavior in response to young, remains relatively insensitive to endocrine facilitation.

In contrast to the male, the female of course is not exposed to androgen during neonatal life and in fact is only first exposed to endogenous estrogen at about day 14 postpartum (Cierciorwska & Russfield, 1968). As a consequence, her maternal control system may escape the kind of "desensitization" to which the male is subjected.

D. Neural Mechanisms in the Control of Maternal Behavior

Where in the central nervous system do the hormones of which we have been speaking act? Where, in other words, does progesterone withdrawal exert its priming effect and estrogen and prolactin their excitatory effect? There have been many attempts to identify the neural mechanisms involved in the mediation of maternal behavior and each can be classified according to the technique employed.

1. Ablation Studies

Beach (1937, 1938) removed various amounts of neocortex from the female rat, systematically studying the effect on a variety of maternal responses. He found that lesions inactivating as little as 10% of the total neocortical surface produced in otherwise normal puerperae deficiencies in both retrieving and nest building. As progressively larger lesions were inflicted, the deficiencies become more severe, "increasing in a manner roughly proportional to the amount of tissue destroyed" (1937, p. 432). With lesions involving more than 50% of the cortex, virtually no maternal behavior was displayed. These results were subsequently confirmed by Stone (1938) and Davis (1939).

Stamm (1955) also lesioned the cortex but confined his lesions to the cingulate-retrosplenial area on the median surfaces of the cerebral hemispheres. Having found previously (Stamm, 1954) that destruction of this area interferes with hoarding behavior, he reasoned that it might interfere also with maternal behavior since, in his view, both were "unlearned." His median lesions, destroying less than

20% of the total neocortex, did affect maternal behavior insofar as the puerperal females on whom these lesions had been inflicted built indifferent nests, allowed their pups to remain scattered, and in general failed to protect litters against experimentally introduced stress (air and heat blast). These results, however, are not unambiguous. When the operates own pups, many of which had died, were removed and replaced by foster young, the foster young survived, becoming in fact as "healthy and as big as their litter mates which had remained with their natural mothers" (p. 351).

Other experimenters have also studied the effects of median cerebral lesions on maternal behavior in the puerperal female. Wilsoncroft (1963), for example, lesioned the anterior and posterior cingulate areas, respectively, taking as his behavioral measure "the picking up and depositing of pups in response to scattering the litter and directing air and heat blasts onto the litter" (p. 835). Compared with females bearing posterior cingulate lesions and with control females bearing lateral lesions, the "anterior group" exhibited more erratic retrieving behavior. For example, in response both to scattering and to "air and heat blasts," they were observed to pick up a pup and, rather than bring it directly to the nest, to carry it around the cage several times before depositing it. Beach observed similar behavior following nonlocalized lesioning of the cortical surface. Unfortunately, Wilsoncroft did not present data on pup survival.

In a more extensive and more carefully controlled study than that performed by either Stamm or Wilsoncroft, Slotnick (1967) compared puerperal females bearing, respectively, full-cingulate, anterior-cingulate, or posterior-cingulate lesions. Again the data revealed erratic and confused retrieving, with pups repeatedly brought into and out of the nest. Here, however, nursing was also affected insofar as the experimental females were frequently observed to crouch over only part of their litters, ignoring the remaining pups lying near by. This was particularly evident among subjects bearing full cingulate lesions; those bearing partial lesions, while displaying aberrant nursing, improved during subsequent observations. In contrast to the data reported by Wilsoncroft, no significant differences were found between the anterior and posterior cingulate groups.

Two points warrant emphasis in the study by Slotnick. First, in allowing the cingulate females to remain relatively undisturbed, Slotnick observed that they were in fact able to keep their own pups alive and did not require substitution of foster litters. Evidently, the erratic nursing and retrieving behavior observed in cingulate-lesioned females is deleterious to the survival of the young only under conditions of environmental stress, or at least under conditions of "air and heat blasts." And second, cingulate-lesioned females are by no means indifferent to their litters; on the contrary, their maternal behavior, as Slotnick observed, is "well motivated," although of course often inappropriate.

Considering the data of the three studies just reviewed, there appears no real justification for regarding the median cortex as playing a unique role in the expression of maternal behavior in particular or of "unlearned behavior" in general. That the median cortex serves an integrative function may well be the case, but that this function is implicated only in the display of those responses variously designated as "unlearned," "innate," or "species typic" is without any substantive foundation.

There have been surprisingly few experiments in which the maternal behavior of puerperal rats has been studied following lesions in brain areas other than the cortex. Kling (1964), for example, inflicted lesions, respectively, in the olfactory stalk, medial olfactory area, and hippocampus in rats which were between 3 and 19 days of age. Following puberty, these animals were impregnated and allowed to deliver normally. Interested primarily in somatic development rather than in maternal behavior, Kling reports only that in each lesion group there was a "slight reduction in mothering capacity" (p. 1398).

More relevant is the study by Kimble, Rogers, and Hendrickson (1967). Bilateral dorsal-hippocampal lesions were inflicted in female rats of about 90 days of age, after which they were observed both for sexual and maternal behavior. With respect to sexual behavior, the hippocampal-lesioned females were essentially similar to nonoperated controls; with respect to most aspects of maternal behavior, they were strikingly different. More specifically, the majority of the hippocampal-lesioned females cannibalized at least part of the litters, built poor nests, and spent only about 20% of the time nursing. They did not, however, show any significant deficit in retrieving. Disturbingly enough, a third group of animals, sustaining lesions over 40 to 50% of the dorsal surface of the neocortex, behaved in a manner indistinguishable from controls. Kimble *et al.* (1967) suggest that the deficits in maternal behavior seen by Beach were the result not of cortical lesions as such but of cortical lesions that invaded the hippocampus. Accordingly, Kimble *et al.* (1967) discount the role previously attributed to the total neocortex in the display of maternal behavior and focus instead on the limbic system, particularly the hippocampus and the anatomically related cingulate region. They decline, however, to speculate as to just how this "neural machinery" characteristically functions in the "execution of maternal behavior in the rat" (p. 406).

Perhaps the most puzzling study of those included thus far under the present rubric is that by Avar and Monos (1967). These investigators placed electrolytic lesions in the lateral (tuberal dorsolateral and infundibular ventrolateral) hypothalamus of female rats during the third trimester of pregnancy. Such lesions were found to raise perinatal mortality to a level above 90% and, in addition, to prevent the expression of virtually all maternal behavior. That lack of adequate stimulation from own young was not responsible for the behavioral deficit observed, is seen in the fact that the lesioned mothers showed no evidence of nest building, retrieving, or nursing even when presented with foster young.

Avar and Monos, to be sure, recognize that lateral hypothalamic lesions produce aphagia and adipsia and in fact attribute the excessive number of stillbirths to a reduction in the fluid content of fetal tissue. But confusingly enough, they still conclude that "the lateral hypothalamic regions play a large part in the integration of maternal behaviour" (p. 260), pointing out that their lesioned animals were maternally indifferent not only to surviving young in their own litters but to healthy foster young as well. However, if even limited aphagia and adipsia resulted from the lateral hypothalamic damage, then there might well have been general debilitation. That females suffering such debilitation would fail to behave maternally is hardly cause for surprise.

Obviously the studies just reviewed have not answered the question of where in the central nervous system estrogen, progesterone, and prolactin act to facilitate the expression of maternal behavior. Instead, they have demonstrated simply that destruction of certain brain regions—the hippocampus or the cingulate-retrosplenial area—will interfere with the effective performance of one or another nurtural response. However, from this we do not know in what capacity the hippocampus and the cingulate-retrosplenial area functions to mediate these responses nor even whether their destruction produces deficits unique to maternal behavior. Indeed, if such deficits as are produced do extend beyond the maternal to effect response systems entirely unrelated to the care of the young, then the areas involved have only indirect relevance to the question raised.

2. *Implantation Studies*

Implantation of fine capillary tubes carrying crystalline sex steroids has proven an effective technique for identifying brain areas involved in the hormonal control of sexual behavior (e.g., Lisk, 1962, 1967). Essentially the same technique of course could be used in studying maternal behavior. Estrogen and prolactin, for example, could be implanted after the withdrawal of systemically administered progesterone. Or conversely, progesterone could be implanted during the time that both estrogen and prolactin are being injected. As long as the progesterone remains implanted, maternal behavior in response to foster pups would not be expected; however, within some 40 hours after removal these same females should build nests, retrieve, and attempt to suckle young.

One of the difficulties involved in carrying out experiments of the kind just described is knowing where in the brain to implant. It would seem that the most propitious choice would be of those areas which, during parturition, selectively incorporate and retain progesterone, estrogen, and prolactin, respectively. The autoradiographic method is ideally suited to the task of such identification, but unfortunately has not been applied to the pregnant female. Moreover, in the nonpregnant female, only estrogen (estradiol ^{3}H) binding has been studied extensively (e.g., Eisenfeld & Axelrod, 1966, 1967; Kato & Villee, 1967a, 1967b; Pfaff, 1965, 1968; Stumpf, 1968), and prolactin binding not at all.

The earliest and most widely cited attempt to induce maternal behavior through direct hormonal stimulation of the brain is that of Fisher (1956; see also Fisher, 1961, 1966), who used male rather than female rats and testosterone (sodium testosterone sulfate) rather than either estrogen, progesterone, or prolactin. When implanted into the medial preoptic area, testosterone was able to facilitate the "complete maternal response," but only, however, in 5 of the more than the 125 animals tested with foster young. Of this low rate of success, Fisher writes "we still do not understand the problems which prevent a greater incidence of replication from animal to animal" (1966, p. 120).

The several reports in which these preliminary data were presented do not contain a complete description of the controls used. We do not know, for example, whether the five males that responded immediately to foster young following testosterone placement in the medial preoptic would have failed to respond maternally following placement in the same area of either a blank cannula or a control substance. Correlatively, we do not know whether or not these same males were of a type that Wiesner and Sheard (1933) called "spontaneous responders," that is, nonpuerperal control animals showing, not the typical 6 to 7 day latency, but near-immediate maternal behavior following the presentation of foster young. This raises the question of whether Fisher actually demonstrated testosterone facilitation of maternal behavior or whether he merely reaffirmed the fact of spontaneous responding. It is not irrelevant that, in the author's laboratory, the spontaneous responder is found in about 2% of the cases, a frequency not significantly different from that which Fisher found in his population of brain-implanted animals.

There are only two additional reports in the literature in which the attempt was made, through brain implantation, to discover those regions critically involved in the hormonal facilitation of maternal behavior. Unfortunately, both studies were presented only in preliminary form, leaving many questions about procedure and results unanswered.

One of these is by Braden (1966), involving bilateral placement of estrogen (estradiol beta) into the medial preoptic area of the pregnant rat. She reports, first, that the mortality rate among the young of the estrogen-preoptic mothers was significantly higher than among the young of the control mothers. She does not, however, specify when the mortalities occurred—whether *in utero,* during parturition, or during the postpartum observation period. After litter size was adjusted to six and supplemented, when necessary, with foster young, observation indicated that the estrogen-preoptic mothers were inferior to control females in nursing, although not in nest building nor in what Braden called "general maternal behavior."

Such results are difficult to interpret for two reasons. First, the endocrine events that facilitate maternal behavior in the puerperal female are most likely different from those which function to maintain the behavior once it is established. If this is correct, and some supporting evidence will be presented below, then studying already established mother-litter interactions is irrelevant to the task of identifying the mechanisms critical for the induction of these same interactions. Thus Braden reports that the estrogen-preoptic mothers were inferior to the control mothers on the "nursing scale," but does not report whether nursing as such was initiated with equal frequency in both groups. And second, we have already mentioned that plasma estrogen slowly increases beginning about day 15 of pregnancy and then rises abruptly on or shortly before the time of parturition. This change in titer is conceived to elevate the concentration of bound estrogen in selected brain areas, resulting near term in a concentration optimal for the facilitation of maternal behavior. To administer estrogen at this time, either systemically or through brain placement, might well have the effect of raising the steroid to a level no longer optimal for facilitation, and perhaps even to a level that interferes with facilitation. In other words, when bound at one concentration, estrogen (together with prolactin and against a background of progesterone withdrawal) may induce immediate maternal responding, while when bound at higher concentrations may actually preclude such responding. Whether Braden's report of perinatal mortality was the result of such interference or, as we have already said, was even due to maternal neglect at all, cannot be determined.

The second of the two studies mentioned above was reported by Roth and Lisk (1968). On day 18 of pregnancy, they implanted progesterone into either the arcuate or the mammillary region of the hypothalamus. In the arcuate region, progesterone caused reproductive failure as evidenced by delayed parturition and a high incidence of stillbirths. It resulted also in failure to respond maternally toward those of the litter that did survive, although this may have been due to the suboptimal stimulation offered the mother by the surviving pups. In any event, when foster young were proffered, the progesterone-arcuate mothers were observed to initiate both retrieving and nursing. Unlike progesterone in the arcuate region, that in the mammillary region did not cause reproductive failure nor did it interfere with the maintenance of own young. It did, however, appear to reduce the incidence of retrieving, since 40% of these females failed to retrieve when tested with foster young.

The authors maintain that their data "begin to show" that different stages of the breeding sequence are governed by different hypothalamic sites—parturition by the arcuate region and maternal behavior by the mammillary region. This conclusion is interesting, but not entirely warranted by the data. Parturition, for example, was affected by progesterone implants in the arcuate region, but in only 50% of the animals tested, the remaining 50% showing no sign of reproductive failure. And while the incidence of retrieving decreased following placement of progesterone in the mammillary region, this decrease was evidenced in only 40% of the subjects and, in comparison with control females, failed to reach statistical significance. Evidently, the present experiment must be repeated, using more animals and, it would seem, more precisely localized progesterone implants.

As the evidence now stands, we have little or no knowledge of which brain regions actually mediate the influence of estrogen, progesterone, and prolactin on the expression of maternal behavior. There are only the studies just reviewed, and none of these, as we have already indicated, is free of criticism.

E. THE SYNCHRONY OF MATERNAL BEHAVIOR IN THE LITTER SITUATION

Thus far we have spoken only of the initiation of maternal behavior, neglecting the fact that after it is initiated the behavior must not only be maintained for a period of time, but within that time must be modulated to the changing demands of the growing litter. In other words, the female has not only to respond immediately after parturition but has also to continue to respond, in synchrony first with helpless, pink-skinned pups and then with active, adult-looking young.

The full course of maternal behavior has been followed systematically in the postpartum female but not, unfortunately, in the nulliparous female induced to act maternally either through continuous association with young or through hormone injection. In the postpartum female, retrieving is maintained at a high level for about the first 9 days following parturition and thereafter begins to decrease progressively (Beach & Jaynes, 1956; Moltz & Robbins, 1965; Rosenblatt & Lehrman, 1963; Wiesner & Sheard, 1933). By day 21 the pups are not retrieved at all. Nest building follows essentially the same functional course as retrieving: a high compact nest is seen for some 9 or 10 days postpartum after which time the nest progressively disintegrates and of course nest building behavior progressively declines (Moltz & Robbins, 1965; Rosenblatt & Lehrman, 1963). In contrast to nest building and retrieving, nursing does not decline during the first 20 or 21 days postpartum (Moltz & Robbins, 1965), although beginning at about day 14 or 15 it is the pups who are likely to initiate nursing rather than the mother (Rosenblatt & Lehrman, 1963). Only by the end of the fourth week does nursing usually cease altogether.

What we have just described is the behavior characteristically displayed toward a single litter allowed to progress to weaning age. This is to be contrasted with behavior displayed toward litters which are repeatedly replaced by newborn pups each time they become 10 or 12 days old (Bruce, 1961; Nicoll & Meites, 1959; Wiesner & Sheard, 1933). Under such conditions of stimulation, maternal attachment is prolonged. That is, the mother continues to respond, maintaining the nest, nursing and retrieving, often for as long as 3 or 4 months and in a manner consistently appropriate to each new litter.

Several investigators have raised the question of whether changes in the stimulus properties of the young act directly to elicit whatever level of maternal behavior is required at each stage of litter development or whether such changes must first act to induce in the mother selected endocrine responses which then serve to elicit behavior consonant with the needs of the young. In relation to these alternatives, we have available only fragmentary evidence.

Through the injection of both prolactin and oxytocin "in doses similar to those previously used to maintain lactation," Rosenblatt (1965) succeeded in maintaining what he called the "maternal condition" in mothers whose young had been removed. From this he suggested that prolactin and oxytocin, normally released by suckling, are essential not only for galactapoiesis but for the maintenance of maternal behavior in the litter situation as well. This is an interesting hypothesis, implying that, at each stage of the nurtural episode, the suckling demands of the litter modulate maternal behavior through variations in the prolactin and oxytocin output of the mother.

That suckling causes the discharge of both prolactin and oxytocin in the postparturient rat has been firmly established. That suckling, however, is essential in synchronizing maternal behavior to the requirements of the young has been contraindicated by the study of Moltz *et al.* (1967, cf. above). There, it will be recalled, mammectomized females, who of course received no suckling stimulation at all, were seen not only to initiate maternal behavior without delay but thereafter to display the behavior in a manner indistinguishable from the normal. This indicates that either prolactin and oxytocin are not essential for response modulation or, if they are essential, that their release can be effected through mechanisms in addition to suckling. Perhaps the sight and sound of the young, or merely physical contact with the young, are alone sufficient to sustain whatever prolactin and oxytocin titers are necessary for the characteristic expression of maternal behavior. If this is in fact the case, then the decrease in maternal responsiveness that occurs as the pups advance in age could be attributed, in part at least, to their changing stimulative properties—properties which after a time become so altered as to no longer induce the appropriate endocrine responses and thus no longer induce maternal behavior. A recent study (Moltz, Levin, & Leon, 1969b), using postparturient females in contact with young but surgically deprived of all opportunity to suckle, indicated that suckling is not essential for prolactin discharge, that the sight, sound, and perhaps "feel" of the young are alone sufficient to maintain the output of the gonadotropin.

Of course, the results of Moltz *et al.* do not establish the essentiality of prolactin in the modulation of maternal behavior. They indicate merely that the hypothesis regarding the role of prolactin and oxytocin is not necessarily contradicted by the normal display of nursing, nest building, and retrieving in the absence of suckling stimulation. To determine whether litter-induced changes in the output of prolactin (or the output of any other hypophyseal hormone) actually underlie the observed changes in postparturient maternal behavior, we would need follow the full functional course of the behavior in animals that have been hypophysectomized shortly after the birth of their young (cf. Obias, 1957). To date, no such study has been performed.

At this point it might be well to raise the question of whether the endocrine manipulations found to delay the onset of maternal behavior will also disrupt the functional course of the behavior once it is established. If such imposed imbalances are seen to have disruptive effects, then some insight might be gained into whatever mechanisms underlie the mother-litter synchrony of which we have been speaking. Unfortunately, there are only two studies available here, neither of which is conclusive.

One study is by Moltz *et al.* (1969a), involving the injection of progesterone into puerperal females for a period of 4 days beginning 48 hours after maternal behavior had been established. Here the steroid, in doses previously found effective in delaying the onset of maternal behavior, was entirely without effect—in each case the experimental females continued to nurse, retrieve, and maintain their nests, all in a manner appropriate to the developmental age of their young. Evidently, injection of progesterone does not influence the progress of maternal behavior; that it does not is hardly surprising in view of the characteristic increase in this steroid beginning soon after parturition (Grota & Eik-Nes, 1967).

Another study relevant to the topic under discussion involved ovariectomy, performed on puerperal females shortly after parturition (Folley & Kon, 1938). The experimenters were concerned primarily with the continuation of lactation and did not report in detail on the maternal behavior of their animals. However, from the limited descriptions they did provide, it appears that neither nursing nor any other nurtural response was disrupted by the ovariectomy. In support of this conclusion is the fact that, when ovariectomy is performed just prior to the expected time of parturition, those females that do not suffer delay in the onset of maternal behavior enter into an entirely normal litter interaction (Moltz & Wiener, 1966).

From the two studies just reviewed, it is tempting to conclude that the hormones involved in the induction of maternal behavior are not themselves involved in the modulation of that behavior once it is established. Such a conclusion, however, would be premature, for actually we know very little as to how the interaction between mother and young is synchronized during the course of a breeding episode. Much additional research involving manipulation of both the stimulus characteristics of the pups and the endocrine profile of the mother must be carried out before the basis of this synchrony can be specified.

II. The Rabbit

For the rabbit, as for the rat, the main source of progesterone during pregnancy is the ovary and in both species removal of the ovary leads to abortion. Like the rat, the rabbit also evidences a marked change in progesterone output during the course of gestation. As measured in ovarian vein blood (Mikhail, Noall, & Allen, 1961), progesterone in the rabbit begins to increase on the second day after mating, reaching maximal values on day 16 of her 31- to 32-day gestation period. From that point the titer decreases, at first gradually but then abruptly beginning a few days before parturition. By day 4 postpartum, progesterone can no longer be detected. Unfortunately, neither estrogen nor prolactin levels have been measured in the pregnant rabbit.

Parturition in the rabbit differs significantly from that in the rat. In the rat, strong uterine motility can often be observed beginning as early as 3 days prior to delivery, with delivery itself frequently taking as long as 2 hours. In the rabbit, however, intrauterine pressure waves (as measured both by radiotelemetry and balloon insertion) show a sudden onset, followed almost immediately by the rapid expulsion of the fetuses (Fuchs, 1964a, 1964b, 1966; Kuriyama & Csapo, 1961; Schofield, 1957, 1968, 1969). This difference in the time of occurrence of uterine contractions and in the duration of labor may reflect, as several investigators have suggested, a difference in the role played by oxytocin. In the rabbit, oxytocin release probably occurs suddenly and, along with progesterone withdrawal, is likely instrumental in initiating delivery; in the rat, delivery is probably initiated by progesterone withdrawal alone, with oxytocin entering only during the second or final stage of the delivery process.

Maternal behavior in the rabbit is in some respects similar to and in other respects different from that in the rat. Depending on the strain, the rabbit begins to build a nest either a few days before parturition or on the day of parturition itself (Ross *et al.*, 1961). In the laboratory, the nest is usually built of hay, straw, or other materials of a similar nature while in the wild, a burrow is first dug and the nest then constructed of natural compost (Deutsch, 1957; Ross, Zarrow, Sawin, Denenberg, & Blumenfield, 1963b). In either case, the female lines the site with hair plucked from her own body, hair that had already started to loosen a few days before parturition. Whether or not the quality of the nest improves with successive breeding experiences has been debated. The rat clearly shows no improvement, the primiparous female being as proficient in nest building as her multiparous counterpart (Moltz & Robbins, 1965). For the laboratory rabbit, in contrast, there are data (Ross, Denenberg, Sawin, & Meyer, 1956) available suggesting a linear increase in nest quality from the first to the fourth parturition. Deutsch (1957), however, questions the generality of these data, maintaining that under natural conditions—when presumably opportunities for nest construction are optimal—the primiparous doe builds a nest that in no way differs from the multiparous doe.

Delivery, of course, whether in the wild or in the laboratory, takes place in the nest. As soon as the young emerge, the female becomes attentive—she quickly licks and cleans her pups and within only a few minutes begins to nurse (Ross, Sawin, Zarrow, & Denenberg, 1963a). But unlike the rat, the postparturient doe nurses only once a day and then for only 2 or 3 minutes at a time (Deutsch, 1957; Zarrow, Denenberg, & Anderson, 1965a). Also retrieving, one of the most interesting components of rodent maternal behavior, has never been observed with any frequency either in wild or in laboratory-reared rabbits (Ross, Denenberg, Frommer, & Sawin, 1959).

Unfortunately, there have been no attempts to determine whether the nulliparous doe, if simply kept in continuous association with young, will come to respond maternally. However, there have been a number of attempts to induce maternal behavior, or more precisely maternal nest building, in the nulliparous doe through hormone injection. Taking ovariectomized females, Zarrow, Sawin, Ross & Denenberg (1962), Zarrow, Farooq, Denenberg, Sawin, & Ross (1963), and Zarrow *et al.* (1965b) administered both estradiol and progesterone for some 4 weeks in one experiment and for about 2 weeks in another. When injection of both steroids was terminated simultaneously, maternal nest building failed to occur. However, when progesterone was terminated 2 or 3 days before the cessation of estradiol, nest building was observed in all experimental animals. Evidently, progesterone withdrawal is as critical for the induction of maternal behavior (or at least for the induction of maternal nest building) in the nulliparous rabbit as it is in the nulliparous rat.

Moreover, it is significant that in both species maternal behavior does not occur immediately after progesterone withdrawal, but only after some period of time: in the rabbit about 2 to 3 days and in the rat, as we have already mentioned, about 40 hours. Unfortunately, Zarrow did not proffer foster young, so we do not know whether the induction of nest building in the nulliparous doe would have been followed also by the induction of licking and nursing.

Attempts at endocrine disruption in the pregnant rabbit have also provided some interesting parallels with the rat. In both species, repeated injections of progesterone beginning near term will delay the onset of parturition and, at a dosage of 4 mg per day, will, in many females, also inhibit the appearance of maternal behavior (Zarrow *et al.*, 1963). However, once again our information is incomplete: we do not know whether the does used by Zarrow were primiparous or multiparous and consequently whether the behavioral inhibition observed following the administration of progesterone can be modified in the pregnant rabbit by previous breeding experience. It will be recalled that, in the pregnant rat, such modification was found to occur readily.

Thus far we have not mentioned specifically the role of prolactin in the induction of maternal behavior in the rabbit. That the nulliparous rat fails to respond to estrogen and progesterone withdrawal alone is clear; superimposed on these steroids must be exogenous prolactin. In contrast, it would seem that the nulliparous rabbit does not require exogenous prolactin, but will respond in most cases simply to the steroids alone. However, as Zarrow points out, the role of prolactin in the induction of nest building in the doe cannot be assessed properly until hypophysectomized animals are tested. In other words, it is possible that the extended estrogen injections found necessary to induce nest building in the rabbit functioned as well to discharge sufficient prolactin from the female's own pituitary (e.g., Welch *et al.*, 1968), making injection of the gonadotropin itself gratuitous. Had a more abbreviated hormone schedule been employed, approaching, for example, that used with the rat, then perhaps prolactin may have been found critical.

One final comparison between the rabbit and the rat remains to be discussed, namely, that relating to maternal behavior in the male. It will be recalled that in the male rat we were able to effect only a limited reduction in the latency to maternal behavior, that even when estrogen and progesterone dosages were doubled, the male still did not respond in a manner comparable to that of the treated female. Zarrow was even less successful with the male rabbit. Their castrated males were administered 10 rather than 5 μg of estradiol and 8 rather than 4 mg of progesterone under the same schedule found effective in inducing nest building in the nulliparous female. Despite these dosages, the male failed to show any signs of nest building at all following progesterone withdrawal.

Perhaps in the rabbit as in the rat, the presence of androgen early in ontogeny critically affects thresholds to endocrine activation within the maternal mediating system. In other words, for the rabbit also we might expect neonatal (or perhaps prenatal) castration—freeing the male from the "desensitizing" effects exerted by endogenous androgen during a critical period in ontogeny—to increase subsequently his reactivity to inductor hormones. Correlatively, androgen injection during the same period in the genetic female might raise her hormone sensitivity to a threshold level characteristic of the normal male.

III. The Golden Hamster

Gestation in the golden hamster lasts but 16 days, a period surprisingly brief for a rodent species. A significant increase in plasma progesterone is not attained until day 8 of pregnancy and maximal values of the steroid are not attained until day 14 (Lukaszewska & Greenwald, 1970). As in the rat and rabbit, peripheral titers of progesterone decrease before parturition, although in the hamster this decrease begins much closer to term. It is noteworthy that at no time during gestation does the hamster show progesterone values significantly higher than those found during the estrous cycle. This is in marked contrast, of course, to both the rat and the rabbit in which an appreciable enhancement of progesterone occurs during gestation.

Estrogen levels have not been determined for the hamster. However, the fact that the pregnant female will mate, especially when near term (Krehbiel, 1952), has been taken as presumptive evidence of an elevation of estrogen during the last trimester of the gestation period (Greenwald, 1964). Unfortunately, no data are available concerning the behavior of prolactin in either the pregnant or the cycling hamster.

Nest building in the golden hamster occurs in both sexes, but with the pregnant female using appreciably more material in nest construction than either the male or the nonpregnant female (Richards, 1965). This "maternal nest" does not appear abruptly at the time of parturition, as it does characteristically in the rat and rabbit. Instead, the hamster begins increasing her use of nesting material on day 5 or 6 of gestation, thereafter adding progressively to the total amount employed (Richards, 1969). By the time parturition approaches, she has a large, well-structured nest. The young, of course, are delivered in the nest and nipple attachment, together with the characteristic arched posture of the nursing female, are often seen before the full litter has emerged (Rowell, 1961).

Nulliparous female hamsters have been tested for maternal behavior, or at least tested for certain nurtural responses. Richards (1966a, 1966b), for example, found that nulliparous females offered pups that were between 1 and 5 days of age invariably attacked and killed them, while those offered pups that were between 6 and 10 days of age were occasionally seen to lick and sometimes to nurse. However, from the descriptions provided by Richards, it is impossible to know—when a nulliparous female did act maternally—just how integrated her behavior was. The impression gained was that, in response to older pups, some females showed licking and that still other females showed nursing—in other words, that only fragments of the maternal complex were exhibited. Why such fragments were displayed more frequently toward older than younger pups remains unanswered.

Richards observed his animals during only a single 15-minute period following the presentation of foster young. The question that obviously arises is whether the nulliparous hamster will behave maternally—displaying the fully integrated nurtural repertoire—if kept in continuous association with young. We do not, at present, have an answer to this question; the closest we approach is a study by Noirot and Richards (1966) in which two 15-minute observations were taken, separated by an interval of 48 hours. During this interval, foster young were not present. The data indicated that initial contact with 1-day-old young increased the percentage of females showing nest building, nursing, and retrieving during a second test with 5-day-old young. Once again, however, it could not be determined just how integrated or even how concordant these responses were. Curiously enough, when pups of the same age were proffered during both the initial and subsequent test, there was not, for the most part, a significant increase in the number of nulliparous females displaying any maternal responses.

It is obvious that the extent to which the behavior of the nulliparous hamster parallels that of the nulliparous rat has not been determined. Although the hamster, to be sure, seems more prone than the rat to attack foster young, the hamster has never been observed under the same conditions as the rat, that is, has never been observed over a period of days in the presence of foster young. Under such enforced association, the tendency to attack may disappear and may be replaced by full maternal behavior, perhaps even with a latency comparable to that of the rat.

There are only two studies addressed to the question of whether the immediate responsiveness characteristic of the puerperal hamster is hormonally facilitated. Unfortunately, both provide only incomplete answers. In the first, pregnant females were observed in their behavior toward foster young during a single 15-minute test period (Richards, 1966b). Each female was within 24 hours of parturition and thus exposed to decreasing levels of progesterone (Lukaszewska & Greenwald, 1970) and presumably to increasing levels of estrogen and prolactin as well. Compared to their nulliparous counterparts, these females showed a significantly higher incidence of both nursing and licking.

In a more recent study (Richards, 1969), pellets of both estrogen and progesterone were implanted under the skin of ovariectomized nulliparous females, following which they were observed to determine whether the amount of straw used for nest construction would increase in a manner parallel to that seen typically in the pregnant female. The expected parallel was obtained insofar as a progressive rise in straw utilization was seen beginning at about six days after implantation. In contrast, nulliparae bearing implants of only one of the two steroids, respectively, displayed no such increase and in fact showed essentially the same use of straw as control females.

This is a curious study, difficult in some respects to interpret. First, no mention was made of just how much estrogen and progesterone were retrieved from the implant site at the termination of the 16-day test period. If one assumes that substantial quantities still remained and, moreover, that each steroid was being actively released during the entire time of implantation, then the hamster is indeed different from both the rat and rabbit. That is, instead of maternal nest building being induced in some part by progesterone withdrawal, it would seem, in the hamster, to be induced by progesterone elevation. This of course is consonant with the fact that the pregnant female initiates the characteristic increase in nest building beginning early in gestation, when progesterone titers are in fact increasing. And second, the nulliparous females tested in the present study were at no time offered foster young. The question that quite naturally arises is would they have licked and retrieved these young and would they have attempted also to nurse? It is difficult to imagine why the opportunity to display these various behaviors was not made available.

One additional part of the present study remains to be mentioned. When castrated male hamsters were implanted with estrogen and progesterone, the use of straw in nest construction did not increase (Richards, 1969). This insensitivity to endocrine facilitation parallels, of course, that exhibited by male rats and rabbits. What is not obvious, however, and what, in fact, is contrary to that expected from the rat and rabbit, is the failure of neonatal steroid manipulations to effect a reversal of male-female differences in straw utilization following adult estrogen and progesterone implants. More specifically, male hamsters castrated at between 2 and 4 days postpartum and female hamsters injected at the same age with testosterone propionate did not differ, respectively, from normal males and females in their response to subsequent steroid administration; each, in other words, showed a level of straw utilization characteristic of its genetic sex (Richards, 1969).

Not only does the hamster appear to be different from the rat and rabbit with respect to the role of progesterone in the induction of maternal nest building, but it appears to be different as well with regard to some of the events governing the continuation of nursing. In the rat, as already mentioned, successive substitution of foster litters will keep the postparturient female lactating for an extended period, and, in addition, will prolong the time over which she will continue to exhibit nursing behavior. Here evidently, the suckling demands of the litter are of primary importance. In contrast, the duration of nursing in the postparturient hamster is determined less by stimulation from young than by the number of days the female has already been nursing (Rowell, 1960). There appears, in other words, to be a fixed period postpartum during which the female hamster will suckle young, this period strictly determined by her physiological condition. Just why the hamster is what may be called a "determinate nurser" and the rat an "indeterminate nurser" remains to be answered.

It is obvious that, for the hamster, there is much to be done toward elucidating the mechanisms underlying maternal behavior. But given the recent spate of interest in this rodent as a subject of psycho-endocrinologic investigation, perhaps rapid progress can be expected.

IV. The Mouse

In examining the rat, the rabbit, and the hamster, we raised the question of just how elicitable was the maternal behavior of each in the absence of endocrine facilitation. For the rat, of course, the answer was clear; for the rabbit and hamster, the available data were less conclusive, although for these species as well it was possible to make a case for endocrine facilitation, since, at the very least, nulliparous rabbits and hamsters are not immediately responsive to young. Along this dimension, the mouse diverges sharply. Here nulliparous females, never having been in contact previously with foster pups, characteristically respond within as brief a period as 5 *minutes* after presentation (Leblond, 1940; Noirot, 1964a, 1964b, 1964c, 1965, 1969). They are seen, specifically, to retrieve, lick, and even assume a nursing posture, particularly if the young offered are only 1 or 2 days of age. Neither hypophysectomy nor ovariectomy significantly affects this immediate responsivity (Leblond & Nelson, 1937). And interestingly enough, the castrate male also behaves with much the same alacrity as the nulliparous female, retrieving, licking, and huddling young in an entirely maternal manner (Leblond, 1940). Evidently, the neuronal mechanisms mediating these several nurtural responses in the mouse do not require endocrine facilitation; they appear, unlike those in rat, rabbit, and hamster, to be immediately excitable by pup-related stimuli alone. Just what adaptive function such excitability serves is obscure.

There is, however, one dimension of maternal behavior along which the nulliparous female does not respond until she has been exposed to pups for some time—she does not build what Koller (1952, 1955) called a "brood nest." In the absence of pups, and when supplied with sufficient nesting material, she, as well as the adult male, will characteristically build a "sleeping nest," which is a small, almost flat structure weighing between 8 and 12 gm. The brood nest, in contrast to the sleeping nest, is at once larger and more compact, and on the average weighs some 40 gm. It appears abruptly on day 5 of pregnancy and it is in such a nest that the female then gives birth and subsequently nurtures young.

This dramatic appearance of the brood nest on day 5 coincides with the onset of nidation and with the mucification of the vaginal epithelium which is indicative of both progesterone and estrogen secretion (Choudary & Greenwald, 1969). From this point until day 15, there is an even further enhancement of ovarian steroid release. Day 16 of pregnancy, however, marks the beginning of luteal regression, and following shortly thereafter the beginning of a progressive decrease in plasma levels of progesterone (Hooker & Forbes, 1947). Estrogen secretion, in contrast, probably remains enhanced until term, as suggested by the characteristic development on day 16 of a squamous but still mucified vaginal epithelium.

It is obvious that the initiation of the brood nest in the pregnant mouse coincides with a rise in progesterone and estrogen release from the ovary. Whether the increase in either one or both of these steroids actually induces brood-nest building was investigated by Koller (1952, 1955) and more recently by Lisk, Pretlow, and Friedman (1969).

Koller reports that "one or two" injections of progesterone in non-pregnant mice (of unspecified breeding history) resulted in the appearance of a brood nest some 24 to 48 hours later. In this regard, neither prolactin, FSH, nor estrone was effective. It is of interest that the adult male, whether intact or castrated, did not respond to progesterone, or at least did not respond to the dosage level used in the female. Unfortunately, there was no attempt here to vary dosage systemically.

Using subcutaneous implants of progesterone rather than discrete injections, Lisk *et al.* (1969) essentially confirmed the results of Koller: intact and ovariectomized females responded to progesterone by building brood nests, while castrate males did not. Interestingly enough, estradiol alone in ovariectomized females not only failed to induce construction of brood nests, but actually decreased the amount of material used in building sleeping nests.

From even the limited data available, it is evident that the control of maternal behavior in the mouse is distinctly different from that in each of the species discussed thus far. First, there is the characteristic tendency of the nulliparous foster mouse to lick, retrieve, and attempt to huddle young some *minutes* after presentation. Second, there is the fact that the pregnant mouse builds a "maternal nest" on day 5, at a time not of progesterone withdrawal but of progesterone enhancement. Here she differs from the rat and rabbit. But with regard to this maternal nest she differs from the hamster as well. Although both mouse and hamster show an increase in nest building early in pregnancy, this increase occurs abruptly in the mouse, while in the hamster the appearance is more gradual. And second, progesterone enhancement in the mouse is all that is needed to facilitate the appearance of the maternal nest, while in the hamster, as already mentioned, both progesterone and estrogen seem equally critical.

One final point must be mentioned in relation to the mouse. Although we have been speaking only of *Mus musculus*, there are numerous and discrete strains within the species, each characterized to some extent by distinctive physiological and behavioral traits. It is conceivable that, as the study of maternal behavior proceeds, strain differences in the mouse will be revealed which may be every bit as divergent as some of the species differences discovered thus far.

From what has already been said, it is obvious that maternal behavior defies dichotomization into categories such as "innate" and "acquired." It is of course neither "innate" in the sense of being encoded isomorphically in the genome nor "acquired" in the sense of being entirely learned or conditioned (cf. Lehrman, 1956; Moltz, 1965). Rather, it can be understood only by reference to a complex of intraorganic events and extrinsic stimulative conditions interacting to excite a mediating substrate whose reactivity thresholds, in turn, have been affected by neonatal steroid levels and probably by previous breeding experience. This now is the picture that emerges, at least for rat, rabbit, hamster, and mouse. It should be gratuitous to add that much additional research, carried out systematically on a much wider variety of mammalian forms will be needed before the full spectrum of interspecific differences and similarities in the mechanisms underlying maternal behavior become apparent. However, such an endeavor, painstaking as it will undoubtedly prove, will richly reward the psychologist and the physiologist alike.

References

Amenomori, Y., Chen, C. L., & Meites, J. Serum prolactin levels in rats during different reproductive states. *Endocrinology*, 1970, **70**, 506-510.

Arai, Y., Hiroi, M., Mitra, J., & Gorski, R. A. Influence of intravenous progesterone administration on the cortical electroencephalogram of the female rat. *Neuroendocrinology*, 1967, **2**, 275-282.

Avar, Z., & Monos, E. Effect of lateral hypothalamic lesion on maternal behaviour and foetal vitality in the rat. *Acta Medica Academiae Scientiarum Hungaricae*, 1967, **23**, 255-261.

Barraclough, C. A., & Cross, B. A. Unit activity in the hypothalamus of the cyclic female rat: Effect of genital stimuli and progesterone. *Journal of Endocrinology*, 1963, **26**, 339-359.

Bartholomew, G. A. Mother-young relations and the maturation of pup behaviour in the Alaska fur seal. *Animal Behaviour*, 1959, **7**, 163-171.

Beach, F. A. The neural basis of innate behavior. I. Effects of cortical lesions upon the maternal behavior pattern in the rat. *Journal of Comparative Psychology*, 1937, **24**, 393-436.

Beach, F. A. The neural basis of innate behavior. II. Relative defects of partial decortication in adulthood and infancy upon the maternal behavior of the primiparous rat. *Journal of Genetic Psychology*, 1938, **53**, 109-48.

Beach, F. A., & Jaynes, J. Studies of maternal retrieving in rats. II. Effects of practice and previous parturitions. *American Naturalist*, 1956, **90**, 103-109.

Beach, F. A., & Wilson, J. R. Effects of prolactin, progesterone, and estrogen on reactions of non-pregnant rats to foster young. *Psychological Reports*, 1963, **13**, 231-239.

Beyer, C., Ramirez, V. D., Whitmoyer, D. I., & Sawyer, C. H. Effects of hormones on the electrical activity of the brain in the rat and rabbit. *Experimental Neurology*, 1967, **18**, 313-326.

Bindra, D. *Motivation: A systematic reinterpretation.* New York: Ronald Press, 1959.

Birch, D. Sources of order in the maternal behavior of animals. *American Journal of Orthopsychiatry,* 1956, **26,** 279-284.

Braden, I. C. Effects of hypothalamically implanted estrogen on the maternal sequence of rats. *American Psychological Association Proceedings,* 1966, **1,** 187-188.

Bruce, H. M. Observations on the suckling stimulus and lactation in the rat. *Journal of Reproduction and Fertility,* 1961, **2,** 17-34.

Choudary, J. B., & Greenwald, G. S. Ovarian activity in the intact or hypophysectomized pregnant mouse. *Anatomical Record,* 1969, **163,** 359-372.

Christophersen, E. R., & Wagman, W. Maternal behavior in the albino rat as a function of self-licking deprivation. *Journal of Comparative and Physiological Psychology,* 1965, **60,** 142-144.

Cierciorwska, A., & Russfield, A. Determination of estrogenic activity of the immature rat ovary. *Archives of Pathology,* 1968, **85,** 658-662.

Cosnier, J. Quelques problèmes posés par le "comportement maternel provoqué" chez la ratte. *Comptes Rendus des Seances de la Societe de Biologie,* 1963, **157,** 1611-1613.

Cross, B. A. Neural control of oxytocin secretion. In L. Martini & W. F. Ganong (Eds.), *Neuroendocrinology.* Vol. 1. New York: Academic Press, 1966. Pp. 217-259.

Davis, C. D. The effect of ablations of neocortex on mating, maternal behavior and the production of pseudopregnancy in the female rat and on copulatory activity in the male. *American Journal of Physiology,* 1939, **127,** 374-380.

Deis, R. P. Mammary gland development by hypothalamic and hypophyseal estrogen implants in male rats. *Acta Physiologica Latino Americana,* 1967, **17,** 115-117.

Denenberg, V. H., Taylor, R. E., & Zarrow, M. X. Maternal behavior in the rat: An investigation and quantification of nest building. *Behaviour,* 1969, **34,** 1-16.

Deutsch, J. A. Nest building behaviour of domestic rabbits under seminatural conditions. *British Journal of Animal Behaviour,* 1957, **2,** 53-54.

Dilley, W. G., & Adler, N. T. Postcopulatory mammary gland secretion in rats. *Proceedings of the Society for Experimental Biology and Medicine,* 1968, **129,** 964-967.

Eibl-Eibesfeldt, I. Das Verhalten der Nagetiere. In J. Helmecke, H. V. Lengerken, & D. Starck (Eds.), *Handbuch der Zoologie.* Berlin: de Gruyter, 1958. Pp. 1-88.

Eibl-Eibesfeldt, I. The interactions of unlearned behaviour patterns and learning in mammals. In J. F. Delafresnaye (Ed.), *Brain mechanisms and learning.* Oxford: Blackwell, 1961. Pp. 53-73.

Eisenfeld, A. J., & Axelrod, J. Selectivity of estrogen distributions in tissues. *Journal of Pharmacology and Experimental Therapeutics,* 1965, **150,** 469-475.

Eisenfeld, A. J., & Axelrod, J. Effect of steroid hormones, ovariectomy, estrogen pretreatment, sex and immaturity on the distribution of ^{3}H-estradiol. *Endocrinology,* 1966, **79,** 38-42.

Eisenfeld, A. J., & Axelrod, J. Evidence for estradiol binding sites in the hypothalamus—effect of drugs. *Biochemical Pharmacology,* 1967, **16,** 1781-1785.

Eto, T., Hsoi, T., Musudo, H., & Suzuki, Y. Progesterone and pregn-4-ene 20α-ol-3-one in rat ovarian venous blood at different stages in the reproductive cycle. *Japanese Journal of Animal Reproduction,* 1962, **8,** 34-40.

Everett, J. W. The control and secretion of prolactin. In G. W. Harris & B. T. Donovan (Eds.), *The pituitary gland.* Vol. 2. Berkeley: Univ. of California Press, 1966. Pp. 166-194.

Fajer, A. B., & Barraclough, C. A. Ovarian secretion of progesterone and 20α-hydroxy-pregn-4-en-3-one during pseudopregnancy and pregnancy in rats. *Endocrinology*, 1967, **81**, 617-622.

Fisher, A. E. Maternal and sexual behavior induced by intracranial chemical stimulation. *Science*, 1956, **124**, 228-229.

Fisher, A. E. Behavior as a function of certain neurobiochemical events. In R. Patton *et al.* (Eds.), *Current trends in psychological theory: A bicentennial program.* Pittsburgh: Univ. of Pittsburgh Press, 1961. Pp. 70-86.

Fisher, A. E. Chemical and electrical stimulation of the brain in the male rat. In R. A. Gorski & R. E. Whalen (Eds.), *Brain and behavior.* Vol. 3. Berkeley: Univ. of California Press, 1966. Pp. 117-130.

Fitzpatrick, R. J. The posterior pituitary gland and the female reproductive tract. In G. W. Harris & B. T. Donovan (Eds.), *The pituitary gland.* Vol. 3. Berkeley: Univ. of California Press, 1966. Pp. 453-504.

Folley, S. J., & Kon, S. K. Effects of sex hormones on lactation in the rat. *Proceedings of the Royal Society, Series B*, 1938, **124**, 476-492.

Friedlich, O. B. A study of maternal behavior of the albino rat as a function of self-licking deprivation. Unpublished master's thesis, Southern Illinois University, 1962.

Fuchs, A. R. The role of oxytocin in the initiation of labour. *Proceedings of the Second International Congress of Endocrinology, 1964*, 1965, 753-758. (a)

Fuchs, A. R. Oxytocin and the onset of labour in rabbits. *Journal of Endocrinology*, 1964, **30**, 217-224. (b)

Fuchs, A. R. Studies on the control of oxytocin release at parturition in rabbits and rats. *Journal of Reproduction and Fertility*, 1966, **12**, 418.

Greenwald, G. S. Ovarian follicular development in the pregnant hamster. *Anatomical Record*, 1964, **148**, 605-10.

Greenwald, G. S. Ovarian follicular development and pituitary FSH and LH content in the pregnant rat. *Endocrinology*, 1966, **79**, 572-578.

Grindeland, R. E., McCulloch, W. A., & Ellis, S. Radio-immunoassay of rat prolactin. Paper presented at the 51st meeting of the Endocrine Society New York, June 1969.

Grota, L. J., & Eik-Nes, K. B. Plasma progesterone concentrations during pregnancy and lactation in the rat. *Journal of Reproduction and Fertility*, 1967, **13**, 83-91.

Hamburg, D. Effects of progesterone on behavior. In R. Levine (Ed.), *Endocrines and the central nervous system.* Baltimore, Md.: Williams & Wilkins, 1966. Pp. 251-263.

Hashimoto, I., Hendricks, D. M., Anderson, L. L., & Melampy, R. M. Progesterone and pregn-4-en-20α-ol-one in ovarian venous blood during various reproductive states in the rat. *Endocrinology*, 1968, **82**, 333-341.

Holland, H. C. An apparatus note on A. M. B. A. (Automatic Maternal Behavior Apparatus). *Animal Behaviour*, 1965, **13**, 201-202.

Hooker, C. W. & Forbes, T. R. A bioassay for minute amounts of progesterone. *Endocrinology*, 1947, **41**, 158-169.

Kato, J., & Villee, C. A. Factors affecting uptake of estradiol-6,7-^{3}H by the hypophysis and hypothalamus. *Endocrinology*, 1967, **80**, 1133-1138. (a)

Kato, J., & Villee, C. A. Preferential uptake of estradiol by the anterior hypothalamus of the rat. *Endocrinology*, 1967, **80**, 567-575. (b)

Kawakami, M., & Sawyer, C. H. Neuroendocrine correlates of changes in brain activity thresholds by sex steroids and pituitary hormones. *Endocrinology*, 1959, **65**, 652-668.

Kimble, D. P., Rogers, L., & Hendrickson, C. W. Hippocampal lesions disrupt maternal, not sexual, behavior in the albino rat. *Journal of Comparative and Physiological Psychology*, 1967, **63**, 401-407.

Kirby, H. W., & Horvath, T. Self-licking deprivation and maternal behaviour in the primiparous rat. *Canadian Journal of Psychology*, 1968, **22**, 369-376.

Klein, M. Uterine distension, ovarian hormones and maternal behaviour in rodents. *Ciba Foundation Colloquim on Endocrinology* [*Proc.*], 1952, **3**, 84-87.

Kling, A. Effects of rhinencephalic lesions on endocrine and somatic development in the rat. *American Journal of Physiology*, 1964, **206**, 1395-1400.

Kobayashi, T., Kobayashi, T., Takezawa, S., Oshima, K., & Kawamura, H. Electrophysiological studies on the feedback mechanism of progesterone. *Endocrinologia Japonica*, 1962, **9**, 302-320.

Koller, G. Der Nestbau der weissen Maus und seine Hormonale Auslösung. *Verhandlungen der Deutschen Zoologischen Gesellschaft, Freiburg*, 1952, 160-168.

Koller, G. Hormonale und psychische Steuerung beim Nestbau weisser Mäuse. *Verhandlungen der Deutschen Zoologischen Gesellschaft*, 1955, 123-132.

Komisaruk, B. R., McDonald, P. G., Whitmoyer, D. I., & Sawyer, C. H. Effects of progesterone and sensory stimulation on EEG and neuronal activity in the rat. *Experimental Neurology*, 1967, **19**, 494-507.

Krehbiel, R. H. Mating of the golden hamster during pregnancy. *Anatomical Record*, 1952, **113**, 117-121.

Kuhn, N. J. Lactogenesis in the rat. Metabolism of uridine diphosphate galactose by mammary gland. *Biochemical Journal*, 1968, **106**, 743-748.

Kuhn, N. J. Progesterone withdrawal as the lactogenic trigger in the rat. *Journal of Endocrinology*, 1969, **44**, 39-54.

Kuhn, N. J., & Lowenstein, J. M. Lactogenesis in the rat. Changes in metabolic parameters at parturition. *Biochemical Journal*, 1967, **105**, 995-1002.

Kuriyama, H., & Csapo, A. Placenta and myometrial block. *American Journal of Obstetrics and Gynecology*, 1961, **82**, 592.

Kwa, H. G., & Verhofstad, F. Prolactin levels in the plasma of female rats. *Journal of Endocrinology*, 1967, **39**, 455-456.

Laumas, K. R., & Farooq, A. The uptake *in vivo* of [1,2-^{3}H] progesterone by the brain and genital tract of the rat. *Journal of Endocrinology*, 1966, **36**, 95-96.

Leblond, C. P. Nervous and hormonal factors in the maternal behavior of the mouse. *Journal of Genetic Psychology*, 1940, **57**, 327-344.

Leblond, C. P., & Nelson, W. C. Maternal behavior in hypophysectomized male and female mice. *American Journal of Physiology*, 1937, **120**, 167-172.

Lehrman, D. S. On the organization of maternal behavior and the problem of instinct. In P. P. Grassé (Ed.), *L'instinct dans le comportement des animaux et de l'homme*. Paris: Masson, 1956. Pp. 475-520.

Lehrman, D. S. Hormonal regulation of parental behavior in birds and infrahuman mammals. In W. C. Young (Ed.), *Sex and internal secretions*. Vol. 2. Baltimore, Md.: Williams & Wilkins, 1961. Pp. 1268-1382.

Lisk, R. D. Diencephalic placement of estradiol and sexual receptivity in the female rat. *American Journal of Physiology*, 1962, **203**, 493-496.

Lisk, R. D. Neural localization for androgen activation of copulatory behavior in the male rat. *Endocrinology*, 1967, **80**, 754-761.

Lisk, R. D., Pretlow, R. A., & Friedman, S. M. Hormonal stimulation necessary for elicitation of maternal nest-building in the mouse (*Mus musculus*). *Animal Behaviour*, 1969, **17**, 730-737.

Lott, D. F. The role of progesterone in the maternal behavior of rodents. *Journal of Comparative and Physiological Psychology*, 1962, **55**, 610-613.

Lott, D. F., & Fuchs, S. S. Failure to induce retrieving by sensitization or the injection of prolactin. *Journal of Comparative and Physiological Psychology*, 1962, **55**, 1111-1113.

Lukaszewska, J. H., & Greenwald, G. S. Progesterone levels in the cyclic and pregnant hamster. *Endocrinology*, 1970, **86**, 1-9.

Martin, R. D. Tree shrews: Unique reproductive mechanism of systematic importance. *Science*, 1966, **152**, 1402-1404.

Meites, J., & Nicoll, C. S. *In vivo* and *in vitro* effects of steroids on pituitary prolactin secretion. In L. Martini & A. Pecile (Eds.), *Hormonal steroids*. Vol. 2. New York: Academic Press, 1965. Pp. 307-316.

Mikhail, G., Noall, W. W., & Allen, W. M. Progesterone levels in the rabbit ovarian vein blood throughout pregnancy. *Endocrinology*, 1961, **69**, 504-509.

Moltz, H. Contemporary instinct theory and the fixed action pattern. *Psychological Review*, 1965, **72**, 27-47.

Moltz, H., Geller, D., & Levin, R. Maternal behavior in the totally mammectomized rat. *Journal of Comparative and Physiological Psychology*, 1967, **64**, 225-229.

Moltz, H., Levin, R., & Leon, M. Differential effects of progesterone on the maternal behavior of primiparous and multiparous rats. *Journal of Comparative and Physiological Psychology*, 1969, **67**, 36-40. (a)

Moltz, H., Levin, R., & Leon, M. Prolactin in the postpartum rat: Synthesis and release in the absence of suckling stimulation. *Science*, 1969, **163**, 1083-1084. (b)

Moltz, H., Lubin, M., Leon, M., & Numan, M. Hormonal induction of maternal behavior in the ovariectomized nulliparous rat. *Physiology and Behavior*, 1971, **5**, 1373-1377.

Moltz, H., & Robbins, D. Maternal behavior of primiparous and multiparous rats. *Journal of Comparative and Phsyiological Psychology*, 1965, **60**, 417-421.

Moltz, H., Robbins, D., & Parks, M. Caesarean delivery and the maternal behavior of primiparous and multiparous rats. *Journal of Comparative and Physiological Psychology*, 1966, **61**, 455-460.

Moltz, H., & Wiener, E. Effects of ovariectomy on maternal behavior of primiparous and multiparous rats. *Journal of Comparative and Physiological Psychology*, 1966, **62**, 382-387.

Nicoll, C. S., & Meites, J. Prolongation of lactation in the rat by litter replacement. *Proceedings of the Society for Experimental Biology and Medicine*, 1959, **101**, 81-82.

Nicoll, C. S., & Meites, J. Estrogen stimulation of prolactin production by rat adenohypophysis *in vitro*. *Endocrinology*, 1962, **70**, 272-277.

Noirot, Elaine. Changes in responsiveness to young in the adult mouse. I. The problematic effect of hormones. *Animal Behaviour*, 1964, **12**, 52-58 (a)

Noirot, Elaine. Changes in responsiveness to young in the adult mouse. IV. The effect of an initial contact with a strong stimulus. *Animal Behaviour*, 1964, **12**, 442-445. (b)

Noirot, Elaine. Changes in responsiveness to young in the adult mouse: The effect of external stimuli. *Journal of Comparative and Physiological Psychology*, 1964, **57**, 97-99. (c)

Noirot, Elaine. Changes in responsiveness to young in the adult mouse. III. The effect of immediately preceding performances. *Behaviour*, 1965, **24**, 318-325.

Noirot, Elaine. Serial order of maternal responses in mice. *Animal Behaviour,* 1969, **17,** 547-550.

Noirot, Elaine, & Richards, M. P. M. Maternal behaviour in virgin female golden hamsters: Changes consequent upon initial contact with pups. *Animal Behaviour,* 1966, **14,** 7-10.

Obias, M. D. Maternal behavior of hypophysectomized gravid albino rats and the development and performance of their progeny. *Journal of Comparative and Physiological Psychology,* 1957, **50,** 120-124.

Pfaff, D. W. Cerebral implantation and autoradiographic studies of sex hormones. In J. Money (Ed.), *Sex research: New developments.* New York: Holt, 1965. Pp. 219-234.

Pfaff, D. W. Uptake of ^{3}H-estradiol by the female rat brain: An autoradiographic study. *Endocrinology,* 1968, **82,** 1149-1155.

Pitters, H. Untersuchungen über angeborene Verhaltensweisen bei Tylopoden, unter besonderer Berucksichtingung der neuweltlichen Formen. *Zeitschrift fuer Tierpsychologie,* 1954, **11,** 213-303.

Raisinghani, K. H., Dorfman, R. I., Forchielli, E., Gyermek, L., & Genther, G. Uptake of intravenously administered progesterone, pregnanedione, and pregnanolone by the rat brain. *Acta Endocrinologica (Copenhagen),* 1968, **57,** 395-404.

Ramirez, V. D., Komisaruk, B. R., Whitmoyer, D. I., & Sawyer, C. H. Effects of hormones and vaginal stimulation on the EEG and hypothalamic units in rats. *American Journal of Physiology,* 1967, **212,** 1376-1384.

Ramirez, V. D., & McCann, S. M. Induction of prolactin secretion by implants of estrogen into the hypothalamo-hypophyseal region of female rats. *Endocrinology,* 1964, **75,** 206-214.

Ratner, A., Talwalker, P. K., & Meites, J. Effect of estrogen administration *in vivo* on prolactin release by rat pituitary *in vitro. Proceedings of the Society for Experimental Biology and Medicine,* 1963, **112,** 12-15.

Richards, M. P. M. Aspects of maternal behaviour in the golden hamster. Doctoral thesis, University of Cambridge, 1965.

Richards, M. P. M. Maternal behaviour in virgin female golden hamsters (*Mesocricetus auratus* Woterhouse): The role of the age of the test pup. *Animal Behaviour,* 1966, **14,** 303-309. (a)

Richards, M. P. M. Maternal behaviour in the golden hamster: Responsiveness to young in virgin, pregnant, and lactating females. *Animal Behaviour,* 1966, **14,** 310-313. (b)

Richards, M. P. M. Maternal behavior in rodents and lagomorphs: A review. In A. McLaren (Ed.), *Advances in reproductive physiology.* Vol. 2. New York: Academic Press, 1967. Pp. 54-110.

Richards, M. P. M. Effects of oestrogen and progesterone on nest building in the golden hamster. *Animal Behaviour,* 1969, **17,** 356-61.

Riddle, O. Physiological responses to prolactin. *Cold Spring Harbor Symposia on Quantitative Biology,* 1937, **5,** 218-228.

Riddle, O., Hollander, W. F., Miller, R. A., Lahr, E. L., Smith, G. C., & Marvin, H. N. Endocrine studies. *Carnegie Institution of Washington, Yearbook* 1942, **41,** 203-211. (a)

Riddle, O., Lahr, E. L., & Bates, R. W. Effectiveness and specificity of prolactin in the induction of the maternal instinct in virgin rats. *American Journal of Physiology,* 1935, **113,** 109. (a)

Riddle, O., Lahr, E. L., & Bates, R. W. Maternal behavior induced in virgin rats by prolactin. *Proceedings of the Society for Experimental Biology and Medicine*, 1935, **32**, 730-734. (b)

Riddle, O., Lahr, E. L., & Bates, R. W. The role of hormones in the initiation of maternal behavior in rats. *American Journal of Psychology*, 1942, **137**, 299-317. (b)

Rosenblatt, J. S. The basis of synchrony in the behavioral interaction between the mother and her offspring in the laboratory rat. In B. M. Foss (Ed.), *Determinants of infant behaviour*. Vol. 3. New York: Wiley. 1965. Pp. 3-41.

Rosenblatt, J. S. Nonhormonal basis of maternal behavior in the rat. *Science*, 1967, **156**, 1512-1513.

Rosenblatt, J. S., & Lehrman, D. S. Maternal behavior of the laboratory rat. In Harriet L. Rheingold (Ed.), *Maternal behavior in mammals*. New York: Wiley, 1963. Pp. 8-57.

Ross, S., Denenberg, V. H., Frommer, G. P., & Sawin, P. B. Genetic physiological and behavioral background of reproduction in the rabbit. V. Nonretrieving of neonates. *Journal of Mammology*, 1959, **40**, 91-96.

Ross, S., Denenberg, V. H., Sawin, P., & Meyer, P. Changes in nest building behaviour in multiparous rabbits. *British Journal of Animal Behaviour*, 1956, **4**, 69-74.

Ross, S., Sawin, P. B., Denenberg, V. H., & Zarrow, M. X. Maternal behavior in the rabbit: Yearly and seasonal variation in nest building. *Behaviour*, 1961, **18**, 154.

Ross, S., Sawin, P., Zarrow, M. X., & Denenberg, V. H. Maternal behavior in the rabbit. In Harriet L. Rheingold (Ed.), *Maternal behavior in mammals*. New York: Wiley, 1963. Pp. 94-121. (a)

Ross, S., Zarrow, M. X., Sawin, P. B., Denenberg, V. H., & Blumenfield, M. Maternal behaviour in the rabbit under semi-natural conditions. *Animal Behaviour*, 1963, **11**, 283-285. (b)

Roth, L. L., & Lisk, R. D. Effects of hypothalamic implants of progesterone on parturition, lactation, and maternal behavior in the rat. *American Psychological Association Proceedings*, 1968, **3**, 267-268.

Roth, L. L., & Rosenblatt, J. S. Changes in self-licking during pregnancy in the rat. *Journal of Comparative and Physiological Psychology*, 1967, **63**, 397-400.

Rothchild, I. Interrelations between progesterone and the ovary, pituitary, and central nervous system in the control of ovulation and the regulation of progesterone secretion. *Vitamins and Hormones* (*New York*), 1965, **23**, 210-328.

Rothchild, I., & Schwartz, N. B. The corpus-luteum hypophysis relationship: The effects of progesterone and oestrogen on the secretion of luteotrophin and luteinizing hormone in the rat. *Acta Endocrinologica* (*Copenhagen*), 1965, **49**, 120-137.

Rowell, Thelma E. On the retrieving of young and other behaviour in lactating golden hamsters. *Proceedings of the Zoological Society of London*, 1960, **135**, 265-282.

Rowell, Thelma E. The family group in golden hamsters: Its formation and break-up. *Behaviour*, 1961, **17**, 81-94.

Schneirla, T. C. Levels in the psychological capacities of animals. In R. W. Sellars, V. J. McGill, & M. Farber (Eds.), *Philosophy for the future*. New York: Macmillan, 1950. Pp. 243-286.

Schneirla, T. C. Interrelationships of the "innate" and the "acquired" in instinctive behaviour. In P.-P. Grassé (Ed.), *L'instinct dans le comportement des animaux et de l'homme*. Paris: Masson, 1956. Pp. 387-452.

Schofield, B. M. The hormonal control of myometrial function during pregnancy. *Journal of Physiology* (*London*), 1957, **138**, 1.

Schofield, B. M. Parturition. In A. McLaren (Ed.), *Advances in reproductive physiology.* Vol. 3. New York: Academic Press, 1968. Pp. 9-32.

Schofield, B. M. Parturition in the rabbit. *Journal of Endocrinology,* 1969, **43,** 673-674.

Seiki, K., Higashida, M., Imanishi, Y., Miyamoto, M., Kitagawa, T., & Kotani, M. Radioactivity in the rat hypothalamus and pituitary after injection of labelled progesterone. *Journal of Endocrinology,* 1968, **41,** 109-110.

Seiki, K., Miyamoto, M., Yamashita, A., & Kotani, M. Further studies on the uptake of labelled progesterone by the hypothalamus and pituitary of rats. *Journal of Endocrinology,* 1969, **43,** 129-130.

Shinde, Y., Ota, K., & Yokoyama, A. Lactose content of mammary glands of pregnant rats near term: Effect of removal of ovary, placenta, and foetus. *Journal of Endocrinology,* 1964, **31,** 105-114.

Slotnick, B. M. Disturbances of maternal behavior in the rat following lesions of the cingulate cortex. *Behaviour,* 1967, **29,** 204-236.

Stamm, J. S. Control of hoarding activity in rats by the median cerebral cortex. *Journal of Comparative and Physiological Psychology,* 1954, **47,** 21-27.

Stamm, J. S. The function of the median cerebral cortex in maternal behavior in rats. *Journal of Comparative and Physiological Psychology,* 1955, **48,** 347-356.

Steinberg, J., & Bindra, D. Effects of pregnancy and salt-intake on genital licking. *Journal of Comparative and Physiological Psychology,* 1962, **55,** 103-106.

Stone, C. P. Preliminary note on maternal behavior of rats living in parabiosis. *Endocrinology,* 1925, **9,** 505-512.

Stone, C. P. Effects of cortical destruction on reproductive behavior and maze learning in albino rats. *Journal of Comparative Psychology,* 1938, **26,** 217-236.

Stumpf, W. E. Estradiol-concentrating neurons: Typography in the hypothalamus by dry-mount autoradiography. *Science,* 1968, **162,** 1001-1003.

Sturman-Hulbe, M., & Stone, C. P. Maternal behavior in the albino rat. *Journal of Comparative Psychology,* 1929, **9,** 203-237.

Taylor, R. E. Hormones and maternal behavior in the rat. Unpublished doctoral dissertation, University of Michigan, 1965.

Terkel, J. Induction of maternal behavior: Cross-transfusion of blood from postpueral to virgin females. Unpublished doctoral dissertation, Rutgers University, 1970.

Terkel, J., & Rosenblatt, J. Maternal behavior induced by maternal blood plasma injected into virgin rats. *Journal of Comparative and Physiological Psychology,* 1968, **65,** 479-482.

Welch, C. W., Sar, M., Clemens, J. A., & Meites, J. Effect of estrogen on pituitary prolactin levels of female rats bearing median eminence implants of prolactin. *Proceedings of the Society for Experimental Biology and Medicine,* 1968, **12,** 817-821.

Wiesner, B. P., & Sheard, N. M. *Maternal behaviour in the rat.* Edinburgh: Oliver & Boyd, 1933.

Wiest, W. G. On the function of 20α-hydroxypregn-4-en-3-one during parturition in the rat. *Endocrinology,* 1968, **83,** 1181-1184.

Wiest, W. G., Kidwell, W. R., & Balogh, K. Progesterone catabolism in the rat ovary: A regulatory mechanism for progestational potency during pregnancy. *Endocrinology,* 1968, **82,** 844-860.

Wilsoncroft, W. E. Effects of median cortex lesions on the maternal behavior of the rat. *Psychological Reports,* 1963, **13,** 835-838.

Yoshinaga, K., Hawkins, R. A., & Stocker, J. F. Estrogen secretion by the rat ovary *in vivo* during the estrous cycle and pregnancy. *Endocrinology,* 1969, **85,** 103-112.

Zarrow, M. X. Gestation. In W. C. Young (Ed.), *Sex and internal secretions.* Vol. 2. Baltimore, Md.: Williams & Wilkins, 1961. Pp. 958-1031.

Zarrow, M. X., Denenberg, V. H., & Anderson, C. O. Rabbit: Frequency of suckling in the pup. *Science,* 1965, **150,** 1835-1836. (a)

Zarrow, M. X., Denenberg, V. H., & Kalberer, W. D. Strain differences in the endocrine basis of maternal nest-building in the rabbit. *Journal of Reproduction and Fertility,* 1965, **10,** 397-401. (b)

Zarrow, M. X., Farooq, A., Denenberg, V. H., Sawin, P. B., & Ross, S. Maternal behaviour in the rabbit: Endocrine control of maternal nest-building. *Journal of Reproduction and Fertility,* 1963, **6,** 375.

Zarrow, M. X., Sawin, P. B., Ross, S., & Denenberg, V. H. Maternal behavior and its endocrine basis in the rabbit. In E. L. Bliss (Ed.), *Roots of behavior.* New York: Harper, 1962. Pp. 187-197.

CHAPTER 8

THE ONTOGENY OF MOTHER-INFANT RELATIONS IN MACAQUES

Leonard A. Rosenblum

I. Introduction

Along with many other scientific achievements, the onset of the second half of the twentieth century marked the beginnings of an enormous and continuing growth in our knowledge of nonhuman primate behavior. In the past 10 years particularly, a sizable portion of this effort has been directed toward studies of behavioral development. In light of the relatively prolonged period of infantile dependency in primates, mother-infant relations have long been recognized as of crucial significance in the development of these psychologically complex organisms and, in consequence, this dyadic relationship has been the object of much research.

As a result of this work, an attempt at summarizing the major landmarks in the early ontogeny of mother-infant relations in a delineated group of nonhuman primates now seems a feasible and valid undertaking. But although no longer in its infancy, the study of primate development is at best at an awkward adolescence: it is at times confusing in its eager search for generalizations, it is easily infatuated with particular approaches and techniques, and, most important, it is unsure of its direction and goals.

In light of this level of maturity and the variably structured, often nonequivalent, data scattered through the literature, the current review will attempt to be selective and integrative rather than encyclopedic. It must be noted as well that although some attempt will be made to describe the course of early development of mother-infant relations in the whole genus *Macaque*, the great bulk of available data, unfortunately, focuses largely on a single species, *Macaca mulatta*, the rhesus monkey.

So that the reader will be able to place the general progression of behaviors described below into a broad developmental context, several basic phylogenetic and ontogenetic factors must first be considered. The macaque genus is extremely large and diversified in its wild habitat, body size and structure, and, though not as yet clearly delineated, undoubtedly in group social structure as well. Depending upon which classification scheme is used, there are at least twelve species and numerous subspecies in the genus and these range in habitat across more than 50 degrees of latitude, from seashore, swamp and cliff to city, farm, and mountain. Of considerable pertinence in the present developmental context, however, is the fact that although mature body weight in the genus varies from approximately 3500 to 18,000 gm in males and 2,500 to 16,000 gm in females, a relatively narrow range of gestational periods averaging approximately 165-170 days (with expected individual variation within species ranging from perhaps 145 to 180 days) seems to characterize the genus. Similarly, sexual maturity is reached in most macaque males at about 4 years, whereas females mature perhaps a year earlier. It must be remembered, however, that these specific figures on comparative maturation are often influenced by the conditions of rearing, with maturation progressing somewhat more rapidly in the laboratory than in the field, and with hand-reared subjects showing the most precocious development of all (Yang, Kuo, Del Favero, & Alexander, 1968).

II. The Onset of Maternal Behavior

A. Maternal Behavior Prior to Parturition

Numerous workers in both the field and laboratory have reported evidence of "interest" and "attention" to newborns by nonmothers in the group, including adult males (Mitchell, 1969). Harlow, Harlow, and Hansen (1963), for example, have suggested that "It is probable that maternal affectional responding is never completely absent in the rhesus female, particularly after the birth of her first baby" (p. 258). However, little precise information has actually been obtained concerning the relation of such behaviors to the age, experience, or hormonal status of the female. Furthermore, there is little clear indication regarding the degree to which such interest in young infants is maternal in character or should be assigned to response systems more generally described as associated with exploration in response to novelty.

In a study designed to assess the appearance of maternal (or, perhaps more precisely, "paramaternal") patterns in adolescent rhesus monkeys, Chamove, Harlow, and Mitchell (1967) tested behavior toward a 1-month-old infant in fifteen males and fifteen females ranging in age from 18 to 30 months. Following a 24-hour adaptation period during which the infant could be seen and heard, but not reached, it was placed into the adolescent's cage for 15 minutes. Using a conglomerate measure of positive social behavior, females were found to manifest four times as many responses to the infant as males; the males, on the other hand, showed significantly more hostile behavior to the infant. Of particular relevance is the fact that, in contrast to the males, all but one female contacted the infant, ten showed "other maternal relevant behaviors" and five achieved ventral contact. Although responses were somewhat variable during this brief exposure (unfortunately no age-related data are presented), and some overlap in male and female responses was evident, the authors seem justified in concluding that sexually immature rhesus females "typically exhibited maternal-like affiliative patterns towards infants, whereas the males exhibited patterns of indifference or hostility" (p. 334). Although in a group setting, young animals may be prevented by the mother or other adults from approaching and contacting young infants, in a setting free from these potentially inhibiting factors, it would appear that in rhesus females the elements of maternal behavior begin to emerge prior to puberty and at least by the second year of life. In a recent study in the author's laboratory, similar evidence of maternal responsiveness was observed in adolescent pigtail (*M. nemestrina*) females. In this instance, the females involved were older siblings of 5- to 6-month-old infants and were continuously present during the infant's life as part of the large social group in which the mothers and infants lived. Some intermittent evidence of maternal (ventral) contacts with one infant by its sister appeared in the presence of the mother after about the second month of the infant's life, an event never previously observed between nonsiblings. However, clearly sustained ventral-ventral contact was observed between this adolescent (2½ years old) female and her infant sibling following the experimental removal of the mother over a 2-week period. It may be noted that some limited "maternal" responding was also observed in this infant's older male sibling as well. The second preadult female in the group, whose younger sibling was similarly treated, was never observed to manifest any clear maternal behavior toward the infant either before or after the maternal separation. These limited observations make it possible to conclude only that response elements of the maternal repertoire can be elicited in preadult female macaques under various conditions, but that the precise nature of these eliciting conditions is not entirely understood.

Maternal responses have also been observed in adult nonmothers in various macaque groups. In particular, we should consider first the available information on maternal responsiveness to infants in older females who are not themselves mothers of young infants at the time of observation. Several investigators, both in the field (e.g., Kaufmann, 1966) and in the laboratory (Hinde, 1965; Rosenblum & Kaufman, 1968) report the attentiveness of nonmothers to young infants. Hinde has referred to such females as "aunts," although Raphael (1969) has recently suggested the use of the Greek term "doula" to describe both the behavior and the fact that the behavior was initiated by a female (doulos would then refer to a male engaged in sustained caretaking activities). As the observations of both Kaufmann and Hinde suggest, nonmothers often show great interest in and attempts to contact young infants, although the full expression of maternal behavior by them is generally inhibited by the biological mother, particularly if the latter is relatively high in the dominance hierarchy. Kaufmann, for example, studying the development of infants in the free-ranging colony of rhesus monkeys on Cayo Santiago Island, indicates that females would not only hesistate to approach infants still in contact with their mother, but would often back away from infants who approached them. We have observed similar interference with attempts at doula behavior in our laboratory-observed pigtail macaques (Fig. 1). In contrast, however, we have observed a striking permissiveness on the part of bonnet macaque mothers (*M. radiata*) in allowing access to their infants (Fig. 2). What is most striking and most pertinent in the present context is the frequent emergence of clear maternal responding in females when the possibility of interference by the biological mother is eliminated, as when

Fig. 1. A pigtail mother preventing another female in the group from contacting her newborn infant.

she is removed from the group. Hinde, Spencer-Booth, and Bruce (1966), for example, report a sharp increase in maternal contacts with separated rhesus infants. In our own laboratory, we also have observed "full adoptions" of separated bonnet infants 2-6 months of age, with such adoptions lasting a month or more. More recently, we have observed a similar, though less sustained, manifestation of maternal behavior in a pigtail adult female in response to a separated 5-month-old infant.

Harlow *et al.* (1963) have also indicated that at least some female rhesus, even 9 months after the birth and immediate removal of their own infants, will adopt a separated 1- to 2-month-old presented to them in isolation. Moreover, these females not only showed "entirely normal mother-infant relationships" but ultimately evidenced copious lactation. Such a juxtaposition of female reproductive status and infant age would be difficult to achieve in the wild, since in the wild most eligible females deliver within a short time of each other.

Fig. 2. A bonnet female grooming the newborn infant of another mother in the group.

Thus, it is not surprising that it is only from the laboratory that we have data regarding the potentialities of maternal responding in nonpuerperal females.

Further discussion of these maternal caretaking activities in nonmothers will be presented below. At this point, however, it is sufficient to conclude that relatively extensive maternal behavior can be elicited, albeit somewhat variably, in nonmother female macaques under the appropriate conditions. Again, the actual internal and external eliciting conditions are not clearly known, although as Harlow *et al.* (1963) suggest, the infant's own approach behavior plays a critical role in determining the degree of maternal responsiveness subsequently elicited.

There are some data, however, which suggest that there are differences in what may be called the "threshold conditions" for maternal responding in both nonmother adult females and in those who currently have or who have recently had offspring. Simons, Bobbitt, and

Jensen (1967), for example, using pigtails, presented tape-recorded infant calls to: (a) two mothers (whose 7-month-old infants were temporarily removed during the several hours of testing); (b) two adult (parous) females without young; and, (c) two adult males. Simons *et al.* concluded from this study that "Mothers and only mothers increased locomotive activity significantly when infant calls were the stimuli; mothers and only mothers increased vocalization when infant calls were the stimuli;" (p. 9). Thus, although there was apparently no selectivity of response for calls of their own infants (also assessed), females currently with young were more responsive to these highly delimited infant stimuli. In a related but perhaps more ambiguous experiment, Cross and Harlow (1963) used a visual exploration apparatus and tested preference for viewing a 10- to 40-day-old infant versus a 1-year-old. The subjects were (a) nulliparous adult females, (b) multiparous females delivered 2–4 months before the start of the experiment but whose infants had been removed immediately upon parturition, and (c) multiparae both during late pregnancy and after delivery. The nulliparous females, and the gravid subjects during the 2 weeks prior to delivery, showed no preference for baby-viewing. In contrast, the multiparae that had delivered some months earlier showed a modest but persistent preference throughout the 49 test days for viewing infants. Most striking, however, was the sharp rise in infant-viewing in the period just *following* parturition in those subjects who had shown no preference while still pregnant, even though these females had their own infants permanently removed from them on day 3. Although the results of this study are certainly more suggestive than conclusive, they do fit with the general view that, when compared to current mothers or to mothers whose infants have just been removed, nonmothers show only variable and transient interest in young infants. These results suggest further that even among females who have borne young in the past, no marked change in maternally related responses is evident as parturition approaches. Regarding this latter point, however, an early report by Tinklepaugh and Hartman (1930) suggests that it may not be until just before parturition that specific behavioral changes in the direction of increased maternal responses appear. In order to observe her reactions to the birth of an infant, a young primipara, some 3 weeks prior to term, was placed in a cage with a female whose own parturition was imminent. When delivery did occur, the primipara (about 26 days prior to term) showed no apparent interest in either the new infant, the placenta, or the birth fluids. However, when this same primipara, now only 3 days prior to term, was housed below another female during the latter's labor and delivery, she was observed to lick ". . . such drops of the (amniotic) fluid as she could recover from the walls of her lower cage. She next licked up some of the blood which came through the floor of the cage above and also succeeded in filching a considerable portion of the afterbirth which she ate with avidity" (p. 77).

Although neither our own research nor that of Tinklepaugh and Hartman have systematically assessed the response to birth by-products at all stages of pregnancy, the fragmentary observations that are available suggest that this initial step in the full maternal behavior repertoire may be dependent upon hormonal changes coincident with the terminal phases of pregnancy, or a particular chain of behavioral events, or perhaps both.

B. Parturition

Although many investigators involved in breeding of macaques, including the present author, have occasionally observed the parturitional process, the actual birth of most infants usually goes unobserved. This is largely a result of the fact that more than 90% of such births occur at night, when personnel of laboratories are generally not in attendance, and when field observations are generally impossible. Jensen and Bobbitt (1967), however, have recently manipulated the timing of parturition in a number of pigtail females (*M. nemestrina*) by reversing both light-dark cycles and associated laboratory routines beginning at least 3 days prior to parturition. Under these conditions, eleven of twelve deliveries occurred during the 10-hour artificial night. Although this suggests that it is relatively easy to obtain laboratory data on the parturitional process as well as on the initiation of formal mother-infant relations in macaques, little has actually appeared.

The most generally detailed descriptions of parturition in macaques was provided a number of years ago by Tinklepaugh and Hartman (1930, 1932) who observed a number of rhesus (*M. mulatta*) and several *M. cynmolqus* births. These superb early papers provide both excellent descriptive and quantitative data in support of what remains as the most reasoned interpretation yet available of the unfolding behavioral processes observed in parturition. It may be noted further that these data also seem quite applicable to the more limited observations we have made on pigtail and bonnet macaque deliveries.

As Tinklepaugh and Hartman indicate, there are few overt changes in general behavior to suggest the approach of labor in most females, except perhaps a reduction in total activity in the final days of pregnancy (cf. Southwick, Begg, & Siddiqi, 1965). A number of hours or occasionally a number of days prior to the onset of labor, touching and manual exploration of the vagina, with subsequent sniffing and licking of the fingers may be observed. The onset of labor itself is signaled by the periodic assumption of a squatting posture, often with the female tightly clasping some aspect of her perch against which to strain (Fig. 3). The labor period can apparently range from less than an hour to more than a day. (Tinklepaugh and Hartman report a female who showed more than 1200 contractions over a 32-

Fig. 3. A pigtail female straining during labor contraction shortly prior to the delivery of her infant.

hour period.) The female alternately paces, squats, and rests, often assuming the relatively unusual posture of lying on her side.

> As mucus begins to be expelled immediately prior to or during labor, the hand is used to remove the substance and carry it to the mouth. Then as labor becomes more intense and the fetus advances in the canal, there is a marked increase in this "removal function" of the hand, as though effort were being made to eliminate the source of irritation within the bulging perineum. Finally, the baby's face appears in the vertical axis. The mothers without exception, explore this with their hands. They then attempt to seize hold of the fetal head. As soon as the baby's head is expelled the mother seizes it and pulls it to one side and forward (Tinklepaugh & Hartman, 1930, p. 88).

After the infant is completely expelled, it is then usually set against the ventral surface to which it adheres by means of strong clasping reflexes of both fore- and hindlimbs. Actually, the active clasp reflexes of the newborn may begin to function before the infant is completely free of the birth canal.

> Activity on the part of the typical monkey baby begins as soon as it is sufficiently free from the birth canal to make any form of movement possible. When the head appears, the eyes begin to blink and the baby gives shrill piping sounds. The arms reach out and seize hold of any object within reach as soon as they are expelled. At times it appears that this behavior actually aids in the process of delivery. (Tinklepaugh & Hartman, 1932, p. 267)

It is of some importance to note that the mother, after manually bringing the infant forward to her ventrum, does not appear to concern herself with or adjust the infant's actual orientation. It is apparently the strong righting reflexes and negative geotropism (Mowbray & Cadell, 1962) of the healthy newborn which function to produce the proper orientation. "The mother may hold the infant to her breast with one arm if she turns to this other activity (i.e., attention to the placenta). She does not otherwise determine the baby's position relative to her, but it *orients itself* toward her in the ventral-ventral relationship and clings tightly with both hands and feet" (Tinklepaugh & Hartman, 1932, p. 267). It may be noted that when the infant is unable to carry out these adjustments, a mother will often clasp her disoriented baby for long periods. In our laboratory, we observed the birth of a pigtail infant to a primiparous female who allowed the infant, in its final expulsion from the birth canal, to fall to the floor from an overhead bar. The infant, as revealed from subsequent autopsy, received a severe brain injury and though it lived for 18 hours, it was apparently uncoordinated although moving. This mother carried her injured newborn both dorsal-ventral and upside down for long periods before and after it died (Fig. 4).

Fig. 4. A pigtail female clasping her newborn, brain-damaged infant. Note the lack of clinging by the infant and its inappropriate orientation toward its mother.

About 5-15 minutes after the delivery of a macaque infant, the female characteristically begins further grasping and manual exploration of the vaginal area into which the cord, of course, still passes. Once the placenta descends into the vagina it is usually withdrawn by the mother. The next phase has been described in rhesus as follows:

> For the time being the baby is largely ignored and there begins one of the most interesting and puzzling features of parturition. This herbivorous creature suddenly becomes carnivorous. After licking the afterbirth, she begins the gruelling task—one common to most if not all subhuman mammals and probably related to human placentophagia—of consuming this tough fiberous mass. Holding the organ in her hands she bites and tears at it with her teeth. Long strands of it are loosened and she tries to swallow them, only to draw them back again and separate them completely from the main mass. Her hands become wet with blood and fluids and she stops from time to time to lick them. Her behavior does not suggest the eating of a delicacy so much as it does the performance of a task under compulsion (Tinklepaugh & Hartman, 1930, p. 89). (Figs. 5 a, b)

Fig. 5a and b. A bonnet female and a pigtail female licking and devouring the placenta shortly following the birth of their offspring.

These early observers report for rhesus and cynmologus (and this is very much in keeping with our own observations on pigtails and bonnets) that, in the great majority of normal deliveries, the placenta is eaten by primiparous and multiparous females alike. There are exceptions, however, which are seen not only after stillbirths or unusually prolonged labors but after normal deliveries as well. Typically, the cord itself is rarely eaten but rather the eating ceases at or close to the junction of placenta and cord and the latter, though licked and handled, is then allowed to dry and fall away.

Fig. 5b.

Although the mother continues to lick amniotic fluid and blood from her own hands and legs following delivery, she very soon turns her attentions to the newborn. She will intermittently lick her infant, often handling and turning it in attempts to lick the fluid from remaining parts of the infant's body. Later, as the infant's coat dries and often after the mother awakens from a brief postparturitional sleep, the licking ceases and is replaced by more commonly sustained manual grooming patterns of the type generally observed in adult macaques. In most instances, the mother intently inspects and picks through the fur of the infant, quite literally from head to toe.

Before leaving the parturitional phase itself, it might be of interest to consider Tinklepaugh and Hartman's interpretation of the sequence of behavior observed at parturition, as that sequence is related to changing physiological events. Although speculative in nature, their hypotheses still provide as reasonable a structure for the sequence as is available:

> At each successive stage in the process of birth-giving, the behavior is nearly always appropriate. The swelling and irritation of the genital region lead to manual exploration. The presence of mucus, urine, and fecal matter calls forth the posture and straining common to eliminatory processes. Later fetal fluids begin to appear. In the course of the manual investigation of the genitals the animals lick these fluids from their hands, and this act, it seems, may determine much of their subsequent behavior. Why these fluids are eaten we do not know, but it does appear that to the monkey subjects they are fluids to be consumed.
>
> As labor progresses with its successive contractions, the eliminatory position is taken more often, straining is more intense and, facilitating the latter, the animals seize hold of the walls or floor of the cage. The fetus advances in the canal, causing the perineum to bulge from the pressure. Then what have been manual exploratory movements, change to manual eliminatory movements—namely, efforts to remove the source of irritation, and the licking of the fluids from the hands continues.
>
> The head of the baby eventually appears at the entrance of the canal and then the manual efforts become affective. The wet, newborn baby is drawn forth and around one side to the mother's breast. The fluid is licked from both baby and maternal hands. During the washing process the genitals are again irritated by the frequent tautness of the cord when the baby is shifted from one position to another. The afterbirth is discharged into the vagina. These events result in further exploration and finally in the drawing forth of the afterbirth—a mass of tissue which like the baby, is covered with fetal fluids. It is licked then eaten.
>
> The question, why does the mother consume the afterbirth and only wash the baby, when both are bathed in the same fluids, is a pertinent one. It might be assumed that the mother, even though inexperienced, reacts to the one as a baby and to the other as an inanimate object. But we may suggest that the baby is only covered with the 'to-be-consumed' fluids, whereas the afterbirth is permeated with them. That the cord, likewise, is only covered with the fluid may explain why it is licked and left structurally intact. Such explanations are naturally, only suggestive and require experimental testing (p. 94–95).

III. Mother-Infant Interactions

Although many papers have reported qualitative observations of mother-infant relations in numerous members of the genus (e.g. *M. speciosa:* Chevalier-Skolnikoff, 1968; *M. sylvana:* Lahiri and Southwick, 1966; *M. silenus:* Sugiyama, 1968; *M. cyclopis:* Yang *et al.*, 1968) systematic, quantitive data are currently available only on the rhesus monkey (*M. mulatta*), the pigtail (*M. nemestrina*), the bonnet (*M. radiata*), and the Japanese macaque (*M. fuscata*). Since there appears to be considerable overlap with respect to the major quantitative dimensions and sequential themes in the ontogeny of mother-infant relations in these species, it is reasonable to treat them together. However, interspecific variations in the precise chronology of development do exist, and variations in the pattern and pace of development within a species have also been reported. Unfortunately, the diversity of social and physical settings within which observations have been made, often make attempts at interpretation of these interspecific differences extremely difficult.

Several authors have attempted to present the ontogeny of mother-infant relations in macaques (Hansen, 1966; Harlow *et al.*, 1963; Rosenblum & Kaufman, 1968) by delineating the characteristics of several successive stages in the relationship. Although making for oversimplification, as Hinde has recently emphasized (personal communication, 1969), use of a stage-by-stage descriptive approach seems useful for integrating and summarizing the multidimensional chronology of infant-mother, infant-group, and mother-group interactions. However, as Hansen (1966) has pointed out, it is necessary to bear in mind that "No precise temporal delineation of these stages can be made and they are typified by considerable overlapping" (p. 116). Furthermore, the actual description of specific stages in any given study is a function of what Harlow *et al.* (1963), focusing on the mother have termed, ". . . such multiple variables as the nature of the physical and social environments; the life history of the animal; previous pregnancies and experiences with infants; endocrinological factors; and even the behavioral criteria chosen and the time of day the data are collected" (p. 258). In light of the great diversity of methodology and behavioral terminology employed in this field, we call the readers attention to this last caveat.

A. Primary Mother-Infant Contact

As indicated earlier, the completion of the birth process, i.e., full emission from the birth canal, is, in most instances, simultaneous with the onset of mother-infant interactions as they are observed for much of the initial phase of the infant's development. After the mother has brought the infant around to her ventrum, tight clinging by the infant begins and often, although not always (Hinde, Rowell, & Spencer-Booth, 1964), the infant, through the so-called rooting reflexes, obtains nipple contact. As we have seen, the mother, in turn, occasionally supports the newborn and frequently licks and grooms it as well. This ventral-ventral contact position, most often accompanied by the infant's oral contact with the nipple and the mother's active clasping or bodily enclosure of the infant, is the primary mode of interaction within the dyad in its earliest phase. Such primary contact, though varying in the degree of total body contact area and the amount of support provided, is maintained both at rest and when the mother moves about the environment. The gradual decline in this form of interaction in the ensuing months largely reflects (but as we shall see does not completely describe) the waning or change in the mother-infant bond as the infant develops. Figures 6 and 7 indicate the change in this primary contact pattern in mother-infant dyads containing male and female infants in bonnet and pigtail macaque groups observed in the laboratory (Kaufman & Rosenblum, 1969; Rosenblum & Kaufman, 1967). These subjects lived in homospecific groups, each containing several mother-infant dyads and a breeding male, and were observed with great regularity during their development. Tape recorded observations followed by computor analysis, allowed computation of the frequency and duration of a wide variety of behaviors (both in absolute terms and relative to the total amount of time spent observing the subjects), (Kaufman & Rosenblum, 1966; Rosenblum, Kaufman, & Stynes, 1964). These curves show a characteristic decline during the first 3–4 months after birth followed by a more gradual decline until, toward the final quarter of the first year of life, low asymptotic values are reached in both sexes and in both species.

Hinde and his co-workers (e.g., Hinde *et al.*, 1964; Hinde and Spencer-Booth, 1967) have studied the ontogeny of mother-infant relations in the rhesus monkey under conditions of group living very similar to those employed for the bonnets and pigtails described above except that indoor-outdoor pens were utilized. Observations in these studies were recorded primarily in terms of the relative frequency of occurrence of a diversity of behaviors during each successive half-minute interval of observation. These half-minute check sheet records afforded ample opportunity to tabulate the changing patterns of mother-infant relations and to propose analyses regarding the relative contribution of mother and infant to the changing rela-

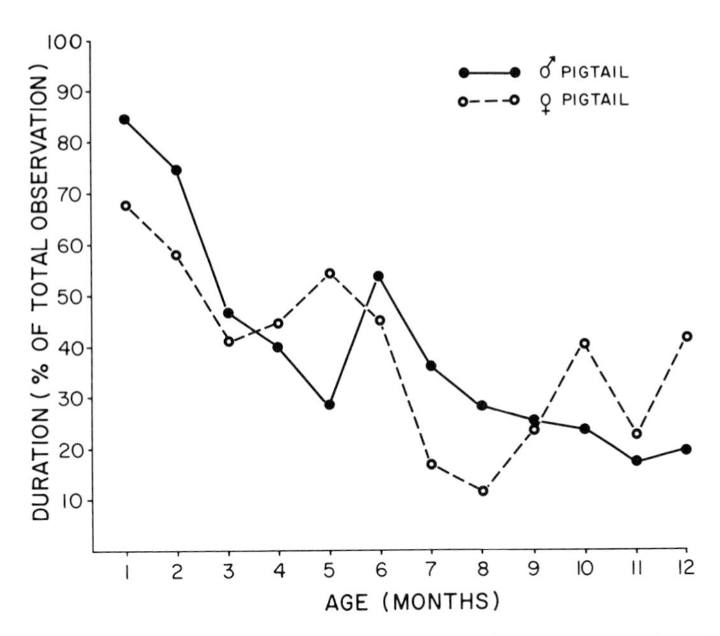

Fig. 6. The median duration of ventral contact with the mother in male and female pigtail infants during the first year of life.

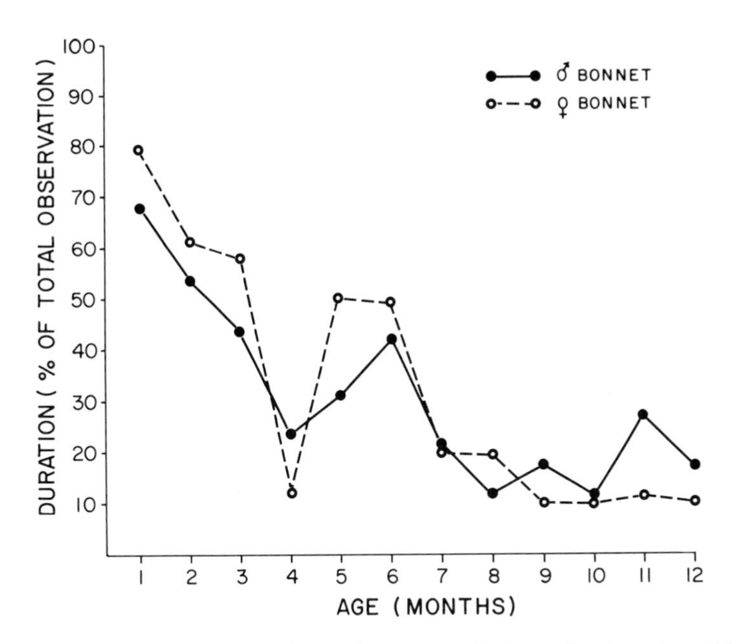

Fig. 7. The median duration of ventral contact with the mother in male and female bonnet infants during the first year of life.

tionship over time. These data on rhesus infants show a decline in mother-young contact similar to that observed in pigtail and bonnets, except that in the rhesus the decline occurs somewhat more gradually, although with the most rapid decline still restricted to the first 16 weeks of life (Hinde *et al.*, 1964, p. 628). The relative plateau in the decrease in mother-infant contact toward the end of the first year was also observed in these rhesus dyads. In addition, the quantitative data of the Hinde studies, quite unique among any of the observations appearing to date, continues until the infants are 125 weeks old and indicates that the final or complete cessation of contact is not reached until these rhesus infants are about 2 years of age.

In still another quantitative assessment of the changes in the mother-infant relationship with age in a complex social setting, Itoigawa (1971) has reported observations covering the first 8 months of life in free-ranging Japanese macaques. The troop observed by Itoigawa ranged over some 24 square kilometers. They were observed regularly at the park site to which they returned each day for food provisioning, the key method through which Japanese investigators have made otherwise wild troops available for systematic study over long periods. Itoigawa's troop had been "provisionized" (sic) and studied in this way for about 6 years prior to his observations. Itoigawa used a checksheet method for half-minutes of observation similar to that used by Hinde and his co-workers. His results also showed the characteristic decline in ventral contact with mother after the period of maximum contact in the first days of life. However, in Itoigawa's data the decrease in ventral contact during the first 3 months of life was extremely sharp, with a rather *abrupt* asymptote.

Similar observations of an extremely sharp decrease in ventral contact with the infant have been reported by Hansen (1966; Harlow *et al.*, 1963) in rhesus dyads kept in a constricted laboratory setting. Hansen's subjects were observed in a "playpen" apparatus in which four dyads were housed in separate cages, each measuring only 3 feet cubed. The dyads, although physically isolated, were in visual and auditory contact with one another. These four dyadic chambers connected to a center play area (5 feet square × 2½ feet high) into which only the infants could move. It was in this play area that the infants were able to interact with one another for a limited period each day. Hansen used a variant of a checksheet method, essentially similar to those described above, except that 15-second time intervals were used. Under these carefully controlled conditions, Hansen observed a dramatic decrease in the incidence of ventral contact during the first 3 months, with only a slight additional decline until such contact was virtually absent after the ninth month of life.

An extended program of observation of pigtail dyads has been carried out by Jensen, Bobbit, and Gordon (1967a, 1967b, 1968a, 1968b), also under well-controlled but even more restricted conditions than that employed by Hansen. Mother-infant dyads were housed and studied separately in individual cages, under conditions designed to manipulate environmental complexity. At one extreme, smooth barren cages set into separate soundproof chambers were used; at the other, mesh wall cages were used containing several manipulanda, set within visual and auditory contact of other monkeys. Jensen and his co-workers used a precise dictated tape-recording system somewhat similar to that used by the present author, and subsequent computer analysis to determine the relative frequency and duration of various behaviors and behavior combinations. Unfortunately, most of these data have been presented only in terms of changes in specific items relative to other groups or classes of items, the latter themselves changing over time. Thus, for example, the frequency of breaking contact between the dyad has been presented relative to the frequency of all changes in dyadic positions, including various changes in types of contact. Nonetheless, although precise comparisons with other data are difficult, the results of Jensen *et al.* (1968a) indicate that infant-mother contacts in their individual pigtail dyads show a sharp decrease during the first 15 weeks of life with a slope similar to that found for group-living rhesus dyads and for group-living pigtail and bonnet dyads. In addition, however, Jensen *et al.* (1967a) found a more rapid decrease in contact with increased environmental complexity. Thus, it seems likely that, despite the similarity in the slope of the decreasing contact, the absolute levels of contact observed in individually caged subjects are not comparable to those reported for any of the dyads studied in group settings.

B. Infant Separation from the Mother

An examination of the data provided by the same investigations described above but relevant to the other extreme of the mother-infant distance continuum, namely, full separation rather than intimate ventral contact, indicates a pattern similar to that depicted in Figs. 8 and 9 for pigtail and bonnet subjects. Although based on the same observations used in describing the decline in contact, the data on maximal separation from the mother speaks to another aspect of the

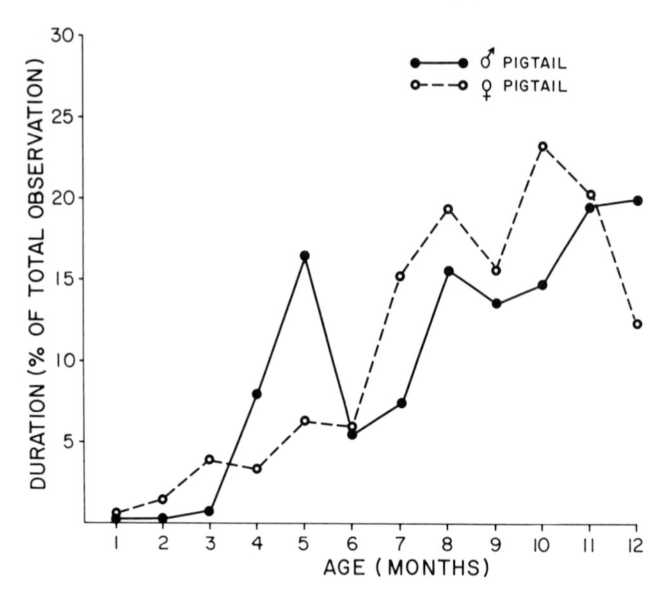

Fig. 8. The median duration of maximal separation between male and female pigtails and their mother during the first year of life.

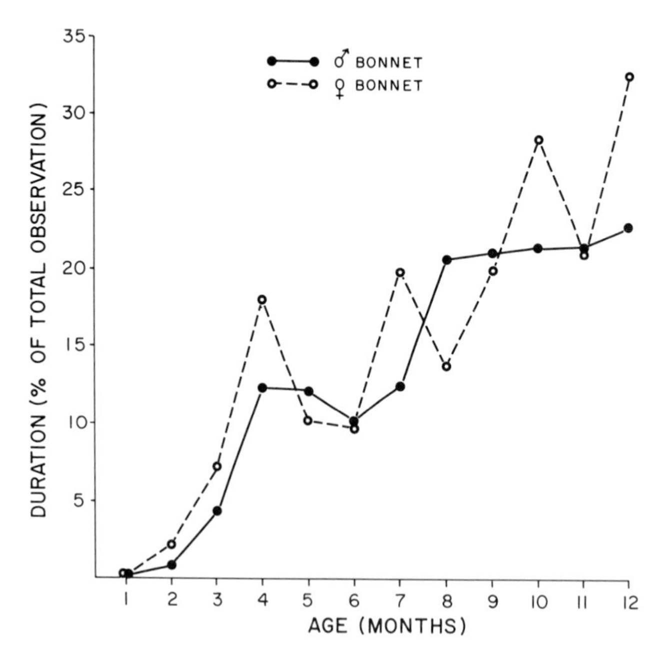

Fig. 9. The median duration of maximal separation between male and female bonnets and their mother during the first year of life.

relationship—the degree to which the infant, having broken physical contact, actually moves independently out into the environment. Initially, breaks in contact with the mother begin as early as the second day of life (Kaufmann, 1966) and are observed with characteristic regularity by the second week of life (Hinde *et al.*, 1964; Lashley & Watson, 1913) in both captive and free-ranging rhesus. This age of first breaking contact is in keeping with our own data on captive pigtail and bonnets. Although the first movements from the mother are generally of short distance and brief duration (usually less than a minute), a gradual increase occurs in both the duration and extent of those excursions. Obviously, the varied restrictiveness of the environments within which macaque infants have been observed set the boundaries for the degree of spatial separation that it was possible to observe; nonetheless, general trends seem relatively comparable in the several species studied.

Hinde *et al.*, (1964) for example, report a gradual but steady increase in the percent of complete half-minute intervals of observation spent more than 2 feet from the mother. This percent climbs most sharply from about the sixth to the twenty-second week, reaching a plateau of about 30–35% of the observation time after that point. The relative frequency of these longer periods away from the mother apparently rises still further in the last quarter of the first year (Hinde & Spencer-Booth, 1967). With respect to extended excursions from the mother, Itoigawa's data on Japanese macaques show a much smoother developmental trend than was true of his data on ventral contact reported above. The Japanese macaque apparently obtains distances greater than 1 m from the mother at levels about comparable to those shown in Hinde's data on the rhesus. Moreover, the Japanese macaque reaches this distance in about 35–40% of the observation periods by the third month. Itoigawa also reports a secondary increase in the frequency of these extended separations from the mother to about 60% of the observation periods by the eighth month of life.

The data on pigtails and bonnets shown in Figs. 8 and 9 reflect the percent of observation time which subjects of both sexes in each species spent maximumly separated from their mothers as they grew older. In the pen situation in which these subjects were raised and observed, maximum separation involved distances of about 3–15 feet, but, most importantly, involved separations between mother and infant at different vertical levels of the pens, i.e. ceiling, bars, shelves, floor. Our observations indicate that these vertical excursions occur later in development than separations of equivalent distances along the same horizontal level (Kaufman & Rosenblum, 1969). This distinction, perhaps not really surprising in a semiarboreal form, has not been articulated by other observers and may account for the generally lower levels and later development of the "maximum" separations achieved in our pigtail and bonnet subjects when compared with the most extensive separations designated and recorded by other observers. Indeed, in contrast with the trends presented above for rhesus and Japanese macaques, even 30% levels of this maximum degree of separation were not observed in our pigtails and bonnets until the very end of the first year of life.

Somewhat in contrast to these data on relatively free-moving subjects, Hansen's data on his more restricted dyads show an extremely sharp increase in a measure of greatest separation from the mother during the first 3 months. Data on the amount of time the infant spends outside the cage and consequently away from the confined mother, increases to about 50% of the observation periods by the third month and then rises to about 60% by the fourth quarter of the first year.

This overview of the changing spatial relations of mother and infant during the first 1-2 years after birth tells us a great deal about mother-infant interaction. Indeed, these relatively discrete measures of contact and distance may be considered as the guideposts for understanding the changing behavior of mother and infant. In other words, these twin parameters may be important indicators of basic differences between species as well as between conditions of observation. Thus, for example, the relatively sharp decline in contact in Japanese macaques described by Itoigawa may be attributable in part to the fact that his observations were made only when mothers were "sitting and resting" and "when neither mother nor infant was in antagonistic relation with others" (p. 5). Such qualifications can easily shift the recorded data in the direction of relatively low contact scores. Similarly, Hansen's observations of a dramatic early decline in intimate mother-infant contact, could be due, as Hinde *et al.* (1964) have suggested, to (a) the mother's inability to retrieve or even approach the infant after it left her cage, or (b) the fact that complete access to peers was possible only during the actual observation period each day, thus perhaps encouraging the infant to leave as much as possible during this relatively curtailed period.

C. Nursing

Often beginning only moments after birth, and of course during those periods when the infant is in ventral-ventral contact with the mother, the infant engages in prolonged periods of oral contact with the nipple. All observers agree that periods in which actual sucking behavior occurs are difficult to distinguish from periods in which the nipple is merely held in the mouth. As a result, precise separations of these two behaviors are not available for any species (cf. Harlow *et al.*, 1963; Fig. 4). Furthermore, as Hinde has pointed out, an infant may be asleep while on the ventrum but may maintain nipple contact nonetheless. The relative time spent in nipple contact when eyes are open or closed may follow different developmental courses, however. Thus, Hinde *et al.* report, for rhesus at least, that after the second month of life the infant retains the nipple during most of the time it is on the mother's ventrum with its eyes closed, presumably asleep. The same, however, is not the case when it is on the mother but awake, and varying periods may be spent without oral contact.

Although nipple contacts do occur outside the ventral-ventral posture, most observers agree that there is a gradual decrease in total level of nipple contact rather congruent with the general decline in ventral-ventral contact with mother. However, the age at which complete termination of nipple contact occurs may be influenced by various factors, including, for example, the birth of siblings, which may terminate nipple contact abruptly (Hinde & Spencer-Booth, 1967; Simonds, 1965). It may be noted, however, that even the birth of siblings has a variable influence on the termination of nipple contact. For example, we occasionally have observed that both bonnets and pigtails obtain brief nipple contacts with their mother some weeks after the birth of a sibling. Notwithstanding the fact that much of the nipple contact observed probably does not include active sucking, the duration of time the mother is capable of maintaining active milk secretion may be another important factor influencing the age at which nipple contact ceases. In this regard, Harlow *et al.* (1963) report that palpation of the nipples of rhesus mothers indicated that lactation could occur for at least 18 months after the birth of a single infant. Furthermore, Hinde and Spencer-Booth, for their group-living rhesus subjects, report that nipple contacts generally terminate between 90–110 weeks of age, but that one female infant continued to obtain such contact until the end of the 2½ years of their observations.

One final note on the developmental course of infant contacts with the mother's nipple concerns the relationship of nipple contact to ventral contact. It appears, at least throughout the course of the first year, that as total contact with the mother decreases, the percentage of time the infant effects nipple contact while on the mother remains rather constant. In rhesus, for example, even at the end of the first year, more than 90% of the time that the infant is in close contact with the mother the nipple is held in the mouth (Hinde & Spencer-Booth, 1967). Comparable data for pigtails and bonnets are presented in Fig. 10. Thus, it seems reasonable to suggest that well after the mother ceases to provide the major source of nourishment, the infant's contact with her is intimately tied to its attempts to obtain the breast. Indeed, it has been observed in older infants that nipple contact is often achieved in the absence of ventral-ventral or even actual body contact with the mother, the infant reaching from her side to take the nipple in its mouth. In light of these data, it would be of interest to determine the impact nonaccessibility of the nipple might have on the infant's continuing orientation and movement to an otherwise accessible mother.

D. Protective Maternal Behavior

Regardless of the setting within which macaque mothers and infants have been studied, mothers have been observed to act in a manner that may be described as protective or restrictive of their infants when confronting apparent sources of threat or danger. Indeed, the coincidence of the onset of this protectiveness with birth itself is reflected in the frequent observation that many mothers will zeal-

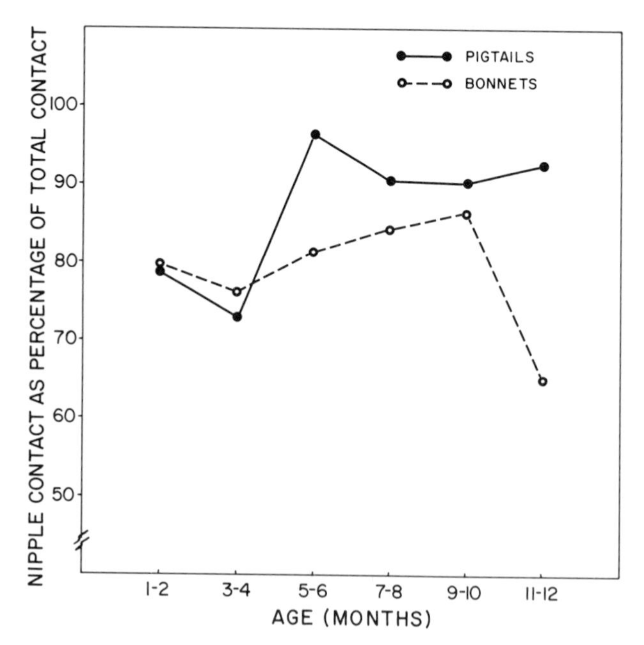

Fig. 10. The percentage of time spent with the nipple in the mouth as a function of time spent in ventral contact with the mother.

Fig. 11. A bonnet female clinging to her dead infant.

ously guard uneaten placenta or the body of a stillborn infant (Fig. 11). Such protectiveness may be manifested in externally directed threatening or aggressive behavior (Fig. 12); of at least equal importance, however, are a series of addyadic behaviors. These behaviors, serving to keep or bring the dyad together, may be interpreted as generally protective of the infant, although they are directed toward regulating the behavior of the infant itself (Kaufman & Rosenblum, 1966). They may occur in the presence of an obvious or subtle social threat or in the presence of perceived danger from an inanimate source. The former might take the form of an overt attack either on the mother or on her infant, the latter may range from a sudden burst of sound in the area to the mere visual presence of a novel stimulus in the home environment (Kaufmann, 1966). On the other hand, just as frequently as an external stimulus for eliciting maternal protectiveness may be discerned, it is very common, particularly in the early period of the infant's life, to observe various forms of protective maternal regulation when no evident external provocation exists. In such instances, it would appear only that an observable but difficult to define increase in maternal agitation instigates the protective behavior.

Fig. 12. A pigtail female threatening another female in the group as the latter approaches her and her newborn infant.

At this point it becomes appropriate to ask about the specific forms that infant-directed maternal regulation takes early in life and about the chronology of its development in those species studied to date. As we have already seen, the very young infant clings tightly to the mother's ventrum, often sleeping and showing little overt interest to the outside world. However, by approximately the second week, the infant begins its attempt to make excursions from the mother's body. It is at this time that the first addyadic regulatory patterns are seen. In general, most of the infant's initial attempts to break contact with the mother are prevented by various forms of maternal restraint, including tight clasping of the infant to the ventrum, or holding tightly the body, leg, or tail of the infant (Fig. 13). These maternal restraints may appear as early as the first week in rhesus, Kaufmann reporting a single instance on the sixth day after birth and Hinde and Spencer-Booth many more instances during the same period. However, maternal restraint is probably exercised more frequently beginning during the second week of life in rhesus (Hinde & Spencer-Booth, 1967; Lashley & Watson, 1913). In any event, there is a peak in maternal restraint in the second and third weeks with a gradual decline during the first 2 months and complete disappearance after weeks

45–46. Kaufmann's observations in the wild indicate also that "most" infants are no longer restricted after the seventh or eighth week. Hansen (1966) on the other hand, indicates that his rhesus mothers' "active interference with their infants activity" shows a rather sharp continuous *increase* during the first 40 days of life, not reaching a peak until the eighth week, but then declining very rapidly to reach virtually zero levels by the twelfth week of life.

Itoigawa's data on the restrictive behaviors of Japanese macaque mothers also show a sharp increase in the second month of life, but with a more gradual decline and with some limited frequency of occurrence even at the end of his eight-month observation period. Similarly, Jensen's data on pigtails indicate a more gradual increase in mothers "retaining" their infants until about the fifth week with a slow decline thereafter until observations ended at 15 weeks. Unfortunately, each of these data samples on the several macaque species is presented using combined categories of maternal protection or restriction. Analyses of the ontogeny of several of the separate maternal patterns in pigtails and bonnets suggest that each of these behaviors may follow its own developmental course.

Thus far we have identified three primary forms of infant-directed

Fig. 13. A pigtail female restraining the departure of her 1-month-old in the presence of another mother and infant of the group.

addyadic maternal patterns. The first to occur developmentally is the type of behavior that restrains an infant from breaking contact with the mother. As the infant's efforts to break free of the mother persist, physical separation is intermittently achieved; the mother, however, keeps close guard on her infant either by moving along next to it, one hand outstretched toward the infant's side, or by stretching herself over the infant, producing a kind of canopy beneath which the infant may move (Fig. 14). From these positions the mother makes frequent and very rapid manual retrievals, bringing the infant back to her ventrum. Finally, as the infant matures and its coordination and speed of movement increase, the extent and duration of the infant's excursions increase and the mother's primary protective tie to the infant is manifested by prolonged watching and apparent attentiveness to any vocalization the infant might make. At this stage, the mother protects and regulates her infant's activity primarily by rushing to it and retrieving it when danger threatens. As can be seen in Figs. 15 and 16, mothers of both bonnet and pigtail species engage in each of these behaviors during the first 6 months of the infant's life. In keeping with most of the data presented for other macaques, the percentage of mothers engaging in each of these behaviors is highest after the first 2 weeks of life, with a decline being evidenced by the end of the second month. However, it is important to note that, although observed under comparable conditions, relatively fewer bonnet mothers engaged in these protective behaviors at each infant age, and, in contrast to pigtails, the general trend shows a much more rapid drop in response after the eighth week; by the end of the third month, less than 10% of the bonnet mothers evidenced any of these behaviors. This trend among bonnets, observed in a freely interacting group situation, resembles that described by Hansen for his more restricted rhesus females. In contrast, as Fig. 17 shows, pigtail mothers engage in these protective behaviors with much greater frequency and continue to show them until considerably later in the life of their infant. Figure 18 shows that the three primary protective regulatory behaviors in pigtails follow discriminatively different time courses, both with respect to relative duration and frequency of occurrence. The initial appearance of restraint and its rapid rise by the third week is followed in the fifth week by the growth and subsequent peak level of "guard." During the same period, retrieving increases rapidly and although retrieving is less frequent after the third week, it is sustained after the first 6 weeks as the most prominent overtly protective behavior.

Fig. 14. A bonnet mother guarding her 6-week-old infant as the latter climbs on a supporting post.

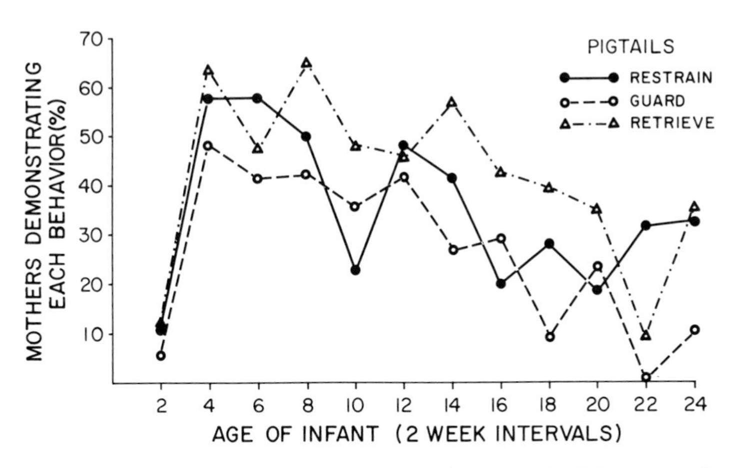

Fig. 15. The percentage of pigtail mothers manifesting each of the maternal protective behaviors during the first 6 months of their infant's life.

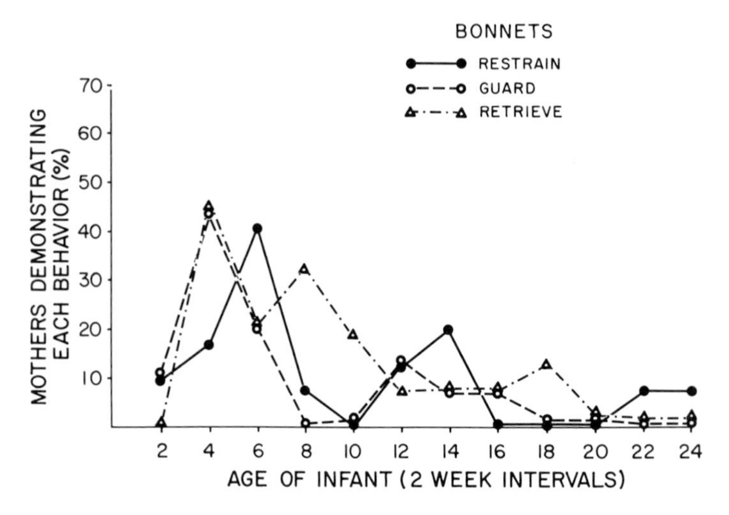

Fig. 16. The percentage of bonnet mothers manifesting each of the protective behaviors during the first 6 months of their infant's life.

It is by no means clear at present what factors are involved in shaping the developmental courses that have been described for these addyadic maternal patterns in macaques. On the one hand, the variously defined conglomerate indexes of maternal protection that have been used as well as the diversity of observation conditions employed make it noteworthy that all authors agree that at least by

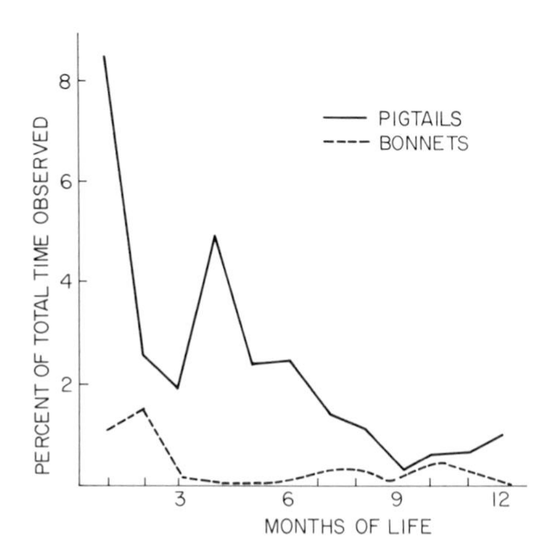

Fig. 17. The mean duration of engagement in protective maternal behaviors in pigtail and bonnet mothers during the first year of their infant's life.

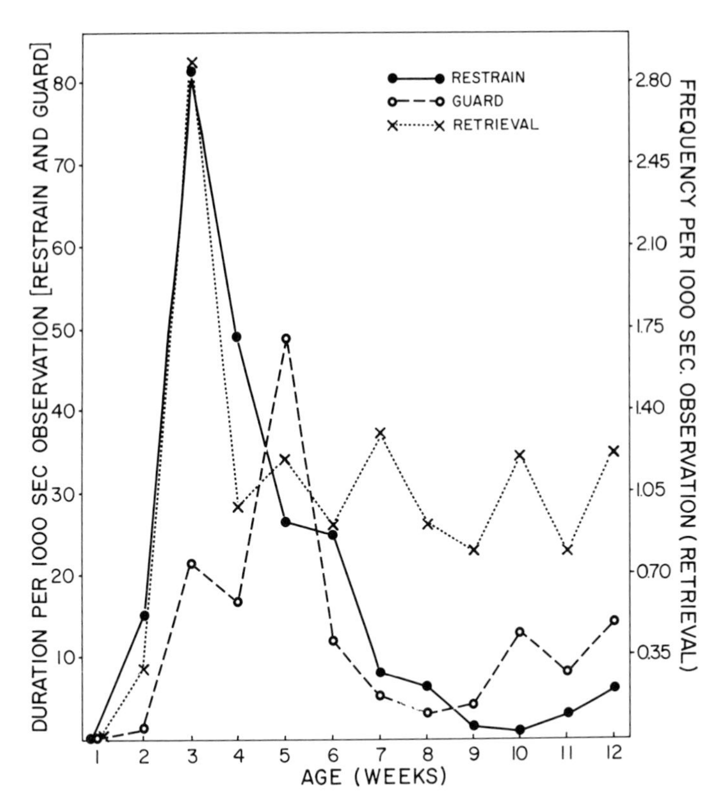

Fig. 18. The chronology of the onset and decline of the major protective behaviors in pigtail mothers during the first 12 weeks of their infant's life.

the end of the third month of life protective patterns either fall to low levels or virtually disappear. This fact suggests the possibility that some endogenous, relatively invariant factors may play a crucial role in the termination of maternal protection. The factors that most readily come to mind include a shift in the hormonal status of the mother and a change in the stimulus characteristics of the maturing infant. The latter might include not only the gradual change in physical size, but a change as well from the so-called "natal" coat color toward the shade of fur characteristic of the adult. Of course, in the final analysis, some coordinated change in both members of the dyad may be involved. One possibility here might be that the infant's increasing consumption of solid foods decreases its nursing demands and thereby facilitates or triggers selected hormonal changes in the mother.

The concordance of data for the onset of these addyadic maternal patterns in rhesus under group living conditions in both field and laboratory and for pigtails under similar laboratory conditions suggests that, under freely interacting and relatively heterogeneous conditions, these maternal patterns assume prominence early in life, usually after the first 2 weeks. When observations are restricted to parts of the day when the group is rather quiescent, as in Itoigawa's observations, the detection of such behaviors may not occur until some weeks later. Similarly, in the variously confined and restricted subjects of the Hansen and Jensen studies, it may well be the early failure of the stimulus environment to evoke outward movement from the mother that accounts for the relative delay in the appearance of maternal restrictiveness, a possibility alluded to by the authors themselves. Some combination of these environmental conditions and the endogenous factors suggested above may account for the rather varied developmental curves for protective behavior that have been described—the more restricted or less complex settings producing a gradual, delayed onset of maternal protection followed by a relatively sharp decline as maturation ensues, whereas the freely interacting, more varied environment results in a very early, intense protective pattern which declines gradually with age.

E. Maternal Rejection

The zealous protective and supportive behavior of the neoparturient mother is gradually interlaced with maternal behavior which represents attempts to discourage rather than enhance the further strengthening or even sustentation of the infant-mother bond. To this rejecting behavior, we give the name "abdyadic." Even the earliest systematic observations of Lashley and Watson (1913) indicated that by the fourth week of the infant's life the otherwise quite protective rhesus mother will occasionally slap her infant, particularly when it attempts to touch or take a desired food from her. By the eighth week, Lashley and Watson report an incident in which the mother, having been given some desired food, then ". . . forced the young monkey out of her arms. He began to eat in a half-hearted way at the corn, but tired of it and approached the mother and reached out for the rolls. This led to severe and continued chastisement. She first pulled him away, holding him on level with her eye and looking fiercely at him all the while. Then cuffed him with a paw and bit him upon the skin of the head and back. The young monkey was forced to keep out of her reach . . . When I made noises designed to frighten them she would go over to the young monkey and place him in position but would not allow him to nurse" (p. 127).

More recent quantitative studies have considerably enlarged upon our knowledge of these rejecting or abdyadic (Kaufman & Rosenblum, 1966) behaviors of the mother. Hansen (1966) has provided a detailed analysis of the appearance and subsequent decline in maternal "punishment" patterns in his restricted laboratory rhesus dyads. He identifies four major components of the pattern, including "mouthing" "clasp-pulling" of the infant's fur, "cuffing" (slapping) the infant and "rejection," that is, preventing the infant's attempts to obtain, maintain, or regain contact with the mother. As a total pattern, these punishment behaviors increase gradually, beginning after the second month and reach peak frequency levels at the fifth month; thereafter they exhibit a gradual decline. Hansen's detailed analysis indicated that whereas the initial form of the behavior was primarily mouthing of the infant, after the fourth month about half of all punishments seen were of the "rejection" type. Unfortunately, Hansen's data do not provide material on the appearance of maternal responses calculated exclusively to prevent nursing. However, as already suggested, such specific "weaning" responses may be the result of a complex of hormonal and external stimulus factors which may or may not be entirely congruent with those mediating the related punishment patterns. Indeed, Hinde *et al.* (1964) indicate that as a result of sore nipples, rhesus mothers may intermittently deter nipple contact (i.e., "wean") even on the first day of the infant's life. Our own observations on both pigtails and bonnets support this finding and indicate that such deterrence may occur for some days without any evidence of other discriminable rejecting or punishment patterns.

Hinde and Spencer-Booth (1967) consider separately the frequency with which mothers "hit or threaten" infants when the latter attempt to gain the nipple and those punishments which occurred at other times. From their data it would appear that rejection outside the context of nipple attainment emerges in about 15% of the mothers by the fifth week of the infant's life and is seen most prominently at about 9 months, at which time half the mothers engage in these punitive responses; thereafter this pattern declines gradually. However, as already mentioned, the long period over which Hinde and Spencer-Booth extended their observations allowed them to detect a subsequent peak level in the frequency of rejecting responses at the end of the first year followed by a gradual increase to quite high levels at the end of the second year. On the other hand, when

Hinde and Spencer-Booth focused their analysis on those rejections directly associated with nursing attempts, they found that these behaviors become "common only after the first two months." When the relative frequency of occurrence across subjects was considered, these behaviors showed a minor peak at approximately 23–24 weeks with their highest level at the end of the first year, at which time they were seen in about 2.5% of the half-minute intervals of observation. There was a sharp decrease after this point, and "rejections became rare once the infants were 18 months old" (p. 175).

Both Hansen and Hinde and Spencer-Booth stress that the relative decline in the absolute levels of rejection after the final portion of the first year are due to a decrease in the number of attempts on the part of the infant to reach the mother and the breast. Indeed, Hansen speaks of a "progressive avoidance of the mother" as being important in influencing the decline. Most important, Hinde and Spencer-Booth indicate that, if one considers the amount of rejection in relation to the actual frequency of attempting to attain the nipple, then a temporary peak in this rejection ratio can be seen at week 20 with a gradual increase by the end of the first year that is then maintained at high levels during the entire second year of life. This sustained level of relative rejection, when compared to the decrease in absolute levels in the second year, supports the contention that fewer attempts to reach the nipple account for the drop in total occurrences. However, it is important to note that this rejection ratio hovers at about only 0.5 in this later period and is less than 0.3 in the earlier, first-year peak. Thus, even at the end of the second year only about half of the attempts to reach the nipple result in rejection. This fact, in turn, lends further credence to the conclusion of Hansen that, in the course of his 15 months of intense observation and even during his subsequent 6 months of viewing his rhesus dyads, the data ". . . did not indicate the existence of a stage of true rejection" (p. 125). The same may also hold during a comparable period for the rhesus dyads observed by Hinde and his colleagues. The conclusion of Hansen and its apparent parallel in the Hinde data is perhaps all the more striking in light of the fact that at peak levels Hansen's females, on the average, were engaging in punishment behaviors (albeit including those both related and not related to nipple contact) at about six times the rate observed by Hinde *et al.*

Jensen and his colleagues (1967a, 1967b, 1968a, 1968b) also describe the onset of maternal abdyadic behaviors in their relatively restricted pigtail dyads. The material on these patterns is difficult to assess precisely, since once again different components of the maternal pattern are presented in several different papers, and the base-lines against which quantitative changes during development are presented differ in each presentation. Thus, "biting," which Jensen *et al.* (1967b) indicate ". . . probably most closely reflects the mothers punitive behavior," (p. 289) is presented "relative to mothers oral mouthing behavior" (p. 6) over time. But, in Jensen *et al.* (1968a) "punitive manipulation," presumably containing biting in addition to other patterns, is presented relative to "total number of mother's manipulations." Again, no apparent differentiation of behaviors directly related to nursing attempts on the part of the infant are presented. Nonetheless, it would appear that, in the most restricted of Jensen's conditions (the "privation environment"), some infants begin receiving maternal punishment by the second week of life. If some of these early punitive behaviors specifically relate to the nursing attempts of the infant, then this would be in keeping with the observations of Hinde's group on rhesus and our own work on pigtails and bonnets. In addition to this early appearance of punishment in some of Jensen's subjects, it is striking that such behavior continued to increase and reached a high plateau at about 6–8 weeks and thereafter was maintained at this level through the 23 weeks, when all observations were concluded (Jensen *et al.*, 1967b). This chronologically early and apparently high-level plateau in Jensen's restricted pigtail dyads contrasts sharply with the development of apparently similar behaviors both in Hansen's somewhat restricted rhesus dyads (which progress steadily to peak levels of punishment at about 5–10 months) and in group-living rhesus dyads studied by Hinde *et al.* It will be remembered that in both cases the appearance of punitive behavior did not occur with any appreciable frequency until after the second month and did not increase to initial peak values until after the fifth or sixth month. Our own data on bonnets and pigtails depart in similar ways from the data presented by Jensen *et al.*

The observations of abdyadic patterns in pigtail and bonnet mothers reveal some striking parallels and differences in comparison to one another and to the related material described above. We have identified a series of patterns through which mothers in these species reject their infants. These primarily include (a) removal of the infant from the mother's body (Fig. 19, a,b) or preventing him from achieving contact (cf. Hansen's "rejecting"); (b) weaning (Fig. 20), including the removal of the nipple from the infant's mouth, pushing or pulling his head from the nipple, and related efforts to prevent further access to the nipple, by such gestures as clasping the infant's

Fig. 19a and b. A pigtail female stripping her 5-month-old infant from her and walking away from the screaming infant.

Fig. 20. A pigtail female pushing her infant's lips from the nipple with the back of her hand.

head beneath the mother's arm; and (c) punitive deterrence (Fig. 21) which primarily includes the biting, mouthing, and manual clasp-pulling described in the work of others. In comparing the bonnets and pigtails, one is immediately struck by the fact that the overall expression of these abdyadic behaviors is considerably lower in bonnet than in pigtail mothers. This species difference is reflected in the comparison of Figs. 22 and 23. Indeed, the overall rates of abdyadic patterns in bonnets is so low that, except for the brief and slight increase in the fourth month, little in the way of developmental trends can be identified. These overall low levels in bonnets are in turn reflected in each of the component responses: if we consider a moderate level of frequency to be one occurrence per 1000

Fig. 21. A pigtail female punishing her infant by biting his hand and pulling back on his fur.

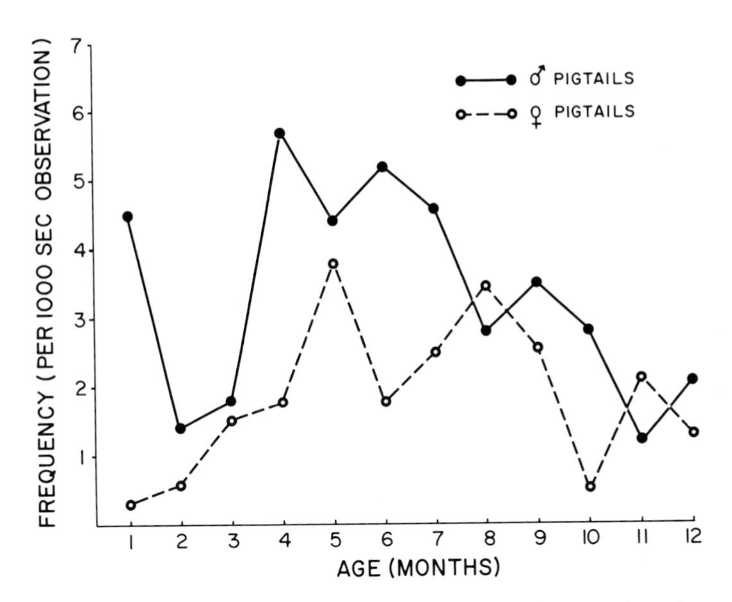

Fig. 22. The median frequency of punishment patterns by pigtail mothers toward their male and female infants during the first year of life.

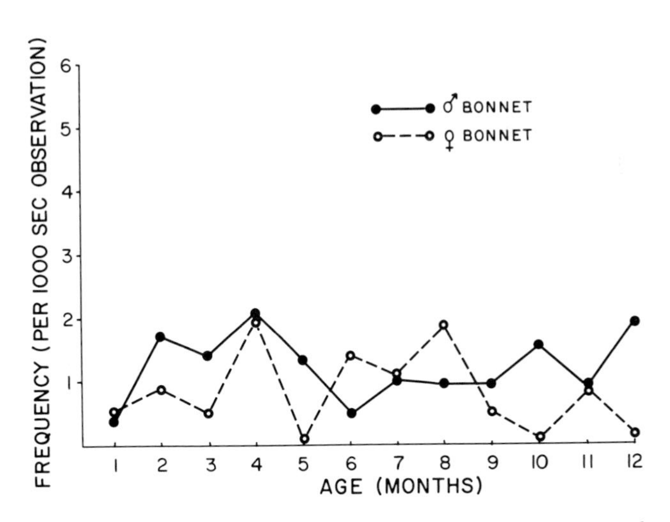

Fig. 23. The median frequency of punishment patterns by bonnet mothers toward their male and female infants during the first year of life.

seconds of observations, then during only the fourth and eighth months did more than half the bonnet mothers show this level of punitive deterrence and in only one month did even a third of these subjects show weaning behavior at this rate. (Infant removal and prevention of contact were at relatively low levels in both species.)

An examination of the pigtail data on the other hand, reveals that considerably higher levels of abdyadic responses are characteristic of pigtails. As seen in Fig. 22, the general curve for the total pattern, after an initial high level due primarily to weaning (one-third of the mothers showed weaning in the first month), indicates that peak levels of abdyadic behaviors are reached in the fourth through seventh month, with a decline thereafter. If we contrast these data on the pigtail with those on the bonnet in terms of moderate frequencies of occurrence, then we see that in each of the months from 4 through 9, 67–88% of pigtail mothers showed moderate or greater levels of punitive deterrence and, in three of the first 6 months, one-third or more also showed similar levels of weaning. In addition, if one looks at the relative number of mothers in each species manifesting high levels of all types of abdyadic behavior, (i.e., the occurrence of one or more abdyadic behaviors of any type per 5 minutes of observation) then it is clear that such behavior was rare in bonnets and relatively frequent in pigtails. In only the third and fourth months did as many as 25% of the bonnet mothers show high rejection frequencies; in 5 of the first 12 months, no bonnet mother at all showed this much rejection of their infants. On the other hand, in the fourth through the ninth months, 44–67% of the pigtail mothers showed high levels of rejection, and in only 1 month of the first year did the pigtail mothers fail to show this active rejection behavior. Moreover, not only was punitive behavior more frequent in pigtails, but the intensity of such punishment as reflected in the duration of each occurrence was consistently greater in pigtails than in bonnets. The mean duration of punishment decreased over each quarter of the first year in both species, but in each successive quarter averaged 9.9, 6.8, 5.7, and 5.3 seconds per rejection in pigtails and only 6.7, 6.6, 3.9, and 3.4 seconds per rejection in bonnets.

These data on the development of abdyadic behavior in pigtails are essentially in keeping with the data presented by both Hansen and Hinde for rhesus, but contrast with the data Jensen *et al.* presented for their environmentally restricted pigtail dyads.

Finally, it should be pointed out that, as reported previously (Kaufman & Rosenblum, 1969), the decrease in total abdyadic behavior after the seventh month in pigtails can be explained in terms of a decrease in weaning behavior rather than a decrease in the total pattern. Punitive deterrence itself continues to decline slowly until the end of the first year. Thus, although at least half of the pigtail mothers manifested moderate or higher levels of punitive deterrence in the ninth through the twelfth month, none was showing similar levels of weaning by the time their infants reached that age.

As in the case of the addyadic patterns described earlier, there is considerable congruence in most of the data on the appearance of abdyadic patterns in the several macaque species studied. Both Hansen's and Hinde's data on the rhesus as well as our own results on pigtails indicate a sustained high frequency of appearance of these patterns in the second quarter of the first year. Even the bonnets, in whom rejection patterns never reach even the moderate levels displayed by rhesus and pigtails, manifest their peak in the fourth month of the infant's life. With respect to the possibility that maternal responses may differ toward male and female infants it may be of importance to note that Hansen's curves for rejection of his subjects (all males) coincide largely with that observed for male pigtail infants in our own studies.

Unfortunately, we lack systematic data on the appearance of abdyadic patterns in Japanese macaques. We must, however, consider the intriguing data of Jensen and his associates on their restrictively reared pigtail dyads. As we have now seen, these pigtails showed a delayed onset of maternal restrictions and somewhat early initiation of maternal rejection. In an attempt to place Jensen's material into some perspective, one may hypothesize that under living conditions of relatively severe restriction there is less social and inanimate "danger" and, furthermore, that infants are less stimulated to leave their mother, hence there is less early restriction. Moreover, if, as Hansen and Hinde have suggested, rejection is in part a function of developmentally inappropriate levels of contact by an infant, then an alternative hypothesis can be advanced, namely, that unusually high levels of contact in the young infant as may occur under relative privation conditions may serve to stimulate early onset of maternal abdyadic patterns. It should be pointed out that the first of these hypotheses is supported by the data of Jensen *et al.* (1968b) in the comparison of the dyadic patterns in their "privation" and "rich" environments. However, the second hypothesis runs directly counter to their conclusion in the same study regarding the behavior of the mother under these two environmental conditions. After indicating that no differences in maternal "disciplining behavior" appeared in their two rearing environments, they conclude that "environment does not affect the basic nature of the mother's role" (p. 261), and that their data, ". . . do not indicate that mothers in either environment respond differently to the qualitative differences in behavior that the infants showed . . ." (p. 262). There can be no doubt that the Jensen data support their conclusions within the context of the conditions they studied. However, it seems likely that even their "rich" environment was in fact restricted along several social and physical dimensions as compared to the conditions which obtained in other studies of macaques. Consequently, their conditions may represent an inadequate range upon which to base general conclusions regarding maternal responsiveness.

It must surely be obvious that this examination of existing material on the quantitative dimensions of the changing relationships of infant and mother macaques in the first year or two of life has necessarily been narrow in the scope of the behaviors considered. To be sure, a wide variety of other maternal activities at each stage has been studied to some degree at least. Thus, we know that the means by which the mother supports her clinging infant changes over time in the direction of decreasing physical support of the infant. We know also that specific elements of maternal caretaking change over time. Grooming of the infant, for example, falls quite noticeably toward the end of the second month of life but remains in evidence for sometime thereafter, perhaps indefinitely (Hinde & Spencer-Booth, 1967). In similar fashion, we have not considered here the developing repertoire of behaviors observed in the infant itself. Much could be said, for example, of exploratory activities and various forms of social and inanimate object-play, all of which increase steadily in the first 8-10 months of life (Kaufman & Rosenblum, 1969) and remain at relatively high levels on into early adulthood. Certainly the development of eating and drinking patterns, sexual behavior, and the changing expression of emotionality are relevant also to a consideration of the unfolding pattern of mother-infant relations in macaques. However, it has been the purpose of this portion of the present chapter to consider in some detail the major dimensions of mother-infant interaction which most readily characterize the changes in their relationship during the earliest period of the infant's life. It may be noted in conclusion that the variation evident in these major thematic responses indicates that the interrelationships between more discrete patterns mentioned above will undoubtedly take many years of additional work to untangle.

F. Discussion

What general themes regarding the changing mother-infant relationship are evident from the data on the four species studied in detail thus far? First it is our judgment that the species variable is not as significant in its influence on the course of the mother-infant relationship as one might expect. The *observed* temporal course of various aspects of maternal and infant behavior seem more a function of environmental conditions and secondarily the methodology of observation than the result of apparently phylogenetically determined differences among the species. Although each species is certainly not identical to the other, it seems reasonable to suggest that, given relatively moderate degrees of environmental complexity or "danger," rather similar overall patterns of maternal behavior are exhibited by rhesus, pigtails and bonnets. Under conditions of extremely marked privation, a rather different trend was observed in pigtails, as was also the case when rhesus were observed under rather constricted conditions and when the data were further confounded by an observational procedure which directly influences the recorded dimensions of mother-infant interaction (Hansen, 1966). A similar variation in the pattern was observed in the Japanese macaque when a period of dyadic and group quiescence was chosen for observation, even though the total environment was quite rich and varied.

In an attempt to look past these variations in each study, but utilizing a number of points from each, we can suggest an hypothesis, depicted graphically in Fig. 24, regarding the changing relationship of mother and infant macaques during the first 2 years of life. Taking the time spent in contact with the mother and the time spent maximumly separated from the mother as the counterposed measures reflecting the level of attachment-independence, in the first days of life, the degree of environmental complexity or danger does not markedly influence the dyadic relationship, that is, the pair remains closely attached. Thereafter, however, if an infant matures under conditions of low complexity, the increase in the independent functioning of the dyadic members occurs slowly. If it matures under conditions of moderate complexity, then there is a facilitation of independent functioning, although at high levels of complexity there may well be a contraction of the dyad, perhaps mutually dependent upon the infant and maternal addyadic responses. With increasing age, this U-shaped relationship between complexity and attachment gradually flattens, with increased complexity even at somewhat high levels facilitating greater levels of independent function when compared with more restricted situations.

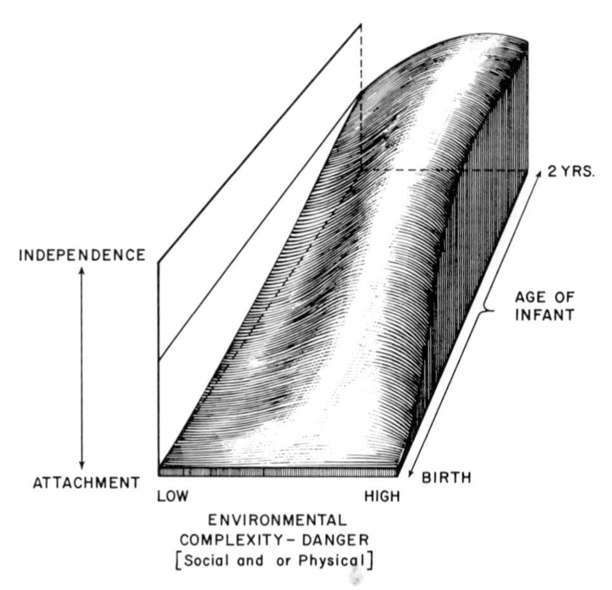

Fig. 24. A theoretical curve depicting the relationship between environmental complexity and the development of infant independence in macaques.

IV. Maternal Rejection and Infant Attachment

We turn now to a consideration of an hypothesis which has received considerable attention in the recent literature on macaque development. This hypothesis suggests that the rejecting and punitive responses of the mother toward her infant are perhaps the primary factor accounting for the growth of infant independence, at least as that independence is defined above. Thus, Hansen (1966), for example, states "One of the primary functions that the mother monkey served was seen in her contribution to the gradual, but definite, emancipation of her infant. Although this process was aided and abetted by the developing curiosity of the infants to the outer world, this release from maternal bondage was achieved in considerable part by responses of punishment and rejection . . ." (p. 122). Jensen *et al.* (1968a) similarly identified the greater frequency of rejection responses toward male as compared with female pigtail infants, as the most likely factor producing the higher levels of independent functioning these males achieved. Mitchell (1968) also observed high levels of independent functioning by 6 months of age in rhesus males compared with rhesus females and also recorded a greater incidence of maternal rejection of male infants in months 3–6 than of female infants (however, absolute levels of rejection, which averaged higher for males than females throughout the 6-month observation period, did not differ statistically). Likewise, Hinde and Spencer-Booth (1967, 1968), after a detailed consideration of rejection patterns in relation both to time in contact with and at some distance from mother, conclude that ". . . after the first few weeks at any rate, the mother plays a large role in the increasing independence of the infant" (1967, p. 190). More specifically, the data of Hinde and Spencer-Booth indicate that, when the group as a whole is considered over time, there is a consistent positive correlation between time off the mother and the relative frequency of maternal rejections. That is, time off the mother and the ratio of rejections to the total frequency of an infant and mother coming together, reaches a significant correlation coefficient of +.69 after week 20, by which time rejections are very much in evidence. These authors stress that the absolute frequency of rejections become negatively correlated with time off the mother after this age due to the fact the infant is avoiding contact with the mother, a point also suggested by Hansen (1966) regarding a similar drop in punitive responses in his data. It is because of this "adjustment" on the part of the infant that relative rather than absolute frequency of rejection is considered most pertinent by Hinde and Spencer-Booth. It should be added that there is also an indication in the data of Hinde and Spencer-Booth that after the fourth week of life the relative frequency of rejections is positively related to the time which the infant spends more than 2 feet away from the mother.

It is evident from these several sources of data ". . . that the importance of the mother's role in facilitating the independence of the infant has been underestimated" (1967, p. 194). However, there seems some indication, at least, that a reverse error in the direction of overestimating the importance of the mother's rejection patterns on the infant's independence may be occurring. Kaufmann (1966), for example, drew this conclusion regarding the development of infant independence after his study of free-ranging rhesus: "On Cayo Santiago rejection of infants seemed insignificant compared to the infant's interest in other monkeys, especially other infants" (p. 25). He suggests that Hansen's emphasis on rejection patterns may have been due to an accentuation of such patterns under the constricted laboratory conditions in which he worked, a possibility that Hansen himself acknowledges.

Other data from the recent literature also raise serious questions regarding the influence of maternal rejection on the development of independence in infants. Lindburg (1969), for example, indicates that normal rhesus infants were consistently rejected or punished by mothers more than were thalidomide-deformed infants, yet body contacts with mother were higher in the normal group throughout the first 6 months of life. Other measures of the infants behavior did not suggest that it was the other infant's deformity that reduced their contact. Similarly, Spencer-Booth (1968), in reporting the development of mother-infant relations in a pair of rhesus twins compared to singletons, indicates that early in life the twins and singletons were with their mothers equivalent amounts of time, yet the singletons were rejected more than the twins.

Jensen and his colleagues (1968a), despite their suggestions concerning the role played by higher levels of maternal punishment toward males in inducing greater male independence, provide some contradictory indications in another aspect of their work. In comparing the impact of environments of different complexity on the growth of independence, their study indicated significantly greater proportions of time spent off the mother by infants in the complex environment even though, ". . . mothers in the two environments did not differ in total number of behaviors directed toward their infants at any point of the development period studied. Also, there were no differences in caring, protecting, or disciplining behavior, such as grooming, restraining, retrieving, cradling, or punishing" (p. 262).

Perhaps the most striking piece of evidence suggesting that maternal punishment may not always play the primary role of producing infant independence is presented in the work of Harlow and his students on the response of infants born to the so-called "motherless-mothers," that is females hand-raised in the Wisconsin laboratory in extreme social isolation. When these females were finally bred they were frequently abusive and punitive toward their newborn infants. However, Arling and Harlow (1967) indicate that ". . . even though an infant was the victim of frequent rebuffs and violent attacks, he persisted in his attempts to gain contact with the mother during each observation session throughout the study (240 days)" (p. 372), and furthermore, that by the fourth month, "The perserverance of the MM (motherless-mother) infant in seeking contact of any sort was partially reflected in the significantly greater number of manual explorations and clasps of the mother by MM infants than by FM (feral reared mothers) infants . . ." (p. 373). These results are in keeping with the conclusion presented in one of the earlier Wisconsin papers on these abusive mothers and their offspring: "A surprising phenomenon was the universally persisting attempts by infants to attach to the mother's body regardless of neglect or physical punishment" (Seay, Alexander, & Harlow, 1964, p. 353). A study of the influence of rejecting cloth-mother surrogate on infant contact carried out by Harlow and the present author some years ago (Rosenblum & Harlow, 1963) suggested a similar finding of "rejection" being responded to by ardent and sustained efforts to cling to the rejecting mother.

The studies of Hinde and Spencer-Booth (1967, 1968), although containing the most reasoned data analysis in support of the maternal-rejection hypothesis, also present some intriguing alternative possibilities. Considering the data on the group studied as a whole, median levels of the rejection ratio upon which Hinde and his colleagues place considerable importance never exceeded about .35 during the first year, indicating that only a third or less of all attempts at reestablishing contact with mother resulted in rejections. Moreover, even throughout the second year, the rejection ratio remained at about .5. Thus, mothers showed these rather ambivalent rejection ratios when their infant's time off increased most dramatically, that is, during the first year, and continued rejecting at these levels when the time off the mother rose progressively to 100%, that is, by the end of the second year. At the very least, these data raise some interesting questions regarding the principles of intermittent reinforcement as they may operate in this critical area of development.

In considering the role of maternal rejection as it affects infant independence, it is of pertinence to note that Hinde and Spencer-Booth report that as the proportion of possible maternal rejections increases (the rejection ratio) the proportion of approaches within the dyad accounted for by the infant increases, and the proportion of "leavings" (i.e., departures) shown by the infant decreases. Whereas this latter correlation may be interpreted as indicating that the more the infant approaches or fails to leave the more likely he is to be rejected, it is feasible to consider the opposite interpretation of the same statistics, namely, the more an infant is rejected, the more likely he is to attempt to attain and maintain contact with the mother. This latter interpretation, of course, might be applied also to the fact that the correlation between the amount of rejection and total time off, although initially positive, becomes negative after week 20. In other words, it is not only that infants stay off more after this age and that as a result of this "adjustment" receive less rejection, but that as an infant is rejected more he stays off less of the time.

However, it is when the analyses of Hinde and Spencer-Booth turn to a consideration of the effect of the maternal rejection behavior within individual dyads that a more pronounced ambiguity of interpretation than that just considered presents itself. The authors carried out the same series of correlations between various measures within each dyad over varying ages as they had employed initially on group median data over approximately the same age periods. In other words, correlations were initially calculated using group median scores for rejection, time off mother, etc., across successive age periods; correlations were subsequently calculated within each dyad separately, using the successive changes in the scores for each behavior at each age within a single dyad, and then averaging the separate correlations obtained. After these analyses, the authors conclude, "Turning to those pairs of measures whose relationships indicate the nature of differences between mother-infant pairs, we see that time off is not significantly correlated with the absolute or relative frequency of rejections at any age. Thus, both differences between mothers and differences between infants play a part in producing the differences in time-off observed" (Hinde and Spencer-Booth, 1968, p. 189). In light of the alternative interpretation which we offered above, it is interesting that, on an individual dyad basis, most correlations between the level of both absolute and relative rejection with time off the mother were negative after 20 weeks of age, leading Hinde and Spencer-Booth to conclude, ". . . that infants which are often rejected are those which tend to spend time off their mothers. . . ." (1967, p. 193).

It should be emphasized at this point that it is not our intention to imply that maternal rejection is completely irrelevant to the development of infant independence. It would be strange indeed if these dramatic maternal patterns did not have an impact on the infant's relationship with the mother. What we wish to emphasize, however, is that the effect of rejection may, at the very least, vary over time both in and between dyads (as Hinde and Spencer-Booth have themselves indicated) and most significantly, may in many instances actually operate in a manner exactly opposite to that hypothesized by several of the authors discussed, including Hinde, Hansen, and Jensen. That is, it is our contention that in much the same way that highly intense stimulation in the environment may retard rather than facilitate independent functioning, maternal rejection may actually motivate the infant toward initially higher levels of contact and association. The mother in essence remains the primary source of fear reduction, and motivation to contact her when frightened or upset is elevated even when she herself, or some stimulus closely associated with her, is the source of the fear (Rosenblum & Harlow, 1963). This does not mean that the mother cannot remove or drive away her young infant for any given period of time. Rather, the question here is what impact such rejection has on the infant's behavior toward the mother during the intervals when overt rejections are not immediately present. Our own analyses of the pigtail and bonnet data we have obtained cannot offer more than some additional suggestive evidence. First, it is obvious that although the data illustrated in Figs. 22 and 23 indicate a striking difference in the frequency of rejection patterns between pigtails and bonnets, an examination of Figures 6 through 9 indicate, at the very least, no concomitant difference in the ontogeny of contact or separation patterns in these species. Within the pigtails themselves, in which rejections were quite prominent, an analysis of the role of rejection within individual dyads further illustrates the complexity of this problem. An examination of Table I reveals that during the months in which rejections were at their peak in this species (months 4–8), the infants receiving

Table I
MATERNAL REJECTION AND INFANT ATTACHMENT IN PIGTAILS (*Macaca nemestrina*)

Contact or Maximum Separation		Months of age				
		4	5	6	7	8
Group median:	Con	44.1	43.8	45.3	27.2	24.4
	Sep	6.0	8.3	5.8	11.3	16.0
Most rejected infant each month	Con	51.1	18.5	45.3	29.3	28.6
	Sep	2.4	16.1	0.0	20.4	15.7
Least rejected infant each month	Con	51.9	40.3	94.9	43.8	11.5
	Sep	8.8	8.3	22.2	6.2	31.9
Most rejected infant (mo. 4–8)	Con	51.1	60.9	35.4	25.2	28.6
	Sep	2.4	0.9	5.8	10.5	15.7
Least rejected infant (mo. 4–8)	Con	44.9	68.3	40.7	20.6	9.6
	Sep	3.4	3.5	5.9	12.2	19.4
Most rejected infant previous month	Con	51.1	60.9	71.3	43.8	38.7
	Sep	2.4	0.9	5.1	6.2	12.9
Least rejected infant previous month	Con	5.0	54.4	11.7	20.6	11.5
	Sep	19.9	6.1	22.2	12.2	31.9

the highest frequency of rejection for that particular month of age, fell below the group medians in maternal contact only once; similarly, the infants receiving the lowest frequency of rejections were above the median only three times. With regard to maximum separations from the mother, the most rejected infants in each month are above the median twice whereas the least rejected are below the median separation only once. An examination of the records for the infant that received the highest and the infant that received the lowest total amount of rejection from the fourth through eighth month age period reveals about the same irregular pattern, except that the most rejected infant was consistently below the group median in maximum separation scores. Finally, in terms of the sequelae of maternal rejection, it might be expected from the hypothesis that rejection produces independence, that if an infant receives a high degree of rejection in a given month, an examination of his subsequent responses to the mother in the following month will show a change over what might otherwise have been expected. During the month following a period of harsh rejection he would have had time to make his "adjustment" toward staying away from the mother to avoid additional punishments. Moreover, an examination of these data from the month following high rejection might illuminate the hypothesis that the infant on the mother most at a given age is the infant punished most at that age. Figure 25 presents the data from months 3–8 in pigtails on a simple composite index of attachment-independence, that is, the score for maternal contact in a given month divided by the comparable score for maximum separation. The curves show the median index for the pigtail group as a whole at these ages and those comprising the scores drawn from the infants most and least rejected in the immediately preceding month. That is, the first point in the open circle curve, i.e., month 3, was that score obtained by that infant in the group most rejected in month 2, the point at month 4 taken from the infant most rejected in month 3, etc. A similar method was used to derive the "least-rejection" curve. The separate data on contact and maximum separation for these groups are also presented in Table I. It can be seen that those infants most rejected in a given month have higher contact scores and lower separations than the average for the group in the following month. On the other hand, those least rejected in a given month have generally lower contact and consistently higher maximum separation scores than the average of the group in the month that followed. Once again, it is obvious that these results are merely suggestive, and ulti-

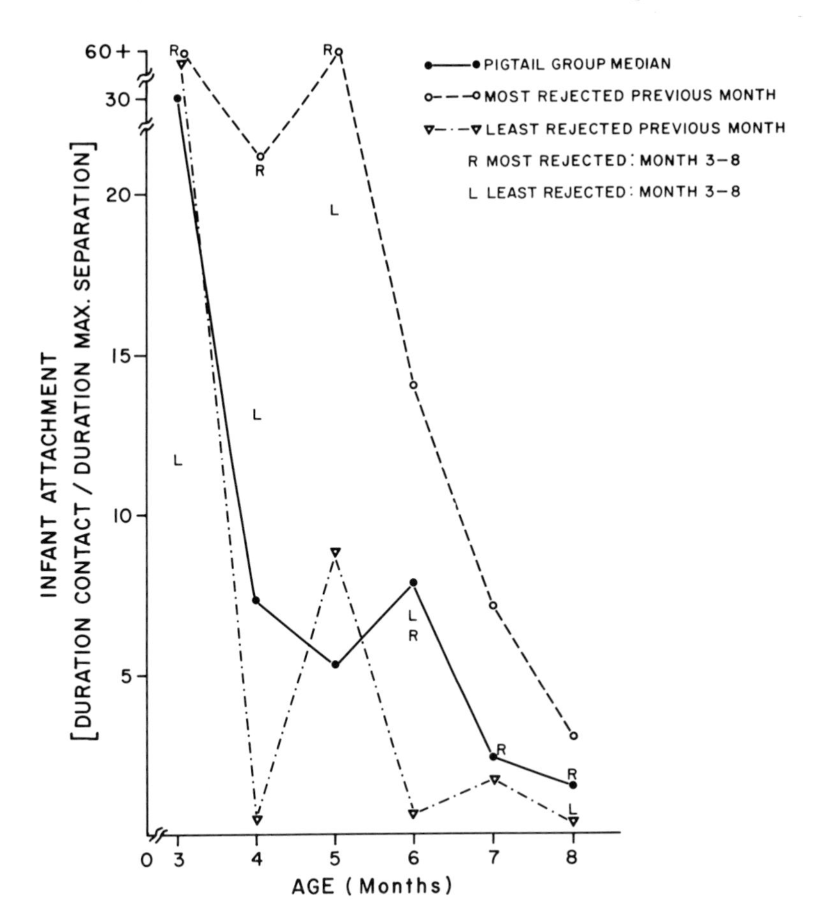

Fig. 25. A comparison of an index of infant attachment in the pigtail group as a whole, and as a function of maternal rejection.

mately will be understood only in the context of a store of information vastly increased over that which we now have on the operation of a host of variables. It is clear that current studies not only are incapable of controlling many pertinent variables, but in all likelihood have not even begun to identify many important ones.

These questions and alternatives are not raised in order to resist productive formulations and hypotheses even when these are quite speculative. They have been raised with the feeling that, in this exceedingly bewildering area, it is best not to close doors in front of us, but only after we have actually passed through them.

References

Arling, G. L., & Harlow, H. F. Effects of social deprivation on maternal behavior of rhesus monkeys. *Journal of Comparative and Physiological Psychology*, 1967, **64**, 371-378.

Chamove, A., Harlow, H. F., & Mitchell, G. D. Sex differences in the infant-directed behavior of a preadolescent rhesus monkeys. *Child Development*, 1967, **38**, 329-335.

Chevalier-Skolnikoff, Suzanne. The ontogeny of communication in *Macaca speciosa.* Paper presented at the annual meeting of the American Anthropological Association, Seattle, 1968.

Cross, H. A., & Harlow, H. F. Observation of infant monkeys by female monkeys. *Perceptual and Motor Skills*, 1963, **16**, 11-15.

Hansen, E. W. The dvelopment of maternal and infant behavior in the rhesus monkey. *Behaviour*, 1966, **27**, 107-149.

Harlow, H. F., Harlow, Margaret K., & Hansen, E. W. The maternal affectional system of rhesus monkeys. In Harriet L. Rheingold (Ed.), *Maternal behavior in mammals*. New York: Wiley, 1963. Pp. 254-281.

Hinde, R. A. Rhesus monkey aunts. In B. M. Foss (Ed.), *Determinants of infant behaviour*. Vol. 3. New York: Wiley, 1965. Pp. 67-71.

Hinde. R. A. 1969. Personal communication.

Hinde, R. A., Rowell, T. E., & Spencer-Booth, Yvette. Behaviour of socially living rhesus monkeys in their first six months. *Proceedings of The Zoological Society of London*, 1964, **143**, 609-649.

Hinde, R. A., & Spencer-Booth, Yvette. The behaviour of socially living rhesus monkeys in their first two and half years. *Animal Behaviour*, 1967, **15**, 169-196.

Hinde, R. A., & Spencer-Booth, Yvette. The study of mother-infant interaction in captive group-living rhesus monkeys. *Proceedings of The Royal Society, Series B*, 1968, **169**, 177-201.

Hinde, R. A., Spencer-Booth, Yvette, & Bruce, M. Effects of 6-day maternal deprivation on rhesus monkey infants. *Nature* (*London*), 1966, **210**, 1021-1023.

Itoigawa, N. Early maternal influence upon infant behavior in a troop of Japanese monkeys. In C. R. Carpenter (Ed.), *Social regulatory mechanisms of primates.* Philadelphia: Univ. of Pennsylvania Press, 1971.

Jensen, G. D., Bobbitt, Ruth A. Changing parturition time in monkeys (Macaca nemestrina) from night to day. *Laboratory Animal Care*, 1967, **17**, 379-381.

Jensen, G. D., Bobbitt, Ruth A., & Gordon, Betty N. The development of mutual independence in mother-infant pigtailed monkeys, Macaca nemestrina. In S. A. Altmann (Ed.), *Social communication among primates*. Chicago, Ill.: Univ. of Chicago Press, 1967. Pp. 43-53. (a)

Jensen, G. D., Bobbitt, Ruth A., & Gordon, Betty N. Sex differences in social interaction between infant monkeys and their mothers. In J. Wortis (Ed.), *Recent advances in biological psychiatry*. New York: Plenum, 1967. Pp. 283-298. (b)

Jensen, G. D., Bobbitt, Ruth A., & Gordon, Betty N. Sex differences in the development of independence of infant monkeys. *Behaviour*, 1968, **30**, 1-14. (a)

Jensen, G. D., Bobbitt, Ruth A., & Gordon, Betty N. Effects of environment on the relationship between mother and infant pigtailed monkeys (*Macaca nemestrina*). *Journal of Comparative and Physiological Psychology*, 1968, **66**, 259-263. (b)

Kaufman, I. C., & Rosenblum, L. A. A behavioral taxonomy for *M. nemestrina* and *M. radiata:* Based on longitudinal observations of family groups in the laboratory. *Primates*, 1966, **7**, 205-258.

Kaufman, I. C., & Rosenblum, L. A. The waning of the mother-infant bond in two species of macaque. In B. M. Foss (Ed.), *Determinants of infant behaviour.* Vol. 4. London: Methuen, 1969. Pp. 41-59.

Kaufmann, J. H. Behavior of infant rhesus monkeys and their mothers in a free-ranging band. *Zoologica*, 1966, **51**, 17-27.

Lahiri, R. K., & Southwick, C. H. Parental care in *Macaca sylvana. Folia Primatologica*, 1966, **4**, 257-264.

Lashley, K. S., & Watson, J. B. Notes on the development of a young monkey. *Journal of Animal Behavior,* 1913, **3**, 114-139.

Lindburg, D. G. Behavior of infant rhesus monkeys with thalidomide-induced malformations: A pilot study. *Psychonomic Science*, 1969, **15**, 55-56.

Mitchell, G. D. Attachment differences in male and female infant monkeys. *Child Development*, 1968, **39**, 611-620.

Mitchell, G. D. Paternalistic behavior in primates. *Psychological Bulletin*, 1969, **71**, 399-417.

Mowbray, J. B., & Cadell, T. E. Early behavior patterns in rhesus monkeys. *Journal of Comparative and Physiological Psychology*, 1962, **55**, 350-357.

Raphael, D. Uncle rhesus, auntie pachyderm, and mom: All sorts and kinds of mothering. *Perspectives in Biology and Medicine*, 1969, **12**, 290-297.

Rosenblum, L. A., & Harlow, H. F. Generalization of affectional responses in rhesus monkeys. *Perceptual and Motor Skills*, 1963, **16**, 561-564.

Rosenblum, L. A., & Kaufman, I. C. Laboratory observations of early mother-infant relations in pigtail and bonnet macaques. In S. A. Altmann (Ed.), *Social communication among primates.* Chicago, Ill.: Univ. of Chicago Press, 1967. Pp. 33-41.

Rosenblum, L. A., & Kaufman, I. C. Variations in infant development and response to maternal loss in monkeys. *American Journal of Orthopsychiatry*, 1968, **38**, 418-426.

Rosenblum, L. A., Kaufman, I. C., & Stynes, A. J. Individual distance in two species of macaque. *Animal Behaviour*, 1964, **12**, 338-342.

Seay, B., Alexander, B. K., & Harlow, H. F. Maternal behavior of socially deprived rhesus monkeys. *Journal of Abnormal and Social Psychology*, 1964, **69**, 345-354.

Simonds, P. E. The bonnet macaque in South India. In I. DeVore (Ed.), *Primate behavior.* New York: Holt, 1965. Pp. 175-196.

Simons, R. C., Bobbitt, Ruth A., & Jensen, G. D. An experimental study of mother M. nemestrina's responses to infant vocalizations. *American Zoologist*, 1967, **7**, 112.

Southwick, C. H., Begg, M. A., & Siddiqi, M. R. Rhesus monkeys in North India. In I. DeVore (Ed.), *Primate behavior.* New York: Holt, 1965, Pp. 111-159.

Spencer-Booth, Yvette. The behaviour of twin rhesus monkeys and comparisons with the behaviour of single infants. *Primates*, 1968, **9**, 75-84.

Sugiyama, Y. The ecology of the lion-tailed macaque (*macaca silenus*) (Linnaeus)—a pilot study. *Journal of the Bombay Natural History Society*, 1968, **65**, 1-10.

Tinklepaugh, O. L., & Hartman, K. G. Behavioral aspects of parturition in the monkey (*Macaca rhesus*). *Comparative Psychology*, 1930, **11**, 63-98.

Tinklepaugh, O. L., & Hartman, K. G. Behavior and maternal care of the newborn monkey (*Macaca mulatta*). *Journal of Genetic Psychology*, 1932, **40**, 257-286.

Yang, C. S., Kuo, C. C., Del Favero, J. E., & Alexander, E. R. Care and raising of newborn Taiwan monkeys (*Macaca cyclopis*) for virus studies. *Laboratory Animal Care*, 1968, **18**, 536-543.

CHAPTER 9

IMPRINTING

P. P. G. Bateson

I. Introduction

Many species of birds which are already feathered and active shortly after hatching will respond socially to a surprising range of visual stimuli. They will approach and attempt to nestle against animals that look nothing like adults of their own species and, even more remarkably, will behave in a similar way toward moving boxes, flashing lights and much else besides. As the result of experience with one object, the young bird comes to prefer this to others and eventually may direct its social behavior exclusively toward the familiar object. Thus to take two famous examples, greylag goslings and mallard ducklings hatched in an incubator and raised by hand for several days avoid their natural parents and do their utmost to stay with their human keepers. Pictures of ethologists leading parties of young birds around are so commonplace that the phenomenon has lost some of its novelty—if none of its charm.

The process by which the birds develop a social attachment for a particular object was called "imprinting" by Heinroth (1911) and

Lorenz (1935), because the birds seem to learn the characteristics of the object so quickly and because the effects of the process were thought to last throughout life. The term does, indeed, suggest an instantaneous, irreversible process. In later years, as more evidence became available, these implications were disputed and the term was held to be misleading (e.g., Hinde, 1962; Moltz, 1960); nonetheless it has been retained in the literature. Possibly "imprinting" is too vivid and too well embedded in the language to be dislodged by purely scientific arguments.

The popularity of imprinting both as a phenomenon and as a concept has had two opposed but equally unfortunate effects. On the one hand, inferences about the nature of underlying mechanisms have been uncritically accepted and used to explain many other processes, such as the development of social attachments in man. On the other hand, the evidence has been tainted by popularization and overcritically rejected by many of those whose primary interest is the experimental analysis of learning. It is a curious position—at one extreme, imprinting is used as the basis for wide-ranging generalizations about early learning, and at the other is excluded from any broad statement about "the learning process." A middle view is both possible and, in the present circumstances, desirable. In this chapter I shall attempt to show that while many of the claims made for imprinting have been extravagant, the evidence *does* pose problems which should not simply be brushed aside and which are central to the analysis of behavioral development.

II. Imprinting Procedure

In considering the imprinting process, it is important to realize that the experimental operations by which learning is demonstrated differ in significant ways from classical conditioning, for example, and from discrimination learning in an instrumental situation. Essentially, the procedure involves a period of exposure to a particular stimulus and a subsequent test in which the reactions of the bird to that same stimulus and to a dissimilar stimulus are measured. Generally, if the majority of birds in a group approach the first stimulus rather than the second or follow the first more than the second, they are said to have been imprinted to the first stimulus. However, the birds might have preferred that stimulus from the outset. Obviously, then, the experimental procedure must be counterbalanced: what was the second stimulus for one group must be made the first stimulus for another group, with each stimulus having been matched previously for initial effectiveness in eliciting social behavior. If both groups prefer the stimulus to which they had been first exposed, then the difference between them can be safely attributed to the imprinting procedure.

The crucial points about imprinting are first, that the acquisition of a preference is not directly observed in the way that changes in behavior during the acquisition of a skill, for example, can be observed (cf. Sluckin, 1962, 1964). Second, the body of knowledge about imprinting is still poor enough that frequently a rigorous demonstration of the sources of the difference in behavior between two groups is needed. The same requirement, of course, should also be imposed on learning phenomena when studied in a relatively unworked context, since the observed change in behavior may not be due to the experimental manipulation but merely to the passage of time. The danger is especially great in studies of learning early in development, since behavioral changes frequently occur whether or not the animal is stimulated in a particular way. For example, the motor-coordination of chicks improves with age so that a 3-day-old bird is better able to peck accurately at grain (Cruze, 1935) and better able as well to follow a moving object (Bateson, 1964b) as compared with a 1-day-old bird. These changes occur whether or not the bird has had an opportunity to peck grain or follow objects beforehand.

The behavior restricted to a specific class of objects by the imprinting procedure is often loosely referred to as "social behavior." Many components of this social behavior occur frequently in other contexts. For example, the characteristic twitter call of domestic chicks can be elicited both by an imprinting stimulus and by the presence of food.

In general, the response measures most often used in imprinting experiments are more restricted and defined more rigorously than is suggested by the term "social behavior." For example, in the case of precocial birds that actively follow the mother (or her surrogate), the time they spend traveling in the same direction as the mother and the distance they maintain in relation to the mother are each recorded. Similarly, in choice tests, the object that is first approached to within a certain distance and the time taken to effect this approach are noted. Debate still continues, however, as to what constitutes the most satisfactory method for measuring the effects of imprinting (e.g., Guhl & Fischer, 1969). Relevant to this debate is the work of Connolly (1968) which indicates that initial choice provides a more sensitive index of imprinting in domestic chicks than amount of time spent following a familiar or unfamiliar object when both are simultaneously visible and moving.

Whether familiar and unfamiliar objects are presented simultaneously or successively depends on the question in which the experimenter is interested. However, on the face of it, simultaneous presentation would seem to provide the more sensitive measure of preference and successive presentation the more rigorous test. In most simultaneous choice tests, once the bird starts to approach one stimulus, it of necessity moves away from the other stimulus. The situation is therefore unstable. Furthermore, the presence of an unfamiliar stimulus may disrupt the response to a familiar stimulus. But there is a disadvantage as well to the procedure of successive presentation, namely, its pronounced order effects. For example, Fischer, Campbell, and Davis (1965) found that, all else being equal, domestic chicks tend to follow the second stimulus presented to them more often than they do the first stimulus. Despite these reservations, no fundamental difficulties are raised by the various options open to the experimenter, providing he realizes that different methods will give slightly different answers, and that there is no absolute measure of imprinting.

In contrast to the test phase of an imprinting experiment, what might be called the "training phase" seems straightforward enough, insofar as it involves simply the repeated presentation of a single stimulus. However, the likelihood of imprinting taking place does, of course, depend on many conditions. For example, just how much the bird learns about the stimulus depends on the characteristics of the stimulus and on the length of time the bird is exposed to the stimulus. Not surprisingly, the bird's responsiveness also plays a major role in determining how much learning actually takes place; indeed, this is so important that many investigators have treated measures of responsiveness, such as the amount of time spent following during training, as measures of imprinting. The responsiveness of the bird, in turn, is affected by a host of other factors, some of which can be traced to events outside the animal and some of which cannot. Since a discussion of the imprinting process must hinge partly on the independent variables that are known to affect it, it is worth dealing with these variables in greater detail.

A. Stimulus Characteristics

Domestic chicks and domestic ducklings, to mention two species customarily used in imprinting experiments, have relatively unstructured social preferences at hatching—at least in the visual modality.

Nevertheless, it would be a mistake to regard them as *tabulae rasae*, since even in these animals, some stimuli are more effective than others in eliciting social behavior. At one time, movement was regarded as essential in "releasing" the following response and hence in initiating the imprinting process. It is now clear, however, that the effectiveness of the many visual stimuli used in the imprinting situation depends on such properties as their size and shape, as well as on the angle they subtend and the intensity and wavelength of light they reflect (review in Bateson, 1966). Moreover, the rates at which these variables change are also important—hence the effectiveness of movement and flicker.

Under the circumstances, it would be difficult to provide a precise and detailed delineation of the most effective visual stimulus for imprinting. In fact, the best that we can apparently offer in this regard would be to say that the more conspicuous a stimulus is to the human eye, the more effective it is in the imprinting situation (Bateson, 1964c). This is, admittedly, a gross way of characterizing the visual stimuli. In particular, it does not take into account the size and wavelength ranges which are most effective in inducing responses (see Bateson, 1966). But in searching for a more refined delineation, we must not lose sight of the fact that when a bird responds to an imprinting object, it is responding to a pattern of stimulation, and that, therefore, any characterization of the most effective stimulus must be cast in terms of clusters of physical variables.

Visual stimuli which are most effective in eliciting approach responses in experimentally naive birds do not necessarily have characteristics which are easily learned. For example, the highly variable pattern coming from a television screen would probably be very effective in eliciting approach in naive day-old chicks, but its characteristics would be extremely difficult to learn. On the other hand, a repetitive form of stimulation, such as a rotating white disc containing a black sector, might not be as attractive as a television screen but might easily be identified after a relatively brief exposure. It is only fair to point out that these matters have hardly been touched on in experimental studies.

Although there is evidence to suggest that young birds learn the characteristics of auditory stimuli played to them shortly after hatching (e.g., Gottlieb, 1965; Klopfer, 1959), and of tactile stimuli with which they have been in contact (Taylor, Sluckin, Hewitt, & Guiton 1967), the great majority of studies have been of visual imprinting. Therefore, the present chapter is devoted primarily to the processes involved in learning characteristics of visual stimuli; only brief consideration will be given to stimuli in other modalities.

One of the effects of intermittent auditory stimulation is probably to alert the bird (cf. Guhl & Fischer, 1969). For example, Pitz and Ross (1961) found that loud bangs caused day-old domestic chicks to follow a visual stimulus more vigorously. Stimulation in other modalities may also have general effects on the bird's alertness. It seems likely, for example, that mild electric shock, which can also cause day-old domestic chicks to follow a visual stimulus more strongly (Kovach & Hess, 1963), operates in the same way. However, not all the interactions between auditory and visual stimuli in their relation to imprinting can be explained in terms of nonspecific effects. Intermittent auditory stimuli presented in the absence of visual stimuli will induce chicks to give the so-called "contentment twitter" and to approach the sound source (e.g., Collias, 1952). Moreover, Gottlieb (1965) found that, in domestic chicks and mallard ducklings, the sounds most effective in eliciting pursuit of a moving visual stimulus are conspecific maternal calls. His results suggest that a highly specific mechanism controlled by particular auditory stimuli is coupled to a parallel mechanism controlled by visual stimuli.

Tactile stimulation also seems to have specific effects on the birds' behavior. A chick or duckling when touched, particularly on the head and back, will emit twitter calls and make characteristic nestling movements; under natural conditions, such movements would bring the bird under the wing or ventral feathers of its parent (e.g., Driver, 1960; Salzen, 1967). Just how auditory or tactile stimuli are involved in the visual imprinting process is not known. However, it would be reasonable to assume that a sound, making a bird more responsive to a visual stimulus, and contact, resulting from movement toward the visual stimulus, would increase the rate at which the visual characteristics were learned.

B. Length of Exposure

The implication of the term "imprinting" is that all aspects of the eliciting stimulus that are going to be learned are stamped in at the same time. Indeed, imprinting has often been conceived as occurring after a single brief exposure or in one "trial." This view was implicitly rejected by Thorpe (1956) when he argued that perception proceeds from the discrimination of gross differences to the discrimination of fine differences. And in their joint and separate writings, Sluckin and Salzen have consistently argued that the ability to discriminate between familiar and unfamiliar objects as a result of imprinting is related to the amount of perceptual experience with the familiar (e.g., Salzen, 1962; Sluckin, 1964; Sluckin & Salzen, 1961). Over the years, considerable evidence has accumulated in support of the view that the longer a bird has been exposed to an imprinting object the better able it is to respond selectively to that object (review in Bateson, 1966; Connolly, 1968; Sluckin, 1964).

Occasionally, attempts have been made to refute the evidence that learning the fine details of the imprinting object is accomplished by "normal conditioning" (e.g., Hess, 1959a). However, the criteria for separating the early phases of learning the characteristics of an object from the later phases have not been made explicit and the assumption that two processes are involved seems gratuitous and unwarranted. Short of a passionate desire to retain the purity of the original concept of imprinting, there seems little justification for postulating two processes. Whether or not the way in which a young bird learns the characteristics of its parent differs from other types of perceptual learning is, of course, a separate issue.

Whatever the mechanisms involved in imprinting, the analogy with the formation of an imprint seems a bad one. It would be more satisfactory to liken it to the painting of a portrait in which the broad outlines are sketched first and the details are then filled in by degrees. This conception correctly implies that the consequences of the process are not all-or-nothing and makes sense of apparently contradictory evidence, particularly in relation to the long-term effects of the process.

III. Activities Connected with Imprinting

Obviously enough, the process of learning requires the active involvement of the animal. However, the concept of imprinting suggests an element of passivity on the part of the animal. It is odd that a concept with such implications should be associated with Konrad Lorenz, who did so much, on other fronts, to attack the notion of the reactive organism and who reinstated the notion of active, appetitive elements in behavior. In contexts other than those in which birds form their social attachments, a general swing away from the push-button view of animal behavior seems to have taken place. Occasionally, however, attempts are still made to account for the development of appetitive behavior in terms of the elaboration of what seem to be simple stimulus-response systems. For example, Schneirla (1965, p. 46) has argued that, initially, precocial birds simply approach appropriate sources of stimulation, and subsequently, the behavior of *seeking* an object in its absence develops as a consequence of previous reinforcement. While there is much to be said for looking at development in these terms, the behavioral organization of the recently hatched chick or duckling seems to be more elaborate than such an explanation would suggest.

When a 12-hour-old mallard duckling, which has been sitting quietly in a dark incubator, is removed and placed in a bare alley at room temperature it soon begins to move about. Before long it starts "distress" calling and it shuffles about in disorientated fashion with its neck extended. If a conspicuous visual stimulus is now presented to the duckling, it turns toward that stimulus and its "distress" calling stops. In many ways, its behavior resembles that of an older duckling that has become separated from its mother and then regains contact with her. Such an observation suggests that, even before they have been imprinted, ducklings will behave in a way that increases the likelihood of their making visual contact with a conspicuous object. If stimuli that are highly effective in the imprinting situation do bring such appetitive behavior to an end and are thereby consummatory, they might be expected to reward the young bird. This is indeed the case. Day-old domestic chicks and wild mallard ducklings taken from a dark incubator will quickly learn to operate a pedal that turns on a flashing light (Bateson & Reese, 1968). Prior exposure to the flashing light, which in itself is a highly effective imprinting stimulus, far from increasing the rate of learning to press a pedal, slows up acquisition.

Now it can be argued that these results could have been produced by any sensory change (cf. Kish, 1966). However, a close correspondence was found between the factors affecting the reinforcing properties of the flashing light and those affecting the elicitation of social behavior by such a stimulus (Bateson & Reese, 1969). Particularly relevant here was the observation that an orange flashing light, approached preferentially by naive day-old chicks in a choice test, was a more effective reinforcer than a green flashing light. Furthermore, Bateson and Reese (1969) showed that, in the course of learning to work for a flashing light, day-old chicks also learned its characteristics and preferred it to a different flashing light when given a choice. So strong are the links between the imprinting situation and the situation in which the flashing light was used as a reinforcer, that it seems valueless to attribute the reinforcing effects to nonspecific stimulus change. Indeed, one can go further and argue that before imprinting occurs the bird will show specific appetitive behavior that is terminated by any one of the broad class of effective imprinting objects.

When a bird is within eye-shot of a conspicuous object, its appetitive behavior will, of course, orient it toward the object. What happens next depends on the species, on age, and on other variables. For example, 18-hour-old mallard ducklings begin to approach an object and upon reaching it may nestle against it or may start a number of other activities such as preening or pecking. Domestic chicks of the same age, having poorer control over their body temperature, are more likely to nestle against the object. If the object then moves away from the bird, its approach response is called "following." Classically, this approach response was given great prominence, but probably it is only one of a number of activities that keeps the young bird within a certain distance of the conspicuous object. What the bird does when it is near the object depends on factors such as temperature, how long it has been deprived of food, and so on. As far as visual imprinting is concerned, the response of greatest functional importance is presumably the response of looking at the stimulus, but there is no evidence that the bird will do this to the exclusion of all else. Such visual input as they experience appears to be embodied in the process of keeping close to the imprinting object.

On commonsense grounds, one would expect that any activity which orients the bird toward the visual characteristics of the object (e.g., approach and following) might be supposed to accelerate the rate at which these visual characteristics are learned. But it is possible that, in imprinting as in other learning situations, an inverted U-shaped relationship exists between responsiveness and rate of learning (cf. Yerkes & Dodson, 1908). Unfortunately, no studies have been concerned explicitly with this possibility. Indeed, analysis has been directed simply toward teasing apart the diverse variables that affect a bird's responsiveness to a particular visual stimulus. Some of these are considered next.

IV. Factors Affecting Responsiveness

Lorenz (1935) drew a vivid analogy between embryological induction and imprinting, and since then the dependence of imprinting on maturational processes has been stressed by many authors. The idea was that endogenous changes, rather than specific forms of experience, are the determinants of both the onset and the termination of the optimal period for imprinting. While a great deal of evidence suggests that imprinting with a novel and conspicuous object usually occurs more readily at one stage of development rather than at another (review in Bateson, 1966), such evidence does not necessarily indicate that endogenous changes determine when that stage occurs. Specific types of experience affecting the behavior of the bird can and indeed do covary with developmental age, and so attempts must be made to break the correlation before a reasonable hypothesis can be advanced. In the case of the optimal period of imprinting, analyses have revealed different causes underlying its onset and termination. Maturational changes, occurring independently of specific experience, have been implicated in the onset. The evidence here comes largely from the work of Gottlieb (1961) who made use of the fact that birds hatch at different stages of embryonic development. This makes it possible to have birds of the same developmental age, but with different amounts of post-hatch experience; equally, it is possible to have birds of the same post-hatch age, but, in having hatched at different stages of embryogeny, of different developmental ages. Gottlieb found that the sensitive period for imprinting in mallard ducklings is not so well marked if age is measured from hatching as it is if age is measured from the beginning of embryonic development. This suggests that experience after hatching has relatively little effect on the onset of the sensitive period.

To what can this increase in sensitivity be attributed? For one thing, locomotor ability improves with age (e.g., Hess, 1959a, 1959b), so the increase could be due to an increase in the ability to orient toward and, if necessary, approach and follow conspicuous objects. Changes in the visual system have also been implicated in the increase in sensitivity. The speed with which the system reacts to a flashing light has been measured by recording evoked responses in the retina and at the optic lobes of Peking ducklings (Paulson, 1965). The time from a flash of light to the first response was found to decline markedly shortly before hatching and, what is more interesting, continued to decline after hatching. In other words, the visual system of the older bird responds more quickly to changes in the environment than does that of the younger bird. Whether the changes on the sensory and motor side provide a sufficient explanation for the endogenous changes involved in onset of the sensitive period cannot be stated at present.

The evidence for growth processes affecting the onset of the sensitive period does not, of course, exclude the possible involvement of environmental events. Indeed, there is evidence to suggest that imprinting occurs at an earlier age in visually stimulated than in visually deprived chicks (e.g., Dimond, 1968; Haywood & Zimmerman, 1964).

For many years the termination of the sensitive period was attributed to the growth of fear occurring independently of specific experience (e.g., Hess, 1959a). To be sure, a great deal of evidence shows that, on the second day after hatching, naive chicks and ducklings are much more likely to avoid a strange moving object than on the first day after hatching (see Bateson, 1966). However, a number of authors have demonstrated that rearing conditions have a marked effect on the age at which avoidance of novel objects first occurs (Guiton, 1959; Moltz & Stettner, 1961; Sluckin & Salzen, 1961). Moreover, Bateson (1964a) has shown that chicks isolated for 3 days following hatching avoided a moving object that resembled in pattern their home pens much less than one that differed in pattern. This suggests that the birds had learned the characteristics of their home pens and avoided objects which they could detect as being different. Thus the sensitive period seems to be brought to an end by a very specific type of experience. Its end, as defined under any given set of experimental conditions, does not of course mark the point at which learning is complete; it merely marks the point at which the animal prefers its familiar environment to the object used by the experimenter for training purposes.

It would seem easy to test the hypothesis that familiarity with a particular input solely determines the end of the sensitive period. In fact, however, it is difficult to deny a bird all visual information, for even darkness constitutes an environment in which the retinal receptors discharge in a characteristic way. Confronted with this difficulty, McDonald (1968) attempted to reduce visual experience in chicks through the injection of sodium pentobarbital during the first 4 days after hatching. When tested on day 5, these chicks showed considerably more social responsiveness to a strange moving object than control birds. Here, of course, there is no evidence for the view that endogenous factors are even partly responsible for the termination of the sensitive period; on the contrary, McDonald's data support the view that specific experience is the primary cause.

In a category different from the factors affecting the sensitive period for imprinting are those that have relatively short-term effects on responsiveness. For example, Polt and Hess (1966) found that domestic chicks given 2 hours of social experience prior to testing followed a moving object more strongly than isolated birds. Similarly, Bateson and Seaburne-May (unpublished data) found that mere exposure to a static light for as little as half an hour increases the speed with which day-old domestic chicks subsequently approach a flashing light. In contrast, stimulation in other modalities reduces responsiveness to visual stimuli. Thus, after gentle stroking, domestic chicks approach a moving object more slowly than unstimulated birds (Graves & Siegel, 1968). Playing "peep-calls" to chicks in a dark incubator for an hour similarly reduces their responsiveness to a flashing light (Bateson & Seaburne-May, unpublished data). Just how these results are to be explained is unclear, but they do make apparent the need to pay as much regard to the animal's state in studies of imprinting as in other studies of learning.

V. Consequences of Imprinting

Before considering some of the theoretical implications of imprinting, it is worth summarizing briefly what happens as a result of the imprinting process. As already mentioned, the range of objects that can elicit social behavior is restricted by imprinting. Thus, when the young bird becomes familiar with one object, the likelihood of it avoiding dissimilar conspicuous objects increases (see Bateson, 1966). We have also seen how the naive bird appears to search for conspicuous objects and how the presentation of such objects can reinforce certain activities. As imprinting takes place, the range of stimuli which terminates the searching behavior of mallard ducklings and domestic chicks is narrowed down to the familiar one (Guiton, 1959; Weidmann, 1958). Furthermore, imprinting restricts the range of stimuli which can be used as reinforcers (Bateson & Reese, 1969; Hoffman, Searle, Toffey, & Kozma, 1966). Precisely because the familiar object has such powerful reinforcing properties, which incidentally, may be enhanced further by its "anxiety-reducing" effect (see Moltz, 1960, 1963), it is not difficult to see how the searching responses which enable the bird to regain contact will be rapidly elaborated.

The long-term effects of imprinting on sexual preferences, a possibility that has attracted so much attention, undoubtedly exist in some species and under certain conditions (see review in de Lannoy, 1967; Immelman, 1969; Klinghammer, 1967). While all birds that show sexual imprinting probably show filial imprinting, the reverse is not true. The sexual performance of the female mallard, for example, is not affected by this early filial preference (Schutz, 1965). Moreover, while filial imprinting does canalize the development of young birds, in the sense that it restricts the range of objects with which they interact socially, Schutz's results suggest that the very early experience determining filial preferences in mallard ducklings is not sufficient to determine sexual preferences. Such preferences are perhaps determined by the performance of precocial sexual responses which, of course, are likely to be directed toward familiar companions.

Controversy continues as to whether imprinting is absolutely irreversible. There are cases, to be sure, of astonishing retention of sexual preferences in the face of considerable sexual experience with other objects (e.g., Immelman, 1969; Lorenz, 1935; Schein, 1963). Criticisms of these cases have largely been off the point, confusing the issue of whether or not birds will *generalize* their responses from the imprinted object with the principal issue of whether or not they *retain* the original preference after experience with other objects. Few systematic attempts have been made to come to grips with the main problem. But since, as we have seen, the amount a bird learns is dependent on the length of time it is exposed to the imprinting object, the stability of the acquired preference is probably a function of length of exposure. Indeed, in those cases in which stable sexual preferences have been demonstrated, the bird's initial experience with the preferred species was never less than a week and often more than a month (e.g., Immelman, 1969; Schein, 1963; Schutz, 1965).

VI. The Nature of the Process

A great deal more heat than light has been generated over whether or not imprinting is a special kind of learning. Much of the controversy stems from the strong stand Lorenz (1935) took when initially characterizing imprinting. As is common in his writings, attractive evidence subtly shades into plausible hypothesis, and it becomes difficult to disentangle observations from assumptions. At the time, though, he was correct in distinguishing imprinting from classical conditioning since the procedures for demonstrating imprinting were then quite different from those involved in associative learning. Subsequently, however, it became apparent that certain procedures for demonstrating the effects of perceptual experience are in principle very similar to the imprinting procedure, involving as they do a period of exposure to a stimulus followed by a test measuring whether the animal has learned the characteristics of the stimulus (e.g., Gibson & Walk, 1956). For this and other reasons, imprinting has been subsumed under the broad category of "perceptual" or "exposure" learning in modern reviews of the subject (e.g., Bateson, 1966; Sluckin, 1964). No adequate counter arguments have been advanced to justify the view that imprinting is a special process of learning, although Lorenz still asserts that his original views have been fully vindicated (e.g., Lorenz, 1970). However, it is worth noting that the distinction between "exposure learning" and "associative learning" implies two underlying mechanisms.

When Sluckin (1964) first proposed that imprinting be regarded as an example of "exposure learning," he wrote that the term ". . . refers unambiguously to the perceptual registration by the organism of the environment to which it is exposed." However, the nature of the process underlying this "perceptual registration" remains vague, and the implication that exposure learning continues regardless of environmental content or motivational context may be misleading. In what precise ways, then, does exposure learning differ from classical conditioning or discrimination learning in an instrumental situation? Certainly not in its outcome, since in each procedure the animal is equally responsive (or unresponsive) to a range of stimuli at the outset, and at the end of training is more responsive to familiar than to unfamiliar stimuli. It is primarily in the conditions necessary for its occurrence that exposure learning, or more specifically imprinting, appears to differ from what may be called associative learning, since, for imprinting to occur, it is not necessary to pair the training stimulus with any other event.

James (1959) showed that chicks could be induced to approach a flashing light in preference to one that was not flashing. He suggested that a flashing light acts as an unconditioned stimulus, with the other characteristics of the light being learned through the process of classical conditioning. As we have seen, the effective element of an imprinting stimulus cannot be characterized simply as "flickering light," but nonetheless the essence of James's view may be correct. It is noteworthy that when Abercrombie and James (1961) placed a stationary object alongside a flashing light, chicks subsequently approached the stationary object in the absence of the light.

The question remains as to whether some property of the visual imprinting stimulus is reinforcing. Here both theoretical and empirical issues are involved and it is necessary to distinguish one from the other. The empirical issue of whether or not a visual stimulus can be used to strengthen a response has been settled: we have already seen, for example, that a flashing light can be used to reinforce the activity of pressing a pedal (Bateson & Reese, 1968, 1969). The theoretical issue is whether one dimension of visual stimulation is necessary for learning the characteristics of another dimension. For example, the increase in visual angle of a striped box consequent upon approach by a young bird might provide the condition necessary for learning that the box was striped. This particular hypothesis, incidentally, does not seem very plausible, since considerable evidence suggests that restrained birds can learn the characteristics of an imprinting object (e.g., Moltz, Rosenblum, & Stettner, 1960; Smith, 1962). Furthermore, Moltz (1963) has presented evidence that Peking ducklings are more likely to follow an object if, during original training, they were confined so that they could only watch the box move *away* than if they could only watch it move toward them (see also Tronick, 1967).

At present it remains an inadequately formulated and inadequately tested hypothesis that an imprinting stimulus consists of two components. The consummatory element, which could act as an unconditioned stimulus, might be that characteristic of the stimulus which makes it conspicuous to the human eye; the initially neutral characteristics of the stimuli may be those special and detailed features that make it different from other conspicuous stimuli. The difficulties involved in testing such a hypothesis lie in finding stimuli which have one set of properties but not the other. Indeed, the essence of any exposure learning situation may be that neutral and significant aspects of stimulation are closely intertwined. However, until further evidence has been obtained, there is little to be gained by arguing strongly as to whether or not the associative/exposure learning distinction implies two separate mechanisms.

VII. Conclusion

The reality of the imprinting process lies somewhere between the views of the enthusiasts and those of the sceptics. On the negative side is the fact that many of the implications stemming from the original concept are simply wrong. The picture so commonly conveyed in popularizations of imprinting is that the image of an object is stamped instantaneously and irreversibly on the brain of the passive young bird. I have attempted to show that this picture is misleading in at least four respects. First, the bird's preferences are already highly patterned before imprinting takes place. Second, the young bird is certainly not passive; it actively engages its environment, working for stimuli that are most effective in eliciting its social behavior. Furthermore, its responsiveness to these stimuli is very much dependent on its internal state. Third, the learning process is gradual. The longer a bird has been exposed to a particular stimulus, the better able it is to discriminate between that stimulus and something else. And finally, the stability of imprinting almost certainly depends on the length of exposure; in those cases in which preferences were maintained throughout the life of the animal, initial exposure was never less than a week.

On the positive side, imprinting does differ from what we have called associative learning since in imprinting the animal learns the characteristics of a stimulus simply as the result of being exposed to that stimulus; no explicit pairing of neutral and biologically significant stimuli is required. Thus imprinting provides a particularly striking case of the familiarization that presumably takes place during exploration, latent learning, and the establishment of many other preferences and habits. For this reason alone, it warrants the close scrutiny of those who are interested in the learning process. Possibly, the mechanism underlying exposure learning is essentially the same as in better studied associative situations. But it may not be, and ignorance is no justification for complacent theorizing.

The imprinting situation is also striking because it provides such an obvious case of how behavior is affected by interactions between the animal and its environment during development. The constraints on what can be learned are fairly severe but are also highly adaptive. First, growth processes ensure that the timing of the onset of learning is coupled with other aspects of the developing bird's biology. Thus, precocial species leaving the nest shortly after hatching learn the characteristics of their parent at an earlier age than altricial species. Second, the bird is selectively responsive to certain types of stimulation even before imprinting has occurred. For example, objects below a certain size are treated as food rather than as social companions (e.g., Fabricius & Boyd, 1954). This differentiation serves to keep apart the processes involved in the development of social attachments from those, among others, involved in the development of food preferences. Finally, the avoidance of novel objects and the absence of social responses to such objects after imprinting has taken place serves to canalize development by restricting the range of objects with which the bird comes into contact.

While claims for an irreversible process have been overplayed, the fact remains that, in certain instances at least, the consequences of imprinting can be remarkably stable. This raises an important question for the study of behavioral development, namely, is the first experience of a certain type more likely to continue to affect behavior than subsequent experience in the same category? Or in other words, are there any general principles that determine whether or not early experience has long-lasting effects? In many ways, the imprinting situation is ideal for investigating this problem.

Mindless generalizations based on the classical concept of imprinting can be wildly misleading and do considerable harm. Nevertheless, investigation of the process has raised issues of broad significance, making it fair to conclude that some but not all the enthusiasm generated by imprinting has been justified.

References

Abercrombie, B., & James, H. The stability of the domestic chick's response to visual flicker. *Animal Behaviour,* 1961, **9**, 205–212.

Bateson, P. P. G. Effect of similarity between rearing and testing conditions on chicks' following and avoidance responses. *Journal of Comparative and Physiological Psychology,* 1964, **57**, 100–103. (a)

Bateson, P. P. G. Changes in chicks' responses to novel moving objects over the sensitive period for imprinting. *Animal Behaviour,* 1964, **12**, 479–489. (b)

Bateson, P. P. G. Relation between conspicuousness of stimuli and their effectiveness in the imprinting situation. *Journal of Comparative and Physiological Psychology,* 1964, **58**, 407–411. (c)

Bateson, P. P. G. The characteristics and context of imprinting. *Biological Reviews of The Cambridge Philosophical Society,* 1966, **41**, 177–220.

Bateson, P. P. G., & Reese, E. P. Reinforcing properties of conspicuous objects before imprinting has occurred. *Psychonomic Science,* 1968, **10**, 379–380.

Bateson, P. P. G., & Reese, E. P. Reinforcing properties of conspicuous stimuli in the imprinting situation. *Animal Behaviour,* 1969, **17**, 692–699.

Collias, N. E. The development of social behaviour in birds. *Auk,* 1952, **69**, 127–159.

Connolly, K. Imprinting and the following response as a function of amount of training in domestic chicks. *British Journal of Psychology,* 1968, **59**, 453–460.

Cruze, W. W. Maturation and learning in chicks. *Journal of Comparative Psychology,* 1935, **19**, 371–409.

de Lannoy, J. Zur Prägung von Instinkthandlungen (Untersuchungen an Stockenten *Anas platyrhynchos* L. und Kolbenenten *Netta rufina* Pallas). *Zeitschrift fuer Tierpsychologie,* 1967, **24**, 162–200.

Dimond, S. J. Effects of photic stimulation before hatching on the development of fear in chicks. *Journal of Comparative and Physiological Psychology,* 1968, **65**, 320–324.

Driver, P. M. A possible fundamental in the behaviour of young nidifugous birds. *Nature (London),* 1960, **186**, 416.

Fabricius, E., & Boyd, H. Experiments on the following reactions of ducklings. *Wildfowl Trust Annual Report, 1952–1953,* 1954, 84–89.

Fischer, G. J., Campbell, G. L., & Davis, W. M. Effect of ECS on retention of imprinting. *Journal of Comparative and Physiological Psychology,* 1965, **59**, 455–457.

Gibson, E. J., & Walk, R. D. The effect of prolonged exposure to visually presented patterns on learning to discriminate them. *Journal of Comparative and Physiological Psychology,* 1956, **49**, 239–242.

Gottlieb, G. Development age as a baseline for determination of the critical period in imprinting. *Journal of Comparative and Physiological Psychology,* 1961, **54,** 422-427.

Gottlieb, G. The question of imprinting in relation to parental and species identification by avian neonates. *Journal of Comparative and Physiological Psychology,* 1965, **59,** 345-356.

Graves, H. B., & Seigel, P. B. Prior experience and the approach response in domestic chicks. *Animal Behaviour,* 1968, **16,** 18-23.

Guhl, A. M., & Fischer, G. J. The behaviour of chickens. In E. S. E. Hafez (Ed.), *The behaviour of domestic animals.* (2nd ed.) London: Baillière, 1969. Pp. 515-553.

Guiton, P. Socialization and imprinting in Brown Leghorn chicks. *Animal Behaviour,* 1959, **7,** 26-34.

Haywood, H. C., & Zimmerman, D. W. Effects of early environmental complexity on the following response in chicks. *Perceptual and Motor Skills,* 1964, **18,** 653-658.

Heinroth, O. Beiträge zur Biologie, namentlich Ethologie und Psychologie, der Anatiden. *Verhandlungen des V. Internationalen Ornithologen Kongresses,* 1911, 589-702.

Hess, E. H. Imprinting. *Science,* 1959, **130,** 133-141. (a)

Hess, E. H. Two conditions limiting critical age for imprinting. *Journal of Comparative and Physiological Psychology,* 1959, **52,** 515-518. (b)

Hinde, R. A. Some aspects of the imprinting problem. *Symposia of the Zoological Society of London,* 1962, **8,** 129-138.

Hoffman, H. S., Searle, J., Toffey, S., & Kozma, F. Behavioral control by an imprinted stimulus. *Journal of the Experimental Analysis of Behavior,* 1966, **9,** 177-189.

Immelmann, K. Über den Einfluss frühkindlicher Erfahrungen auf die geschlechtliche Objektfixierung bei Estrilden. *Zeitschrift fuer Tierpsychologie,* 1969, **26,** 677-691.

James, H. Flicker. An unconditioned stimulus for imprinting. *Canadian Journal of Psychology,* 1959, **13,** 59-67.

Kish, G. B. Studies in sensory reinforcement. In W. K. Honig (Ed.), *Operant behavior: Areas of research and application.* New York: Wiley, 1966. Pp. 5-42.

Klinghammer, E. Factors influencing choice of mate in altricial birds. In H. W. Stevenson, E. H. Hess, & H. L. Rheingold (Eds.), *Early behavior: Comparative and developmental approaches.* New York: Wiley, 1967. Pp. 5-42.

Klopfer, P. H. An analysis of learning in young Anatidae. *Ecology,* 1959, **40,** 90-102.

Kovach, J. K., & Hess, E. H. Imprinting: Effects of painful stimulation upon the following response. *Journal of Comparative and Physiological Psychology,* 1963, **56,** 461-464.

Lorenz, K. Der Kumpan in der Umwelt des Vogels. *Journal fuer Ornithologie,* 1935, **83,** 137-213, 289-413.

Lorenz, K. *Studies in animal and human behavior.* Vol. 1. London: Methuen, 1970.

McDonald, G. Imprinting: Drug-produced isolation and the sensitive period. *Nature (London),* 1968, **217,** 1158-1159.

Moltz, H. Imprinting: Empirical basis and theoretical significance. *Psychological Bulletin,* 1960, **57,** 291-314.

Moltz, H. Imprinting: An epigenetic approach. *Psychological Review,* 1963, **70,** 123-128.

Moltz, H., Rosenblum, L. A., & Stettner, L. J. Some parameters of imprinting effectiveness. *Journal of Comparative and Physiological Psychology,* 1960, **53,** 297-301.

Moltz, H., & Stettner, L. J. The influence of patterned-light deprivation on the critical period for imprinting. *Journal of Comparative and Physiological Psychology,* 1961, **54,** 279-283.

Paulson, G. W. Maturation of evoked responses in the duckling. *Experimental Neurology*, 1965, **11**, 324–333.

Pitz, G. F., & Ross, R. B. Imprinting as a function of arousal. *Journal of Comparative and Physiological Psychology*, 1961, **54**, 602–604.

Polt, J. M., & Hess, E. H. Effects of social experience on the following response in chicks. *Journal of Comparative and Physiological Psychology*, 1966, **61**, 268–270.

Salzen, E. A. Imprinting and fear. *Symposia of the Zoological Society of London*, 1962, **8**, 199–217.

Salzen, E. A. Imprinting in birds and primates. *Behaviour*, 1967, **28**, 232–254.

Schein, M. W. On the irreversibility of imprinting. *Zeitschrift fuer Tierpsychologie*, 1963, **20**, 462–467.

Schneirla, T. C. Aspects of stimulation and organization in approach/withdrawal processes underlying vertebrate behavioral development. In D. S. Lehrman, R. A. Hinde, & Evelyn Shaw (Eds.), *Advances in the study of behavior*. Vol. 1. New York: Academic Press, 1965. Pp. 1–74.

Schutz, F. Sexuelle Prägung bei Anatiden. *Zeitschrift fuer Tierpsychologie*, 1965, **22**, 50–103.

Sluckin, W. Perceptual and associative learning. *Symposia of the Zoological Society of London*, 1962, **8**, 199–217.

Sluckin, W. *Imprinting and early learning*. London: Methuen, 1964.

Sluckin, W., & Salzen, E. A. Imprinting and perceptual learning. *Quarterly Journal of Experimental Psychology*, 1961, **13**, 65–77.

Smith, F. V. Perceptual aspects of imprinting. *Symposia of the Zoological Society of London*, 1962, **8**, 171–191.

Taylor, A., Sluckin, W., Hewitt, R., & Guiton, P. The formation of attachments by domestic chicks to two textures. *Animal Behaviour*, 1967, **15**, 514–519.

Thorpe, W. H. *Learning and instinct in animals*. London: Methuen, 1956.

Tronick, E. Approach response of domestic chicks to an optical display. *Journal of Comparative and Physiological Psychology*, 1967, **64**, 529–531.

Weidmann, U. Verhaltensstudien an der Stockente (*Anas platyrhynchos* L.). *Zeitschrift fuer Tierpsychologie*, 1958, **15**, 277–300.

Yerkes, R. M., & Dodson, J. O. The relation of strength of stimulus to rapidity of habit formation. *Journal of Comparative Neurology*, 1908, **18**, 459–482.

CHAPTER 10

VOCAL LEARNING IN BIRDS

Peter Marler and Paul Mundinger

I. Introduction

A variety of experiments by psychologists have tended to confirm Skinner's impression (1957) that animal vocalizations are refractory to the effects of operant conditioning, as compared with other motor activities. Several attempts have been made to modify animal vocalization by conditioning. It proved possible to modify the duration and rate of sound production in a number of cases, but the actual acoustical structure or morphology of the vocalizations was changed little or not at all by the procedures (e.g., Ginsburg, 1960, 1963; Grosslight, Harrison, & Weiser, 1962; Grosslight & Zaynor, 1967; Lane, 1961). Ginsburg (1963) was successful in achieving stimulus control over the production of two words from human speech by a mynah bird. However, these words had already been established in the bird's repertoire by other means. The speech training included not only a conditioning procedure but also repeated presentations of normal and recorded speech without any extrinsic reinforcement. The role of conditioning in establishment of the sounds in the repertoire is uncertain.

Lane and Shinkman (1963) reported that they were successful in changing the morphology of the calls of domestic chicks by conditioning. However, careful inspection of the results suggest that they may be equivocal. Sound spectrograms are provided of the vocalizations given during continuous reinforcement and during extinction. A comparison with sound spectrograms of natural vocalizations of chicks (Collias & Joos, 1953) reveals that all of these sounds are present in the normal repertoire. Thus it seems likely that Lane and Shinkman modified the normal frequency of sounds already present in the repertoire without necessarily changing their acoustical structure. The sounds given during reinforcement are those described by Collias and Joos as "pleasure calls." One of the sounds given during extinction, presumed to have novel morphology as a consequence of conditioning, is in fact the normal "distress call" of a chick, given when it is deprived of food. The other sound given during extinction is a normal "fear trill."

One might hope that nonhuman primates would be a more fertile source of examples of changing vocal behavior by learning. Although chimpanzees and an orangutan have been taught to utter a few words of human speech, it seems clear that the processes involved were quite different from those one customarily thinks of as vocal imitation. The training required direct manipulation of the animal's tongue, lips, jaw, and nose by the experimenter, and there is general agreement among the investigators of the extraordinary effort required both of them and of their animal subjects (Furness, 1916; C. Hayes, 1951; K. J. Hayes, 1950; K. J. Hayes & Hayes, 1954, reviewed in Kellog, 1968). Thus, so far there is no evidence of any great facility for vocal imitation among nonhuman primates.

It seems reasonable to draw the inference from investigations by psychologists on vocal learning in animals that animal sounds are indeed resistant to effects of learning. Nevertheless, there is a body of zoological evidence that argues otherwise. The intent of this review will be to establish that, given the appropriate subjects and techniques, it is possible to demonstrate experimentally that some birds have great facility in vocal learning, and that in some species, learning is the normal means of transmitting sounds from one generation to the next, as in man.

II. The Nature of Avian Vocal Behavior

A. Acoustical Properties of Vocalizations

The problems of behavioral description, always present in ethological study, are especially acute in ontogenetic investigations. If changes imposed by experimental manipulation are to be detected, the behavioral measures must be sensitive and reliable. In this regard, vocal behavior presents some advantages. Recording and physical analysis of sounds gives a rather complete record of the motor patterns involved. An analysis into frequency, amplitude, and their changes with time gives a complete acoustical description of the sound. Extensive use has been made in bird studies of the Kay Electric sound spectrograph and the illustrations to this paper are so derived. It is the most convenient method of portraying the physical characteristics of the biologically relevant properties of sounds of species which are themselves frequency-sensitive.

It is important to bear in mind that no single method of analysis can give a complete portrayal of the characteristics of a sound. This is especially true of sounds whose properties are changing rapidly with time. Special care is needed in interpreting sound spectrograms of rapid pulse-trains and tones which are frequency or amplitude-modulated, all of which often occur in animal sounds (Marler, 1969b; Watkins, 1967).

One drawback with the sound spectrograph is the short duration of each portion of sound analyzed (2.4 seconds). For investigation of the longer term organization of vocalizations, a useful alternative is to relay the recorded sounds into a frequency meter, the output voltage of which fluctuates according to frequency. This output is fed to a voltage-sensitive recorder which provides a permanent paper record (Fish, 1953; Mulligan, 1966). This method has the advantage that large numbers of recordings can be surveyed in a short period of time. It lends itself especially to analysis of the bout structure of the singing behavior of birds (Fig. 1).

B. Extent of the Vocal Repertoire in Representative Species

If we record the sounds that a species uses in the course of its life cycle and analyze them by the methods described above, the sounds can be arranged in categories according to their physical structure. From such an analysis it is possible to estimate the size of the acoustical repertoire for a species. Although our knowledge is still very limited, some generalizations are beginning to emerge. The largest repertoires found among birds and mammals are considerably greater than those of lower vertebrates and invertebrates (Marler & Hamilton, 1966). As far as one can tell, there is no great difference between birds and mammals in this regard. For example, an estimate of the number of basic vocalizations in the adult repertoires of animals might reveal a range of 5–14 sounds in birds and 5–17 in various nonhuman primates (Marler & Hamilton, 1966).

In fact, such comparisons can be deceptive, since in some species the sounds are organized in discrete, nonoverlapping categories while in others the sounds grade into one another with a variety of subtle distinctions which probably have significance to the animals themselves. In such cases, it is impossible to derive a meaningful estimate of repertoire size by descriptive analysis alone.

Within the repertoires of many bird species there is often a basic distinction between a variety of sounds in which the fundamental acoustical unit is short, the calls, and another set in which trains of sounds are given in a more or less highly organized pattern, the song (Fig. 2). The latter is often, though by no means always, the prerogative of the male. It is often the loudest sound in the repertoire. Learning plays an especially prominent role in the development of this song.

C. Contexts of Sound Production

As with other behavior, the vocalizations of birds occur in more or less well-defined contexts. Some are characteristic of a particular age. Among adults, some vocalizations are heard the year around while others are restricted to birds in breeding condition. Thus, chaffinches (*Fringilla coelebs*) in nonreproductive condition use only two

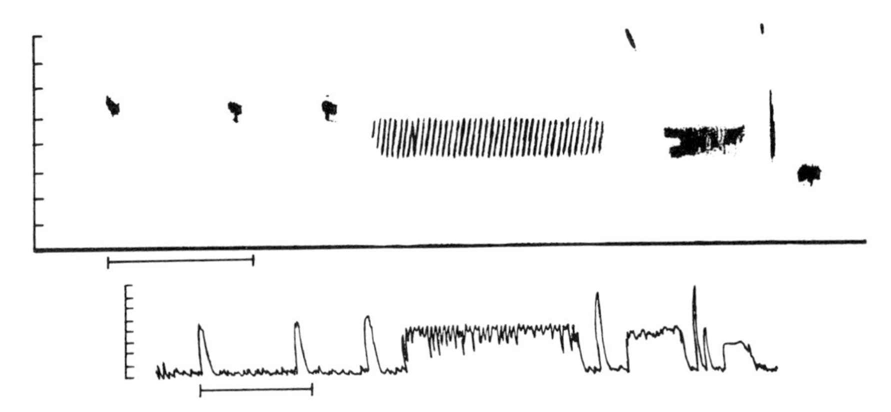

Fig. 1. Comparison of a sound spectrogram (above) and an oscillographic record (below) of the same song sparrow song, illustrating the differences between the two methods of analysis. Ordinate, a frequency marker 0-8 kHz; abscissa, a time marker of 0.5 second. (After Mulligan, 1966).

vocalizations at all commonly, the flight call and the social call, the latter having several variations that serve as aggressive and alarm signals. These evidently suffice for adequate organization of behavior in the winter flock (Marler, 1956).

In the spring the flock breaks up, males stake out territories, each is typically joined by a female, and they raise a brood. There is an underlying cycle of growth of the gonads, which are minimal in size in the winter and maximal in spring and early summer. As the male's testosterone level rises, several new vocalizations enter his repertoire.

Some chaffinch vocalizations—courtship and alarm calls, for example—are normally triggered by an environmental change such as proximity of the mate or appearance of a predator. Others, such as the location call of the fledgling, may be evoked by perception of the approaching parent, but are often triggered by endogenous changes, such as those following food deprivation. The male song is in this latter category, sometimes occurring in immediate response to social stimulation, but more often endogenously triggered.

The male song serves both to attract and retain a mate, and to maintain the spacing of territorial males. This function would not be efficiently served if song were completely contingent upon close proximity of rival or a mate (cf. Marler, 1969a). The male song serves for communication over long distances. In line with these functions, it is perhaps not surprising that in the breeding season of many bird species, the sound with the greatest volume and the highest frequency of utterance is the song of the male.

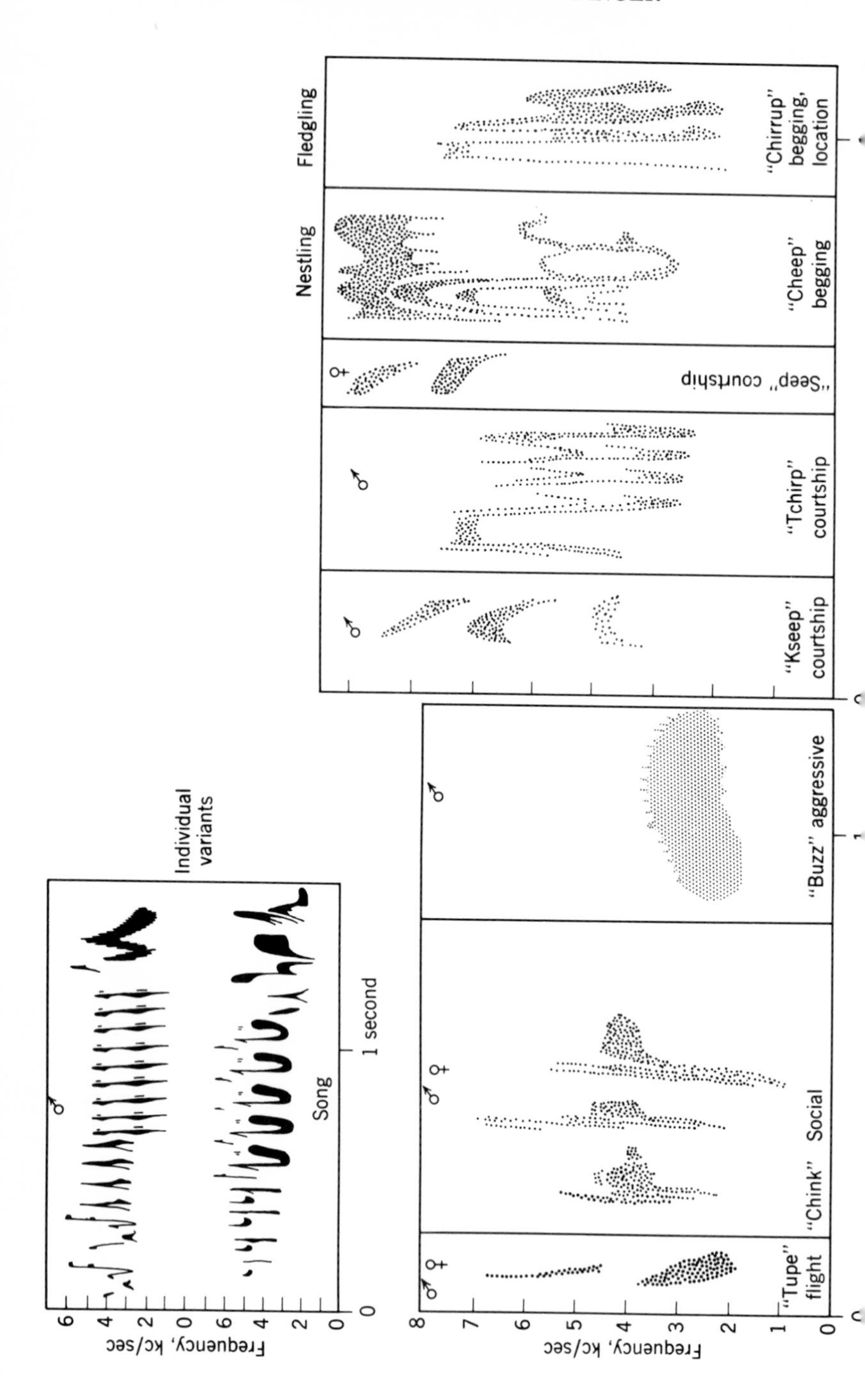
Frequency, kc/sec
Song
Individual variants
1 second
"Tupe" flight
"Chink" Social
"Buzz" aggressive
"Kseep" courtship
"Tchirp" courtship
"Seep" courtship
Nestling
"Cheep" begging
Fledgling
"Chirrup" begging, location

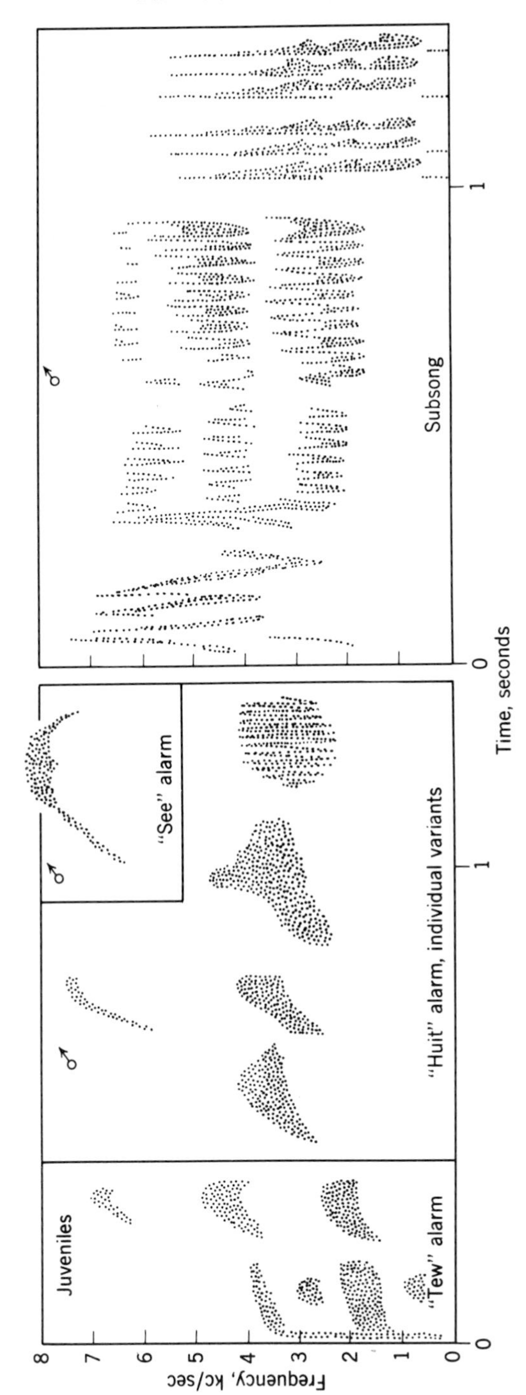

Fig. 2. Repertoire of vocalizations of the chaffinch. In addition to song there are 13 basic call patterns. One call, a squeal given when birds are taken by a predator, is not shown. Some of these sounds are stereotyped, and others such as the "chink" are graded according to context. Individual differences occur in male song patterns and in the "huit" alarm call. (After Marler & Hamilton, 1966.)

D. Degrees of Species Specificity within the Repertoire

A survey of the different functions served by bird vocalizations can lead us to speculate about the rates of change that might be required in the course of phyletic history, if the functions are to be served efficiently. For example, the alarm calls used by various woodland species in Britain are very similar, and this resemblance surely correlates with their frequent use in the interspecific communication of danger (Fig. 3).

There are other less striking cases of vocalizations for which efficient functioning requires only a minimum of specific divergence in the course of evolution (Marler, 1957).

Quite the reverse is true of the male song. This vocalization carries the major burden of reproductive isolation in many bird species, and it is surely no accident that even close relatives usually have highly divergent song patterns. The classic example is provided by three morphologically very similar species of European warbler which were first distinguished on the basis of differences in their song (Fig. 4).

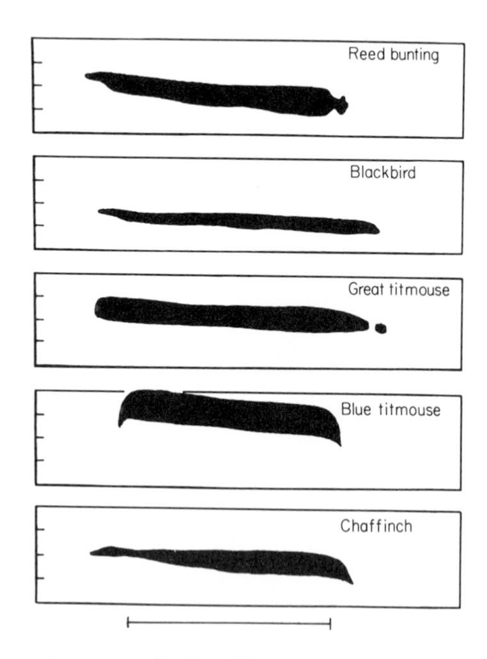

Fig. 3. Sound spectrograms of calls of five species of British birds used when a hawk flies over. These sounds have characteristics, such as pure tones and gradual fade in and fade out, that make them difficult to locate. Frequency markers, 5-9 kHz; time marker at bottom, 0.5 second. (After Marler, 1957.)

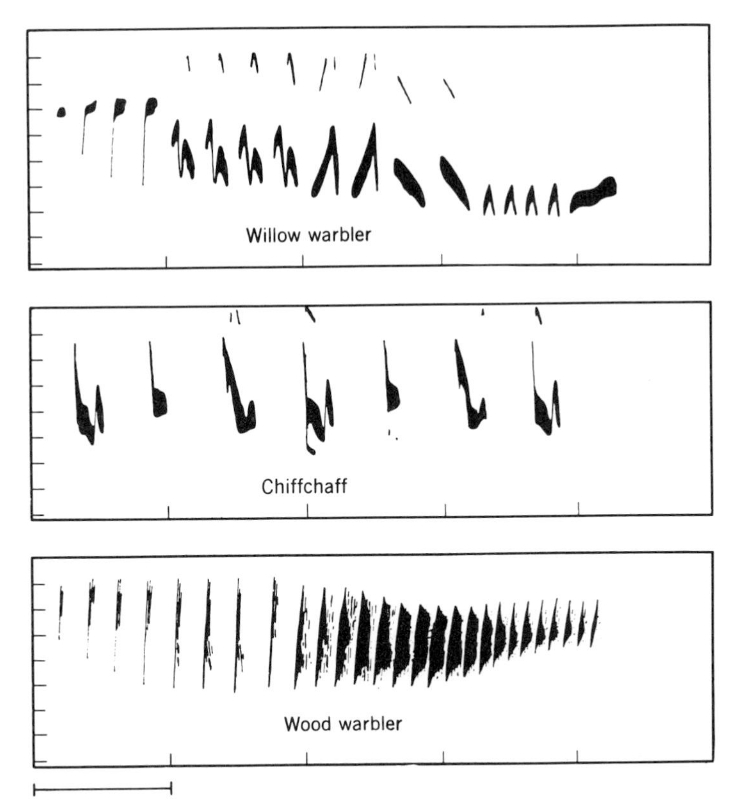

Fig. 4. Sound spectrograms of songs of three European warbler species. The differences between the songs were the first characters used to distinguish between these species. For example, the willow warbler and chiffchaff are morphologically very similar but whereas the willow warbler repeatedly sings a regular song of about 2 seconds' duration, the chiffchaff (*Phylloscopus collybita*) song rambles on continuously, improvising on two basic notes. Frequency markers, 0-8 kHz; time marker, 0.5 second. (After Marler, 1957.)

When we reflect on the richness of adaptive radiation among songbirds, such that many species are to be heard in any given area, it becomes clear that the function of reproductive isolation requires a high rate of change in the structure of song in the course of evolution. This may be important to maintain isolation not only from other species, but even from separate populations of the same species. With different rates of change required for different vocalizations within the repertoire, the stage is set for some radical innovation in the mechanisms of ontogeny. The dominant role of learning in the development of the male song of some birds may be viewed as a response to this kind of evolutionary demand for rapid change.

III. Methods of Study and Experimentation

A. Acoustical Analysis

A sound spectrographic analysis of a tape recording of an animal sound gives a remarkably full description of its physical properties. The Kay Electric Sound Spectrograph is provided with two band widths, 40 Hz and 300 Hz. For most purposes the 300 Hz, or wide band pass filter, is the most convenient for the analysis of bird sounds and this was used in most of the illustrations for this paper. Exceptions are so indicated. Some further comment is required on the methods of sampling a recording of bird vocalizations. The very completeness of the description provided by a sound spectrogram is also an embarrassment, since it is difficult to give a quantitative account of the patterns of frequency change which are often so obvious and distinctive on the record. Statistics on the duration of notes and the intervals between them, or the range of frequencies represented, do little justice to the subtleties of the pattern which, as we show later, are of significance to the birds themselves.

The method adopted in many studies of avian vocalizations is to analyze an array of recordings and to select some as typical representatives of the sample and to present them as evidence. Although there is an element of subjectivity here, this is less dangerous than it sounds. Extensive analysis of vocalizations of many bird species in captivity and in the field has demonstrated that at any given stage of development the repertoire of an individual is usually limited. An experienced investigator can derive a reasonably accurate qualitative picture of that repertoire without necessarily resorting to exhaustive quantification. As experiments become more refined, it will be necessary to devise new methods for quantitative comparison of the patterns of frequency change which are present in avian vocalizations. Meanwhile, it is not unreasonable to assume that the sound spectrograms of vocalizations presented as evidence are in fact representative of the utterances at that particular stage of development. This is particularly true of the species which have been the main subjects of experimental investigation so far, such as the chaffinch and the white-crowned sparrow, which were selected in part because the limited nature of the repertoire of male songs made the task of description easier than in birds with a larger song repertoire. Thus an adult male white-crowned sparrow has a single song pattern, subject only to minor variation from utterance to utterance; a male chaffinch, from one to six song types, with an average repertoire size of 2.3 songs (Marler, 1952, 1956).

B. Isolation and Training Methods

Most of the evidence for learning as a factor in the development of bird vocalizations comes from songbirds. Unlike the young of chickens and other gallinaceous birds, which are precocial, the young of songbirds are altricial. That is to say, they are hatched at an early developmental age, incapable of locomotion. They remain within a nest for one or more weeks, during which time they are fed by the parent. The feeding continues after they leave the nest, with an overall period of dependence on the parent as long as one month in the songbirds discussed here.

In some species, learning is known to occur at an early age. Thus it is a prerequisite for experimentation that birds be taken from the nest at an early age and raised by the experimenter by hand. Ideally one would take the eggs from nests of wild birds and hatch them in an incubator. This has proved difficult in practice, although small numbers have been raised successfully (Blase, 1960; Lanyon & Lanyon, 1969; Messmer & Messmer, 1956; Sauer, 1954). Instead one can compromise by taking birds from the nest a few days after hatching, when they are much easier to raise successfully. Foods used to raise young white-crowned sparrows by hand include hard-boiled eggs, bird seed which has been soaked in water for 24 hours and then crushed, a proprietory canary nestling food mixture, together with small amounts of mealworms, crickets, wax moth larvae, fly maggots and a vitamin supplement. The birds are weaned onto a diet of mixed seed previously soaked in water, and finally onto dry mixed seed. In our studies, nestlings were kept in their original nests with a cleansing tissue lining that was replaced frequently. Food was presented with tweezers or on the tip of a slender wooden spatula.

For some purposes it is of interest to catch young birds after they have fledged and become independent from their parents. This is accomplished by driving them into fine nylon nets. Permits from state and federal authorities are required to operate such nets.

When the birds are brought in from the field, they are placed in some degree of acoustical isolation. In recent years investigators have commonly made use of the small audiometric chambers made by the Industrial Acoustics Corp. These provide an attenuation of frequencies within the range of the birds' vocalizations (500-7000 Hz) of something between 40 and 65 db. The boxes are provided with a microphone and a loudspeaker, and the birds are recorded at intervals. Some details of training procedures will be given later.

It is difficult to determine the sex of many songbirds at an early age by external morphology. If the study is concerned with song learning

in males, it is useful to determine the sex directly by inspection c
the gonads. The techniques for laparotomy are already well worke
out (Bailey, 1953).

C. The Relationship between Field Observation and Experiment

As with many ethological studies, there is a close interrelationshi
between field observation and laboratory experimentation in studie
of song development. An intimate knowledge of natural behavior i
properly regarded as a prerequisite to experimental analysis. De
scriptions of the natural behavior provide a baseline with which vo
calizations that develop under various experimental conditions ca
be compared. Field study can be a valuable guide in some cases t
the selection of suitable subjects for the experimental study of learn
ing, as will be demonstrated in the next section.

Knowledge of the natural behavior is also relevant when trainin
experiments are conducted. The evidence suggests that some spe
cies manifest a predisposition to learn sounds of their own species
Thus at least some experiments should provide an opportunity fo
birds to imitate conspecific songs. The experiments are then mor
likely to be biologically revealing than if attempts at training are re
stricted to sounds such as recordings of human speech, which ar
utterly alien to the species.

The ability of captive mynah birds to imitate human speech i
discussed below. However, knowing that they possess this abilit
helps little in understanding normal vocal development, becaus
wild mynahs do not imitate sounds of human speech or of any othe
bird species. Instead, imitation is restricted to calls of close neigh
bors of the same sex and there are dialects as a result. Matching cal
types are subsequently used in countercalling with neighborin
individuals (Bertram, 1970).

IV. Evidence of Learning from Observational Studies

Field observations can be a deceptive guide to judgments abou
the mechanisms involved in behavioral ontogeny. We are prone t
assume that uniformity of a trait within a population implies a min
imum of environmental contribution to its development (Konishi
1966). The inference would be correct in the case of doves, wher

genetic contributions to singing behavior have been established (Lade & Thorpe, 1964). With the white-crowned sparrow (*Zonotrichia leucophrys*) the inference would be quite wrong, for in this species the more prominent characteristics of the natural song are transmitted by learning from generation to generation.

Study of variation in natural populations on a broader scale can, however, give some reliable cues. Thus, in the white-crowned sparrow, each local population is characterized by a particular song dialect. If such dialects exist without any accompanying morphological distinctions between populations that are separated by relatively short distances, then it is a reasonable guess that learning plays a role in song development.

If an occasional individual within a species produces an unusual and unequivocal rendition of a sound of some other species with which it is known to be in contact, it is reasonable to infer that the sound has probably been learned. We do not usually require proof that renditions of human speech by a mynah bird were learned. Similarly, there are bird species that mimic other birds in nature, and the evidence is strong that learning is the basis for this mimicry.

A. Song Dialects

The existence of dialects in the song of the male chaffinch (Marler, 1952; Promptoff, 1930) proved to be a good clue that learning plays a role in its development. Similarly, the song of the male white-crowned sparrow provides a clear demonstration of dialects (Fig. 5). Further work has shown that these dialects are transmitted by learning (Marler, 1970; Marler & Tamura, 1962, 1964). There is some evidence of dialects in the song of the male yellow bunting (*Emberiza citrinella*) (Kaiser, 1965) and again there is reason to think that learning plays a role in its development (Thorpe, 1964). Cardinals (*Richmondena*) have a repertoire of a dozen or so notes arranged in a variety of song types and some of these are distributed in the form of dialects. These variations in the male song are learned (Dittus & Lemon, 1969; Lemon, 1965, 1966; Lemon & Scott, 1966). The Carolina chickadee (*Parus carolinensis*) (Ward, 1966) and the short-toed tree creeper (*Certhia brachydactyla*) (Thielcke, 1961, 1964, 1965, 1969), with well-marked song dialects, are obvious candidates for ontogenetic study. However, the absence of song dialects does not necessarily imply that learning is without any role in song development (Rice & Thompson, 1968; Thompson, 1968).

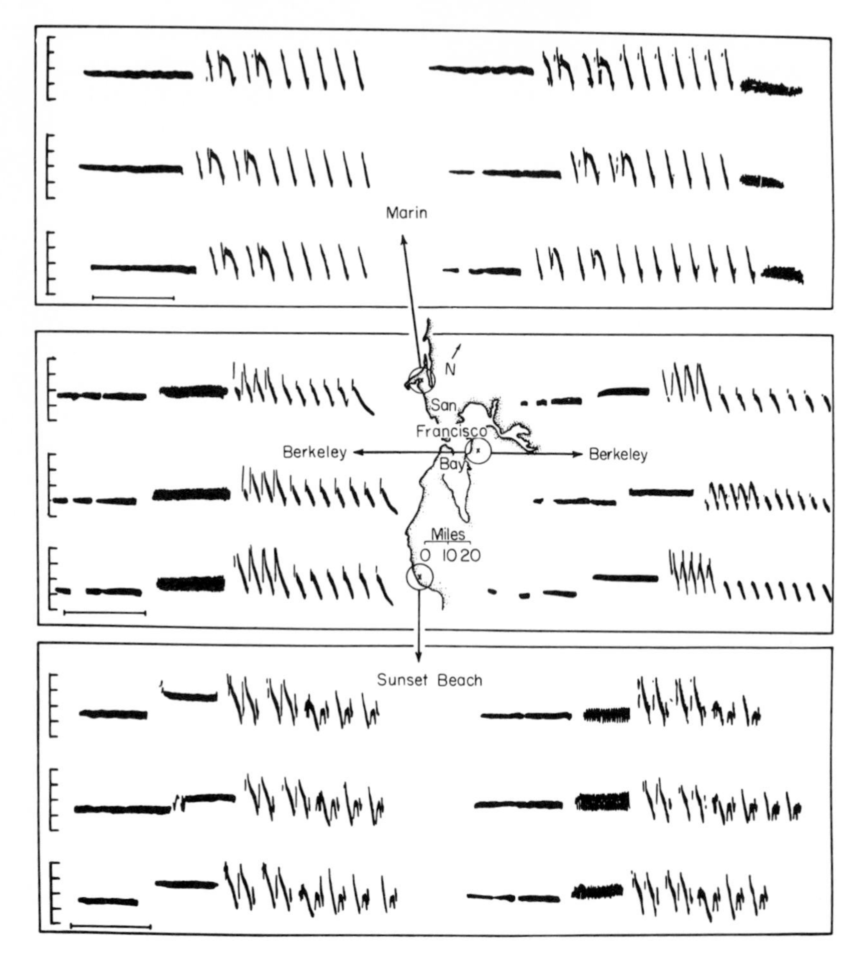

Fig. 5. Sound spectrograms of songs of 18 male white-crowned sparrows from three localities in California. The structure of the second part of the song varies little within an area but is consistently different between populations, illustrating the dialect differences in this species. Frequency markers, 2–6 kHz; time marker at bottom of each panel, 0.5 second. (After Marler & Tamura, 1964.)

B. Interspecific Mimicry

A number of birds are known to produce imitations of other species in the course of their natural singing behavior. The classical North American example is the mockingbird. In Australia, some of the bower birds are renowned mimics, both of other bird sounds and of nonbiological sounds such as the twanging of a wire fence. In

Europe, the starling is a well-known example, as are some of the warblers. Among the most remarkable are some of the viduine finches of Africa which, like the European cuckoo, habitually lay their eggs in other birds' nests. There is strong evidence that they learn the song and some of the calls of their hosts (Nicolai, 1964). Several cases of interspecific mimicry, illustrated by sound spectrograms, have been reviewed by Tretzel (1965a).

The renditions of sounds of other species are often sufficiently novel and accurate as to leave little doubt that they have been acquired through learning. However, we should note that the confusion that a bird watcher might experience on hearing such mimics is usually only brief, for they usually impose some organization of their own upon the imitation. Tretzel (1966) discovered a population of garden warblers in Bavaria whose songs typically incorporate imitations of chaffinch song. Sometimes renditions of complete chaffinch songs are found, but often they are divided into fragments, used within a matrix of other sounds more typical of the garden warbler (Fig. 6). Thus by temporal reorganization, the interspecific imitations acquire some properties species-specific to the mimic. The same tendency has been observed in mockingbirds (Laskey, 1944).

Although there are many anecdotes about interspecific mimicry in nature, few examples have been submitted to acoustical analysis. Tretzel has made this subject his special interest. A crested lark (*Galerida cristata*) in Bavaria has been found imitating four different whistle commands given by a shepherd to his dog, two of which are shown in Fig. 7. The dog responded correctly to tape recordings of the crested lark imitations. Similar though less perfect imitations were detected in the songs of another lark 3 km away, presumed to have learned its versions from the first bird (Tretzel, 1965b).

In another part of Bavaria, Tretzel discovered a group of European blackbirds (*Turdus merula*) singing what appeared to him to be an imitation of human whistling. Some of the imitations were more precise than others. By tracking them down, he was led to the house of a man who had been in the habit of whistling this tune to his 8-year-old cat since it was a kitten. Analysis of recordings of his whistling revealed that he was no musician, and the pitch and rhythm varied considerably. The blackbirds transposed the entire motif to a higher pitch, with some additional notes, and maintained it in a more regular fashion than the man's original versions (Tretzel, 1967).

Although a number of investigators have been concerned with the imitation of speech by captive birds, there has been surprisingly little effort at detailed acoustical comparison between the imitations

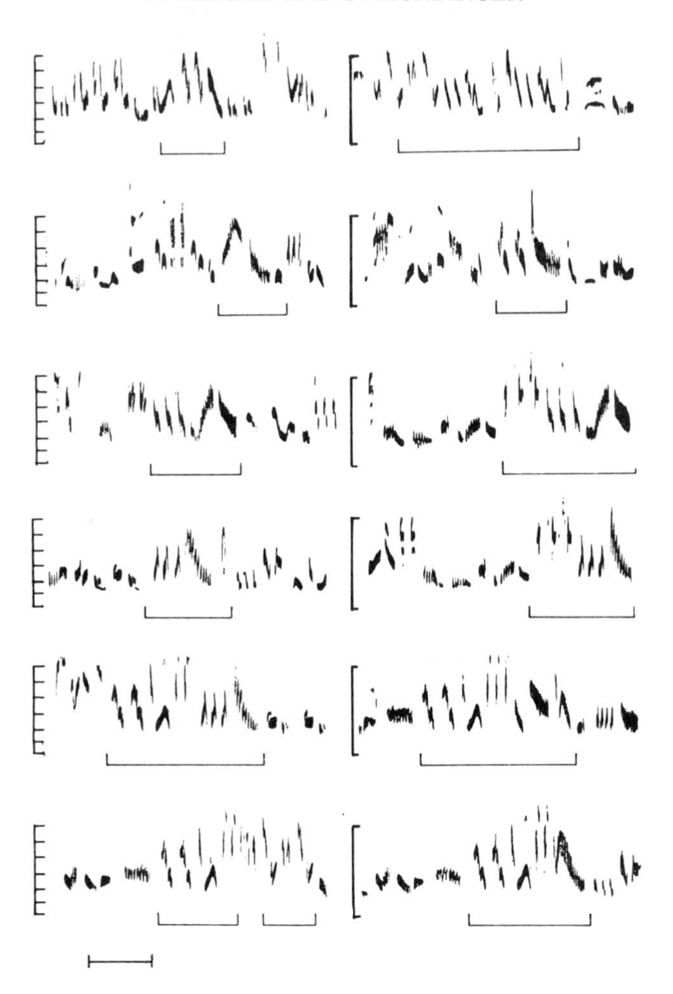

Fig. 6. Portions of 12 songs of a garden warbler (*Sylvia borin*) in which appear fractions of chaffinch song, indicated by the brackets. Frequency markers, 0-6 kHz; time marker at bottom, 0.5 second. (After Tretzel, 1966.)

and their models (Ginsburg, 1963; Grosslight *et al.*, 1962; Grosslight & Zaynor, 1967). Much of the older work on this subject was reviewed by Bierens de Haan (1929), Mowrer (1950), and Thorpe (1959). By subjecting the speech imitations of mynah birds to sound spectrographic and other types of acoustical analysis, Thorpe (1959) and Greenewalt (1968) showed that mynah birds can generate sounds with structures resembling the performance of human speech, although by entirely different mechanisms.

Cases of speech imitation in captive birds, interspecific mimicry in nature, and the occurrence of song dialects all provide evidence that learning has played a role in vocal development. A major puzzle about birds which can be trained to talk in captivity is that no one has yet found evidence that they mimic other species in nature.

C. Learning in the Development of Calls

The distinction in the vocal repertoire of most birds between the song, which is a more or less complex vocal utterance, usually the prerogative of the male, and the calls, which are shorter and simpler, has been noted. Abnormalities reported in the vocalizations of birds raised in social isolation usually concern the male song. Other vocalizations generally seem to develop normally under such conditions, although it must be admitted that few careful comparisons of the calls of normal and isolated birds have been made. Lanyon (1957, 1960) demonstrated that several of the calls of Eastern meadowlarks (*Sturnella magna*) raised in social isolation develop normally (Fig. 8). A similar result is reported for the European blackbird (Messmer & Messmer, 1956; Thielcke-Poltz & Thielcke, 1960) and the white throat (*Sylvia communis*) (Sauer, 1954). The calls of hand-raised song sparrows (*Melospiza melodia*) are normal, and the same is true of house finches and of all but one of the calls of the chaffinch (Marler, 1956; Miller, 1921; Nice, 1943).

Heinroth (1924) summarized the impression of many investigators in his conclusion that the development of bird vocalizations other than the song is generally relatively independent of acoustical influence from other birds. Nevertheless, there are some notable excep-

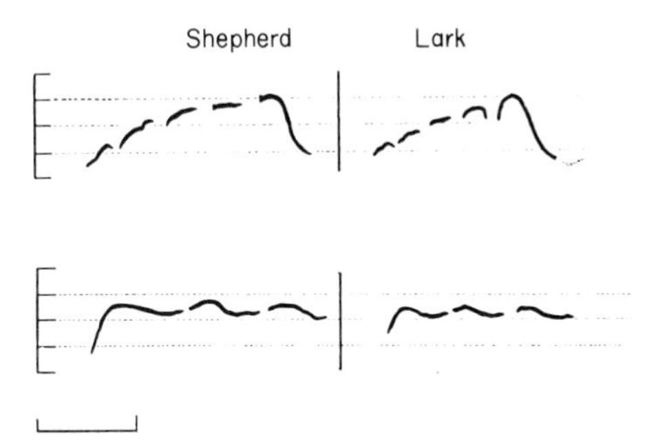

Fig. 7. Narrow band sound spectrograms of two whistled commands used by a Bavarian shepherd and their imitations by a crested lark. Frequency markers, 1-5 kHz; time marker at bottom, 0.5 second. (After Tretzel, 1965b.)

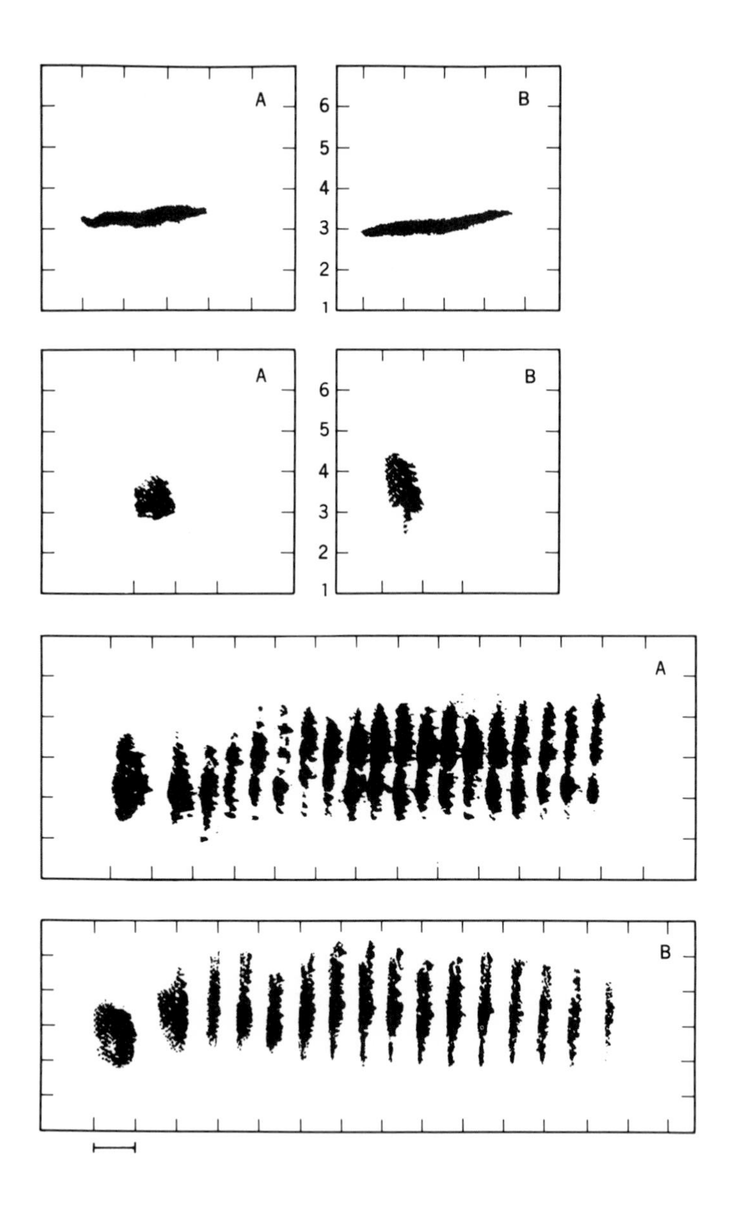

Fig. 8. Some vocalizations of a hand-reared and acoustically isolated Eastern meadowlark (A) are essentially the same as those of a free-living adult (B). Frequency markers, 1-7 kHz; time marker, 0.1 second. (After Lanyon, 1960.)

tions. The common "chink" call of the chaffinch develops abnormally in social isolates (Marler, 1956). Although there is no proof, it seems likely that learning from adults plays a role in normal development of this call. The same may be true of the so-called rain call of the male chaffinch. We know from the work of Sick (1939) and Thielcke (1969) that there are local dialects in the form of this call (Fig. 9).

The same seems to be true of the social call of the European bullfinch (*Pyrrhula pyrrhula*) which develops abnormally in birds raised by hand in isolation from adults of their species (Nicolai, 1959). It is perhaps no accident that this call of a bullfinch, and the two chaffinch calls already mentioned, seem in some respects to be taking over functions which are served only by song in other species of finch.

A more recent example concerns the development of flight calls in several other finches. Mundinger (1970) finds that there are individual differences among the flight calls of wild American goldfinches (*Spinus tristis tristis*). However, a male and female that have formed a stable pair share identical flight call patterns. Playback experiments reveal that a paired female goldfinch recognizes her mate's flight call. Her responsiveness to his call is particularly evident when she is sitting on the nest, and seems to facilitate his visits to the nest to feed her. Thus, it appears that the structure of these flight calls becomes modified as a pair bond is formed, so that they maintain a common pattern for the breeding season which serves to sustain the bond between them. One is reminded of the duets sung

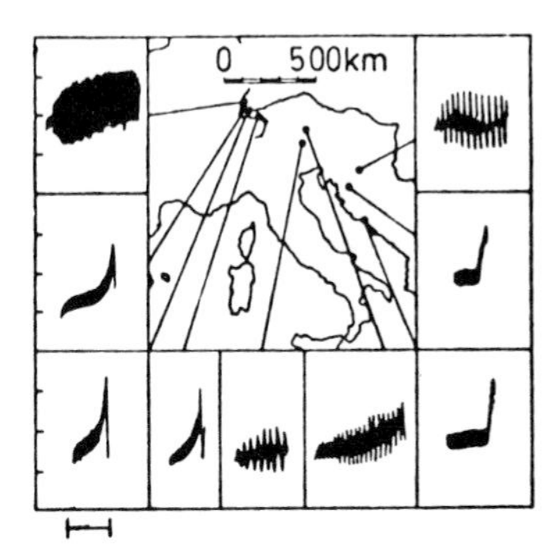

Fig. 9. Mosaic distribution of dialect differences in the chaffinch rain call. The form of the call is very uniform within a given population. Frequency markers, 2-6 kHz; time marker at bottom, 0.1 second. (After Thielcke, 1969.)

by the mated pair in African shrikes and various other species (Grimes, 1965, 1966; Gwinner & Kneutgen, 1962; Hooker & Hooker, 1969; Thorpe, 1963; Thorpe & North, 1965, 1966), which also seem to be learned, and which function in sustaining the pair bond.

Further work reveals a similar change in the flight calls of two related species, the pine siskin (*Spinus pinus pinus*) and the European siskin (*Carduelis spinus*). Again there is some modification of calls within the pair, including an instance of interspecific vocal imitation within a mixed species pair (Fig. 10). Detailed study of these species has also revealed that vocal imitation occurs within social units other than the mated pair, such as the winter flock (Fig. 11).

The modification of these flight calls through learning is subtle, and could easily be overlooked. Although none of these birds has yet been raised in social isolation, it seems unlikely that a social isolate's flight call would be recognized as grossly abnormal. Thus it is not inconceivable that more vocalizations in addition to male song will prove to be subtly modifiable through learning.

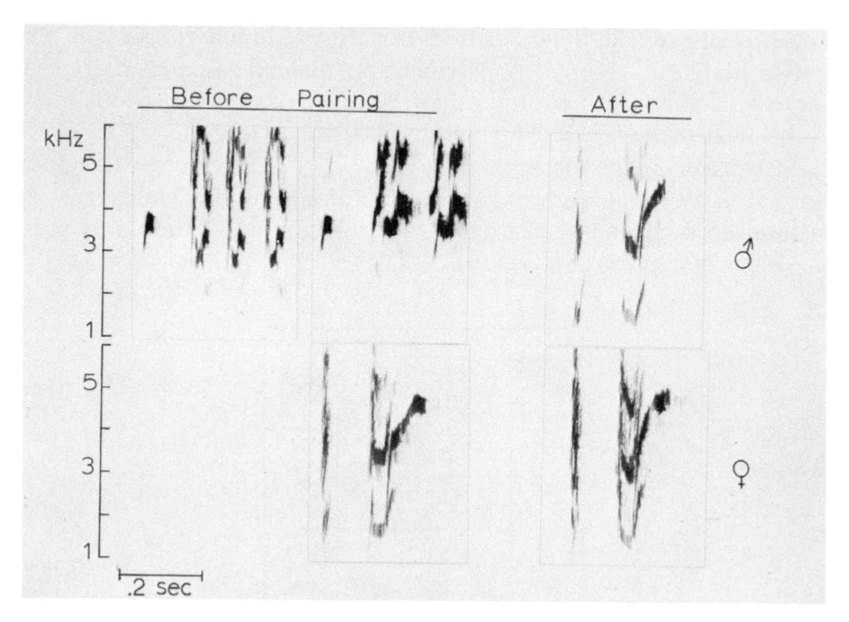

Fig. 10. Vocal imitation of a flight call during the pairing of two closely related finches. Before and after pairing only one type of call pattern was recorded from the female, a European siskin. This pattern was not initially a part of the flight call repertoire of the male, a pine siskin, but it appeared in his repertoire after he was paired with the female. (After Mundinger, 1970.)

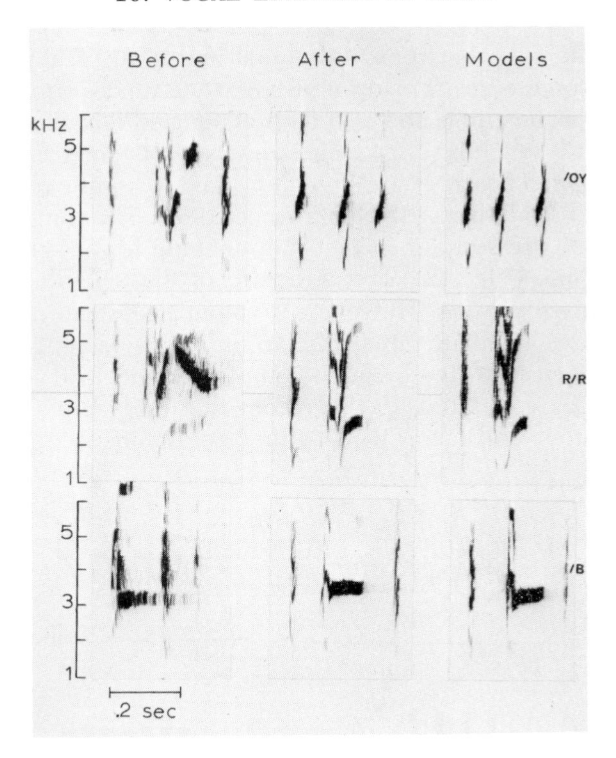

Fig. 11. Vocal imitation of finch flight calls during flock formation. Flight calls recorded from an individual male European siskin before and after his being caged with three strange male conspecifics compared to flight calls recorded from the three new flock mates (the "models"). (After Mundinger, 1970.)

V. Vocal Abnormalities in Social Isolates

If a bird raised out of hearing of adults of its own kind manifests any abnormality in its vocalizations, this suggests that learning is involved in its normal development. The precise interpretation to be drawn depends on the isolation conditions used. Naturalists have often raised young birds out of contact with their own species, but within hearing of a variety of other birds. There are many examples of abnormal development in birds raised under such conditions. Often sounds of other species are learned, even by birds which do not normally mimic other species in nature (review in Marler, 1963).

As a further step, young birds may be isolated from adults of their own and other species, but kept with their own age mates. We know that the difference between individual and group isolation can be critical in some cases. For example, Arizona juncos (*Junco phaeonotus*) taken as nestlings at 4 or 5 days of age and raised in individual isolation will develop songs which differ consistently from those of normal adults. The latter are known to have a highly complex song with many distinct parts (Marler & Isaac, 1961). The songs of an individual isolate are simpler in that they include fewer syllable types, often only one (Fig. 12). The structure of the syllables is simpler than that in wild birds. However, if young males are raised in the company of an age mate rather than in individual isolation, there is a dramatic increase in the syllabic complexity and in the number of syllable types generated, to the extent that the sample is indistinguishable from one taken in a natural environment (Marler, 1957)

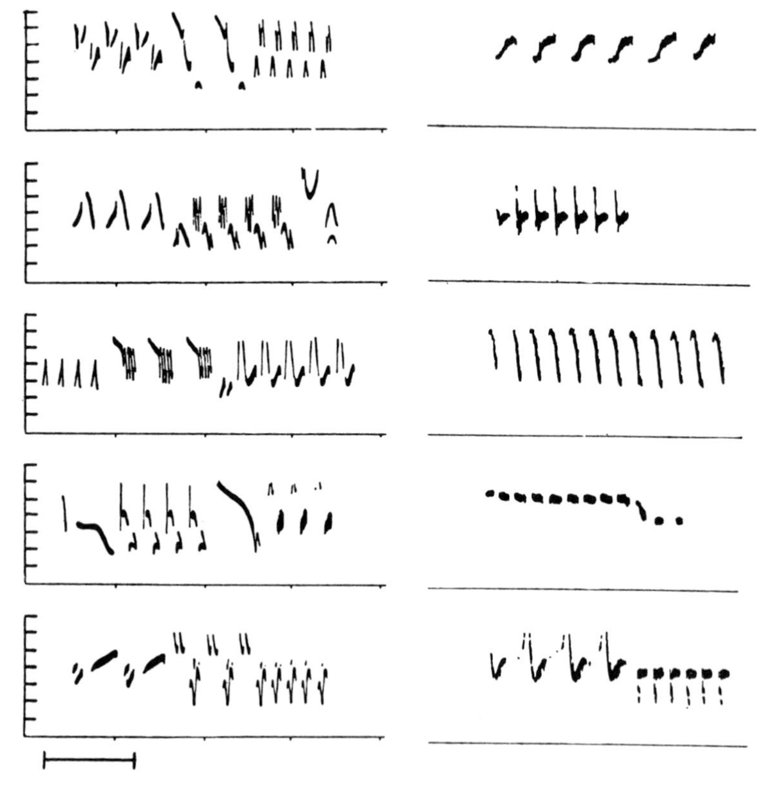

Fig. 12. Left column, song of five wild male Arizona juncos recorded in their natural environment. Right column, five songs simpler than those of wild birds from an Arizona junco raised in acoustical isolation from the nestling stage. Frequency markers, 0–7 kHz; time marker at bottom, 0.5 second. (After Marler, 1967.)

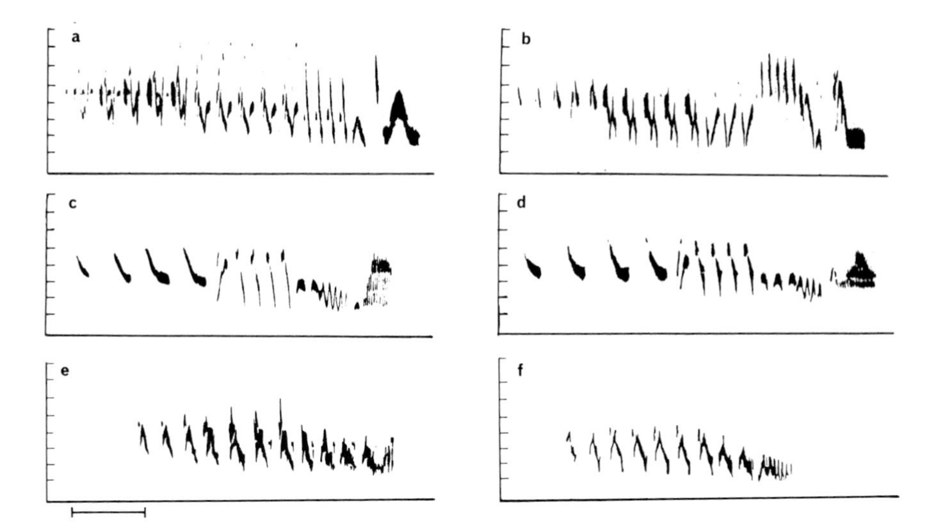

Fig. 13. (a) and (b) Normal songs of two wild-caught chaffinches. (c) and (d) Somewhat simpler songs recorded from two unrelated hand-reared chaffinches isolated from the fourth day of life but permitted to hear each other during the critical period for song learning. (e) and (f) Abnormally simple songs of two hand-reared individual isolates. Frequency markers, 0–8 kHz; time marker at bottom, 0.5 second. (a–c from Thorpe, 1963; d and e from Thorpe, 1958.)

(Fig. 27). Evidently the interchange that occurs between members of such a group is sufficient for normal song patterns to develop without the need for exposure to songs of adult birds. The birds seem to stimulate each other to a greater degree of vocal invention or improvisation, as Dittus and Lemon (1969) have noted in cardinals.

In the ideal study of effects of individual isolation, one would hatch eggs in an incubator and raise birds by hand from the first day. In practice this is difficult with songbirds, although Lanyon and Lanyon (1969) have had remarkable success. More commonly, birds are taken a few days after hatching and placed in individual isolation. Abnormalities in male song have been detected in a number of species raised in this fashion.

The song of a wild male chaffinch includes a wide array of syllable types within the song, organized in two basic parts, a series of introductory trills and an end phrase (Marler, 1952; Thorpe, 1958, 1961). A male taken from the nest at 4 days of age and reared in individual isolation will develop a song which although normal with respect to its duration, is highly abnormal in the small number of syllable types represented, and in their simplified structure (Fig. 13). If such hand-reared isolated male chaffinches are placed together before the period in which song can be modified is terminated, they will stimulate one another to produce more complex songs. These are divided into phrases and may end with a terminal flourish, although the component syllables are still more or less abnormal. Thus, as in the Arizona juncos mentioned above, chaffinches raised in group isolation develop very differently from those raised in individual isolation.

The white-crowned sparrow is another species in which male song develops abnormally in birds taken from the nest at about 1 week of age and placed in social isolation (Marler, 1970; Marler & Tamura, 1964). In this case, birds raised in group or individual isolation develop in essentially the same fashion (Fig. 14). Figure 14 shows the songs of nine males placed together in a large soundproof room between 5 and 9 days of age and left there to develop song in the following year. The birds were taken from three different areas, in each of which a different local dialect prevailed, as illustrated. In no case did the birds develop the characteristics of the local dialect from the

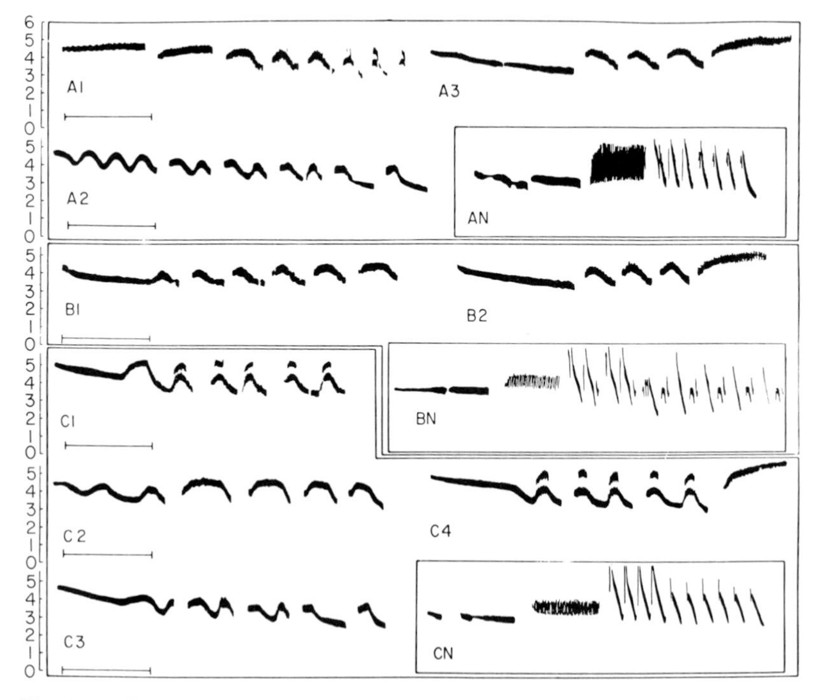

Fig. 14. The full songs of nine male white-crowned sparrows taken as nestlings from three different dialect areas and raised in a large soundproof room. The three inserts, AN, BN, and CN, illustrate the song dialects in the area where the birds were born. Frequency markers, 0–5 kHz; time marker, 0.5 second. (After Marler, 1970.)

area in which they were born. In addition to the characteristics of the dialect, a number of other, more widely shared traits were lacking from their songs, particularly the highly structured trill portion. Nevertheless, we may note the presence of some normal characteristics in the overall duration of the song and in the tendency for sustained whistles which are a basic component in the song of wild male white-crowned sparrows. This point will become significant in later discussion. Figure 15 illustrates the sequence of development in the song of one member of this group, together with parallel displays of development in two other individual males that were raised in individual isolation. The two latter birds developed no signs of the local dialect and are equally lacking in the trill portion of the song, although one bird generated a trill in an interesting way by incorporating into the song a sequence of one of the call notes, which develops normally in isolation.

We should not fail to mention that an equivalent degree of social isolation in some other bird species results in no signs of abnormality. This is the case with domestic chickens (Schjelderup-Ebbe, 1923). Perhaps the best example comes from the work of Mulligan (1966) on the song sparrow, three of which were foster-reared by canaries from the egg, within soundproof chambers. These birds, which had no opportunity to hear adults of their own species at all, developed songs indistinguishable from those of wild song sparrows. This result is all the more interesting when we reflect that the song sparrow is a close relative of the white-crowned sparrow. While adult males of the latter species have only one song type, the male song sparrow may have 16 (Mulligan, 1963). Although these fostered birds developed somewhat smaller repertoires, the songs were quite normal in other respects. Even though a song sparrow is capable of vocal imitation and may copy some songs of neighbors in the wild, exposure to adult song is not necessary for normal development to occur. The same is true of the indigo bunting (*Passerina cyanea*) (Rice & Thompson, 1968) and the cardinal (Dittus & Lemon, 1969; Lemon & Scott, 1966).

VI. Selective Learning of Conspecific versus Alien Song by Social Isolates

In nature, exposure to adult song would normally occur at some phase of the male's early life. Many species of birds are present and generating sound in most natural environments, sounds to which the young male might well be exposed. Reliance upon learning from

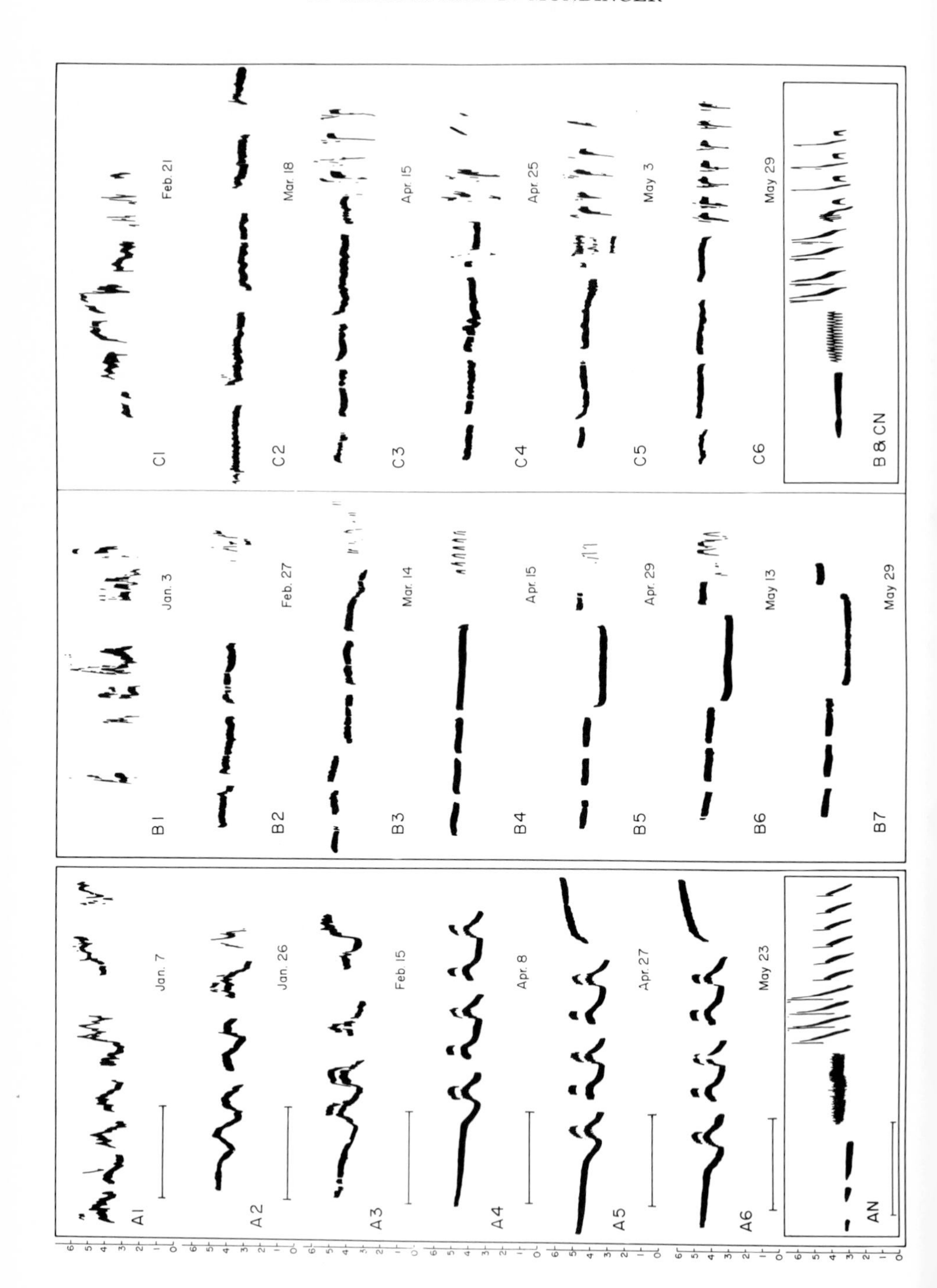
A1
Jan. 7
A2
Jan. 26
A3
Feb. 15
A4
Apr. 8
A5
Apr. 27
A6
May 23
AN
B1
Jan. 3
B2
Feb. 27
B3
Mar. 14
B4
Apr. 15
B5
Apr. 29
B6
May 13
B7
May 29
C1
Feb. 21
C2
Mar. 18
C3
Apr. 15
C4
Apr. 25
C5
May 3
C6
May 29
B8 & CN
6
5
4
3
2
1
0

adults as the basis for normal song development thus involves certain hazards. Birds in which song is learned, song that is distinctively and consistently different from those of other species vocalizing in the same area, must possess some means of restricting the learning to conspecific sounds. Experiments in which young males raised in social isolation are given an opportunity to learn conspecific song come closer to nature if the male is also required to make a choice between the song of his own species and that of another which might be heard in the same environment. The few experiments of this type that have been conducted suggest that males of some species are indeed capable of selective learning.

A. Chaffinch

A young male chaffinch raised in social isolation from 4 to 5 days after hatching develops an abnormal song. If such a bird is exposed to playback of a recording of normal chaffinch song during its youth, the song that it subsequently develops will be closer to the normal condition in a number of respects. Attempts were made to train male chaffinches to sing alien songs by tutoring them during the same period in their first spring with tapes of sounds ranging from a tune on a bird flageolet to a tonally similar bird song. A rearticulated normal song, with the end phrase placed in the center, was learned fairly accurately (Fig. 16). Similarly, a hand-reared bird would also imitate a recording of tree pipit (*Anthus trivialis*) song, which resembles chaffinch song in a number of respects. Models having less resemblance than this were not imitated at all (Thorpe, 1958).

Thus, although the normal song must be learned, there is some selectivity in what the male chaffinch will imitate. As a further point in this direction we may note that a male chaffinch is capable of learning over a period of about 300 days. If it has heard normal song early in this period, subsequent learning becomes still more specific, and only normal chaffinch song will then be imitated.

Exposure to normal song is only effective in achieving normal song development in a socially isolated male chaffinch if the exposure occurs sometime during the first 10 months of its life. At the end of

---

Fig. 15. Song development in three male white-crowned sparrows raised in isolation from 5 days of age. Bird A was acoustically isolated with a group of nine males (cf. Fig. 14), B and C were in individual isolation. AN is the local dialect for bird A, B and CN for birds B and C. Frequency markers 0-6 kHz; time markers 0.5 second. (After Marler, 1970.)

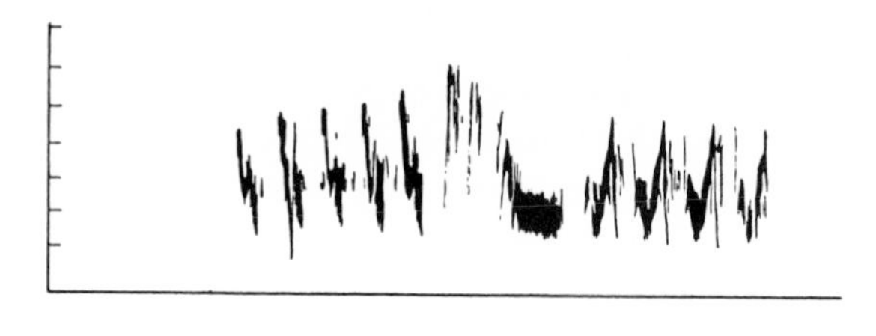

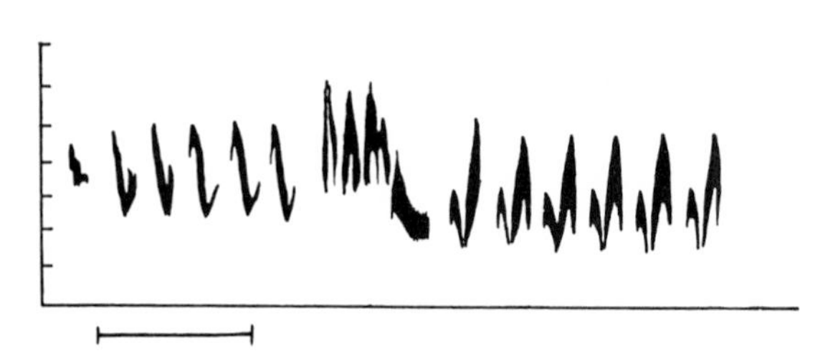

Fig. 16. A rearticulated chaffinch song used as a tutor (above) compared to the song of a hand-reared male chaffinch (below), acoustically isolated from nestling stage but permitted to hear the rearticulated chaffinch tutor song for 12 days, late in the critical period. Frequency markers, 0-7 kHz; time marker at bottom, 0.5 second. (After Thorpe, 1961.)

this period, the bird develops the themes of its full song in a crystallized form and once this is accomplished, exposure to song will have no further effect. In subsequent years the song repertoire will remain virtually unchanged, however abnormal the themes may be. As we shall see, this phenomenon of a critical period of life for song learning recurs in other songbirds as well.

B. White-Crowned Sparrow

If young male white-crowned sparrows taken as nestlings are placed in individual isolation and exposed to 4 minutes of normal white-crowned sparrow song at a rate of 6 songs per minute for a period of 3 weeks somewhere between 7 and 50 days of age, their song will subsequently develop normally, and will constitute a reasonable copy of the model to which they were exposed. Similar exposure between 50 and 71 days of age had little or no such effect. One bird developed a song like that of an untrained isolate. Another had a more normal song, with a trill section, but it bore no other resemblance to the model than this. Training around 100 days of age has only very slight effects, and at 200 days and after, none at all. Thus the critical period for song learning is even more narrowly defined in the white-crowned sparrow than in the chaffinch.

As in the chaffinch, song learning in the white-crowned sparrow is selective in the sense that males exposed to playback of both a normal white-crowned sparrow song and to a recording of another related species during the critical period will learn only the conspecific song (Marler, 1970; Marler & Tamura, 1964) (Fig. 17). Young males exposed to alien song and nothing else during the critical period developed songs resembling those of untrained social isolates.

C. Zebra Finch

One advantage of using the semi-domesticated grassfinches in studies of song development is that some species can readily be fostered upon other members of the same family. Immelmann (1965, 1967, 1969) has exploited this possibility in studies of song learning in the Australian zebra finch (*Taeniopigia guttata*), the African silver bill (*Eudice cantans*), and the Bengalese or society finch, a domesticated form of the Asian striated finch (*Lonchura striata*). Male zebra finches raised by hand or fostered by a female Bengalese finch, with no access to a male song in either case, developed songs which were abnormal in a number of respects, including a smaller number of rather simple and uniform syllable types, considerably longer in duration than in the wild type song. If a male foster parent was also present, however, the young male zebra finch would develop an accurate imitation of his song (Fig. 18). The same result was obtained with male silver bills fostered by Bengalese finches and with Bengalese finches fostered by zebra finches. By separating young male zebra finches from their foster parents at various ages, Immelmann found the learning to be complete by the eightieth day of life. A male zebra finch separated after this age would develop a song identical with that of the foster father. Males isolated between the thirty-eighth and sixty-sixth days developed songs which were made up only of song elements from the foster father, but with a somewhat different total duration and sequence of elements. Males isolated from the foster father before the fortieth day of life developed a song with some elements of the foster father but differing in many other respects such as length, number, and sequence of elements as well as their structure. The song tended to be slow and uniform and thus resembled the song of males raised without a tutor.

It may be noted that the song of the foster father is learned even if members of the young male's own species are within earshot. Thus social bonds ensure some selectivity in the learning process in this species. However, there is also evidence for selectivity at another level, insofar as young males raised only by females under condi-

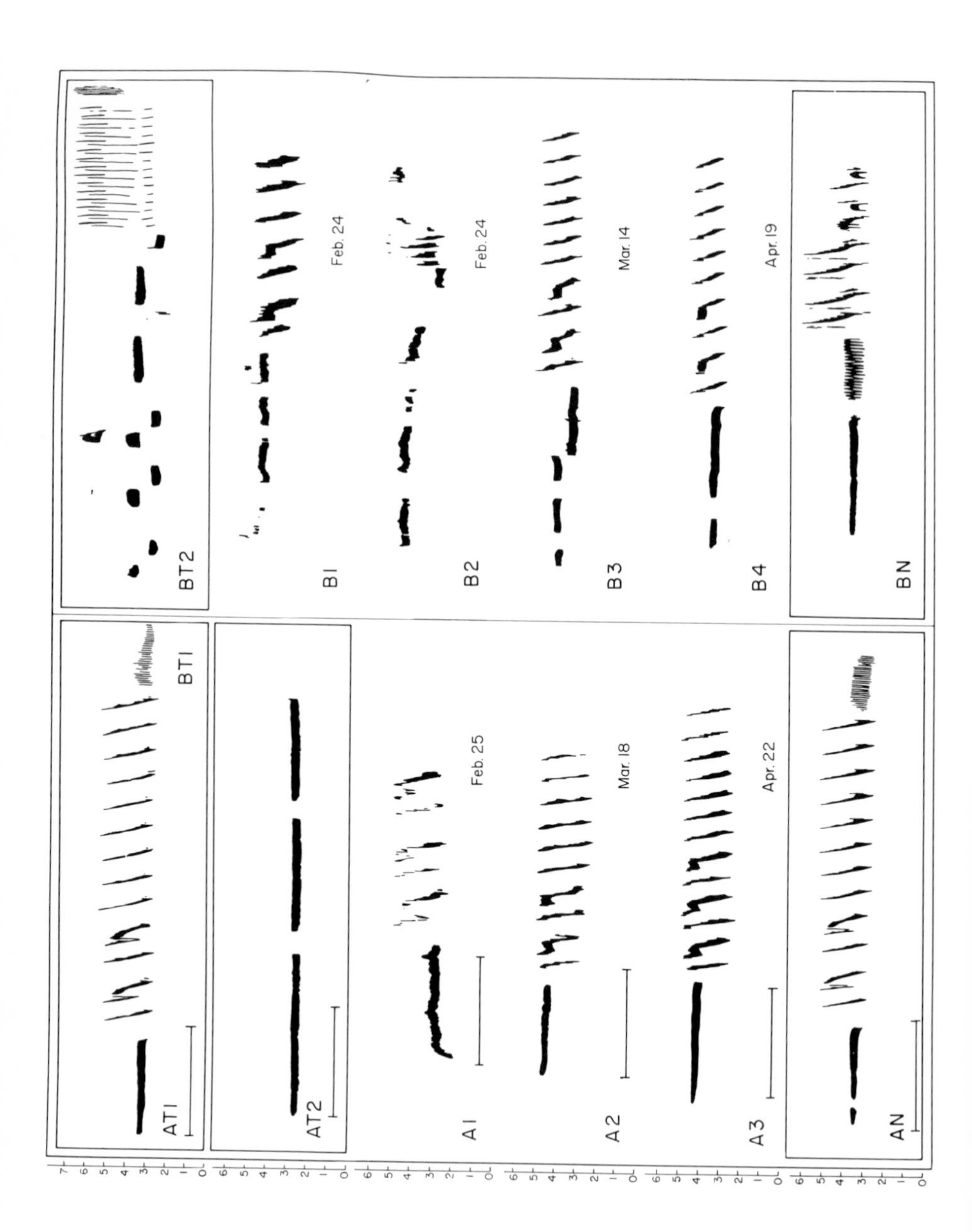
AT1
BT1
AT2
A1
Feb. 25
A2
Mar. 18
A3
Apr. 22
AN
BT2
B1
Feb. 24
B2
Feb. 24
B3
Mar. 14
B4
Apr. 19
BN

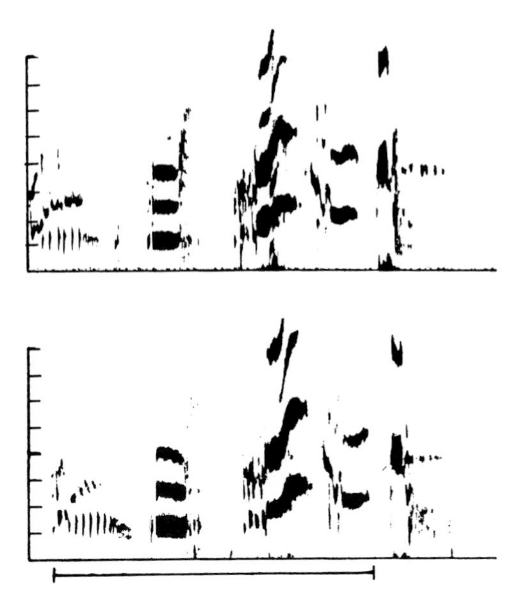

Fig. 18. Song of a male Bengalese finch foster parent (above) and song of a zebra finch raised by the Bengalese finch (below). Frequency markers, 0-8 kHz; time marker at bottom, 0.5 second. (After Immelmann, 1969.)

tions in which they could see and hear both conspecific males and those of other species of grassfinches in neighboring cages, developed songs consisting only of species-specific elements. In contrast to normally raised males, however, they failed to imitate the song of a particular male. Instead they incorporated elements from a number of neighbors in their songs.

Exposure to song after the eightieth day of life seems to have no further effects on subsequent song development. Thus, male zebra finches raised by Bengalese finches, having acquired an imitation of the foster father's song, retained this for many months afterward even though they were housed in an aviary in which many conspecific males were living. As in the white-crowned sparrow, the critical period for zebra finch song learning is quite narrowly defined.

Fig. 17. The emergence of song in two male white-crowned sparrows individually isolated from the fifth day of age and tutored during their critical periods with conspecific and alien songs. Bird A heard 2 minutes of white-crowned sparrow song (AT1) and 2 minutes of Harris' sparrow (*Zonotrichia querula*) (AT2) in the morning and afternoon from the thirty-fifth to the fifty-sixth day of age. B was similarly treated but with song sparrow song (BT2) as the alien pattern from the eighth to the twenty-eighth day of age. AN and BN illustrate the home dialects of these two birds. Frequency markers, 0-6 kHz; time markers, 0.5 second. (After Marler, 1970.)

D. Arizona Junco

A male Arizona junco raised in social isolation develops an abnormal song. If a male is trained for 2 months with 23 minutes of song per day, half Arizona junco song, half Oregon junco song, at a normal rate, beginning about 3 weeks after fledging, the male's songs will subsequently develop quite normally. There are none of the simple single-syllabled songs that occur in an untrained isolate (Marler, 1967). Evidently exposure to the recorded songs diverted development into a more or less normal pathway.

There is evidence of imitation of the Arizona junco song presented as a model. However, most of the song patterns bear no detailed relationship to those presented in training. It seems that exposure to the model has two rather different effects. On the one hand it provides an example of the very general properties of the species-specific pattern which the young male may imitate, and there may be some detailed imitation of syllables heard. However, the training seems to achieve its effect as much by stimulating the male to invent new, more complex syllable types as by providing a model for direct copying. This conclusion is in line with the fact mentioned earlier that a young male raised together with age-mates in group isolation will also develop quite normal song. Thus the interpretation of training experiments such as these is not always as straightforward as it seems.

VII. Mechanisms in Song Learning

A. Effects of Deafening on Vocal Behavior

1. Normal Development in Deaf Birds

We have seen that learning from adult individuals plays a major role in the ontogeny of the vocalizations of some birds. On the other hand, there are many species in which this type of learning is not required for normal development. Audition has a vital function in the former, in permitting young birds to hear the sounds of older individuals. There is another rather different role it might play in enabling birds to hear their own voices, a role which could be equally important in birds that do not require exposure to adult sounds for normal development. Experiments on the effects of deafening on vocal development show that auditory feedback is significant in some bird species, although of little relevance in others.

We know from the work of Konishi (1963) and Nottebohm (1971) that both domestic chickens and ring doves (*Streptopelia risoria*), deafened within a day or two after hatching, will develop all of the normal species vocalizations that have been studied. There are of course some abnormalities in the behavior of deaf birds. Some of the sounds become rare and difficult to elicit, presumably because their occurrence is favored or triggered by acoustical stimuli. But in general, both deaf chickens and ring doves were found to utter their vocalizations under normal stimulus situations. Thus, roosters produced the aerial alarm call upon seeing flying objects. Their food calls were effective in attracting normal hens to them. Some of the chicken vocalizations were graded according to the intensity of external provocation and this also occurred in the deaf birds (Konishi & Nottebohm, 1969) (Fig. 19).

Thus, doves and chickens are able to produce a normal vocal repertoire not only in isolation from sounds of adults of their species (e.g., Schjelderup-Ebbe, 1923) but also when deprived of the ability to hear their own voice, at least after the age of a few days. It is necessary to withhold judgment on this last point, since birds of both species had the opportunity to hear their own sounds for a day or two before deafening occurred. The possible role of this early experience in subsequent vocal development remains to be explored. However, Gottlieb is studying the developmental role of very early vocalization in the behavioral development of ducklings. In unpublished work he finds that ducklings devocalized in the egg and then allowed to produce sounds a week or two afterward seem to develop a normal repertoire. Thus, it is likely that experience that the young bird may have of its own vocalizations in its first day or two of its life does not affect subsequent vocal development.

2. *Abnormal Vocal Development in Deaf Birds*

A young male white-crowned sparrow usually begins to sing when a little over 100 days of age, at irregular intervals, becoming more continuous until full song emerges, usually after 200 days of age. If such a bird is deafened between 40 and 100 days of age, before singing behavior has developed, he will subsequently begin to sing on a more or less normal schedule, but the structure of the song produced will be quite abnormal.

A young male raised in social isolation develops a song which lacks the characteristics of the local dialect and a number of species-specific properties, but nevertheless retains some normal traits, in-

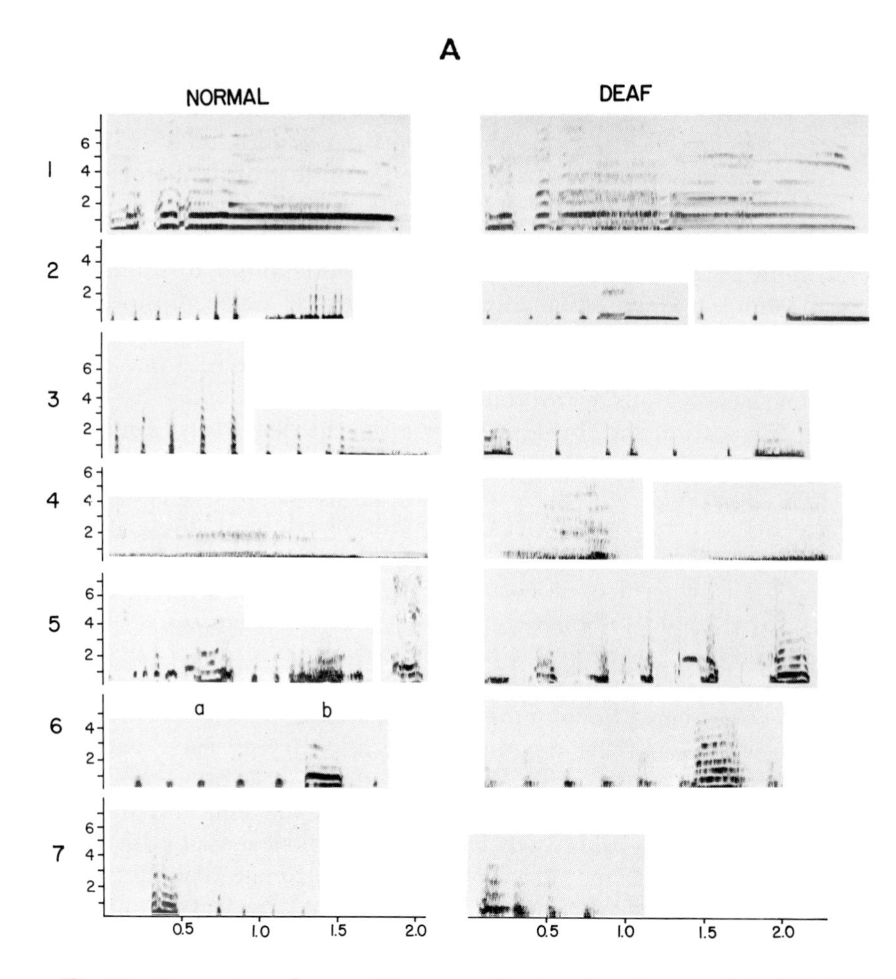

Fig. 19. Comparison of seven different vocalizations from intact and deaf chickens (A) and two different calls from deaf and intact ring doves (B). Ordinate, frequency in kHz; abscissa, time in seconds. (After Konishi & Nottebohm, 1969.)

cluding the sustained pure tones which are a characteristic part of the normal song. In the deaf bird, the song is still more abnormal, even these pure tones tending to disappear. Only the overall duration remains as a normal trait (Konishi, 1965b) (Fig. 20).

Similarly, Oregon and Arizona juncos and black-headed grosbeaks (*Pheucticus melanocephalus papago*) develop abnormal songs if deafened before the onset of singing, although the discrepancy between the songs of deaf and intact birds is less great than in the

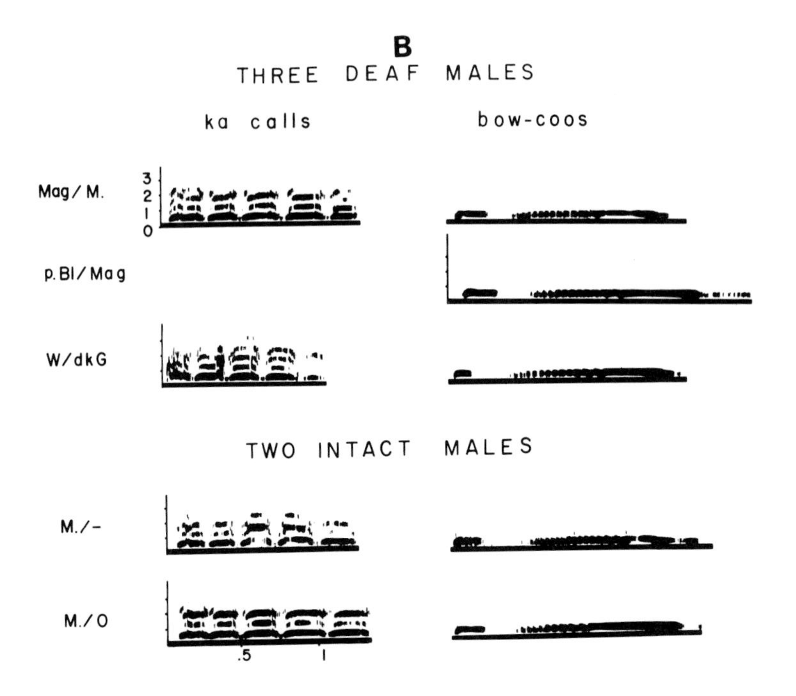

Fig. 19b.

white-crowned sparrow (Konishi, 1964, 1965a). Mulligan (1966) found that a male song sparrow deafened before song had developed sang in a highly abnormal fashion. The songs of early-deafened white-crowned sparrows, Oregon juncos, and song sparrows show fewer striking species differences than those that characterize the natural songs of this group of closely related species. Their variability makes detailed comparison difficult. Figure 21 presents three examples selected for their similarity. Others are more distinct. Nevertheless, it would be impossible to find natural songs of these three species as similar as those portrayed in Fig. 21 from early-deafened birds.

Konishi and Nottebohm (1969), noting that no passerine bird has yet produced a perfectly normal song if deafened before the completion of song crystallization, summarize the abnormalities of their singing as follows.

a. Absence of Unit Organization. The most extreme effect of deafening is the disappearance of all the recognizable structural entities of song.

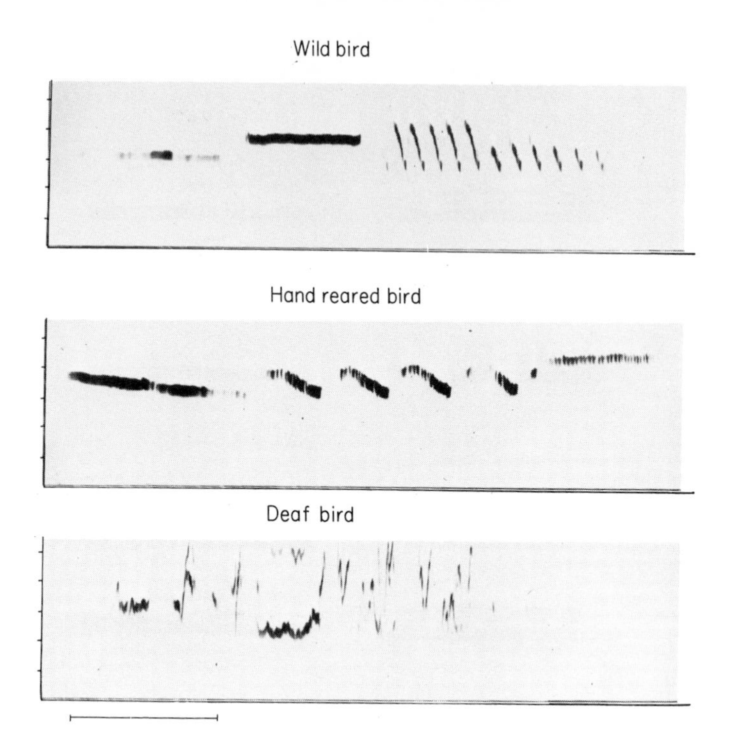

Fig. 20. Selected songs from three male white-crowned sparrows, one wild, one hand-reared in acoustical isolation and one deafened early in life. Frequency markers, 0–5 kHz; time marker, 0.5 second. (After Konishi, 1965b.)

b. Abnormal Notes and Syllables. If deaf birds produce distinct notes and syllables at all, these tend to contain abnormal patterns of frequency modulation. Notes and syllables produced by deaf birds usually appear irregular and fuzzy on the audiospectrogram.

c. Instability in the Form of Notes and Syllables. In sharp contrast with the sounds of intact birds, the notes and syllables of deaf birds are not repeated in exactly the same form either from song to song or within a song, even though their general patterns are maintained (Fig. 22). This type of fluctuation is seldom observed among intact birds. Despite this short-term instability, deaf birds can maintain to a considerable extent the individual characteristics of their songs in successive years.

3. *Invalidation of Learning by Deafening*

Refinement of the deafening experiment is possible in the white-crowned sparrow as a result of the temporal separation between learning the song, between 10 and 50 days of life, and reproduction of vocal renditions of that sound at a later age. The deafening operation can be conducted after learning is complete but before singing has developed. When this is done, a male white-crowned sparrow develops a song indistinguishable from that of a male deafened early in life, without any training. Thus the deafening seems to erase the memory trace which persists after learning or makes it inaccessible.

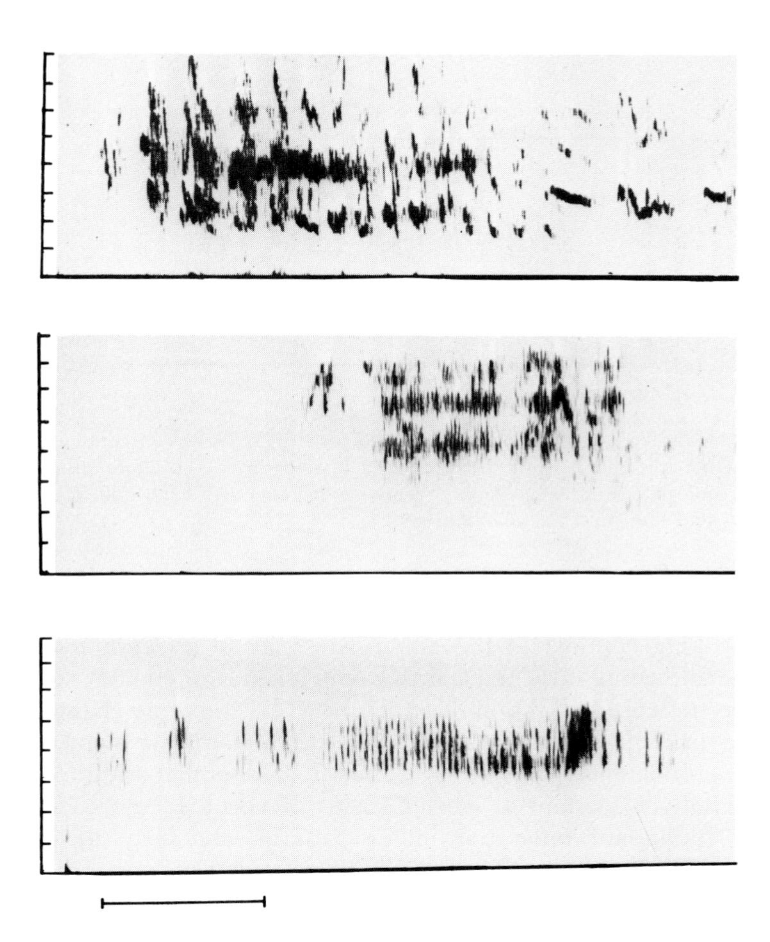

Fig. 21. Comparison of songs from three closely related species of early-deafened birds. Song sparrow (above), Oregon junco (middle), white-crowned sparrow (below). Frequency markers, 0–8 kHz; time marker, 0.5 second. (Oregon junco and white-crowned sparrow compliments of Dr. M. Konishi; song sparrow after Mulligan, 1966.)

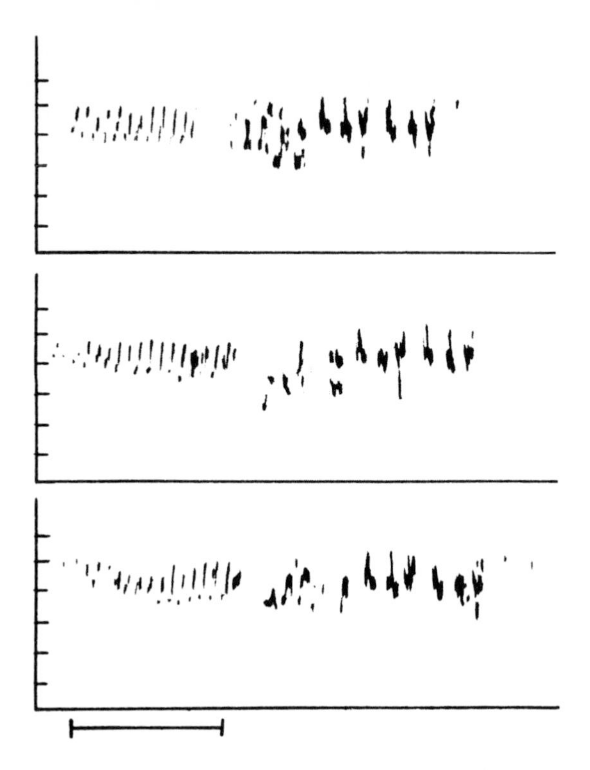

Fig. 22. Instability in the song of deafened passerines. Three consecutive songs of a deaf white-crowned sparrow illustrating fluctuations in fine structure although the general song pattern remains stable. Frequency markers, 0–6 kHz; time marker, 0.5 second. (After Konishi & Nottebohm, 1969.)

4. *Reduced Effect of Deafening at a Later Age*

In striking contrast to the abnormal songs of an early-deafened bird, a male white-crowned sparrow which is deafened after song has fully crystallized will retain its structure with little or no change. The same is true of a male chaffinch, which also manifests many abnormalities if deafened early in life. Thus it appears that auditory feedback, vitally important at earlier stages of vocal development, becomes redundant once the motor patterns are fully developed (Konishi, 1965b; Nottebohm, 1967, 1968) (Fig. 23).

B. Mechanisms Underlying Selectivity in Song Learning

Various lines of evidence point to the existence of constraints imposed on the process of song learning, both in time, such that there are critical periods when learning takes place most readily, and in the acoustical patterns which are most readily learned. In some species, social constraints are involved. However, in the white-crowned sparrow and the chaffinch the restrictions on the type of sound that will be learned are not attributable to social influences, since they are manifest in a bird raised in isolation from a few days after birth, with selection made from sounds coming through a loudspeaker. In these cases the constraints must be in some sense endogenous to the male bird. Two rather different kinds of neural or neuromuscular mechanisms, motor or sensory, might impose conspecific constraints on the learning process.

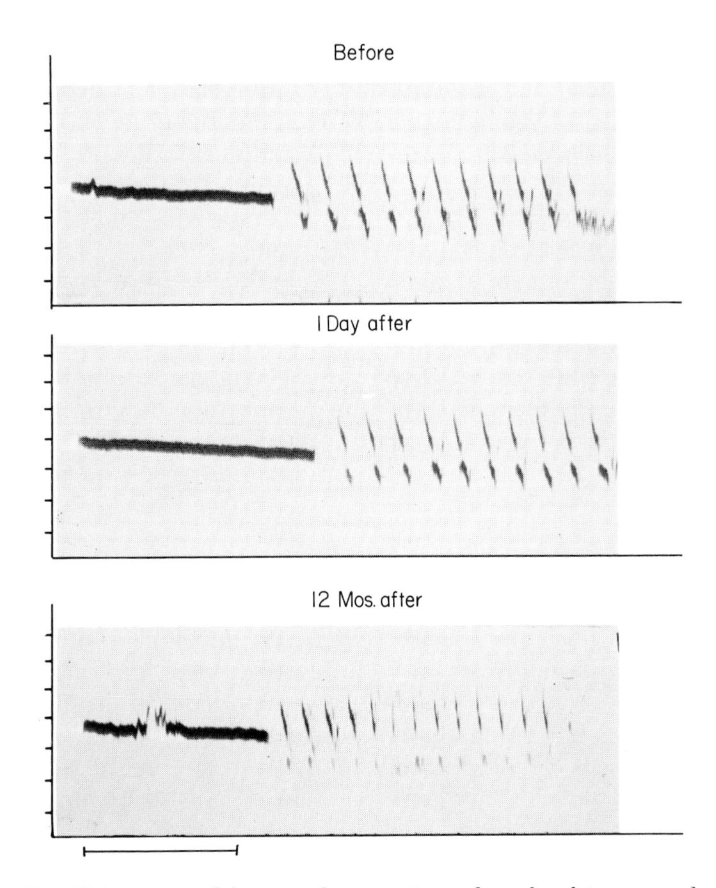

Fig. 23. Maintenance of the normal song pattern of a male white-crowned sparrow deafened when adult. Song pattern given before deafening operation (above) compared with postoperative song patterns. Frequency marker, 0–7 kHz; time marker, 0.5 second. (After Konishi, 1965b.)

1. Motor Mechanisms

The structure of the sound-producing equipment, in this case the respiratory machinery, the syrinx and its associated membranes, muscles, and resonators must impose some restrictions on what sounds can be produced. Could these restrictions be sufficiently specific to provide a basis for the selection of conspecific song for imitation, and rejection of the songs of close relatives?

There are several reasons for doubting that such motor constraints would be sufficient. The sound-producing equipment will of course set limits, but the evidence suggests that these limits are very broad. No difference has been detected between the syringes of closely related birds such as the white-crowned sparrow and the song sparrow, and the structure of the bird syrinx is in fact known to be a very conservative trait that changes only slowly in the course of evolution. It is widely used as a taxonomic character at the higher levels of phylogenetic classification. There are many examples of species with similar syrinx structure producing very different vocalizations. Thus a chaffinch and a bullfinch have very similar syringes, but a bullfinch raised by hand and imprinted on its human keeper will imitate a great variety of unnatural sounds, including musical instruments (Thorpe, 1955). As we have seen, a chaffinch is very restricted in what it will imitate. Furthermore, the resemblance noted between the songs of a white-crowned sparrow, a song sparrow, and a junco deafened early in life suggest that the output from the syrinx of these three species is very similar when freed from auditory constraints. Thus, it seems likely that limitations on the sensory side are involved.

2. The "Auditory Template" Hypothesis

Experiments demonstrating abnormalities in the song of an early-deafened bird, even in species which can develop a normal song when raised in social isolation with hearing intact, lead naturally to the invocation of a sensory mechanism to guide vocal development. The hypothesis has gradually developed over a period of years that certain songbirds develop vocalizations by reference to something we may think of as an "auditory template" (Konishi, 1965b; Konishi & Nottebohm, 1969; Marler, 1963, 1970). This template is visualized as a mechanism or mechanisms residing at one or more locations in the auditory pathway which provide a model to which the bird can match feedback from its vocalizations. Accomplishment of this matching is presumed to take place over a period of time, as the bird acquires sufficient skill in the control of its sound-producing equipment to generate a perfect match with the dictates of the template.

In some species, such as the song sparrow, an individual raised in complete isolation from species members seems to possess a sufficiently well-specified template that it can generate normal song, as long as the individual can hear itself (Mulligan, 1966). However in other species, such as the white-crowned sparrow or the chaffinch, the initial specifications for the template are less complete.

A male white-crowned sparrow raised in social isolation from a week or so of age will develop a song which, although abnormal in its overall pattern, does nevertheless possess certain natural characteristics, particularly the sustained pure tones which are one basic element in the natural song (see Fig. 15). These pure tones are largely absent from the song of a bird deafened early in life (Fig. 20). Thus a male white-crowned sparrow already seems to possess at the age of 10–15 days a crude auditory template of the song. Although this does not suffice to generate normal song by itself, it may be adequate to focus the young male's attention on sounds of his own species rather than those of others to be heard in the same area, such as the song sparrow. We presume that, as he listens to conspecific song, the properties are somehow incorporated in the template, which thus becomes more highly specified, embodying properties both of the species song in general, and of the particular local dialects which the male has experienced. When the male subsequently comes into song, this more refined template will guide development along a normal pathway.

The essential notion exemplified by this "template" is that of active filtering of incoming sensory information, not unrelated to what is implied by the "innate schema" of von Uexküll and Lorenz and the "innate release mechanism" of Tinbergen. The same template can serve both as a kind of filter for focusing attention on sounds that match its crude specification, and as the vehicle for retaining information about the more detailed characteristics of sounds, and for subsequently translating that information into vocal activity.

3. *The Developmental Basis of the Auditory Template*

Beyond the fact that the "template" seems to exist already as a young male white-crowned sparrow enters the critical period for song learning, from about 10 to 50 days of age, we know little of its developmental basis. It must be remembered that the experiments on this species related here were all conducted with birds taken as nestlings. They had several days' potential experience of normal song before coming into the laboratory. Although this experience was evidently insufficient to generate normal song development, it might have sufficed to establish a crude template, the effects of which would then be manifest in a bird raised in isolation after the nestling phase. Until birds of this species are raised from the egg in the laboratory, the developmental basis of the so-called song template remains unexplored. It is conceivable that the male's auditory experience of his own very early vocalizations may play a part. There is evidence from chaffinches that this is indeed the case.

An intact normal chaffinch raised in social isolation from a week or so of age develops a song which, like that of a white-crowned sparrow, is abnormal, yet possesses certain traits in common with that of a wild bird. In studies of the effects of deafening at various stages of the vocal development in chaffinches, Nottebohm (1967, 1968; Konishi & Nottebohm, 1969) demonstrated that a male deafened early in life, at about 90 days, before song development had proceeded very far, eventually came to possess a song with very rudimentary structure, much more abnormal than that of an intact isolated bird. The process of song development in chaffinches extends over several weeks, passing through various stages of subsong, plastic song, and finally full song. Chaffinches deafened at an age at which they were presumed to have had ample experience of their own subsong, but nothing more, resulted in songs which began to approximate those of an intact isolate. One interpretation of this result is that the auditory template which guides song development in an untrained male chaffinch develops at least in part as a consequence of the male's auditory experience of the earlier stages in his own song development. Deprived of such auditory feedback, song development is diverted to a still more elementary form (Fig. 24). Alternatively, performance of subsong, and plastic song, with feedback, may be necessary to accomplish a match between song performance and a preexistent auditory template. According to this interpretation, delaying the age of deafening allows a closer and closer match between vocal output and the specifications of an auditory template.

Similarly, in the process of song development in a chaffinch which has been trained, auditory feedback seems to be important. The effects of training first become clearly manifest in the early stages of so-called "plastic" song. A bird deafened after only a day or two of experience of this stage subsequently reverted to a somewhat simpler song pattern; if the operation was postponed for another 10 days, so that the bird had had more experience of this plastic stage of de-

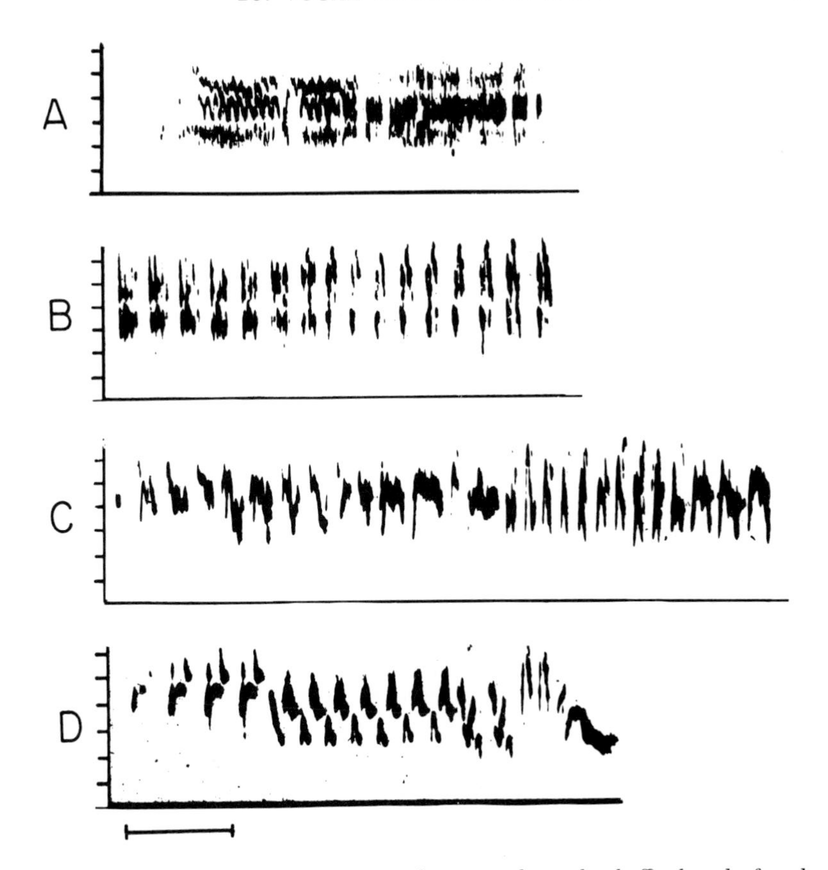

Fig. 24. Comparison of songs sung in the spring by male chaffinches deafened at various stages of the development of song. (A) Deafened at about 90 days of age, (B) deafened at about 200 days of age with some subsong experience, (C) deafened at about 300 days of age with 3 days "plastic" song experience, (D) deafened as an adult bird. Frequency markers, 0–6 kHz; time marker, 0.5 second. (After Nottebohm, 1967.)

velopment, then the song persisting afterward was much closer to the normal wild type. Thus, auditory feedback is important in the chaffinch in the realization of a complete match of song with the dictates of a template that has been refined as a result of exposure to normal song.

4. Social Constraints on Song Learning

Although social restrictions on the direction taken by song learning can be dispensed with in the chaffinch and the white-crowned sparrow, they are critically important in some other species. Nicolai

(1959) has evidence that young bullfinches selectively learn both the song and one of the other vocalizations from their father. He discovered this when a male bullfinch raised by canaries learned some canary phrases and in turn transmitted them to his offspring. Similarly, the sons of a male bullfinch that had been raised by hand and taught to whistle a tune learned the same abnormal song from him, even though normal songs could be heard nearby. We have already noted that young zebra finches will show a preference for learning the songs of the father, or the foster father, even though they also possess an endogenous mechanism for guiding song development as well (Immelmann, 1965, 1967, 1969).

No doubt song learning is guided along social channels in other bird species which have enduring bonds between parent and young during the period at which learning takes place. However, such bonds are of only brief duration in many small birds, as in the chaffinch and the white-crowned sparrow, whose family breaks up as the young are weaned. Thus these species require some other mechanism for ensuring the selectivity of song learning.

5. *The Nature of Reinforcement in Song Learning*

In his autism theory of language learning, Mowrer (1950, 1958) proposed that each sound acquires secondary reinforcing properties as a result of association with primary reinforcement of other kinds from the parent. A similar process may underlie song learning in species in which social constraints are operating to impose selectivity on the learning. Since many songbirds feed their young by mouth, it is not inconceivable that direct reinforcement by parental feeding might be involved in some species. Alternatively, some other kind of social stimulation from parents consequent upon utterance of sounds resembling their own might serve as a reinforcement.

Neither of these processes can be invoked to explain song learning in the white-crowned sparrow. In the first place, learning precedes utterance of copies of the sounds heard by a period of many weeks. In the second place, training is accomplished effectively by presenting recorded sounds through a loudspeaker, with no other concomitant stimulation. One can only assume that production of a sound resembling that heard previously during the critical period is in some sense an intrinsically reinforcing act.

One gathers that some investigators are also skeptical of the adequacy of Mowrer's autism theory as a complete explanation of speech learning in children. Even with talking birds, where Mowrer felt this hypothesis to be especially relevant, Foss (1964) found that mynah birds learned a whistle sound played through a loudspeaker just as well without simultaneous food reinforcement as with it, there being no social stimulation in either case.

6. *The Significance of Subsong*

In many of these species we have been considering, the male song does not suddenly appear in complete form. It develops over a period of weeks, passing through several distinct stages. The first is known to ornithologists as subsong (Lanyon, 1960; Thorpe & Pilcher, 1958), and differs from full song in a number of respects. It is not as loud as full song, and the components are often given in long irregular sequences. In addition to fragments of song, other sounds from the repertoire are often included. The frequency of its components tends to fluctuate erratically, giving subsong a wider range of sound frequencies than occurs in full song. The variability is such that it is hard to discern units which are repeated with any great precision.

As development proceeds, the subsong becomes louder and in many cases more clearly segmented. Finally, the patterns crystallize into stereotyped themes, thus completing the development of full or primary song.

The significance of subsong is uncertain. On the one hand, it may be regarded as a by-product of increasing androgen levels, without any intrinsic significance. On the other hand, it is conceivable that subsong serves to aid the bird in acquiring skill in the use of its sound-producing equipment as a necessary stage in the accomplishment of full song. There are reasons for thinking that the latter may be true in at least some cases. Male chaffinches and white-crowned sparrows coming into song in their second or third years of age pass through the stages from subsong to full song more quickly than young males. This might be expected after the adult song pattern has been fully established, although it could also be argued that rates of hormone production or sensitivity to hormonal effects might be enhanced in older birds that have already experienced one annual cycle.

A comparison of development in trained and untrained male white-crowned sparrows is illuminating in this regard (Marler, 1970). Males that have been trained with a normal song usually pass through the subsong stage rather quickly. Full song is generally crystallized within a few weeks after the onset of subsong. On the other hand, birds raised in social isolation, or unsuccessfully trained either with presentation of conspecific song outside the critical period or exposure to alien song, often remain much longer in this transitional stage of song development. It may take months for full song to become completely crystallized. Thus, a bird which is destined to produce abnormal song may persist with variable patterns for a major part of the singing season. If we think of effective training as providing the bird with a clear goal toward which vocal development should then proceed, it is perhaps understandable that the transitional stages of subsong should be passed over more quickly than in a bird which is lacking such specific instruction.

Careful study of the structure of the components of subsong reveals another difference between white-crowns that have and have not heard normal song during a critical period. Although the subsong of the former shares many formal properties with that of untrained birds, the resemblance between the structure of the earliest components produced and those of the training song is apparent. Thus, some of the basic acoustical components are established early in the course of song development. However, these fragments are produced in a very disorderly fashion at first, and the complete match with the training song is only achieved gradually as the syllables become organized into the appropriate order and timing.

These results are not inconsistent with the auditory template hypothesis advanced earlier to explain song development. Although a bird trained with normal song during the critical period is assumed to possess a highly specified "template" already, the accomplishment of a perfect match between this template and the vocal output will take time. Novel and complex operations of the sound-producing equipment are required. The sequence of development observed in a trained bird is what one might expect if, by engaging in subsong, it thus acquires more proficiency in matching vocal output to the specified pattern of auditory feedback.

To explain development in a male white-crowned sparrow that has been prevented from hearing normal song during the critical period, we have postulated a cruder template which provides only a rough specification for song development. This will allow considerable latitude, and it might be anticipated that in development guided by this crude template, crystallization of song patterns will take longer than in a trained bird which possesses more highly specified instructions. Untrained birds show every sign of vacillation, sometimes changing major patterns several times in the course of development. It will be recalled that an adult male white-crowned sparrow possesses only one song type. Some untrained males had two song themes at intermediate stages of development, subsequently rejecting one of them. This was never recorded in birds that had been subjected to effective training.

If we carry this line of argument still further, one might predict that birds deafened early in life, thus deprived of access even to the crude template of an intact untrained bird, would show an even more attenuated and variable sequence of song development. This is precisely what happens. Apart from their abnormal quality and pattern, the most striking characteristic of the songs of male white-crowned sparrows deafened in youth is their instability (Konishi, 1965b). The same is true of other species deafened early in life (Konishi & Nottebohm, 1969). Some deaf birds even pass through their entire first singing season without a stereotyped theme emerging. Many show some degree of crystallization, but nevertheless vary much more in successive repetitions of a theme than a normal bird. Even successive units within a single song vary widely. The deaf birds seem to have difficulty in maintaining a steady tone. Many of these traits are reminiscent of an early stage of normal subsong development. These birds behave almost as though deafening arrests their development at this stage, in at least some respects, even though the rhythm of song delivery and the singing posture do continue to develop in the normal fashion. We have already noted that a male chaffinch deafened in youth, but after some subsong experience, develops a more advanced song than a male deafened still earlier and lacking such experience (Nottebohm, 1968).

It thus seems probable that the subsong of young males does have developmental significance, perhaps providing opportunity both to learn new operations with the sound-producing equipment, and to employ trial-and-error learning to match the output with a specified pattern of auditory feedback. The recurrence of subsong at the start of each singing season in adult males remains unexplained.

7. Critical Periods for Song Learning

The capacity for song learning tends to be confined to a certain period of life in many bird species. A male chaffinch is able to learn song during the first 10 months of its life, at the end of which the song themes crystallize into their final stereotyped patterns (Thorpe, 1958). In this case we have some notion of the factors that control the timing of this critical period.

Nottebohm (1967, 1969) prevented a male chaffinch from coming into song at the normal age by castration at 7 months (Fig. 25). This bird was then induced to sing at 2 years of age by implantation of a testosterone pellet. While still at the plastic stage of song development, it was exposed to playback of field recordings of two song themes chosen because they differed strikingly from the bird's own developing song. The themes that subsequently developed matched

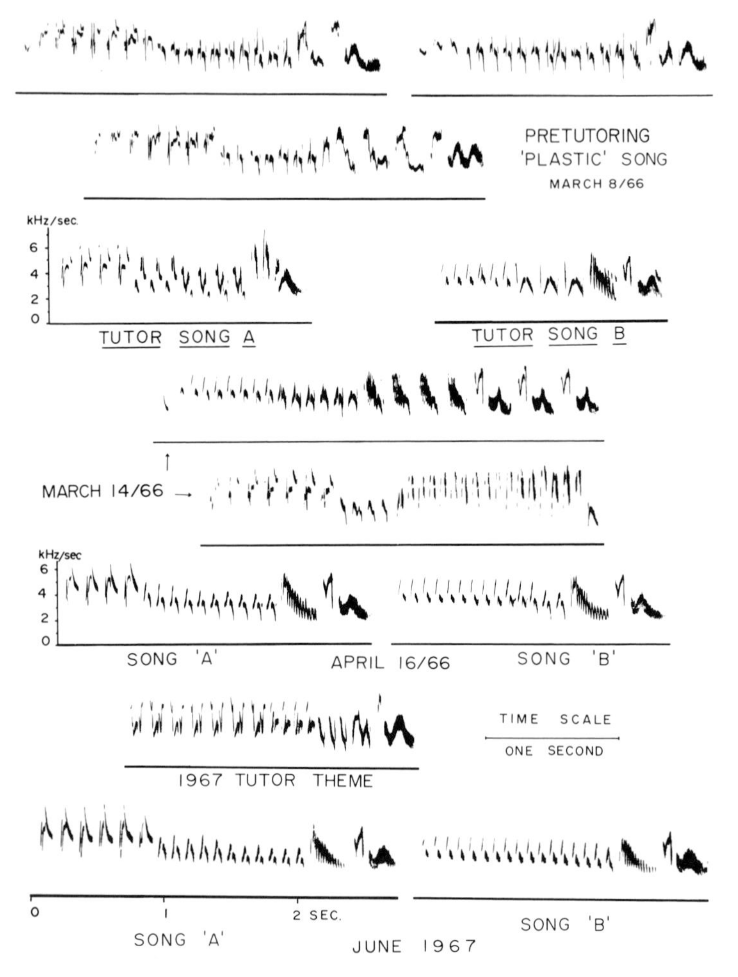

Fig. 25. Delayed critical period in a castrated chaffinch. Castration at 7 months prevented normal song development during the male's first year of life. After testosterone therapy in the male's second winter, song development progressed to the "plastic" song stage (March 8, 1966). The bird was then tutored with song patterns (tutors A and B) different from the developing pattern. A capacity to learn from this tutoring in the second year of life was demonstrated by the inclusion of elements of one tutor song into the final song patterns (April 16, 1966). After crystallization of these song patterns tutoring in a subsequent year (1967) had no demonstrable effect. (After Nottebohm, 1969.)

the training songs in many of their details. Thus this bird demonstrated a capacity for song learning more than a year after the critical period would normally have terminated.

The ending of the critical period for song learning in the chaffinch is not strictly an age-dependent phenomenon. Rather, it seems to follow the crystallization of song, at whatever age this occurs. Whether it results from the crystallization of song patterns as such, or whether it is determined by some other correlate of the high testosterone levels at this time cannot be determined.

In other species, such as the white-crowned sparrow and the zebra finch, there is a critical period which terminates before the onset of full song. In this case motor fixation cannot explain the termination of the critical period. It may be, as Konishi and Nottebohm (1969) have suggested, that the critical period for song learning in birds is not a unitary phenomenon. Rather it may result from the interaction of two critical periods, one for the establishment, or elaboration of the auditory template, and another for its transformation into vocal activity. Although in some species the two periods are separate, in others, such as the chaffinch, they may terminate simultaneously. There is a fertile field for further investigation of the physiological determinants of such critical periods for song learning.

VIII. Genetic Contributions to Song Development

We can discern many points in the ontogeny of bird vocalizations at which genetic influences might readily intrude. The structure of the sound-producing equipment, the syrinx and its associated resonators and respiratory apparatus, must vary under the control of genetic factors. Variations in the timing of critical periods might be under genetic control, as they are in embryological development.

Species differences in the structure of the so-called "auditory template" might be under genetic control. We have drawn attention to the greater similarity of the songs of early-deafened juncos, song sparrows, and white-crowned sparrows, as compared with the normal songs of these species which are radically different in many respects.

In species that undergo normal vocal development when deafened early in youth, such as domestic chickens and ring doves, it seems likely that genetic factors have a direct influence upon species differences in the pattern of motor outflow to the sound-producing equipment from the central nervous system, although the possible significance of patterned proprioceptive feedback must also be borne in mind.

Few attempts have been made thus far to explore the mechanisms involved. Lade and Thorpe (1964) studied the song of hybrid doves. They found that the relationship between the structure of their vocalizations and those of the parental species varied from one combination to another. In some hybrids, the song patterns in the F_1 generation were intermediate. In other crosses, the characteristics of one parent seemed to be dominant. Occasionally, the temporal pattern of the song broke down altogether in the hybrid so that its song lacked any consistent rhythm, even though the parent species had marked rhythms in their songs. They conclude that the specific rhythms of the songs of doves are probably coded at the central nervous system level. The tonal quality of the songs, not radically changed in any of their hybridization studies, is probably a function of the structure of the syrinx and respiratory tract. Wallace Craig noted many years ago (1908, 1914) that if doves of the genera *Streptopelia* and *Geopelia* were incubated and raised by another species, they nevertheless developed normal patterns of song.

Similarly, we know that a cockerel raised in isolation in an incubator will crow normally as an adult (Schjelderup-Ebbe, 1923). There is an ornamental strain of cockerel in the Orient, bred for a very attenuated pattern of crowing. These birds are matched in crowing contests in much the same way as roller canaries in Europe and North America. Nevertheless, there has been no systematic study of the role that genetic factors play.

In one study of the structure of crowing induced in young domestic cockerels by testosterone treatment, comparisons were made between three hybrid strains, but no consistent differences were found. Further study of three highly inbred lines of white leghorns of different parentage revealed crowing patterns that were less variable than those of the hybrid strains, but consistent differences between lines could not be found. It was concluded tentatively that, while genetic factors seem to control such characteristics as the duration of crowing and the period of frequency oscillation within crows, which is similar in young and old birds, the various types of inbreeding for egg and meat production have had little consistent effect on the crowing patterns. The great variability of crowing patterns within a strain might result from heterozygosity in the genes governing crowing or from variable expressivity of the genes concerned, or a combination of both (Marler, Kreith,& Willis, 1962b).

Domestication has had an impact on the vocalizations of several bird species. As a result of direct selection for voice characters, the song of the roller canary is now very different from that of the wild canary (Marler, 1959). Around the turn of the century a special type of domestic pigeon, the trumpeter pigeon, was bred for an unusual pattern of song, although this habit was subsequently neglected by breeders and lost (Levi, 1951). As a result of the ease with which they can be bred, domestic species may prove ideal subjects for further analysis of the genetic control of vocalization.

IX. Improvisation as a Factor in Vocal Development

A. Individual Improvisation

Although the songs of most birds possess certain species-specific traits, there is usually sufficient latitude within the basic framework of each species for a high degree of individuality. The songs of male white-crowned sparrows within one area share many characteristics of the second or trill portion of their song, but there are often strong individual differences in the early part. In other species, such as the Oregon junco, individual differences are much more striking (Fig. 26).

What is the ontogenetic basis of this individuality? We can discern several possibilities. It might result from exposure to different environmental sounds or from variations in the accuracy with which sounds are copied (see Dittus & Lemon, 1969). The characteristics involved might be under genetic control, and the species might be polymorphic for the genes involved. Yet another possibility, which seems the most plausible in many cases, is that there is an element of vocal invention or improvisation in the ontogenetic process which is likely to lead to individual differences.

The song of the European blackbird is very complex and varied, and has been studied both in the laboratory (Messmer & Messmer, 1956; Thielcke-Poltz & Thielcke, 1960) and in the field (Hall-Craggs, 1962; Tretzel, 1967). Whether raised in social isolation or not, even when deafened, all blackbirds studied showed a tendency to recombine song elements in different ways to produce a large number of patterns. Some were stable while others persisted for a time and then changed, so that new sound sequences appeared repeatedly during the life of an individual male.

Although some of the changes are attributable to new environmental sounds, others can only be explained as inventions or refinements imposed by the bird. A wild blackbird studied by Hall-Craggs (1962) began one season with 26 song phrases which were then gradually transformed as a result of recombination, loss, or repetition of

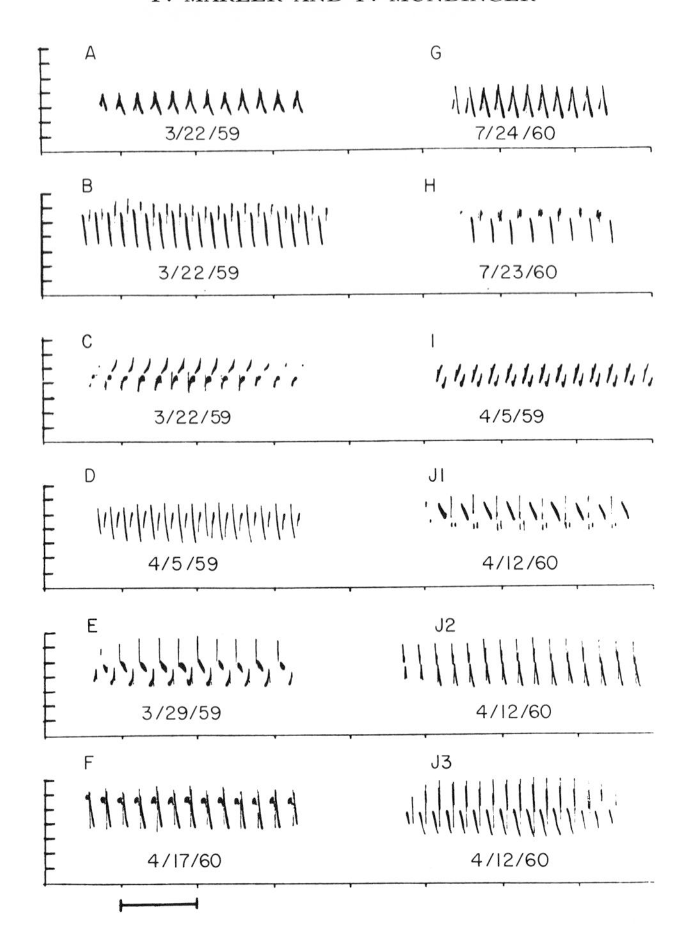

Fig. 26. Songs of nine wild Oregon juncos illustrating individual differences in song pattern. J1-3 are from a single individual. Frequency markers, 0-7 kHz; time markers, 0.5 second. (After Marler *et al.*, 1962a.)

some elements and the addition of new material. There seems no question that invention of new motor patterns takes place. One is tempted to strike an analogy with play behavior.

A similar process seems to underlie the development of structure of the syllables which make up the song of Oregon juncos and Arizona juncos (Marler, 1967; Marler, Kreith & Tamura, 1962a). Birds raised in social isolation developed syllables with a simpler structure than those of wild birds. Birds subjected to training with recorded natural songs developed syllables that were more elaborate, and thus closer to normal. However, they rarely bore any close resemblance to sounds to which they had been exposed. The inference was drawn from these studies that the elaboration of syllable structure results not from imitation as such but rather as a consequence of unspecific stimulation to improvise novel and more elaborate vocal patterns. Such vocal inventiveness is probably a significant factor in the development of certain elements in the song of many bird species, with effects that must often supplement or perhaps even overwhelm the effects of learning from other members of the species.

B. Group Improvisation

As a special case of improvisation, there are two well-documented cases in which young birds placed in group rather than individual isolation, together with age-mates, subsequently produced songs more elaborate and closer to the natural condition than those developed by individual isolates. Thorpe (1958) has described such a case in group-isolated chaffinches, which developed songs with many more normal traits than those of individual isolates. The Arizona junco is another example (Marler, 1967), all the more dramatic since the natural song of this species is extremely varied even within a single population (Fig. 27).

A rough tabulation of the structural characteristics of songs of wild juncos permits a comparison with those of five group-isolated males. The correspondence is remarkable, and provides a striking contrast with the much simpler songs of an individual isolate (Fig. 12).

There is a notable difference between these two cases. While the group-raised chaffinches all shared the same song that was developed, the Arizona juncos all had different songs. Thus, in the chaffinches, we may assume that each bird builds in turn upon improvisations generated by the other. In the juncos, the interchange is less specific, so that the birds stimulate each other to improvise much more freely than they would in isolation, without imitation being necessarily involved.

X. Discussion and Summary

Variation in behavior must ultimately be attributable to two sources, genetic and environmental, and to the outcome of their interaction. This is true whether variation appears as a difference between the behavior of individuals of the same or different species, or as a difference between the behavior of the same individual at different times. Much effort in behavioral research aims to disentangle

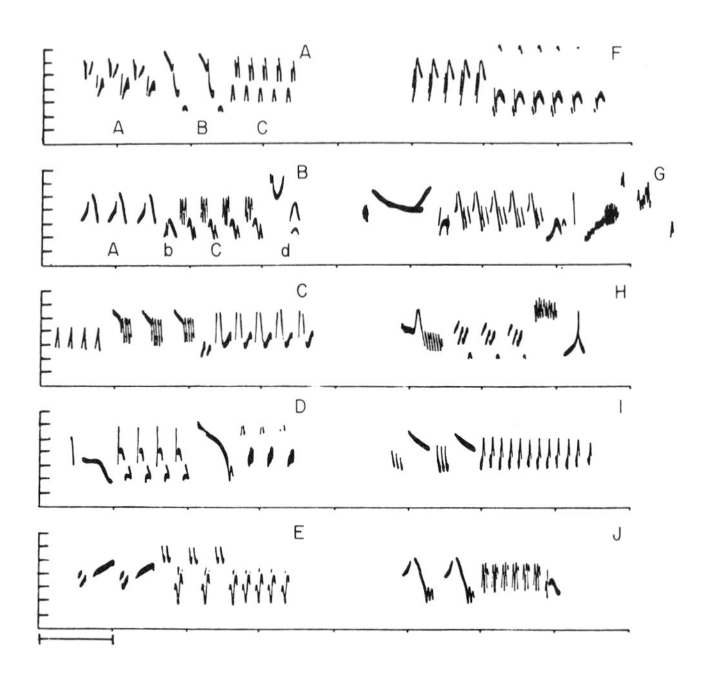

One-trill songs W		E	Two-trill songs W		E	Three-trill songs W		E
4	aBc	1	13	AB	9	5	ABC	7
3	Ab	1	12	ABc	3	2	AbCD	2
	Abc	1	9	AbC	6	2	ABCd	
	A	1	7	AbCd	2	1	ABcD	
			4	aBC	3	1	AbCDe	
			2	aBcD		1	ABcDe	1
			1	aBCd			ABCdE	1
Totals 7		4	48		23	12		11

Fig. 27. Comparable song complexity in group-isolated and wild Arizona juncos. Sonograms are songs of ten wild male Arizona juncos from the same population illustrating the high degree of individual variation in wild populations. The table compares the complexity of the songs of wild (W) juncos and of five experimental (E) juncos raised as a group in acoustical isolation. In the notation used for tabulating songs according to structure, a sequence of repeated syllables, a trill, is represented by a capital letter and any unrepeated unit, a phrase, is represented by a lower case letter. Letters run in alphabetical order from the start of the song. Thus the first and second song in the left-hand column are designated ABC and AbCd respectively. Frequency markers, 0-7 kHz; time marker, 0.5 second. (After Marler & Isaac, 1961.)

the roles of these two types of variables in behavioral development (Marler & Hamilton, 1966). The experimental methods used are complex and difficult in practice but relatively simple in underlying theory. One set of variables is held constant while the other set is systematically changed.

The effects of systematic variation of environmental variables on developing motor patterns are of several types. Certain environmental factors have an all-or-nothing relationship to a given behavior pattern. Experimental variation of these factors causes the behavior to be either present or absent, but does not affect the detailed form of the behavior patterns that develop. On the other hand, there may be graded effects on a behavior pattern, as when change in some environmental variable modifies the frequency or vigor of certain action patterns. The repercussions of experimental handling on the manifestation of "emotionality" in rats (Denenberg, 1962, 1964; Levine, 1962) would be one example. Change in the rate and duration of sound production in birds and mammals accomplished by operant conditioning would be another, vocalization being rewarded by food on various kinds of reinforcement schedules (Ginsburg, 1960, 1963; Grosslight *et al.*, 1962; Grosslight & Zaynor, 1967; Lane, 1961; Salzinger & Waller, 1962).

Perhaps most study has been directed toward the reorientation of actions as a result of environmental experience, as in mazes or problem boxes. This may occur as a consequence of individual exploration or after observing the performance of another animal. Abnormalities of social behavior in animals raised in isolation often seem to result from the lack of opportunity to orient actions appropriately through environmental experience with objects or companions (Eibl-Eibesfeldt, 1961; Mason, 1965).

In most of this work, the manipulation of environmental variables has left the basic patterns of motor coordination unchanged. Although fine adjustments of the motor pattern often occur, at the gross level radically new patterns of motor activity rarely develop except in studies of human motor skills and speech development. Animals often acquire new motor coordinations in the course of operant conditioning. But the shaping procedures involved and the motor patterns generated are often only anecdotally reported and most attention is focused instead on the effect of reinforcement procedures on preestablished operant behavior. When, on rare occasions, operant procedures are expressly aimed at generating new motor coordinations, they sometimes meet unexpected obstacles in the form of preexisting action patterns that are difficult to change or erase (Breland & Breland, 1961).

Actions that are refractory to operant conditioning are often prominent elements in the natural behavior of the species and one suspects that many species-specific behavior patterns would exhibit a similar refractoriness. This is not to deny that learning plays a vital role in development of such motor patterns, affecting their rate and orientation and the stimulus situations which elicit them. But the basic coordinations of neuromuscular activity underlying species-specific actions usually seem resistant to change by learning. In this sense, the songs of some birds are unusual in the extent to which the gross pattern of motor activity can be changed by learning.

There is substantial evidence that normal singing behavior of several songbirds develops by a form of cultural transmission, such that young birds learn vocalizations from adults. The learning that takes place has a number of distinctive characteristics. The capacity for learning is often restricted to a particular period of life, sometimes measured in weeks rather than months or years. There are other species which retain the capacity to modify vocal behavior by learning throughout life. In addition to the existence of critical periods, another characteristic of this learning is the degree of selectivity that it often manifests. There is evidence of a predisposition to imitate certain sounds rather than others, those that are characteristic of members of the bird's own species. The individual's ability to hear its own voice is often of critical importance in development, whether or not the bird learns vocal behavior from other individuals. Another process involved in vocal development in some species has been termed "improvisation." It seems that an element of vocal invention must sometimes be invoked in order to explain the direction taken in development by individual birds. In other species, social facilitation involving a degree of mutual improvisation seems to take place, siblings engaging in a higher degree of invention than that appearing in a single individual.

Factors underlying the selectivity in vocal learning are of special interest. There are constraints imposed on the motor side, as a result of limitations on the potential output from the sound-producing organs. We have also presented evidence for constraints on the sensory side, requiring the postulation of what has been termed "an auditory template" to which the bird seeks to match its own vocal output. In addition, there is evidence from some species that social constraints can guide the direction of vocal learning, by restricting it to an individual relationship of a particular type such as that prevailing between father and son.

Thus, although there is ample evidence that some birds engage in vocal learning in the course of normal development, the capacity for learning is by no means unlimited. Rather, it takes place within a set of constraints which seem designed to ensure that attention will be focused on a set of sounds that is biologically appropriate, and at a certain period of the life cycle. This interplay between learning and the predispositions that the bird brings to the learning situation has been explored in some detail as a manifestation, in rather sharp relief, of what must be a widespread phenomenon in animal learning as it takes place under natural conditions.

Acknowledgments

Research was supported by the National Science Foundation (GB 16606) and the National Institute of Mental Health (MH 14651). Dr. J. Mulligan helped in the preparation of illustrations. We are indebted to Dr. M. Konishi and Dr. F. Nottebohm for extensive discussion of the problems considered, as well as for criticism of the manuscript. Requests for reprints should be sent to Peter Marler, Rockefeller University, New York, New York 10021.

References

Bailey, R. E. Surgery for sexing and observing gonad condition in birds. *Auk,* 1953, **70,** 497-499.

Bertram, B. The vocal behavior of the Indian Hill Mynah, *Gracula religiosa. Animal Behavior Monograph,* 1970, **3,** 2, 79-192.

Bierens, de Haan, J. A. Animal language in relation to that of men. *Biological Reviews of the Cambridge Philosophical Society,* 1929, **4,** 249-268.

Blase, B. Die Lautäusserungen des Neuntoters (*Lanius c. collurio* L.) Freilandbeobachtungen und Kasper-Hauser-Versuche. *Zeitschrift füer Tierpsychologie,* 1960, **17,** 293-344.

Breland, K., & Breland, M. The misbehavior of organisms. *American Psychologist,* 1961, **16,** 681-684.

Collias, N. E., & Joos, M. The spectrographic analysis of sound signals of the domestic fowl. *Behaviour,* 1953, **5,** 175-187.

Craig, W. The voices of pigeons regarded as a means of social control. *American Journal of Sociology,* 1908, **14,** 66-100.

Craig, W. Male doves reared in isolation. *Journal of Animal Behavior,* 1914, **4,** 121-133.

Dennenberg, V. H. The effects of early experience. In E. S. E. Hafez (Ed.), *The behaviour of domestic animals.* Baltimore, Md.: Williams & Wilkins, 1962. Pp. 109-138.

Dennenberg, V. H. Critical periods, stimulus input and emotional reactivity: A theory of infantile stimulation. *Psychological Review,* 1964, **71,** 335-351.

Dittus, W. P. J., & Lemon, R. E. Effects of song tutoring and acoustic isolation on song repertoires of cardinals. *Animal Behaviour,* 1969, **17,** 523-533.

Eibl-Eibesfeldt, I. The interactions of unlearned behaviour patterns and learning in mammals. In J. F. Delafresaye (Ed.), *Brain mechanisms and learning.* Oxford: Blackwell, 1961. Pp. 53-73.

Fish, W. R. A method for the objective study of bird song and its application to the analysis of Bewick Wren songs. *Condor,* 1953, **55**, 250-257.

Foss, B. M. Mimicry in mynahs (*Gracula religiosa*): A test of Mowrer's theory. *British Journal of Psychology,* 1964, **55**, 85-88.

Furness, W. H. Observations of the mentality of chimpanzees and orangutans. *Proceedings of the American Philosophical Society,* 1916, **55**, 281-290.

Ginsburg, N. Conditioned vocalization in the budgerigar. *Journal of Comparative and Physiological Psychology,* 1960, **53**, 183-186.

Ginsburg, N. Conditioned talking in the mynah bird. *Journal of Comparative and Physiological Psychology,* 1963, **56**, 1061-1063.

Greenewalt, C. H. *Bird song: Acoustics and physiology.* New York: Random House (Smithsonian Inst. Press), 1968.

Grimes, L. Antiphonal singing in *Laniarius barbarus* and the Auditory Reaction Time. *Ibis,* 1965, **107**, 101-104.

Grimes, L. Antiphonal singing and call notes of *Laniarius barbarus. Ibis,* 1966, **108**, 122-126.

Grosslight, J. H., Harrison, P. C., & Weiser, C. M. Reinforcement control of vocal responses in the mynah bird (*Gracula religiosa*). *Psychological Record,* 1962, **12**, 193-201.

Grosslight, J. H., & Zaynor, W. C. Verbal behavior and the mynah bird. In K. Salzinger and S. Salzinger (Eds.), *Research in verbal behavior and some neurophysiological implications.* New York: Academic Press, 1967. Pp. 5-20.

Gwinner, E., & Kneutgen, J. Über die biologische Bedeutung der "zweckdienlichen" Anwendung erlernter Laute bei Vögeln. *Zeitschrift füer Tierpsychologie,* 1962, **19**, 692-696.

Hall-Craggs, J. The development of song in the blackbird *Turdus merula. Ibis,* 1962, **104**, 277-300.

Hayes, C. *The ape in our house.* New York: Harper, 1951.

Hayes, K. J. Vocalization and speech in chimpanzees. *American Psychologist,* 1950, **5**, 275-276.

Hayes, K. J., & Hayes, C. The cultural capacity of chimpanzees. *Human Biology,* 1954, **26**, 288-303.

Heinroth, O. Lautäusserungen der Vögeln. *Journal füer Ornithologie,* 1924, **72**, 223-244.

Hooker, T., and Hooker, B. I. Duetting. In R. A. Hinde (Ed.), *Bird vocalizations.* London and New York, Cambridge Univ. Press, 1969. Pp. 185-205.

Immelmann, K. Prägungserscheinungen in der Gesangsentwicklung junger Zebrafinken. *Naturwissenschaften,* 1965, **52**, 169-170.

Immelmann, K. Zur ontogenetischen Gesangsentwicklung bei Prachtfinken. *Verhandlungen Deutschen der Zoolologischen Gesellschaft,* 1967, Suppl., 320-332.

Immelmann, K. Song development in the zebra finch and other Estrildid finches. In R. A. Hinde (Ed.), *Bird vocalizations.* London and New York: Cambridge Univ. Press, 1969. Pp. 61-74.

Kaiser, W. Der Gesang der Goldammer und die Verbreitung ihrer Dialekte. *Falke,* 1965, **12**, 40-42, 92-93, 131-135, 169-170, 188-191.

Kellog, W. N. Communication and language in the home-raised chimpanzee. *Science,* 1968, **162**, 423-427.

Konishi, M. The role of auditory feedback in the vocal behavior of the domestic fowl. *Zeitschrift füer Tierpsychologie,* 1963, **20,** 349-367.

Konishi, M. Effects of deafening on song development in two species of juncos. *Condor,* 1964, **66,** 85-102.

Konishi, M. Effects of deafening on song development in American robins and black-headed grosbeaks. *Zeitschrift füer Tierpsychologie,* 1965, **22,** 584-599. (a)

Konishi, M. The role of auditory feedback in the control of vocalization in the white-crowned sparrow. *Zeitschrift füer Tierpsychologie,* 1965, **22,** 770-783. (b)

Konishi, M. The attributes of instinct. *Behaviour,* 1966, **27,** 316-328,

Konishi, M., & Nottebohm, F. Experimental studies in the ontogeny of avian vocalizations. In R. A. Hinde (Ed.), *Bird vocalizations.* London and New York: Cambridge Univ. Press, 1969. Pp. 29-48.

Lade, B. I., & Thorpe, W. H. Dove songs as innately coded patterns of specific behavior. *Nature (London),* 1964, **202,** 366-368.

Lane, H. Operant control of vocalizing in the chicken. *Journal of the Experimental Analysis of Behavior,* 1961, **4,** 171-177.

Lane, H., & Shinkman, P. G. Methods and findings in an analysis of a vocal operant. *Journal of the Experimental Analysis of Behavior,* 1963, **6,** 179-188.

Lanyon, W. E. The comparative biology of the meadowlarks (*Sturnella*) in Wisconsin. *Publications of the Nuttall Ornithological Club,* 1957, **1,** 1-67.

Lanyon, W. E. The ontogeny of vocalizations in birds. In W. E. Lanyon and W. N. Tavolga (Eds.), *Animal sounds and communication.* Washington, D.C.: Amer. Inst. Biol. Sci., 1960. Pp. 321-347.

Lanyon, W. E., & Lanyon, V. H. A technique for rearing passerine birds from the egg. *Living Bird,* 1969, **8,** 81-93.

Laskey, A. R. A mockingbird acquires his song repertoire. *Auk,* 1944, **61,** 211-219.

Lemon, R. E. The song repertoires of cardinals (*Richmondena cardinalis*) at London, Ontario. *Canadian Journal of Zoology,* 1965, **43,** 559-569.

Lemon, R. E. Geographic variation in the song of Cardinals. *Canadian Journal of Zoology,* 1966, **44,** 413-428.

Lemon, R. E., & Scott, D. M. On the development of song in young Cardinals. *Canadian Journal of Zoology,* 1966, **44,** 191-197.

Levi, H. *The pigeon.* Columbia, S. C.: Brian, 1951.

Levine, S. Psychophysiological effects of infantile stimulation. In E. L. Bliss (Ed.), *Roots of behavior.* New York: Harper, 1962. Pp. 246-253.

Marler, P. Variations in the song of the chaffinch, *Fringilla coelebs. Ibis,* 1952, **94,** 458-472.

Marler, P. The voice of the chaffinch and its function as a language. *Ibis,* 1956, **98,** 231-261.

Marler, P. Specific distinctiveness in the communication signals of birds. *Behaviour,* 1957, **11,** 13-39.

Marler, P. Developments in the study of animal communication. In P. Bell (Ed.), *Darwin's biological work.* London and New York: Cambridge Univ. Press, 1959. Pp. 150-206.

Marler, P. Inheritance and learning in the development of animal vocalizations. In R. G. Busnel (Ed.), *Acoustic behavior of animals.* Elsevier: Amsterdam, 1963. Pp. 228-243, 794-797.

Marler, P. Comparative study of song development in sparrows. *Proceedings of the 14th International Ornithological Congress* 1966, 1967, 231-244.

Marler, P. Aggregation and dispersal: Two functions in primate communication. In P. Jay (Ed.), *Primates: Studies in adaptation and variability.* New York: Holt, 1969. Pp. 420-438. (a)

Marler, P. Tonal quality of bird sounds. In R. A. Hinde (Ed.), *Bird vocalizations.* London and New York: Cambridge Univ. Press, 1969. Pp. 5-18. (b)

Marler, P. A comparative approach to vocal development: Song learning in the white-crowned sparrow. *Journal of Comparative and Physiological Psychology,* 1970, **71,** No. 2, Part 2, 1-25.

Marler, P., & Hamilton, W. J. III. *Mechanisms of animal behavior.* New York: Wiley, 1966.

Marler, P., and Isaac, D. Song variation in a population of Mexican juncos. *Wilson Bulletin,* 1961, **73,** 193-206.

Marler, P., Kreith, M., & Tamura, M. Song development in hand-raised Oregon juncos. *Auk,* 1962, **79,** 12-30. (a)

Marler, P., Kreith, M., & Willis, E. An analysis of testosterone-induced crowing in young domestic cockerels. *Animal Behaviour,* 1962, **10,** 48-54. (b)

Marler, P., & Tamura, M. Song dialects in three populations of white-crowned sparrows. *Condor,* 1962, **64,** 368-377.

Marler, P., & Tamura, M. Culturally transmitted patterns of vocal behavior in sparrows. *Science,* 1964, **146,** 1483-1486.

Mason, W. A. The social development of monkeys and apes. In I. DeVore (Ed.), *Primate behavior.* New York: Holt, 1965. Pp. 514-543.

Messmer, E., & Messmer, I. Die Entwicklung der Lautäusserungen und einiger Verhaltensweisen der Amsel (*Turdus merula merula* L.) unter natürlichen Bedingungen und nach Einzelaufzucht in schalldichten Räumen. *Zeitschrift füer Tierpsychologie,* 1956, **13,** 341-441.

Miller, L. The biography of Nip and Tuck. *Condor,* 1921, **23,** 41-47.

Mowrer, O. H. On the psychology of "talking birds"—a contribution to language and personality theory. In O. H. Mowrer (Ed.), *Learning theory and personality dynamics.* New York: Ronald Press, 1950. Pp. 688-747.

Mowrer, O. H. Hearing and speaking: An analysis of language learning. *Journal of Speech Disorders,* 1958, **23,** 143-152.

Mulligan, J. A. A description of song sparrow song based on instrumental analysis. *Proceedings of the 13th International Ornithological Congress,* 1962, 1963, 272-284.

Mulligan, J. A. Singing behavior and its development in the song sparrow, *Melospiza melodia. University of California, Berkeley, Publications in Zoology,* 1966, **81,** 1-76.

Mundinger, P. C. Vocal imitation and individual recognition of finch calls. *Science,* 1970, **168,** 480.

Nice, M. M. Studies in the life history of the song sparrow. II. Behavior of the song sparrow and other passerines. *Transactions of the Linnaean Society of New York,* 1943, **6,** 1-328.

Nicolai, J. Familientradition in der Gesangsentwicklung des Gimpels (*Pyrrhula pyrrhula* L.). *Journal füer Ornithologie,* 1959, **100,** 39-46.

Nicolai, J. Der Brutparasitismus der Viduinae als ethologisches Problem Prägungsphänomene als Faktoren der Rassen und Artbildung. *Zeitschrift fuer Tierpsychologie,* 1964, **21,** 129-204.

Nottebohm, F. The role of sensory feedback in the development of avian

vocalizations. *Proceedings of the 14th International Ornithological Congress,* 1966, 1967, 265-280.
Nottebohm, F. Auditory experience and song development in the chaffinch (*Fringilla coelebs*). *Ibis,* 1968, **110**, 549-568.
Nottebohm, F. The "critical period" for song learning in birds. *Ibis,* 1969, **111**, 386-387.
Nottebohm, F. Vocalizations and breeding behavior of surgically deafened ring doves. *Animal Behaviour,* 1971, in press.
Promptoff, A. Die geographische Variabilität des Buchfinkenschlages (*Fringilla coelebs* L.). *Biologisches Zentralblatt,* 1930, **50**, 478-503.
Rice, J. O., & Thompson, W. L. Song development in the indigo bunting. *Animal Behaviour,* 1968, **16**, 462-469.
Salzinger, K., & Waller, M. B. The operant control of vocalization in the dog. *Journal of the Experimental Analysis of Behavior,* 1962, **5**, 383-389.
Sauer, F. Die Entwicklung der Lautäusserungen vom Ei ab Schalldicht gehaltener Dorngrassmücken (*Sylvia c. communis* Latham) im Vergleich mit später isolierten und mit wildlebenden Artgenossen. *Zeitschrift füer Tierpsychologie,* 1954, **11**, 10-92.
Schjelderup-Ebbe, T. Weitere Beiträge zur Sozial- und Individual-psychologie des Haushuhns. *Zeitschrift füer Psychologie,* 1923, **92**, 60-87.
Sick, H. Über die Dialektbildung beim Regenruf des Buchfinken. *Journal füer Ornithologie,* 1939, **87**, 568-592.
Skinner, B. F. *Verbal behavior.* New York: Appleton, 1957.
Thielcke, G. Stammesgeschichte und geographische Variation des Gesanges unserer Baumläufer (*Certhia familiaris* L. und *Certhia brachydactyla* Brehm). *Zeitschrift füer Tierpsychologie,* 1961, **18**, 188-204.
Thielcke, G. Zur Phylogenese einiger Lautäusserungen der europäischen Baumläufer (*Certhia brachydactyla* Brehm und *Certhia familiaris* L.) Zeitschrift für Zoologische Systematik und Evolutionforschung, 1964, **2**, 383-413.
Thielcke, G. Gesangsgeographische Variation des Gartenbaumläufers (*Certhia brachydactyla*) im Hinblick auf das Artbildungsproblem. *Zeitschrift füer Tierpsychologie,* 1965, **22**, 542-566.
Thielcke, G. Geographic variation in bird vocalizations. In R. A. Hinde (Ed.), *Bird vocalizations.* London and New York: Cambridge Univ. Press, 1969. Pp. 311-339.
Thielcke-Poltz, H., & Thielcke, G. Akustisches Lernen verschieden alter schallisolierter Amseln *Turdus merula* L. und die Entwicklung erlernter Motive ohne und mit kuntstlichem Einfluss von Testosteron. *Zeitschrift füer Tierpsychologie,* 1960, **17**, 211-244.
Thompson, W. L. The songs of five species of *Passerina. Behaviour,* 1968, **31**, 261-287.
Thorpe, W. H. Comments on the *Bird Fancyer's Delight,* together with notes on imitation in the sub-song of the chaffinch. *Ibis,* 1955, **97**, 247-251.
Thorpe, W. H. The learning of song patterns by birds, with especial reference to the song of the chaffinch, *Fringilla coelebs. Ibis,* 1958, **100**, 535-570.
Thorpe, W. H. Talking birds and the mode of action of the vocal apparatus of birds. *Proceedings of the Zoological Society of London,* 1959, **132**, 441-455.
Thorpe, W. H. *Bird song: The biology of vocal communication and expression in birds.* London and New York: Cambridge Univ. Press, 1961.
Thorpe, W. H. Antiphonal singing in birds as evidence for avian auditory reaction time. *Nature (London),* 1963, **197**, 774-776.

Thorpe, W. H. The isolate song of two species of *Emberiza*. *Ibis*, 1964, **106**, 115-118.

Thorpe, W. H., & North, M. E. W. Origin and significance of the power of vocal imitation: With special reference to the antiphonal singing of birds. *Nature* (*London*), 1965, **208**, 219-222.

Thorpe, W. H., & North, M. E. W. Vocal imitation in the tropical Bou-Bou shrike *Laniarius aethiopicus major* as a means of establishing and maintaining social bonds. *Ibis*, 1966, **108**, 432-435.

Thorpe, W. H., & Pilcher, P. M. The nature and characteristics of subsong. *British Birds*, 1958, **51**, 509-514.

Tretzel, E. Über das Spotten der Singvögel, insbesondere ihre Fähigkeit zu spontaner Nachahmung. *Verhandlungen der Deutschen Zoologischen Gesellschaft*, 1965, **28**, 556-565. (a)

Tretzel, E. Imitation und Variation von Schäferpfiffen durch Haubenlerchen [*Galerida c. cristata* (L.)]. Ein Beispiel für spezielle Spottmotivprädisposition. *Zeitschrift füer Tierpsychologie*, 1965, **22**, 784-809. (b)

Tretzel, E. Spottmotivprädisposition und akustische Abstraktion bei Gartengrasmücken [*Sylvia borin borin* (Bodd.)]. *Verhandlungen der Deutschen Zoologischen Gesellschaft*, 1966, suppl. *30*, 333-343.

Tretzel, E. Imitation und Transposition menschlicher Pfiffe durch Amseln (*Turdus m. merula* L.) Ein weiterer Nachweis relativen Lernens und akustischer Abstraktion bei Vögeln. *Zeitschrift füer Tierpsychologie*, 1967, **24**, 137-161.

Ward, R. Regional variation in the song of the Carolina Chickadee. *Living Bird*, 1966, **5**, 127-150.

Watkins, W. A. The harmonic interval; fact or artifact in spectral analysis of pulse trains. In W. N. Tavolga (Ed.), *Marine bioacoustics*. Vol. 2. Oxford: Pergamon, 1967. Pp. 15-43.

CHAPTER 11

THE ONTOGENY OF LANGUAGE

Joseph Church

There is growing agreement that we shall not be able to solve the riddle of human behavior until we have come to terms with that most human of functions, language. There is further agreement that linguistic functioning can be understood only if we consider it developmentally, in terms of the individual's transformation from an unspeaking infant into a more or less fluent human being, together with all the other transformations of perceiving, thinking, feeling, and acting which may be connected in some way with linguistic development. Let me hasten to say that this is a new conviction only in American psychology. Generations of European psychologists, and, in this country and elsewhere, philosophers, linguists, and literary critics have wrestled with the problems of language and language development.

This is not to say that language was neglected in American psychology. Quite to the contrary, vast energies were spent in trying to tame language, to make it conform to the prevailing behavioristic doctrines, most fundamentally to demonstrate that language had no special emergent character requiring a revision of our reductionistic conceptions. This approach had its last gasp in Skinner's *Verbal Behavior* (Skinner, 1957) and a long-delayed death rattle in Staats' *Learning, Language, and Cognition* (Staats, 1968). Since then, behaviorists have come to view language more seriously and respectfully (e.g., Osgood, 1963), and have filled the once-empty organism with a multidimensional maze of mediating machinery. Needless to say, they have not acted without provocation. The behavioristic approach to language has come under slashing attack from a number of quarters. From Pavlovians, for instance, whom one might naively think of as the behaviorists' natural allies. But Pavlov, it will be remembered, supplemented his reductionism with a "second signal system" which enabled human beings to transcend the limitations of simple conditioning (Chauchard, 1956, pp. 59-66). The second signal system was largely ignored in American associationism, and the more traditional Pavlovians regarded us as backward or wayward. Most of all, though, it has been the "new" generation of cognitive psychologists who have wrenched behaviorism out of doctrinal mindlessness and into a new orientation to language.

Dominating the linguistic landscape at this moment are the figures of Noam Chomsky and Jean Piaget, interlocked half in struggle and half in embrace. They share a structuralistic, biologistic, rationalistic conception of language, but are sharply at odds on two issues. One is the part played by learning in the development of language, Piaget taking more of an empiricist and Chomsky more of a nativist stand. The second issue has to do with the relative primacy of language and thinking, Piaget holding to the opinion that language is molded to fit an underlying "logic" or "intelligence" (Piaget & Inhelder, 1966), whereas Chomsky feels that thought is shaped by more basic linguistic structures.

Considering that he published in the 1920's a seminal work called *The Language and Thought of the Child* (Piaget, 1955), Piaget has had, until recently, remarkably little to say about language and its acquisition. His well-known concepts of *nominal realism, collective monologue,* and *egocentric speech* in effect take language for granted and refer to more basic cognitive processes. In his more recent writings, how intrinsic to thought he considers language varies according to whether he is collaborating with Bärbel Inhelder (Inhelder & Piaget, 1964; Piaget & Inhelder, 1969), in which case language plays an important role, or with Hermina Sinclair (Piaget, 1968), in which case language is of no great consequence. Piaget and Inhelder (1969) propose a sequence of linguistic development from dealing directly with perceived objects, to dealing with "symbols" of absent objects [symbols here including deferred imitation, symbolic (what Americans call *dramatic*) play, drawing (although in fact drawing ordinarily comes after speaking), and images], to "signs," the words which represent objects or images. Although Piaget and Inhelder speak of the logical operations, such as relations and classifications, built into established language, they have little to say about how the individual progresses into abstract thought or, more especially, how language can become instrumental in original or creative thought. In general, Piaget's ideas about the developmental psychology of language are still in a somewhat primitive stage—like the behavioristic psychology of some years back, he seems hardly to go beyond the psychology of concrete nouns—and should detain us no longer.

Chomsky's position is more highly elaborated but, to at least one member of the audience, largely unintelligible. Indeed, at odd moments I have had the conviction that Chomsky is playing an intricate practical joke on the intellectual community. However, we have in hand a document bearing a 1968 imprint (Chomsky, 1968), and from this and other sources we can undertake to reconstruct the Chomskyian thesis.

On the linguistic side, Chomsky is relatively straightforward—up to a point. The essence of language lies in its grammatical and syntactical structure. Let me say that the rediscovery of grammar and syntax came to psycholinguistics like a breath of fresh air. The students of language had been preoccupied so long with vocabulary, phonology, and phonemics that they seemed to have forgotten that there is such a thing as connected discourse. One might suppose that, having rediscovered the sentence, we could go on to the semantics of sentences and other utterances, but Chomsky, despite a few allusions to semantics, is content to dwell at the level of grammar and syntax, which for him largely determine semantics.

An important feature of the Chomsky position is the idea of a universal grammar. Since the grammatical structures of particular languages obviously differ (case endings for nouns in some languages but not others, the past participle remaining attached to the auxiliary in some languages but sliding to the end of the sentence in German, tenses built into verb flexions in some languages but marked by a separate time word in Chinese, English possessive pronouns agreeing in gender with the possessor while in French they agree with the thing possessed, and so forth), Chomsky postulates a distinction between *surface structure,* in which languages differ, and *deep structure,* which all languages have in common. The simple declarative sentence is taken to be the prototype, with a basic structure of noun (or noun phrase) subject plus verb (or verb phrase) plus, in some cases, an object or complement or qualifier. The deep structure also contains a set of *transformation rules* by which the basic sentence can be converted into various surface structures expressing such things as negation, the passive voice, time indices, and interrogation.

Another important Chomskyian concept is the distinction between *competence,* what the individual is capable of saying, and *performance,* what he actually says. This distinction becomes vital to Chomsky's view of language development, which says (Chomsky, 1968, p.

76) that the child does not infer a set of grammatical and syntactical rules from the "meager and degenerate" linguistic data available to him in the speech of people around him, but rather chooses from several sets of innately given rules the one that best accords with the available data. These innately given sets of rules, of course, are all variants of the deep-lying, innate universal grammar. This implies that the very young child is linguistically competent far beyond what his actual speech would lead us to suppose (Chomsky seems to be only at the beginning of recognizing the possible role of maturation in the development of organic structures). Chomsky, of course, is perfectly correct in saying that an individual can be capable of more than he actually does—performance on an intelligence test or in the classroom may give us a serious underestimate of a child's abilities—but he offers little guidance on how to go about demonstrating competence. Some of his followers test children's ability to act out instructions phrased in various ways (e.g., "The cow kisses the horse" vs. "The cow is kissed by the horse") or to imitate various grammatical constructions embedded in sentences. While such studies reveal age and individual differences, their general theoretical implications are unclear.

Chomsky, in discussing deep structure, and Miller and McNeill, in discussing competence, rely most heavily on logical demonstrations of the necessity for these constructs through the analysis of sentence types. Chomsky, for instance, argues that the ambiguity of sentences like "I know a taller man than Bill" can be resolved only with reference to the rather different structures involved in the unspoken "is" or "does." We need mention only in passing that this is a spurious ambiguity. The correct phrasing for the more obvious "is" meaning is "I know a man taller than Bill." But there are ambiguous sentences; take "Movies are more easily adapted from novels than plays," which seemingly could mean either "from plays" or "are plays." But in fact the ambiguity does not reside in the construction of the sentence (stylistically, it would have been better to include the semantically redundant particle). The ambiguity comes about from the sentence having been lifted out of context, which is a discussion of how the play form translated to the screen has a static, artificial quality quite alien to the fluidity inherent in film techniques. And in general most "ambiguities" are not ambiguous in the context of continuing discourse, so that the problem becomes an artifact of the method, and there is no need to invoke a second set of constructs to explain a nonexistent problem. All sentences are of necessity elliptical, but monologs, which have to create their own context, permit less ellipsis than dialogs. It should also be apparent that the context can be supplied by the concrete situation which occasions an utterance. Before dinner, "I don't like John's cooking" is likely to mean "the fact that John is cooking," whereas at table it is more likely to mean "the meals that John prepares." But in any case, the first sense would probably be expressed by "I don't like John to cook" and the second by the intonation pattern of the utterance.

G. A. Miller and McNeill (1969) using a similar technique to demonstrate underlying competence, compare such sentences as "They are drinking highballs," "They are drinking glasses," and "They are drinking companions," all alike in superficial structure but greatly different in meaning. Miller and McNeill argue that we discern these differences by virtue of the very different structures that appear when we paraphrase the sentences. For instance, "They are drinking highballs" can be recast in the passive ("Highballs are being drunk by them") but the others cannot ["Glasses are being drunk" or "Companions are being (not *becoming*) drunk"]. One can say that "They are glasses for drinking [from]" (as contrasted with "for looking through" or "for looking in") but not "They are companions for drinking." And so on. The argument is ingenious but unnecessary. It overlooks in this case the reciprocal context supplied by the words in the sentence (in the *highballs-glasses* contrast, incidentally, we have an intonational difference, the emphasis falling on *highballs* in the first case and on *drinking* in the second).

Here, I believe, is one of the key weaknesses of the entire approach, the attempt to do everything in terms of grammar and syntax without appeal to semantics. We do have a few grammatical imperatives that serve no particular semantic function: the command to "Brush three times daily" is ungrammatical but clear; the claim that "More dentists use mouthwash X" is not likely to evoke such completions as "than LSD" or "than barbers"; advertisers err in promising that a cigarette "travels the smoke farther" or that a girdle "vanishes away the pounds," but they make their point. By and large, though, grammatical and syntactic forms serve a semantic function and are otherwise irrelevant.

The other key weakness is that the Chomsky camp cannot make up its mind whether it is studying linguistics or psychology (and to invent something called "psycholinguistics" does not help). Chomsky repeatedly refers to linguistics as a branch of psychology, but in fact the principles he talks about are linguistic and not psychological. That is, he is analyzing languages as entities abstracted from the utterance of many speakers, and not the behavior of people talking and listening. Even if universal grammar could be shown to be a useful tool for the description of diverse languages—something that has yet to be established; indeed, it is far from a complete description even of English—its psychological implications would still remain obscure.

The reader is free to make what he will of Chomsky's flirtations with mind-body dualism, solipsism, innate ideas, and rationalism, his disregard of the role of stimuli in energizing and mobilizing the organism, his apparent ignorance of the facts of human psychological development, preverbal and verbal, and his seeming yearning after computer-compatible models of analysis [although he repeatedly dismisses the possibility of language-simulating automata, his discussion of vocabulary (Chomsky, 1968, pp. 33–38) sounds very much like a design for a computer-stored lexicon]. I do not feel that they need further comment.

Lenneberg (1967) has set out to provide the biological underpinnings for Chomsky's innate linguistic structures. Lenneberg seems so intent on demonstrating the inconsequential role of experience in linguistic development that he is drawn into a distortion of the facts. He states that Dennis and Najarian's (1957) institution-reared subjects suffered a delay, but no lasting impairment, of language development. In fact, it was precisely in the sphere of language that these children showed the most marked signs of permanent impairment. Lenneberg, in support of the Chomskyian thesis, tries to set an end point to the development of language, and while we can agree that some important foundations have been laid down by age 3 or age 5, it seems obvious that linguistic development goes on lifelong. For instance, some of the manifestations of the middle years of childhood (ages 6–12), following what Lenneberg takes to be the completion of language development, are, besides reading and writing, which may come earlier, puns, double entendre, figures of speech, codes and ciphers, and some ability to test formulations against observations or general principles. In further support of his thesis that experience plays a minor role in language development, Lenneberg (1969), referring to the hearing children of deaf parents, asserts that "In no case have I found adverse effects in the language development of standard English in these children." This assertion leads Friedlander (1971) to comment wryly, "Barring magic, one must suppose that these children found language models *somewhere* . . ." There is, of course, the problem of defining what we mean by normal language. Apart from the absence of aphasia or dysarthria, the only criterion I know of is performance on tests of verbal intelligence, where "normal" means a score in the 85-115 IQ range. But countless studies of institutionalized children have found that they are impaired in verbal intelligence as compared with home-reared children. Unsatisfactory as intelligence tests may be, they at least give us a gross picture of linguistic competence. It is not taking sides in the renewed nature-nurture controversy to say that there is abundant evidence of the impact of experience—good and bad—on linguistic development.

One should not conclude from these strictures that the Chomsky position is barren in the area of research. Various experimental and longitudinal studies are under way which should yield information interesting to psychologists of whatever persuasion. A brief account of one line of research may illuminate the difference between a linguistic and a psychological approach. Bellugi-Klima (1968) is studying the development of transformations, with special attention to interrogation and negation. She reports that young children (ca. age 3) typically use a single transformation for interrogation ("May I can have some?" "Is it is a dog?") but multiple markers for negation ("I don't not want none"). She gives this finding a structural interpretation, that the child "has" one set of rules for interrogation and another set for negation. Heinz Werner (1937) has proposed a distinction between *process* and *achievement*. Achievement corresponds closely to Chomsky's *performance*, but process, unlike competence, refers to active psychological operations which the child applies in transforming a situation. A process interpretation of Bellugi-Klima's data would say that these manifestations are only incidentally related to interrogation and negation. More generally, they represent stages in the child's learning the redundancy rules of English: at first, as illustrated by interrogation, any small change is seen as sufficient; later, as illustrated by negation, one feels the need to use as many markers as possible to make one's point. In fact, these two manifestations do appear in the postulated order, which would seem to support a process interpretation. More generally, the child's errors are revealing. When the child says "He pick it ups" or "I walk homed," it indicates that his competence includes a couple of inflectional rules which, from a process point of view, he misapplies.

While the behaviorists and the structuralists have been battling in the limelight, another tradition has been unobtrusively at work offstage. This is the anthropological linguistic tradition, well exemplified in the work of Hymes (1962, 1964, 1968). Anthropological linguistics attempts to relate language to the ideas, perceptions, beliefs, values, activities, and thought patterns characteristic of various cultures.

Linguistics as practiced by anthropologists has often been static, barren, and reductionistic in several ways. Some anthropologists have been content with the compiling of lexicons while others have been preoccupied with special domains such as kinship terms or status relations. One variety of reductionism apparently took place under the influence of behaviorism, confining the study of languages to their phonologies, so that a description of language development in a given society would be little more than an account of the progressive mastery of speech sounds. At its best, however, anthropological linguistics shows the subtle interplay of forces determining the form of utterances (Frake, 1964; Weinreich, 1963).

Another, more complex kind of reductionism is to be found in the widely influential Whorfian hypothesis (Carroll, 1956). Briefly, the Whorfian hypothesis says that a culture is embodied in the linguistic forms used by the members of the culture, and that these forms impose inescapable categories that determine perception and thought. Anthropologists find it convenient to speak of linguistic *domains,* those aspects of experience which cohere logically or functionally—body functions, food-getting, space, time, the supernatural, and so on ad infinitum. In different languages, different domains are highly *terminologized,* that is, they have elaborate and highly systematic vocabularies. Kinship terms and color names are highly terminologized in all known languages (Weinreich, 1963), while rocks, for instance, may be poorly terminologized except for the specialist. There are striking cultural differences in degree of terminologizing: as everyone knows by now, snow is highly terminologized by Eskimos, and coconuts by Tahitians and probably by other dwellers in the tropics. But beyond phonology and vocabularies, and beyond grammar and syntax, there are also rules of composition, of making meaningful, coherent, intelligible statements. Subtly intertwined with all these levels is the "logic" of any particular language, the kinds of relations it expresses easily, as opposed to those on which the language is silent or which can be expressed only tortuously. For instance, as many observers have pointed out, English tends to be a two-valued language, with many aspects of experience expressed dichotomously, as antonyms: good-bad, male-female, child-adult, and so forth. The spokesmen for the movement known as General Semantics have noted that the two-valued logic of English can lead us into fallacious thinking (specifically, the fallacy of the excluded middle). All of the examples given illustrate this. It requires a certain amount of wisdom to appreciate that a person can be both good and bad. Biology has taught us that male and female are only the anchorage points of a continuum, with many possible degrees of intersex, and homosexuals teach us that it is possible to have a body of one sex and a mentality of quite another. Anyone who had had any dealings with adolescents knows that there is a condition of life which cannot be characterized as either childish or grown-up. In other words, the logic built into English can be a built-in impediment to tolerance of ambiguity. Thus, the propagandist finds it easy to partition people into true-blooded patriotic Americans and communist scum. [Social science jargon converts linguistically discrete categories into continua by a simple device, as in *ethnicity*, *orphanicity*, *schizophrenicity*, and *futurity* (not a horse race but a degree of orientation to the future) (cf. the geologist's *seismicity*); the anthropologist's *nuptiality*, however, seems to mean only marriage customs.] It is extremely difficult to disentangle the implicit logic found in the way people talk from the framework of cultural assumptions, likewise largely implicit, within which they operate. It turns out that one cannot resolve this problem by comparing the behavior of people who speak the same language but have different cultural patterns, e.g., the British and the Americans. For when one looks more closely, it appears that it is only a convenient fiction to say that they speak the same language. There are innumerable dialects in English (as in most languages), overlapping in many pragmatic domains but diverging sharply in the conceptual and evaluative realms. We can see dialectical divisions along religious, social class, educational, ideological, and numerous other lines. But we have no way to determine whether fundamental differences in figure-ground organization—the dress-shop window that is a sharply articulated display to my wife is nothing but background blur to me—can be related to differences in dialects and their logics. The military strategist who juggles megadeaths is obviously operating within a cognitive, linguistic-logical frame very different from that of the nonviolent resister.

The Whorfian view seems to be correct in two senses, one of which negates the other. Many individuals seem indeed to be the prisoners of language, limited in their thinking, perceiving, and feeling by the deadly conventions embalmed in the slogans, cliches, platitudes, formulas, and unquestioned assumptions of the culture. On the other hand, for some number of people language appears to be an instrument of liberation, freeing them from orthodoxy and enabling them to venture into new realms of experience. In Bernstein's (1960, 1964) phrasing, some people learn to speak in *restricted codes* and others in *elaborated codes*. Bernstein, however, distinguishes between two kinds of restricted code. One is the shorthand speech used familiarly among people communicating about standard sorts of situations, which can be made more elaborate as necessary. The other is simply a matter of limited resources, which can be expanded horizontally to include more of the same, but which do not lend themselves to higher conceptual elaborations.

It is only the second kind of restricted code that we are concerned with here. The speaker of a restricted code in ordinary circumstances may also, as in connection with his work, know one or more specialized elaborated codes. But it is one of life's curiosities that a person can be highly articulate in one realm and next to aphasic in the remaining spheres. Thus, there is no reason to suppose that a highly qualified highway engineer understands anything of the ecological relationships or neighborhood values through which he plans to lay a concrete ribbon. The restricted codes in general use leave unexplicated—and apparently unnoticed—the categories and causal conceptions of modern science. They do not lend themselves to original humor [although, according to Riessman (1962) and other observers, the language of black adolescents is highly developed in the domain of witty insult and ridicule], imagination, and unorthodox formulations. American folk humor, for instance, sounds penetrating and fresh when first encountered but becomes stereotyped and tedious upon continued exposure.

More interesting, if more elusive, are the origins and nature of those linguistic styles which transcend the commonplace, which seek to take apart and recombine experience in novel ways, which take note of contrasts and similarities, which generate and test principles, and, most important, reflect self-consciously back upon the nature of language itself, its possibilities for capturing, ordering, rearranging, and extending experience. Notice that at this point we draw no distinction between the brilliant, original user of symbols and the schizophrenic whose formulations of experience may be anything but ordinary. Indeed, we take it as axiomatic that the symbolic innovator, judged by the prevailing standards of his times, will inevitably be ruled insane by most people. We can also recognize that there are those who—the orator, for instance—use symbols brilliantly and persuasively but who, on closer examination, prove to have very little to say.

The distinction between restricted and elaborated codes seems to correspond to two approaches to education. In one, the scholarly, the student is expected to discipline himself and subjugate himself to traditional knowledge and wisdom and to become its perpetuator. In the other, the intellectual, the student is taught to take what is useful from established knowledge but, most of all, to develop his own resources for the expansion and communication of understanding. This contrast is seen clearly in the study of music. On the one hand, the student is expected to master a body of literature so that he can be a faithful interpreter of the work of the Masters. On the other, he is expected to master the resources of an instrument so that he will be able to make his own musical statements. In Piaget's terms, we can describe a contrast between accommodation, adapting oneself to the demands of the material, and assimilation, adapting the materials to one's own purposes. Although assimilation is generally considered the more primitive process, it seems to be essential to what Polanyi (1958) calls "personal knowledge," and represents the style of language usage that confers the greatest conceptual freedom on the individual. Obviously, as in the learning of arithmetic or auto repair, the materials always impose some accommodational constraints, but we must beware of bowing to implicit constraints that the logic of the situation does not really require. For instance, in thinking about the effects of automation and increased leisure, most people are brought up short by the implication that in the future we may have to pay people for doing nothing, which stands in horrifying contrast to our usual convention that it is psychologically destructive not to be gainfully employed. The recurring dilemma is how to have one's roots in history without becoming a prisoner of the past.

What has been said so far seems to boil down to the view that a psychology of language cannot stop with linguistic facts but must examine the uses to which the individual puts his linguistic skills. That is, people talk not for the sake of exercising linguistic forms but in order to say something. This further implies that the child's learning of linguistic forms is merely incidental to his mastery of a whole new, emergent style of behavior, the linguistic. I have proposed elsewhere that we think of the end product of this process not as an organism that can talk but as a verbal organism, transformed by language into a new kind of creature with a special kind of orientation to experience.

As scientists, we are obliged to translate our views into empirical research, and a later part of this chapter will be devoted to describing a new approach to the classification of speech acts in terms of communicative intent. Two things need to be said. This approach cannot be wholly independent of a grammatical-syntactical analysis, both because certain constructions correspond very closely to some semantic types and because we are inevitably interested in the linguistic means by which the child achieves his cognitive goals. Second, the scheme outlined here is of necessity sketchy and tentative, since only longitudinal observations of children early in language development will tell us what actually happens. Thus, what follows is a guide to what we can look for rather than a confident prediction of what we will observe. At a later, less *ad hoc*, stage of the investigation I hope to look for relations between aspects of language development and other, less obviously verbal, aspects of cognition. I have proposed elsewhere (Church, 1970) some tasks suitable to this purpose. I also hope to be able to find a relationship between the linguistic styles of parents and offspring.

Before going on to describe future research, we need to summarize the present state of knowledge about language development and its relation to other aspects of early cognition. A few empirical generalizations are possible about the ontogenesis of language. But note that these generalizations are based almost entirely on observations of white, middle-class children learning Western European languages. We are only beginning to have data on the acquisition of Japanese and of Russian, and we are for all practical purposes totally ignorant with respect to language development in our own underclass and with regard to the more exotic tongues and their dialects. It remains an act of faith that the same generalizations will apply. It is clear that passive understanding of language spoken by other people precedes the child's active speaking (Friedlander, 1971). Most children speak first in one-word utterances (which may in fact be fusions of several words). Some unknown proportion of children speak in expressive jargon, gibberish that simulates the intonations of adult speech, sometimes before any standard words are used and sometimes as a stream in which real words are embedded. It is as though the child attempts to leap magically from babbling to full-blown speech. The growth of sentences goes from two-word juxtapositions ("Baby crying," "Fly bite") to ever longer conglomerations [I am particularly charmed by W. Miller and Ervin's (1964) specimen, "Baby other bite balloon no"]. These conglomerations make no use of articles, prepositions, or the standard inflections for person, tense, or number (although the he-she-it gender discrimination comes early, and children learn, somewhat miraculously, the proper use of I and you). Much of what the very young child says appears to be said mainly for the purpose of practicing new constructions. Two processes have been identified in the transition from agrammatical speech to standard constructions. One is the coupling of a high-frequency "pivot" word, which can belong to any part of speech, with other words, as in

"*Daddy* shoe," "*Daddy* pipe," etc., or "Fix *it*," "Throw *it*, " and so on. The other, similar process is the completion of a stable group of words, or sentence frame: "*I have a* doll," "*I have a* truck," and so forth.

It needs to be emphasized that the great bulk of the child's language learning takes place without explicit instruction. People talk to him in an ongoing context of environmental interactions and the child begins to get the idea of how one operates linguistically. Note that direct imitation plays only a partial role in the process. Obviously the words the child speaks are imitations (at first grossly approximate) of the words he hears others speak. But the use to which he puts the words is rarely an echo of what other people say. This is another way of saying that the child learns language intelligently. Even his errors reveal the misapplication of principles, as in "They're hugging their chother" [This fairly common form is inflected "our chother(s)," "your chother(s)," "their chother(s)."] But we must conceptualize a level of intelligent learning somewhere between the mindlessness of the learning-theory child and the rationalism of the Chomskyites which often seems to imply that the child has an explicit knowledge of the rules of grammar. It seems simpler to say that the child learns one or more linguistic schemata according to which a formulation sounds (or feels) right. In general, the notion of schemata learning helps us find an alternative to reductionism (with its implicit atomism) and mentalism which makes of the child (or animal) an analytic philosopher. The schema can be seen as an organizer of behavior, eliminating the chaining required by an associationist view but providing more flexibility than an organic structuralist view.

Various theorists have felt impelled to ask why the child learns language, what it is that motivates him. Obviously, the structuralist position has no need of a motivational component, since the structuralist child has no choice in the matter—language simply happens to him the same way respiration, circulation, and the like happen to him. But the conventional psychological assumption is a homeostatic one, that without some imbalance in the system the organism drifts blissfully through Nirvana. The first thing that springs to mind is that the baby first learns to speak instrumentally as a way of satisfying organic needs such as hunger and thirst, with the implication that his first words should refer to need states and their satisfiers.

The facts are thoroughly ambiguous. Some generations of observation have failed to identify the first words, or even to yield criteria for knowing when the baby has begun to use words communicatively.

One is usually obliged to disregard many equivocal utterances and rely on the baby's pointing to an object and saying its name, which rules out of consideration all parts of speech except concrete nouns. In his choice of things to name, the baby seems altogether capricious: *dog, car, bath, ball, baby,* proper or pet names, *paper,* along with occasional references to food and drink. *Mama* and *dada* as applied to the chief sources of need gratification tend to appear late, even though they are prominent in the babbling of all babies, and designations of body states much later still. In general, need states seem to inhibit language in favor of crying, whimpering, screeching, pointing, or direct action. An alternative to the need theory is White's (1959) notion of "competence motivation" or "effectance."

In fact, however, one does not need any sort of drive construct. We must acknowledge, however, that Mowrer (1960) is probably correct in saying that if the language directed at the child is largely disagreeable in content and feeling tone, the child will resist learning to speak. The mere fact of speaking generates its own reward which one can then, if one so chooses, think of as a learned motive. But there are at least two ways in which speaking can be seen as gratifying. First of all, communion with other human beings seems to be intrinsically pleasurable, and language elaborates and enhances the possibilities of shared experience—gossiping is prototypical, whereas reticence is hard to come by. The baby activates his parents with numerous commands, teases them, makes and plays jokes (as when the toddler tells his father returning from work that "Mommy go store" when in fact she is in another part of the house), and supplements physical contact with words like *love, hug,* and *kiss.* The second kind of satisfaction intrinsic to language is implicit in the first. It is the power to use language instrumentally, to produce results, whether in the form of rewards or of informative feedback.

Even these intrinsic motivators, however, do not get to the heart of the matter. Language has a fundamental psychological characteristic to which Jean Piaget has given the name of *symbolic realism.* In brief, language—almost totally for the child and at least residually for the adult—creates its own sphere of reality which in some experiential sense is at least as real as the perceived concrete world. Thus, for the child, the name of an object seems to express its essence, and to talk about the world is to manipulate it magically. We see numerous symptoms of word realism in adult language. It is word realism that allows us to participate vicariously in works of fiction, even to the point of endocrine and autonomic involvement. The effectiveness of insults, justified or not, testifies to the way words work upon us. To be moved by the orator's slogans, even when they have no particular content, implies word realism. Indeed, the thing designated need not even exist for us to talk about it "realistically." Not only do we people the world with supernatural entities, such as ghosts, demons, and werewolves, but in our protoscientific gropings we generate phlogiston, caloric, and the ether, animal magnetism, instincts, intelligence, and personality. Notice how, in each era, such naming of supposed causative agents bestows a sense of closure even though, in practical fact, nothing has changed. We see symbolic realism at work in our reluctance to give explicit voice to taboo topics, resorting instead to euphemism, allusion, and circumlocution. Something closely akin to symbolic realism seems to be at work in the feelings we attach to national and religious emblems. We see symbolic realism at its most magical in chants, incantations, curses, spells, and in glossolalia, the speaking in tongues that betokens communion with some divine figure. The jargon and the pompous prose of much scientific writing seem magically intended to infuse the data with an extra dimension of significance (and may also contribute to a high incidence of errors of usage, as in the phrase "ontogenetic development").

The important question is how people outgrow symbolic realism in favor of a more objective, abstract linguistic style. Part of the answer is easy: many do not. But most people, voluntarily or involuntarily, engage in a certain amount of reality-testing. That is, hard experience teaches them that facts do not always fit formulations, as in the discovery that political practice jibes hardly at all with what is said in civics texts. Still others acquire a style—sometimes as part of the identification process, sometimes spontaneously, and sometimes from explicit teaching—of rather systematic symbolic formulation, analysis, and categorical synthesis. Here we must take note of the distinction between habitual and original speech. Habitual utterances are repetitions of or variations on things the speaker has said before—at the extreme, formulas, cliches, and platitudes. Original speech, by contrast, consists of formulations that have never previously occurred in the history of the person speaking. How effective a person's symbolic ordering of experience is, is another question. Obviously, many people described as thoughtful produce some wholly psychotic versions of what the world is like. As we have come to believe, the preventive for disordered or irrational thinking is scientific empiricism.

Now it is obviously nonsense to talk about language and experience and reality without also talking about perception. Restated, what the previous paragraph says is that there is never any fresh, clear, pure, direct apprehension of the world. We seem to move from the primitive world of babyhood, characterized as physiognomic, egocentric, participative, dynamistic, phenomenalistic, and labile yet rigid (we shall return to these terms in a moment) through a world confused by symbolic realism to a world which is stable and orderly and predictable and manageable but only by virtue of the conceptual systems that lock it into place. This is logically so, but psychologically, the scientific explication of experience seems to liberate our senses and, eventually, to let us experience the world as it is, in all its richness. The sense of knowing what lies beneath the surface of things, the interplay of hidden causal forces, enhances rather than detracts from the clarity of our perceptions and the intensity of our feelings. In sum, the symbolic articulation [what I have elsewhere (Church, 1966) called thematization, following the philosopher Merleau-Ponty] seems to bring with it a perceptual articulation.

Now we must recapitulate perceptual development, at least to the extent of contrasting the primitive sphere into which the baby is born and the more sophisticated world experienced by the educated adult. We must stress that there are a few—but only an enumerable few—givens in naive perception. It seems clear, from everyday observations and from Fantz's work (1965), that the neonate is differentially sensitive to the human face. Figure-ground organization seems to be a given, although certain of the gestalt laws, such as good continuation and prägnanz, do not hold for the baby. The visual cliff effect (Walk & Gibson, 1961) seems to be present from the time the child can creep, from the time the eyes open in some other mammalian species, and virtually from hatching in precocial birds. It seems likely that Schiff's (1965) looming effect is a universal. Siqueland and Lipsitt (1966) and Papousek (1967) have both demonstrated conditioning in the neonate, which implies perception of the CS. Lipsitt and associates (1963) have found differential reactions to odors in the newborn, and it is clear that the neonate is sensitive to pain. Certain of the baby's more complex "reflexes" might be taken as perceptual indicators, as in rooting or sucking. The fact that a crying neonate becomes calm when picked up and snuggled tells us something about his perceptions.

There is evidence of perceptual learning early in life. By age 1 month, the baby, held in the feeding position, opens his mouth and strains toward the nursing bottle, suggesting that he has come to recognize the sight of the bottle. However, it may be several months before he distinguishes between the nursing bottle and various other forms (I have used a crumpled cleaning tissue, a gray sphere, a gray cube, a gray cone, a gray cylinder, and a white cylinder, all the solids being roughly equal in volume to a standard 250 ml nursing bottle), or among different orientations of the bottle.

By age 3 months, the baby reacts differentially to familiar people and strangers, indicating that he is learning a framework of the familiar against which he recognizes novelty. Around age 2 months, the baby first turns his head to look at someone who is speaking, which implies a coordination of auditory and visual space, although he does not orient to the sources of inanimate sounds until about age 4 months. At least some babies as young as 3 months produce a crooning sound in response to music. We can assume that, during infancy, a great deal of intersensory development is going on. That is, the child is learning which sights, sounds, flavors, odors, tactile qualities, etc., belong together, that a metal dish striking the table top makes a sound different from a plastic dish. By the same token, the child learns pragmatically a great many causal sequences.

We cannot infer, however, just because the baby perceives some of the same things that we do that he perceives them in the same way. The baby's experience is said to be *physiognomic*, on the analogy of the way we perceive and recognize faces. (The study of releasers by the ethologists amounts to a psychophysics of physiognomic perceiving.) The demand qualities of objects seem to be contained in their physiognomies, and from a few months of age the baby reaches out to grasp the graspable, to probe small openings with his forefinger, and to palpate novel textures. In early stages of speaking, the child learns the names of innumerable shapes and objects, but apparently still perceives them physiognomically, since he can recognize all sorts of triangles in all sorts of orientations without knowing that a triangle has three sides. When he first begins recognizing makes of cars, it is apparently on the basis of their physiognomies, although at a later age the child can begin to spell out the distinctive features of various makes.

The baby's (and young child's) experience is said to be *egocentric* in that he takes no account of perspectives other than his own. This is rather hard to demonstrate in infancy (unless we count the baby's anger when a parent fails to understand from the baby's vocalizings and pointings what it is that the baby wants), but egocentrism is so pervasive in the postinfancy years that it is hard to imagine that baby began as a relativistic creature and became an egocentric one. In general, relativism calls for an act of thinking and only after much practice does it become habitual. For instance, a 4-year-old can learn left and right with respect to his own body, but it is not until age 7 or later than he can correctly designate left and right on a person facing him.

The baby's experience is described as *participative* in that self-world boundaries are very tenuously drawn (which is another way of saying that the body is incompletely schematized and the self is diffusely localized), so that those events that are perceived have an organic impact on the body. One of the more striking manifestations of participation is empathic reactions. A wave of crying can sweep like a contagion down a row of baby carriages in the park; the toddler watching someone on a diving board bobs up and down in rhythm with him, and even the adult watching a fight on television may writhe and grunt empathically with the combatants. It is assumed that empathy is the basis for imitation, although this solves the mystery of imitation only by introducing the more basic mystery of participation.

Dynamism is a more general term used to subsume other ways of describing immature perception of causal relations: animism, magicalism, artificialism, supernaturalism, and so forth. Again, dynamism is hard to demonstrate in infancy, but it is a useful way of accounting for children's incurious acceptance of events, as though they are activated by some self-evident general energy. Protoscientific causal constructs like the caloric or sorcery can be seen as dynamism made explicit.

Phenomenalism refers to the fact that the child accepts things in terms of their surface appearances, without wondering about underlying structures. All children in the first few years of school, and many adults, when asked to make a profile drawing of an island, showing what is under the water as well as what is above, reveal that they think of islands as floating rather than as mountain tops thrusting up from the bottom. It does not occur to most young children to wonder how the wall switch is connected to the light, or where the water coming out of the faucet originates.

Dynamism can be seen as one aspect of phenomenalism, and both can be seen as aspects of more inclusive *realism*, which assumes that all experiences take place in a common, unstratified sphere of reality, where objects mingle with images, words, thoughts, dreams, and feelings. Most children take it for granted that the characters (including cartoon characters) they see on television are actually contained within the set, and it is the beginning of the end of realism when the child starts to worry about how the people can get out. Similarly, the child listening to a record player may begin looking to find "where they have the musicians." Babies' first reactions to pictures or printed forms are to try to pick them up, suggesting that two-dimensional forms are first perceived as three-dimensional solids. In the same way, children may be observed trying to hear the ticking of a pictured watch. Our emotional reactions to the fictional events of a story or play can be seen as manifestations of realism—"The suspension of disbelief."

Considering the foregoing characteristics of the world as perceived by the baby, it is easy to accept that his experience is *labile,* that he lives in a world that is subject to sudden, unpredictable, arbitrary change. His response is *rigidity,* an intolerance of change and novelty and an inability to vary behavior flexibly.

Having defined, however imperfectly, the perceived world of the preverbal baby, we find that we have also described the life space of a great many thoroughly verbal adults. This of course, poses a riddle for the post-Whorfian, who wants to talk about the relevance of learning language to the development of a differentiated, articulated, penetrating, stable yet flexible world-view, one that sees the world as governed by the principles known to science rather than by magic, spirit forces, demons, and their whims and caprices.

The remainder of this chapter is devoted to a research approach which may help solve this riddle. What is proposed is a way of thinking about language in which the linguistic means (phonology, vocabulary, grammar, and syntax) are subordinated to the semantic ends of saying something (whether to other people or oneself). The method is longitudinal and observational. Only longitudinal research enables us to feel confident about sequences of emergence, since the wide range of individual differences in the age span birth-3 years makes cross-sectional comparisons almost meaningless. It is observational because we do not yet have any good standard stimuli for eliciting talking. However, in a complex home environment we can usually be sure that something, even if we do not know in advance exactly what, will provoke the child to speech. This research is an extension and elaboration of an earlier study (Church, 1968).

What we are seeking is a way of classifying utterances into *semantic operations,* ways of ordering experience symbolically to communicate a message. Probably the first identifiable operation produced by the baby is naming, or *denomination.* But there are at least three forms of denomination: asking (by whatever means) for the name of something, supplying something's name as an automatic by-product of encountering it, and naming as a request for verification (Is that a) "Dog?", rather than (That is a) "Dog." Asking for verification is indicated either by an interrogatory tone or by an inquiring glance from object to adult. We should take note that both tone of voice and expressive behavior appear very early as vehicles of shades of meaning.

A fundamental category of speech act is the *imperative,* which appears early and often in the vocabulary of children (and their parents). Although there is very little need to document the appearance of second-person imperatives, we know nothing, as far as I am aware, about the emergence of first-person plural ("Let's Go") and third-person imperatives ("May he stew in his own juice"). Nor do we know anything about children's acquisition of variant forms ("We'd better leave now," "Let him do as he likes," "Would his lordship be so kind as to"). *Please* is a relatively easy form to master, as are *I want* and *I need,* but we have still to learn about such indirect forms as "It would be nice if" or "Wouldn't you like to" or "I'd appreciate it if."

Predicative statements are taken, as by the Chomskyites, as core units in linguistic behavior, and in fact a great many utterances can be classified thus. Predications have the special psychological quality that they are attempts to represent verbally real or imaginary states of affairs. However, it turns out on closer scrutiny that predication is too broad a category. Consider the differences among "There is a dog" "That's a dog," "Dogs bite," "The dog bit his master," "The dog is in the garden," and "My dog is brown." The distinction among different kinds of predication may become clearer if we consider the variety of *question types,* question being a kind of reverse predication whereby one attempts to establish the existence or nature of something. Bellugi (1965) has made a beginning classification of questions, but in terms of grammatical forms (e.g., the use of *wh* words—what, who, when, etc., and of auxiliaries like *have* and *do*) rather than of the kind of information being sought.

A small amount of thought suggests that there are 15-20 kinds of questions, which may appear differentially in the understanding or speaking of children. Perhaps most basic is the identity question—"Who is it," "What is it?" There are several sorts of function questions. "What is it for?" can mean either the abstract "What purpose does it serve?" or the concrete "What can I do with it?" At a higher level the child may ask "How do I make it work?" or "Describe its working to me." There are simple questions of information—"What are we having for dinner?" or "Do rabbits bite?" The child may ask questions of origin: "Where does it come from," "Who made it," or "How did it come into existence?" He can ask about substance:

"What is it made of?" or, more fundamentally, "What is bread?" which verges on a question of origin. A question can ask about justification: "Why that?" or "Why that way?" Many early *why* questions refer to internal or external motivation: "To what end did he do it?" or "In response to what did he act?" Note that motivational questions can be applied to natural events as well as to human-caused ones. The child early asks for permission to do something and for approval after he has done it, although such questions are only tangentially predicative. Questions obviously deal with space (*where*) and time (*when*), both past and future. At about age 3 in bright children hypothetical questions begin to appear in the form *what if* (there were only men in the world, or people had wings and could fly, or I fell through the ice—the possibilities are almost endless and often dire). There are questions of competence ("Does he know how?" or "Is he strong enough?") There are also activity and event questions: "What is he doing?" "What happened?" "What are they waiting for?" I make no pretense to have exhausted all possible question types, but it should be apparent that question-asking is not necessarily a psychologically homogeneous function. Obviously, all these question types have their declaratory counterparts, as in *generalization, description,* and *explanation.*

A subcategory of predication is a group of operations known as deixis. Deixis locates the "subject of discourse" with respect to the speaker and the listener in relation to space, time, person, feelings, attitudes, and values. Expressions of uncertainty and doubt belong to the realm of deixis—we know that children learn to shrug their shoulders as a gestural equivalent of "I don't know," but we do not know at what age or at what point in the linguistic sequence. Various status markers from "sir" to "my respected opponent" to "those people" are deictic features. Deixis is not always coded morphologically but may be carried in tones of voice expressing sarcasm, contempt, disdain, amusement, and so forth. All temporal markers, including adverbs, prepositions, designations of particular times, and tenses, are deictic forms, as are comparable spatial indicators. Various grammatical devices for specifying the object of attention—"this one," "the yellow one," "the one we saw when"—likewise serve a deictic function. Disparaging and belittling can be seen as deictic operations.

Arithmetical operations from counting to summing to adding, subtracting, multiplying, and dividing can be viewed as a special class of predicative acts. We know something about the emergence of numerical competences, although not about how they are related to other aspects of linguistic development. We know, for instance, that counting precedes summing, in that the child can correctly count a collection of objects without recognizing that the last number also represents the total. We know that children familiar with numbers develop a low-level intuitive arithmetic enabling them to answer such problems as the number of pieces produced by cutting an apple in half or what would be left if one began with four blocks and then took one away.

Drawing contrasts is another operation embedded in many acts of predication. Long before he uses standard antonyms, and even longer before he understands the antonymy relation in general, the child finds ways to depict perceived contrasts. Some examples that I have observed are (all at about age 4): "This has an upper part and a downer part," "They're sure not slowpokes, they're fastpokes," and "Sometimes nobody eats with me and sometimes lots-of-bodies."

There has been a good deal of research on the child's ability to *make exceptions,* to respond to instructions like "Give me all the blocks except the red ones" or to say things like "I ate everything except my carrots." Nevertheless, we know little or nothing about how these important competences fit into the total stream of cognitive development.

Taking exception, of course, is an operation quite different from making exceptions. In two children whose language development I have followed closely, taking exception has first appeared in the form *but* followed by an appropriate word. For instance, the child asks whether people can fly and is told they cannot, to which he offers the rejoinder "But airplanes." Or, having asked if he can have a piece of candy and having been told that he can, he objects "But Mommy," implying that his mother has already told him not to eat any candy.

A great many relational statements given formal recognition by logicians have their everyday counterparts, and it would be worth knowing about their emergence in the course of development. One such is *conjunction,* the co-occurrence of events or conditions. For instance, we are accustomed to saying that biological sex is defined by the co-occurrence of certain primary and secondary sex characteristics, so that male genital structures go with facial hair, deep voices, broad shoulders, chest hair, and so forth. Some of the child's conjunctions may surprise us, as when he takes it for granted that any group of adults and children assembled under the same roof are all members of the same nuclear family. Logicians also recognize the relation of *disjunction,* or incompatibility of elements, and here we can expect an even greater lag in the child's development, since it seems that he is largely impervious to paradoxes, contradiction, logical inconsistencies, and cognitive dissonance in general. He does develop, from an early age, a set of expectations about sequences, and when these are disappointed he shows surprise, dismay, anger, or amusement. But there is no evidence that he feels that any logical imperatives have been violated. Much of the current controversy about the development of conservation (Beilin, 1968; Mehler & Bever, 1967; Mehler, Bever, & Epstein, 1968) hinges around the appearance of an intuitive conviction that quantities cannot be changed simply by their spatial rearrangement. At early ages, the child never doubts that the volume of a liquid can be altered simply by pouring it into a container of different proportions. In the same way, the child has a pragmatic understanding of a great many causal sequences, but he lacks—as do many adults—any grasp of general causal principles. It is worth repeating that his early pragmatic causal sense includes some notion of psychological causation, as in the use of imperatives, jokes, and teasing. We know little about the young child's ability to *draw inferences* from concrete or symbolically represented situations, or to *extrapolate* from concrete or symbolic series. We know that children spontaneously commit every known logical fallacy and, in all probability, invent new varieties of their own, but our knowledge is far from systematic. This gap results partly from an overfond attachment to a few experimental paradigms, to the neglect of the child's reasoning in everyday situations. For instance, a recent compilation by Sigel and Hooper (1968) called *Logical Thinking in Children* covers the following topics: conservation of quantity, number, spatial and geometric concepts, and classification. But what are we to make of the logic of a 4-year-old who asks, a propos of a paper cut-out boat, "Would this really float on pretend water?" or who announces, "I'm sorry to have to tell you that I had to take out Jimmy's [a stuffed animal] tonsils. It's all right, though—I put in some new ones"?

Evaluation is an important form of predication, phrased most basically in terms of "good" or "bad" (obviously, there are still more concrete, preverbal expressions of liking or aversion). Adults can usually express approval along a number of more or less independent dimensions, such as the moral, esthetic, utilitarian, rational, and emotional (Church & Insko, 1965), but these appear to be hopelessly jumbled in the young child's judgments. Adults may also be able to distinguish between the value attributes of objects and the emotional reactions these objects produce in the observer, but this distinction seems unclear in young children and uneducated adults. That is, the ability to designate an object as good or bad appears earlier in development than the ability to define its emotional impact on oneself. We also know that verbal evaluation proceeds from the absolute ("good," "bad") to the graduated ("better," "best," "worse," "worst"), but the details are still lacking.

Closely related to expressing valuative reactions is the expression of subjective states in general, from emotions to hunger and thirst, to fatigue, to pain and sickness, to, eventually, more subtle states such as numbness or nostalgia. Here again the child is likely at first to leave subjective feelings implicit, speaking instead of objects ["Meat," "Water" "Fly" (indicating a bite or sting) or, more often, of behavior ("Hug," "Sleep," "Eat"). Still earlier, of course, he expresses his feelings and needs in the concrete "language of behavior." At its most primitive, the language of behavior consists of affective manifestations analogous to those found in nonhuman species—the dog's wagging tail, the cat's bristling fur, the drooping ears and tail that signify depression in the goat, the mooing of cows at sundown. At a more advanced level the child, like the pet dog, communicates with tokens—to climb into his feeding table signifies hunger, to hand an adult a record means that he wants to hear it played, just as the dog's bringing its food dish is a request to be fed, or bringing its leash is to ask to be taken for a walk. Quite young babies point to things they want (although it is not until much later that they comprehend the adult's pointings) and toddlers and chimpanzees move the adult human's hand to perform manipulations of which they are incapable. But we still do not know whether there is a stable order (within or across cultures and languages) for the explicit naming or description of subjective states. The enormous individual variation among adults in their ability to express subjective experience verbally leads us to assume that this is one area in which the routine-original distinction is very marked. Furthermore, the self-discovery that is common through the reading of literature suggests that there is an especially intimate tie between linguistic formulation and self-experience.

To perceive and construct *analogies* is considered one of the more advanced human capacities, even though analogical reasoning often leads to some bizarre outcomes. We can see a concrete drawing of analogies from an early age. For instance, a 3-year-old, seeing her mother holding a newly purchased garment against father's body to judge its fit, asks, "Is Daddy a paper doll?" A 4-year-old, watching a cow being milked, observes, "It's like a water pistol." Simple analogies are included in many tests of early intellectual functioning, but we lack systematic knowledge of how the child moves from the paper-doll level to being able to perform well on the Miller Analogies Test or to reason about human affairs on the basis of analogies. As William Empson (1951) has pointed out, analogy is closely linked to simile, metaphor, and what he calls "equation." Many analogies are embedded in language as dead metaphors, like the *saddle* of a mountain or the *leg* of a table. Asch (1955) has found that a number of metaphors used to describe persons (*warm, cold, sweet, bitter*) are found in many diverse languages. However, Asch and Nerlove (1960) have shown that middle-years children use and understand such figures concretely, without seeing their metaphorical origins, indicating that they are probably dead metaphors. A number of studies, including one by Richardson and Church (1959), show that young school-age children understand metaphorical statements like proverbs literally, and that older children, when they try to grapple with the double meaning of metaphors, distort them beyond recognition. An equation, according to Empson, is an indissoluble fusion of a concrete fact with its metaphorical description, leading to serious misconceptions. Szasz (1961), among others, has pointed out the peculiar consequences of labeling psychological disturbances as "mental illness." The use of analogies shades into the ability to specify *similarities*, as in verbal concept-formation tasks.

Not all predications are factual, of course. The child practices a variety of deceptions, from denial of events or responsibility for events to falsifications or misrepresentations of things, whether in earnest, as lying, or playfully, as joking. We must regard as a serious topic in cognitive development the child's ability to say things that are not so—just as we need to study people's ability to learn and to believe things that are not so.

An important domain of language is *metalanguage*, talking about language itself. Linguists, of course, have developed elaborate metalanguages, but even common discourse recognizes such metalinguistic distinctions as word, sentence, phrase, definition, meaning, and so forth. One metalinguistic function that has been studied developmentally is that of giving word meanings. Traditionally, the task of defining words has been seen primarily as a means of assessing the breadth and depth of the child's vocabulary, the words he has at his disposal. This is a legitimate interest, and indeed has been shown to be one of the better indicators of general intellectual ability. It is somewhat different, however, from the developmental problem of becoming able to talk about words and their meanings. Part of the difference lies in the fact that for a definitions task, as opposed to a vocabulary test, one would want to work with words the child is likely to know. A good beginning in this direction has been made by Feifel and Lorge (1950).

Obviously, the kinds of communicative acts we have been talking about work most effectively when the child can bring into play the appropriate grammatical and syntactic mechanisms. But our attention has been focused on the psychological purposes of the speaker, with only incidental attention to the means by which he realizes his intentions. There are, however, a number of standard grammatical forms and features whose emergence in the child's speech is of interest in its own right. Some of the earliest students of child language concerned themselves with the use of compound and complex sentences. One can, of course, extend the analysis further, as in the use of compound or complex sentences to convey temporal or causal relationships or to arrange a complex situation into background and foreground components: "Although Robin Hood was technically a criminal, he was a true patriot and a friend of the common people."

English is not as rich as some other languages (notably Italian) in *diminutives* and *magnifiers*, but they do exist and diminutives at least are common in baby talk, as in *lambykins*. Nowadays we hear a good deal about miniskirts, we have dinettes and even superettes (which seem to be miniature versions of self-service markets), and *micro-* exists both in common parlance and as a technical term for one millionth. Super is widespread, both as a distinct word and in compounds, and *maxi-* as the opposite of *mini-* as well as *mega-* (one million), are becoming standard.

Standard Average European uses *impersonal construction,* from "It is cold" to "It is raining" to "It is time to . . ." to "It is probable that . . ." The child obviously uses some of these forms from an early age, but others appear we know not when. Notice that many impersonal constructions are linked to the *subjunctive:* "It is essential that . . . ," "It is impossible that . . . ."

Although purists fight it, English words *change part of speech* easily, sometimes with morphological transformation (to anthologize) and sometimes without (child psychology). We do not know whether such shifts represent any particular problem in the child's comprehension or speaking, but it might be worth finding out.

A phenomenon in English is that we often *express two-sided relationships* one-sidedly. We are more likely to say that gloves are too small than that hands are too big. We sometimes have an option, and it has yet to be defined when we say that something is too heavy to lift or that we are too weak to lift it. Once more, we are forced to say that we know nothing of when such constructions appear developmentally.

Obviously, this is not a finished system but a sketch of a method. It will be refined on the basis of empirical observations. Two babies are already being followed by their mothers according to the present scheme, and other subjects will be added to the study as they become available. Progress is slow, but I believe that this approach offers hope that we may yet come to grips with the Whorfian hypothesis.

References

Asch, S. On the use of metaphor in the description of persons. In H. Werner (Ed.), *On expressive language.* Worcester, Mass.: Clark Univ. Press, 1955. Pp. 29-38.

Asch, S., & Nerlove, H. The development of double function terms in children: An exploratory investigation. In B. Kaplan & S. Wapner (Eds.), *Perspectives in psychological theory.* New York: International Universities Press, 1960. Pp. 47-60.

Beilin, H. Cognitive capacities of young children: A replication. *Science,* 1968, **162,** 920-921.

Bellugi, U. The development of interrogative structures in children's speech. In K. F. Riegel (Ed.), *The development of language functions.* Ann Arbor: University of Michigan, Center for Human Growth and Development, Language Development Program, 1965. Pp. 103-148.

Bellugi-Klima, U. Linguistic mechanisms underlying child speech. Paper presented at Joint Study Group, Mechanisms of Language Development, Ciba Foundation and the Centre for Advanced Study in the Developmental Sciences, May, 1968.

Bernstein, B. Language and social class. *British Journal of Sociology,* 1960, **11,** 271-276.

Bernstein, B. Elaborated and restricted codes: Their social origins and some consequences. *American Anthropologist,* 1964, **66,** No. 6, Part 2, 55-69.

Carroll, J. B. (Ed.) *Language, thought, and reality: Selected writings of Benjamin Lee Whorf.* New York and Cambridge: Wiley and Technology, 1956.

Chauchard, P. *Le Langage et la pensée.* Paris: Presses Universitaires de France, 1956.

Chomsky, N. *Language and mind.* New York: Harcourt, 1968.

Church, J. *Language and the discovery of reality.* New York: Vintage, 1966.

Church, J. (Ed.) *Three babies.* New York: Vintage, 1968.

Church, J. Techniques for the differential study of cognition in early childhood. In J. Hellmuth (Ed.), *Cognitive studies.* Vol. 1. New York: Brunner/Mazel, 1970. Pp. 1-23.

Church, J., & Insko, C. A. Ethnic and sex differences in sexual values. *Psychologia,* 1965, **8,** 153-157.

Dennis, W., & Najarian, P. Development under environmental handicap. *Psychological Monographs,* 1957, **71.**

Empson, W. *The structure of complex words.* Norfolk, Conn.: New Directions, 1951.

Fantz, R. L. Visual perception from birth as shown by pattern selectivity. *Annals of the New York Academy of Sciences,* 1965, **118,** 793-814.

Feifel, H., & Lorge, I. Qualitative differences in the vocabulary responses of children. *Journal of Educational Psychology,* 1950, **41,** 1-18.

Frake, C. O. How to ask for a drink in Subanun. *American Anthropologist,* 1964, **66,** No. 6, Part 2, 127-132.

Friedlander, B. Z. Receptive language development in infancy. *Merrill-Palmer Quarterly,* 1971, in press.

Hymes. D. The ethnography of speaking. In T. Gladwin & W. C. Sturtevant (Eds.), *Anthropology and human behavior.* Washington, D. C.: Anthropological Society of Washington, 1962. Pp. 15-53.

Hymes, D. Introduction: Toward ethnographies of communication. *American Anthropologist,* 1964, **66,** No. 6, Part 2, 1-34.

Hymes, D. On communicative competence. Paper presented at Joint Study Group, Mechanisms of Language Development, Ciba Foundation and the Centre for Advanced Study in the Developmental Sciences, May, 1968.

Inhelder, B., & Piaget, J. *The early growth of logic in the child.* New York: Harper, 1964.

Lenneberg, E. H. *Biological foundations of language.* New York: Wiley, 1967.

Lenneberg, E. H. On explaining language. *Science,* 1969, **164,** 635-643.

Lipsitt, L. P., Engen, T., & Kaye, H. Developmental changes in the olfactory threshold of the neonate. *Child Development,* 1963, **34,** 371-376.

Mehler, J., & Bever, T. G. Cognitive capacities of young children. *Science,* 1967, **158,** 141.

Mehler, J., Bever, T. G., & Epstein, J. What children do in spite of what they know. *Science,* 1968, **162,** 921-924.

Miller, G. A., & McNeill, D. Psycholinguistics. In G. Lindzey & E. Aronson (Eds.), *Handbook of social psychology.* Reading, Mass.: Addison-Wesley, Vol. III, 1969. Pp. 666-794.

Miller, W., & Ervin, S. The development of grammar in child language. *Monographs of the Society for Research in Child Development,* 1964, **29,** No. 1.

Mowrer, O. H. *Learning theory and the symbolic processes.* New York: Wiley, 1960.

Osgood, C. E. On understanding and creating sentences. *American Psychologist,* 1963, **18,** 735-751.

Papoušek, H. Conditioning during early postnatal development. In Y. Brackbill & G. G. Thompson (Eds.), *Behavior in infancy and early childhood.* New York: Free Press, 1967. Pp. 259-274.

Piaget, J. *The language and thought of the child.* New York: Meridian, 1955.

Piaget, J. Quantification, conservation, and nativism. *Science,* 1968, **162,** 976-979.

Piaget, J., & Inhelder, B. The psychology of the child. New York: Basic Books, 1969.

Polanyi, M. *Personal knowledge.* Chicago, Ill.: Univ. of Chicago Press, 1958.

Richardson, C., & Church, J. A developmental analysis of proverb interpretations. *Journal of Genetic Psychology,* 1959, **94,** 169-179.

Riessman, F. *The culturally deprived child.* New York: Harper, 1962.

Schiff, W. Perception of impending collision. *Psychological Monographs,* 1965, **79** (11, Whole No. 604.

Sigel, I. E., & Hooper, F. H. (Eds.) *Logical thinking in children.* New York: Holt, 1968.

Siqueland, E. R., & Lipsitt, L. P. Conditioned headturning in human newborns. *Journal of Experimental Child Psychology,* 1966, **3,** 356-376.

Skinner, B. F. *Verbal behavior.* New York: Appleton, 1957.
Staats, A. W. *Learning, language, and cognition.* New York: Holt, 1968.
Szasz, T. S. The uses of naming and the origin of the myth of mental illness. *American Psychologist,* 1961, **16,** 59-65.
Walk, R. D., & Gibson, E. J. A comparative and analytical study of visual depth perception. *Psychological Monographs,* 1961, **75.**
Weinreich, U. On the semantic structure of language. In J. H. Greenberg (Ed.), *Universals of language.* Cambridge, Mass.: MIT Press, 1963. Pp. 114-171.
Werner, H. Process and achievement. *Harvard Educational Review,* 1937, **7,** 353-368.
White, R. W. Motivation reconsidered: The concept of competence. *Psychological Review,* 1959, **66,** 297-333.

AUTHOR INDEX

Numbers in italics refer to the pages on which the complete references are listed.

C

D

E

F

G

H

I

J

L

M

N

O

P

Q

R

S

T

Y

Z

SUBJECT INDEX